THE PHYSICAL GEOGRAPHY OF WISCONSIN

MANITOU FALLS, THE HIGHEST WATERFALL IN WISCONSIN.

One-hundred-and-sixty-foot cascade on the Black River, a tributary of the
Nemadji, south of the city of Superior.

LAWRENCE MARTIN

The Physical Geography
of
Wisconsin

THE UNIVERSITY OF WISCONSIN PRESS
Madison and Milwaukee, 1965

Published by
The University of Wisconsin Press
Madison and Milwaukee
P.O. Box 1379, Madison, Wisconsin 53701

First edition published 1916 and second edition published 1932
as Bulletin No. XXXVI, Educational Series No. 4, Wisconsin
Geological and Natural History Survey

Printed in the United States of America
Library of Congress Catalog Card No. 65-14538

FOREWORD

This volume is a reproduction of the second edition of Lawrence Martin's classical work on the physical geography of Wisconsin, published in 1932. It differs only slightly from the first edition, published in 1916. Why, it might be asked, is such a book being reprinted with virtually no changes after almost half a century? Have there been no changes in statistical data, no advances in scientific knowledge?

The answer is that the timeless quality of this work transcends statistics which, after all, are often outdated even before they are printed. Changes in scientific information have been remarkably few; for the most part they consist of a fuller knowledge of details rather than changes in basic concepts. It matters little whether the Lower Magnesian limestone is now called the Prairie du Chien Group, for names of individual rock units are always prone to change but the rocks themselves remain the same.

The most noteworthy scientific advances of the past thirty years have been in the field of glacial geology. New dating techniques show that many of the glacial events are younger than was previously thought. Studies of modern glaciers and ice sheets have led to a fuller knowledge of glacial processes, which may demand some major revisions of our concept of the Ice Age in Wisconsin; but speculation in this regard would be premature at present, since much work yet remains to be done.

The relief map originally included as Plate I in both previous editions has reluctantly been excluded because of technical difficulties. The printed version did not lend itself to further reproduction, and deterioration of the original three-dimensional model made rephotographing impossible. The omission is pointed out by footnotes at key points of reference in the pages following.

We particularly regret that newly established Menominee County (formerly the northeastern portion of Shawano County) does not appear in maps showing county areas. While some attempt was made in the second edition to revise the old city name of Kilbourn to the present name of Wisconsin Dells, this attempt was not wholly successful and the reader is asked to consider the two names as interchangeable. Minor typographical errors proved to be too numerous for correction.

A final word of precaution specifically concerns the list of maps in Appendix E. Availability of maps, prices quoted, and sources of procurement should all be considered as historical curiosities, surpassed only by the instructions for procuring lantern slides once available from the University of Wisconsin Extension Division. Information about modern maps can be obtained by writing the office of the Geological and Natural History Survey, The University of Wisconsin, Science Hall, Madison, Wisconsin 53706.

The state of Wisconsin is blessed with a variety of scenic features equaled in few, if any, other states. Its citizens and visitors are indeed fortunate in having available once more Martin's magnificent treatise on the state's physical features, their origin, and their role in the historical development of the state.

GEORGE F. HANSON
State Geologist and Director
Geological and Natural History Survey

The University of Wisconsin
June 30, 1964

PREFACE

First Edition

This discussion of the physical geography of Wisconsin is an attempt to describe the surface features of the state in such a way that any man or woman may read and understand, that any grammar school teacher may read and select items for verbal presentation to her pupils, and particularly that any high school, normal school, or college instructor may read and assign to students for reading either the whole book, or the chapters dealing with the home region.

The author has lived in Wisconsin ten years. During the summers of 1907 and 1908 he devoted nearly all of the college vacation to field work on the physical geography and glacial geology of the state. The expense of the first season's work was met by the United States Geological Survey, the second by the Wisconsin Geological and Natural History Survey. In subsequent years he has spent many weeks in field work in connection with his teaching at the University of Wisconsin, taking parties of students for half-day, all-day, and longer excursions, including two periods of four weeks each during the summers of 1913 and 1914. In addition three days were spent in the field in 1913 with Samuel Weidman, W. O. Hotchkiss, and E. F. Bean, five days in 1914 with E. O. Ulrich, W. O. Hotchkiss, and F. T. Thwaites, three days the same year with W. O. Hotchkiss, E. F. Bean, and O. W. Wheelwright, and two weeks in 1915 with F. T. Thwaites and Walter Schoewe. The writing of this book was undertaken in the winter of 1908-9, resumed during the summer of 1912, and completed during the summer and autumn of 1915. Much of the subject matter has been tried out by presentation to students in a university course on The Geography of Wisconsin, first offered in 1910 and repeated during several semesters and summer sessions by the author and, more recently, by Prof. E. F. Bean.

The author has tried to make the book simple. Most of the discussion is without technicality. Nevertheless there are ideas and phenomena which will be new to some. The plan of presen-

vii

tation involves the explanation of all unfamiliar terms and phrases the first time they are used. Accordingly a person who reads the book through from beginning to end should not encounter any terms he cannot understand. Many of the explanations are repeated, but it has not been possible to do this every time. Hence the reader who takes up a chapter without having read those which precede it may sometimes find it necessary to turn to the index at the end of the book and look up the first page listed under moraine, metamorphic rock, cuesta, outwash, peneplain, or whatever the word may be.

The book does not aim to discuss all of the features of the geography of Wisconsin. It deals with the physical geography or physiography of the state. The Wisconsin Geological and Natural History Survey has already published "The Geography and Industries of Wisconsin." This is Bulletin XXVI, now out of print, but republished in the Blue Book for 1915. Regional studies of "The Geography of the Region about Devils Lake and the Dalles of the Wisconsin," on "The Lakes of Southeastern Wisconsin," on "The Geology of North Central Wisconsin (Glacial Geology and Physiographic Geology)," on "The Abandoned Shorelines of Eastern Wisconsin," on "The Inland Lakes of Wisconsin," and on "The Geography of the Fox-Winnebago Valley" have already appeared. Although this discussion of the physical geography of Wisconsin deals primarily with the surface features of the state it has not seemed advisable to omit all references to human geography. Brief incidents in the history of the state and relations of resources and industries to topographic and hydrographic features have been included here and there. Each item introduced has, however, been carefully considered in order to see whether its inclusion adds interest to the discussion of the purely physiographic features and justifies the space given it.

If it were usual to dedicate the publications of the Wisconsin Geological and Natural History Survey to individuals the author would be inclined to inscribe this volume to the memory of Dr. Increase A. Lapham—Wisconsin's first geographer.

Many workers have contributed to the explanation of the physiographic features of Wisconsin. It has not been possible to mention the names of all the geologists, geographers, and other observers who have participated in this work and to tell exactly what each one contributed. The author has omitted all mention of names from the text, because of fear of acknowledging this debt imperfectly and because he believes that names in the text and footnotes at the ends of pages distract the reader and interrupt the continuity of presentation. In the bibliographies at the ends of chapters he has included the names of his fellow workers and the full titles of their books and papers. From a great many of these he has gleaned valuable information. Except in the chapter on The Discovery and Explanation of the Driftless Area, the only names mentioned in the text are those of authors quoted. It is hoped that the youth of Wisconsin may receive from these quotations an inspiration to find beauty and charm in the scenic features of the state and to learn to express it in adequate language.

Certain schools may wish to use this book as a supplementary reader in a course in physical geography. It is conceivable that some school may even try the experiment of using it in place of a textbook of physical geography. In most states this would be impossible, but Wisconsin has such a variety of features that nearly every important physiographic process and topographic form in connection with the physical geography of the lands is here represented. Thus we have weathering and the work of underground water in the Driftless Area of southwestern Wisconsin; we have wind work in the Driftless Area and on the shores of the Great Lakes; the Mississippi River and the other rivers of the state give us nearly all the essential features of stream erosion and stream deposition; in northern and eastern Wisconsin the work of glaciers in erosion and deposition is conspicuously represented; the shores of Lake Michigan, Lake Superior, and the inland lakes include most phases of wave work; the lakes and swamps bring in the relationships of plants and animals; the lava flows of northwestern Wisconsin represent one type of vulcanism; Wisconsin includes several kinds of plains, an area of low plateaus, and a vast expanse of worn-down mountains. If "The Physical Geography of Wisconsin" were used as the chief text and a small amount of supplementary readings were carried on in a reference book, like one of the ordinary textbooks of physical geography—where the ocean, the atmosphere, youthful mountains, volcanoes, and earthquakes were studied briefly—Wisconsin boys and girls might profit quite as much as in the usual high school course. Another plan would be to devote a half year to a study of physical geography in an ordinary textbook, followed by a half year's study of this book on the physical geography of the home state. It is hoped that the publication of the present discussion will result in the introduction of a course on the Geography of Wisconsin in every normal school and college in the state as well as in some of the larger high schools. Such a course might use this book as a text, following it by a consideration of resources and occupations as described in "The Geography and Industries of Wisconsin," or Merrill's "Industrial Geography of Wisconsin" or the United States Census supplements for Wisconsin on Population, Agriculture, Manufactures, Mines and Quarries, or the Wisconsin number of the Journal of Geography.

In any case there should be laboratory and field work. The diagrams and maps in this book furnish some of the material for such laboratory work. The maps listed at the ends of chapters include desirable additional material. Appendix E explains where and at what cost these maps may be obtained. The appendices at the end of the book also furnish material for laboratory work. Lantern slides on the physical geography of Wisconsin may be borrowed from the Extension Division of the University of Wisconsin.

For field work, there is the great Outdoors. No school in Wisconsin, unless possibly in the center of Milwaukee, is so situated that local field work is not possible and profitable. Even if only forty minutes or an hour be available, it is nevertheless distinctly worth while to take students on field excursions. Of course a two-hour or a half-day field trip is better. The teacher who reads this volume will think at once of nearby features similar to those here described.

Similarly a study club might possibly consider making this book the basis of a winter's discussion. If this were done without planning to have one or more outdoor meetings in the spring or autumn a great opportunity would be lost. The layman who reads "The Physical Geography of Wisconsin" and fails to look about the home region for evidences of the processes here described and illustrations of the forms here explained, or who travels about the state without beginning to see and explain for himself has done only half what he might for his own pleasure and enlightenment.

The author has thought it wise to take up the arguments in connection with a few matters that are not yet thoroughly understood or agreed upon. It is felt that this book should be not only informational but disciplinary. It will lead to better habits of thought and reasoning if the readers are taken into the confidence of the author and allowed to realize that by no means every question regarding the physical geography of Wisconsin is to be regarded as settled. One set of facts may favor one interpretation, other facts another, but any person may go into the field and find additional phenomena which support either of the suggested explanations, or lead to an entirely new one.

Much time and thought has been given to the preparation of the illustrations in the book and especially the text figures. It is believed that they are as well worth consideration and study as the text itself. The relief map of the state—the large folded map in a pocket at the end of the volume—may be profitably kept in sight during the reading of nearly every page.* A small number of extra copies has been printed. These have not been folded but are mounted as rolled maps to hang on the walls of school rooms, libraries, and offices. They may be obtained at cost of mounting and mailing by writing to the Wisconsin Geological and Natural History Survey at Madison. It would be well if every high school in the state would take this folded map out of a copy of the book at once and mount it under glass in a wooden frame. The students in manual training might enjoy making and painting the frame. Thus mounted, the Relief Map of Wisconsin would be permanently preserved for daily reference, as well as an ornament to the wall of some class room or corridor in the school. Beside it should be hung the most detailed map of the region near the home town. By looking at the index maps and descriptions in Appendix E the teacher may see what is the best map of the region near the school, and how it may be obtained. It may be one of the topographic maps of the United States Geological Survey, or one of the charts by the Mississippi River Commission, or the United States Lake Survey, or one of the lake maps or soil maps or geological maps issued by the Wisconsin Geological and Natural History Survey. The cost of the map will be only ten cents, or twenty-five cents at most, the glass for the frame will cost little, and the making of the frame as a permanent piece of school equipment will lend interest to the work. The location of the school building might be added to the map in red ink.

The material in the appendices at the end of the book is for general information, but may also be adapted to special uses.

*Regretfully excluded from the 1965 reprinting. See Foreword.

Thus a school class in civil government might profitably study part or all of Appendix F—The Land Survey in Wisconsin. A history class might take up Appendix B—The Boundaries of Wisconsin—also reading the paragraphs in the body of the book, referred to there. It will be of interest in any school or any community to look up the elevation of the home region as given in Appendix H.

The author is under obligations to many persons for help in writing this volume. His first acknowledgment must be to the authors whose names are listed in the bibliographies at the ends of the chapters and in Appendix G. He realizes that their contribution to the physical geography of Wisconsin is greater than his own, and regrets that it could not be acknowledged in connection with their individual ideas which he has used. At the same time he does not feel that they should be held responsible for all the interpretations which the author has introduced and with which they might not in every case agree. Such matters as the relation of the Mississippi terraces to tilting, the amount of glacial erosion in eastern Wisconsin, the presence of cuestas rather than peneplains in western Wisconsin, the burial, warping, and exhumation of the Northern Highland, and the origin of the basin of Lake Superior—to cite a few of the many cases—are not to be wholly charged to earlier writers.

The manuscript of the book has been read by Mr. F. T. Thwaites, curator of the geological museum at the University of Wisconsin. He has given freely from his broad knowledge of the geology and physical geography of the state and has compiled Appendix H and several of the text figures. Prof. E. F. Bean has likewise read the whole volume and made many valuable suggestions. Mr. W. O. Hotchkiss, the state geologist, criticized much of the text. Dr. W. C. Alden of the United States Geological Survey has supplied many facts and made many suggestions to the author, both in the field and by correspondence. Dr. Warren Upham, formerly of the Minnesota Geological Survey and the United States Geological Survey, and Mr. Frank Leverett of the latter bureau have advised on several points. President C. R. Van Hise of the University of Wisconsin, Prof. W. M. Davis of Harvard University, and Prof. Bailey Willis of Leland Stanford University, each made important suggestions to the author. Mr. Eric Miller of the United States

Weather Bureau kindly read the section on climate. Dr. Samuel Weidman of the State Geological Survey supplied valuable data. Each of the author's colleagues in the Geological Department at the State University has been good enough to look over and help with some part of the discussion in manuscript or in proof. Mr. Walter Schoewe aided materially in the compilation of the map material listed in Appendix E.

Many persons have supplied photographs for use in the half-tone plates, as indicated in the list on a later page. A number of the photographs were taken by the author. A majority of the text figures were drawn by Mr. F. W. Gillis. Others were taken directly from geological reports or modified from published diagrams. The authorship of these text figures has been acknowledged in the legends. Special acknowledgment should be made of two groups of text figures, which are based upon earlier publications of the Wisconsin Geological and Natural History Survey. The revised and amended maps of river systems (Figures 58, 100, and others) are based upon figures compiled by Prof. L. S. Smith and published in Bulletin XX on "The Water Powers of Wisconsin." Many of the geological cross-sections (such as Figures 12, 14, and others) are taken from the colored structure sections on W. O. Hotchkiss and F. T. Thwaites' "Map of Wisconsin, showing Geology and Roads."

If the book helps a little in making clear the nature, variety, and beauty of the surface features of Wisconsin, in leading people to look with interest and inquiry at the immediate localities in which they live, and in encouraging them to try to discover the relation between the geographical environment and the resources, industries, transportation, and history of the state, the author will feel amply repaid. LAWRENCE MARTIN

Madison, Dec. 23, 1915.

PREFACE TO SECOND EDITION

The first edition (6,000 copies) of this bulletin, published in 1916, was so popular with the public that the supply was exhausted by 1922. This popularity was due to the fact that Dr. Martin was very successful in presenting scientific facts in an interesting fashion.

In 1915 the total number of automobiles licensed was 115,645. In 1930 the number was 679,613. The improvement in highways and the large increase in the number of automobiles has meant that the scenery and physical geography of the state is today of interest to a much larger public than it was in 1916. The second edition is published to assist this public in understanding outdoor Wisconsin.

In preparing the manuscript for the second edition only such changes were made as are necessary to bring the scientific and statistical data up to date. The borders of the Driftless Area in Wisconsin and Iowa are revisable in the light of new data, but have not been revised on the maps in the book. As was the case in the preparation of the first edition, many persons have aided in the preparation of the second edition. Dr. Martin, now Chief, Division of Maps, Library of Congress, has served as editor. F. T. Thwaites made a critical study of the text and revised the lists of references. D. H. Kipp, Superintendent of Education and Publication for the Conservation Commission, has devoted much time to the preparation of text and the selection of illustrations which will modernize the State Park and Forest Reserve descriptions. Miss Louise P. Kellogg, Research Associate, Wisconsin Historical Society, has edited all of the historical references. Charles E. Brown, Director, State Historical Museum, made many suggestions regarding trails and sites of historic interest. Charles E. Halbert, State Chief Engineer, reviewed engineering data and revised the discussion of water powers. Walter H. Ebling, Agricultural Statistician, Wisconsin Department of Agriculture, revised the agricultural data. Others who have assisted in the preparation of this edition are Prof. W. M. Davis, Prof. Kirk Bryan, Dr. W. C. Alden, Dr. V. Stefansson, Dr. W. A. Johnston, Prof. J. Harlen Bretz, Justice W. C. Owen, Prof. N. M. Fenneman, Prof. Wellington Jones, Prof. Albert Perry Brigham, Prof. William S. Cooper, and Prof. Robert S. Platt. The Survey is deeply appreciative of this valuable assistance. We believe that the public will find the new volume of great interest as a guide to the study of the physical geography of the state. We hope that the second edition will be as popular as was the first.

E. F. BEAN

Madison, May 1, 1931.

CONTENTS

ILLUSTRATIONS

*Excluded from the 1965 reprinting.

ACKNOWLEDGEMENT OF PHOTOGRAPHS.

The author gratefully acknowledges his indebtedness to the following persons for photographs used as illustrations in this book. The photographs were either taken by these authors or reproduced in their publications. Others were purchased from commercial photographers. A few photographs have been published in earlier bulletins of the Wisconsin Geological and Natural History Survey. Many appear here for the first time. The authorship of one or two photographs could not be ascertained. The numbers refer to the plates in this volume.

Alden, W. C.—Plates XVI; XIX, A.
Aldrich, H. R.—Plate XXXI, B.
Bennett, H. H.—Plate XXIV.
Christie, photographer.—Plate X, B.
Detroit Photographic Co.—Plate XX, B.
Fish, J.—Plate VII.
Hotchkiss, W. O.—Plates V, A; VIII, A.
Johnson, D. W.—Plate XX, A.
Kerschner, H. M.—Plates IX, B; X, A.
Kipp, D. H.—Frontispiece; Plates XXIII; XXIV; XXXIII.
Salisbury, R. D., and Atwood, W. W.—Plate II, A.
Steidtmann, Edward.—Plate V, B; XI, B.
Thwaites, F. T.—Plates VIII, B; XII, A; XVIII, A; XXXII; XXXV.
Underwood & Underwood.—Plates XXV; XXVI.
U. S. Geological Survey.—Plates XVI; XIII, B; XIX, A.
Weidman, Samuel.—Plates III; XXII; XXIX; XXX, B.
Whitbeck, R. H.—Plates II, B; XI, A; XIII, A.
Whitson, A. R., and others.—Plates IV; XV; XVII, B; XVIII, B; XXI; XXXVI.

LIST OF TEXT FIGURES

Illustrations   xxiii

TABLES IN TEXT.

THE PHYSICAL GEOGRAPHY OF WISCONSIN

CHAPTER I

THE STATE OF WISCONSIN

GENERAL GEOGRAPHY

Wisconsin is larger than England, yet it contains only about one-twelfth as many people. This is clearly because of the more favorable position of England and its longer period of settlement, since the topography, soil, and climate of the two regions are not widely dissimilar. The greater mineral resources of England and her trade advantages in relation to Europe and to the British Empire also help to account for the difference.

On the other hand, Wisconsin is only about half as large as Colorado, yet it supports nearly three times as many people as Colorado. This difference is partly explained by topography and climate. About half of Colorado is too mountainous to support a dense population. The other half of Colorado is a high plain or plateau. Its climate is too dry for agriculture without irrigation. Wisconsin is largely a plain (Fig. 1). Very little of it is too hilly for cultivation, and its soil and climate are favorable to a successful agricultural and dairying industry. Wisconsin's general position may be somewhat more advantageous than that of Colorado, but Colorado's present mineral output is greater.

The position of Wisconsin as one of the states on the Great Lakes and the Mississippi River is its chief geographical asset, for, as was well said by Theodore Roosevelt, these states are "destined to be the greatest, the richest, the most prosperous of all the great, rich, and prosperous commonwealths which go to make up the mightiest republic the world has ever seen."

The capacity of this state for population depends chiefly upon topography, soil, and climate, hence it will be of advantage to study the nature of the surface features of Wisconsin and their origin.

The state of Wisconsin is located about a third of the way from the Atlantic to the Pacific Ocean, near the northern boundary of the United States (Fig. 1). It lies in what is commonly called the Middle West, between Lake Superior, Lake Michigan, and the Mississippi River. Its greatest length is 320 miles, its greatest width about 295 miles, and its area 56,066 square miles.

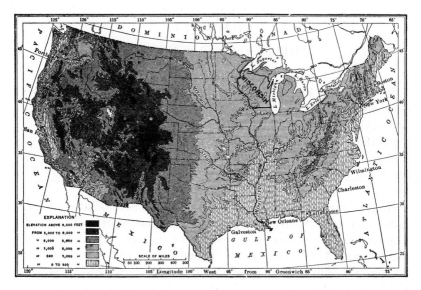

Fig. 1. Map of the United States, showing the location of Wisconsin.

GEOLOGY

The Geological Column. Wisconsin includes a large area of the oldest of rocks, called pre-Cambrian, and a still larger area of rocks of later, though very ancient, formation, called Paleozoic. The rocks of middle age, called Mesozoic, are not represented here but the state includes rocks of Cenozoic age. The latter is represented by widespread, unconsolidated surface deposits made (a) by the decay, or weathering, of older rocks, (b) by river, wind, and wave deposition, and (c) by the ice sheet of the Glacial Period.

The following tables, which in order of time read upward from the bottom, show the whole geological column and the part of it represented in Wisconsin. The first table gives the eras and periods of geological time, essentially as now accepted by the United States Geological Survey, together with the probable duration and the characteristic life of each period. The second table gives the local names of the geological systems and series as now proposed by the State Geological Survey, and the character and thickness of the several rock formations in Wisconsin (Fig. 2). The first table is copied, with minor modifications, from Bulletin 769 of the United States Geological Survey.

TABLE SHOWING THE GEOLOGICAL COLUMN

Era	Period or Epoch	Characteristic Life	Duration, in millions of years, according to various estimates
Cenozoic (recent life).	Quarternary Recent Pleistocene or Glacial Period or Great Ice Age.	Age of man. Animals and plants of modern types.	55 to 65
	Tertiary Pliocene Miocene. Oligocene. Eocene.	Age of mammals. Possible first appearance of man. Rise and development of highest orders of plants.	
Mesozoic (intermediate life)	Cretaceous. Jurassic. Triassic.	Age of reptiles. Rise and culmination of huge land reptiles (dinosaurs), of shell-fish with complexly partitioned coiled shells (ammonites), and of great flying reptiles. First appearance (in Jurassic) of birds and mammals; of cycads, an order of palmlike plants (in Triassic); and of angiospermous plants, among which are palms and hardwood trees (in Cretaceous).	135 to 175
Paleozoic (old life)	Carboniferous. Permian. Pennsylvanian. Mississippian.	Age of amphibians. Dominance of club mosses (lycopods) and plants of horsetail and fern types. Primitive flowering plants and earliest cone-bearing trees. Beginnings of backboned land animals (land vertebrates). Insects. Animals with nautilus-like coiled shells (ammonites) and sharks abundant.	350 to 470
	Devonian.	Age of fishes. Shellfish (mollusks) also abundant. Rise of amphibians and land plants.	
	Silurian.	Shell-forming sea animals dominant, especially those related to the nautilus (cephalopods). Rise and culmination of the marine animals sometimes known as sea lilies (crinoids) and of giant scorpion-like crustaceans (eurypterids). Rise of fishes and of reef-building corals.	
	Ordovician.	Shell-forming sea animals, especially cephalopods and mollusk-like brachiopods. Culmination of the bug-like marine crustaceans known as trilobites. First trace of insect life.	
	Cambrian.	Trilobites and brachiopods most characteristic animals. Seaweeds (algae) abundant. No trace of land animals found.	
Pre-Cambrian	Algonkian.	First life that has left distinct record. Crustaceans, brachiopods. and seaweeds.	700 to 1000
	Archean	No fossils found.	

The geological record consists mainly of sedimentary beds—beds deposited in water. Over large areas, long periods of uplift and erosion intervened between periods of deposition. Every such interruption in deposition in any area produces there what geologists term an unconformity. Many of the time divisions shown above are separated by such unconformities—that is, the dividing lines in the table represent local or widespread uplifts or depressions of the earth's surface

Geologic Column

	System	Formation		Character, Use, Thickness	Cities on or near Formation Outcrop
CENOZOIC	Quaternary	Glacial Drift		Sand, clay, gravel, bowlders marl and peat. 0-600 ft	
PALEOZOIC	Mississippian	Kenwood		Black shale with blue clay bands Sparsely fossiliferous. 0-55 ft	MILWAUKEE
	Devonian	Milwaukee		Gray, brown or white dolomite, Calcareous shale in upper part Fossiliferous. 120-180 ft	MILWAUKEE LAKE CHURCH
	Silurian	Waubakee		Light gray to white thin-bedded dolo Local. 0-25 ft	WAUBAKEE
		Niagara Group		Dolomite, light gray, cherty in some layers, massive to thin bedded Locally coral reefs. Road material building stone & lime. 300-719 ft	RACINE WAUKESHA CHILTON STURGEON BAY
		Neda		Oolitic iron ore in local patches 0-55 ft	IRON RIDGE
	Ordovician	Richmond		Gray to blue and brown dolomitic shales. Fossiliferous Brick clays 150-540 ft	GREEN BAY
		Galena		Gray, buff-weathering dolomite, thick-bedded, cherty bands. Sparsely fossiliferous. Lead and zinc. 125-150 ft	BLUE MOUNDS
		Decorah	Black River	Calcareous and dolomitic blue shale. Fossiliferous. 15-20 ft	CASSVILLE
		Platteville		Shaly, blue-gray limestone and buff dolomite. Fossiliferous. Road material. 40-110 ft	MINERAL POINT
		St Peter		Sandstone moderately coarse; white, yellow, red. 0-330 ft	VIROQUA
		Shakopee	Lower Magnesian	Thin-bedded, light gray dolomite sandy at base. Cryptozoa. 0-50 ft	ARGYLE
		Oneota		Dolomite gray, thick-bedded, with layers of chert and oolite. Cryptozoa. Sparsely fossiliferous. Road material. 0-200 ft	LA CROSSE MADISON
	Cambrian	Madison		Dolomitic, fine-grained sandstone Locally fossiliferous. Building stone. 5-35 ft	MADISON
		Trempealeau		Sandstone, coarsest at top passing below to dolomitic shaly sandstone with dolomite locally at base. Fossiliferous. Road mat'l. 50-160 ft	MINDORO BLACK EARTH
		Mazomanie		Fine grained sandstone and green sand (locally rather coarse) with limy and shaly zones. Fossiliferous. Road shale. 100-170 ft	MAUSTON
		Franconia			DURAND
		Dresbach		Medium grained, thick bedded, white to yellow sandstone 100-170 ft	MENOMONIE
		Eau Claire		Very coarse grained, cross-bedded, thick-bedded, yellow to white sandstone (Mt Simon), passing upward gradually to fine-grained, fossiliferous sandstone and shale (Eau Claire). Building stone and road shale. 250-350 ft	COLFAX BLACK RIVER FALLS
		Mt Simon			EAU CLAIRE KILBOURN
PRE CAMBRIAN	Keweenawan			Ancient dark colored lava flows, conglom. and sandstone. Copper ore and crushed stone. 40 000-55 000 ft.	ST CROIX FALLS ASHLAND SUPERIOR
	Huronian			Quartzite, slate, marble, iron formation Iron ore and ganister. 8000-12 500 ft.	HURLEY ABLEMAN
	Archean			Granites, greenstones, schists, used for monumental and crushed stone.	CHIPPEWA FALLS WAUSAU RHINELANDER

Fig. 2. Geological Column for Wisconsin. W. H. Twenhofel, F. T. Thwaites, G. O. Raasch, and R. R. Shrock collaborated in the preparation of this column.

Minerals and Rocks. The table showing the geological column for Wisconsin indicates in one of the columns the kind of rock deposited in each period. Most rocks are aggregates of minerals. The composition and form of crystallization of the minerals helps to determine the nature of the rock, as is explained in the description of the three main classes of rocks.

Igneous Rocks. The granite and greenstone of the Archean and the trap rock of the Keweenawan are igneous rocks, formed by the cooling of masses of molten material. The individual minerals in the rapidly-cooled trap are not always distinguishable to the naked eye. In granite and some of the coarser greenstones, made by slow cooling, the minerals can often be seen plainly. Depending on the minerals which may be present, igneous rocks are weak or resistant. In respect to durability of different rocks, some knowledge of the geology of Wisconsin is of great importance in our study of the physical geography, or physiography, of the state.

Sedimentary Rocks. The rocks from the Keweenawan sandstone to the Niagara dolomite and Milwaukee shale are of an entirely different origin. They are known as sedimentary rocks. The sandstone and shale are made up of fragments of other rocks, worn and transported by rivers, waves, or the wind. Most sandstones and shales are assorted and deposited in the sea, as we know from the fossil shells of marine animals, preserved in them. Another type of sedimentary rock is conglomerate. This consists of cemented bowlders, cobblestones, and pebbles, like those in a gravel bank. Conglomerate occurs near the bases of certain of the Wisconsin sandstones and altered sandstones or quartzites.

In sandstones the chief mineral is quartz, the commonest and one of the most durable or resistant of minerals. The cementing of the loose sand into sandstone, by the deposition of quartz or lime or iron between the quartz grains, does not produce as resistant a rock as granite. Sand grains are more easily broken apart than mineral crystals. The latter interlock during the cooling of an igneous mass like granite or trap. On the other hand, all of the material in sandstone is resistant and fairly uniform, while the different minerals in granite are of varying durability. Some minerals decay more rapidly than others when exposed to the weather. Chert or flint, consisting chiefly of silica, like the mineral quartz, is exceedingly resistant. It occurs in certain of the limestone formations and in the soils derived from their decay.

Shale, as in the Richmond or the Milwaukee formation, is a relatively weak rock, being consolidated clay or mud. It is made up of disintegrated feldspar; in Wisconsin many shale beds contain impurities in the form of tiny fragments of mica, quartz, and other minerals.

Limestone may be a chemical precipitate of the mineral calcite —calcium carbonate—or dolomite—calcium magnesium carbonate. It may also be made up of the shells of sea animals. Some limestone is weak, some resistant. It is weak in the sense of being easily dissolved by underground water. The sedimentary rocks of this type in Wisconsin, however, are more durable than in some other parts of the United States. Most of them are dolomite rather than pure limestone, and, hence, are less easily dissolved. Part of the Galena limestone is more soluble and more porous, so that it is weaker than the Niagara limestone.

Metamorphic Rocks. The rocks of the pre-Cambrian, aside from the granite, greenstone, and trap, are metamorphic rocks. They have been cemented and recrystallized by water action and pressure, sometimes with heat. Shale has been made into slate or into schist, sandstone into quartzite, limestone into marble, and granite, greenstone, or conglomerate into gneiss or schist. This process of change, or metamorphism, may have produced banding. It may have developed new minerals. It has nearly always made the rock more resistant. Another type of metamorphism, however, weakens and destroys the rocks (p. 7). All the igneous and most of the metamorphic rocks, being made up of crystals, are spoken of as crystalline rocks.

Geological Structures. An exceedingly important geological feature is the arrangement of the rocks. Igneous rocks may be deposited originally in lava flows, as in northwestern Wisconsin and at Berlin and other points in the valley of the Fox River. Other igneous rocks occur in great underground masses. This was the case with the granites of northern Wisconsin, now exposed at the surface by the wearing away of the overlying rocks. Sedimentary beds or strata are originally flat-lying. Either one may be folded into arches and troughs, or broken by faults or joints. Faults are breaks along which there is movement and displacement. They are not very common in Wisconsin. Joints are similar breaks along which there is no displacement. They are to be seen in nearly all our rocks, as closely-spaced, vertical cracks, often in two sets at right angles (Pl. III).

In Wisconsin some of the metamorphic rocks, originally flat-lying sediments, have been folded in the most complex manner. Subsequently erosion has cut off the folds, so that now we have the bases of arches or anticlines, troughs or synclines, or faulted portions of either one. The ledges reveal rock layers in every position from vertical to horizontal. On the other hand, the flat-lying sedimentary rocks of Wisconsin have merely been warped or arched into an extremely-broad, anticlinal fold. The axis of this fold trends north and south; it is inclined, or pitches, southward. Consequently the sedimentary rock layers in central Wisconsin slope, or dip, southward. Those near Lake Michigan and near the Mississippi River, dip southeastward and southwestward respectively, in each case at a very low angle. This slight angle of inclination, however, has been a factor of vast importance in determining the topographic features of Wisconsin.

PHYSICAL GEOGRAPHY OR PHYSIOGRAPHY

Physiographic Processes. The rocks of the lands of the earth's crust, no matter what may be their composition or structure, are being attacked by two dominant processes—weathering and erosion. The first of these is a process of disintegration, partly chemical, partly mechanical. It weakens the rocks, causing them to crumble and decay. The second is largely a mechanical process, accompanying the transportation of weathered materials by gravity, streams, waves, glaciers, and the wind. The most important result of weathering and erosion is the modification of the land by disintegration and stream cutting, so that the upper layers of rock are removed and lower layers exposed. This laying bare of rock layers originally concealed is called denudation.

Weathering. The decomposition and disintegration, or weathering, of rocks is accompanied chiefly by (a) the oxidation—sometimes rusting—of minerals exposed to air and water, (b) by hydration and other chemical action, (c) by expansion of freezing water in joint cracks—frost action, (d) by expansion and contraction of rocks alternately heated and cooled, and (e) by the work of plants and animals. The removal of soluble parts of igneous, sedimentary and metamorphic rocks by underground water results in the destruction of the earth's surface. This is, in one sense, a part of the process of weathering; and weathering, indeed, is one type of metamorphism.

Weathering breaks down the strongest of rocks, so that gravity may cause them to fall or slide or creep, and streams, wind, waves, and glaciers may easily erode and carry them away. The surface of the land becomes mantled by the rock waste or soil, upon which our forests and orchards, and our agricultural and dairying industry depend. Soil, subsoil, and weathered material mantle the surface of the earth, because weathering goes on rapidly there. At some depth beneath the soil, however, there is always solid rock.

Underground Circulation in Relation to Hard Water and Ore Deposits. Underground water carries dissolved material, partly from weathered, surface rock and partly from deeper within the earth. A familiar evidence of this is the hardness of our drinking water and the deposit of lime in tea kettles and boiler tubes. In southern Wisconsin, there is abundant, soluble limestone in rock ledges, in glacial bowlders, and in the soil. Accordingly the solid content of drinking water—largely lime—is 300 to 550 parts per million. In northern Wisconsin, where there is not much limestone or limy soil, it is often as little as 22 to 41 parts per million.

The deposition of this dissolved material modifies all rocks. Sand is cemented to sandstone by this process. Layers or veins of minerals are deposited in crevices in the rock. The lead and zinc deposits of southwestern Wisconsin were made through solution and deposition by underground water. The iron ores of northern and eastern Wisconsin have been enriched and modified by the work of underground water, though the rocks containing these ores already had a large iron content.

The value of the mineral resources of Wisconsin in 1913 was 12½ million dollars, including zinc and lead, iron, limestone and granite, clay products, mineral water, and sand and gravel.

Erosion, Transportation and Deposition. The process of transporting weathered materials by streams results also in the erosion of valleys. Since the land is cut up by valley erosion, the process is sometimes called dissection. The valleys cut by streams are classified as young, mature, and old. The age is not measured in years, however, but in the stage of erosion in relation to what remains to be done. Thus a valley in solid rock may still be steep-sided, while a valley in unconsolidated surface materials may have gently-sloping sides, though both have existed the same number of years. The steep-sided valley or gorge is young. The valley with gently-sloping sides is mature or old. In young, mature, and old valleys there may be stream deposits temporarily laid down on their way to the sea.

The ultimate level toward which all streams are cutting their valleys is that of the sea. It is spoken of as baselevel. The sea level is the baselevel for the Mississippi and its tributaries in Wisconsin. Lake Superior and Lake Michigan are the local baselevels for streams in northern and eastern Wisconsin.

Other physiographic processes, which have eroded the surface in Wisconsin, are the waves in lakes, the wind in sandy and dry regions, and the former glaciers. The ice of glaciers differs from the waters of streams in being able to erode below baselevel, as in the basins of Lake Michigan and Lake Superior, which descend below sea level.

The ice which formerly existed in Wisconsin was part of an ice sheet or continental glacier which overrode all of northeastern North America (Fig. 27). It made vast changes in Wisconsin, both by erosion and by deposition. The deposits left by the former ice sheets are called glacial drift. They covered all but about a fourth of the state,—the Driftless Area. The glacial drift, and the contemporaneous deposits of the Driftless Area, occasionally yield the bones and tusks of the huge mammoth and mastodon, which inhabited Wisconsin before the Glacial Period.

Stages in the Erosion Cycle. An erosion cycle is the period of time during which a block of the earth's crust remains stationary with respect to sea level. During a completed cycle the streams pass successively through the stages of youth, maturity, and old age; and the land mass is denuded till it is reduced to low relief. Not all cycles, however, are completed. In Wisconsin few of the streams are yet in the stage of old age. Some of those in the Driftless Area are in late youth or maturity. Nearly all the rivers and creeks in the glaciated parts of the state are in the stage of extreme youth. The shorelines, likewise, are mostly youthful.

Soils,—the Product of Weathering, Erosion, and Deposition. The soils of Wisconsin are of the two main classes: (1) residual, (2) transported (Pl. II). Residual soils are largely confined to the Driftless Area, being made by the weathering of limestone, sandstone, and crystalline rocks (p. 7). The Driftless Area also contains a little transported soil, such as (a) dune sand and wind-blown silt, or loess, (b) stream-laid sand and gravel, and (c) hillside wash which creeps down the slopes under the influence of gravity.

The soil throughout the larger part of Wisconsin is transported material, brought to its present position by the former ice sheet. The glacial soils include (a) till or unassorted clay and bowldery sand deposited directly by the ice, (b) assorted glacial gravel and sand, laid down by ice-born rivers, (c) stratified clay and sand of the borders and the bottom of former glacial lakes, (d) loess, and (e) deposits made by weathering since glaciation.

Both driftless and glaciated areas also have (a) deposits of modern alluvium laid down by rivers, (b) deposits due to vegetable accumulation, including peat, muck, and meadow, (c) marl deposits, made up of the shells of small animals, and of plant secretions, and (d) modern beach deposits and sand dunes on the shores of Lake Michigan, Lake Superior, and some of the inland lakes.

The Soil as a Resource. In what follows in other chapters of this book it should be recognized that all soil is produced by natural agencies, such as have made the physiographic features of Wisconsin, that hillside-gullying and creep are continually at work carrying the upland soils to the valley bottoms; that, even though the subsoil is fresh and unaltered, weathering and underground water have leached away a notable portion of the lime from the glacially-transported soils—the youngest of our widespread surface accumulations. Least of all should we overlook the soil of Wisconsin as our greatest natural resource.

In recent years Wisconsin has produced annually between 400 and 500 million dollars' worth of crops, livestock, dairy products, fruit, lumber, and other commodities directly or indirectly dependent upon the soil. Dairying is the state's leading industry, the annual value of milk sold from farms being close to 200 million dollars. Cattle and calves valued between 40 and 50 million dollars and hogs valued at approximately the same amount have been marketed from the farms of the state in recent years. Poultry and eggs valued between 20 and 30 million dollars have also been marketed annually. Crops having a total farm value of between 250 and 300 million dollars annually are largely used on the farm and marketed through livestock. The annual crop in-

A. TRANSPORTED SOIL.

Glacial till in the Baraboo region. Below this transported glacial material, but not shown in this picture, is firm unweathered glacial rock ledge.

B. RESIDUAL SOIL.

Weathered limestone in the Driftless Area, grading downward into solid rock.

Pl. III.

JOINT PLANES IN METAMORPHIC ROCK, NORTHERN WISCONSIN.

come has been between 50 and 60 million dollars per year, or about one-fifth of the farm value of the total crop production. Over eighty per cent of the farm income in the state is now obtained from livestock and livestock products, leaving only about one-sixth of the total as coming directly from crops. Only a little over sixty per cent of the state's land area is in farms and less than a third of it is in crops and plowable pasture.

Two centuries and a half ago Wisconsin's great resource was its fur-bearing animals. Forty years ago the pine forest was our greatest asset. Indeed raw lumber and farm produce then approached equality. Each was worth 70 million dollars a year. Wheat raising came and went. Hay, oats, and corn are now the most valuable crops; and dairying is the dominant business of today. Yet even at the last census, the manufacturing industries employed nearly 250 thousand wage earners, and the value added to raw materials by manufacturing has in recent years exceeded 700 million dollars annually.

Throughout our progress from the fur trade to agriculture, from agriculture to dairying, and from the agricultural-dairying stage to the industrial stage, we have been dependent upon the soil. The most valuable of our manufactured articles are the products of the wood-working industry, which come from the soil. Water power for use in our industries is available (see Chapter XVI), thanks to the effects of glaciation upon drainage. The utility of the fertile soil depends on (a) favorable topography, (b) favorable climate, (c) markets. The productivity of these soils results from the operation of three physiographic processes,—weathering, erosion, and deposition, for long periods of time.

Present Occurrence of Rock Formations. The geological map of Wisconsin (Fig. 3) shows the portions of the state where the different rock layers now come to the surface, or outcrop. Where the pre-Cambrian is shown, its igneous and metamorphic rocks extend an unknown distance downward. The pre-Cambrian underlies the whole state, however, even where the map shows Cambrian sandstone or Silurian limestone on the surface. Likewise, the Cambrian lies below the Ordovician and Silurian. Although the whole series of Paleozoic rock layers seem to overlap each other like shingles on a roof, they are not at all the same length, as shingles are. They resemble a pile of boards laying one above the other, with the edges of the lowest ones projecting farthest.

The Quaternary formations—Pleistocene and Recent—cover the whole state, but to avoid confusion are not shown on the geological map. Of these, the glacial drift is of vast importance because it completely mantles a large area. In many places the underlying rocks are deeply buried. The stream deposits, or alluvium, cover

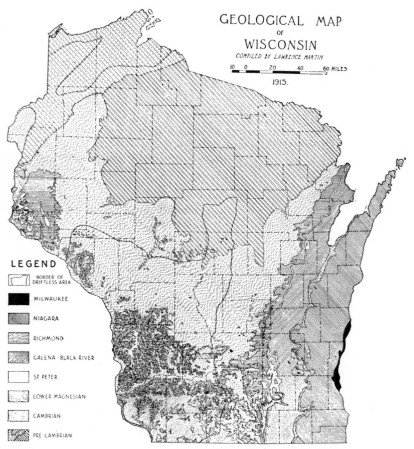

Fig. 3. The areas where the several rock formations are exposed at the surface in Wisconsin.

less of this state. Still less important in Wisconsin are the wind-blown deposits—dunes and loess—and the wave-worn deposits—beaches. The largest, modern, lake deposits are hidden beneath the waters of Lake Michigan, Lake Superior, and the inland lakes of Wisconsin. In some places, however, are deposits of extinct lakes.

History of Physiographic Features. The geological history of the state has included the following:

(1) all Wisconsin was mountainous, after the pre-Cambrian rocks were deposited and folded;

(2) before the Paleozoic sediments were deposited, the whole state was a low plain, or peneplain, with a few isolated hills rising above the general level; thus a long period in the ancient history of Wisconsin was devoted to erosion rather than to deposition;

(3) all of the state may have been alternately submerged beneath the waters of the ocean and slightly elevated, while the sedimentary rocks of the lower Paleozoic were being deposited;

(4) Wisconsin was again dry land, and was being fashioned into something similar to its present form; this was another period of erosion, but, because the land lay low, less was accomplished during this period than during the second one referred to above;

(5) all but the southwest corner of the state was buried beneath an ice sheet like those now found in Greenland and Antarctica;

(6) the ice sheet has melted away, leaving Wisconsin somewhat as before the Glacial Period, but with notable modification of topography, soil, and drainage. The Glacial Period may have lasted over a million years. It ended in the geological yesterday, perhaps only 35,000 to 50,000 years ago.

CLIMATE

General Position. The climatic features, essential to an understanding of the physical geography of the state, may be stated briefly. Weather and climate are to be treated in detail in a later publication of this Survey.

The state lies between 42° 30' and 47° north latitude. It is, therefore, never heated by the vertical rays of the sun. Although free from the extreme conditions of the tropics, the state is far enough south to escape the polar extremes and to have a year divided into four seasons. It receives sufficient heat from the sun to give a temperate climate. The position of the state, 900 to 1,000 miles from the Atlantic Ocean and the Gulf of Mexico, results in its having a continental climate,—that is, in having very cold winters and rather hot summers. Modifying this is the influence of the water in Lake Superior and Lake Michigan. This makes the range of temperature in Wisconsin somewhat less than in the states away from the Great Lakes, like Minnesota and North Dakota. Wisconsin lies in the belt of prevailing westerly winds. It has the variable climate incidental to the passing of a succession of cyclonic storms, determined by areas of high barometric pressure and low barometric pressure.

Temperature.   The mean annual temperature of Wisconsin is about 43° Fahrenheit. The average temperature ranges from 48°, in the southwestern part of the state, to 39°, in the northern portion (Fig. 4). The extreme range is from 111° above to 50° below zero. There is a slight difference of temperature with altitude,

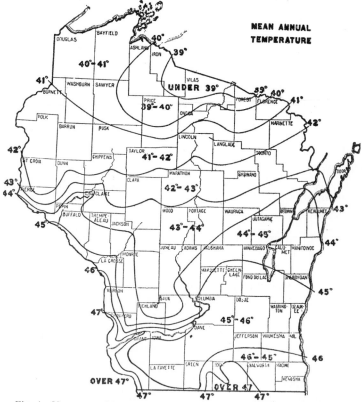

Fig. 4. Mean annual temperatures in Wisconsin. (Whitson and Baker.)

as from La Crosse to Viroqua. The latter place is $1\frac{1}{10}°$ colder, because it is 730 feet higher. The daily range of temperature is about 18° in summer and 14° in winter. The state has a summer temperature similar to that of France, Germany, and southeastern

England, and a winter temperature comparable to that of northern Sweden and central Russia.

Temperature may be thought of as important in relation to the physical geography of the lands,—first, in regard to daily extremes of heat and cold, which bring about weathering of rocks by expansion and contraction; and secondly, in the limitation of such work to the spring, summer, and autumn, for in the winter of a snowy region much weathering, and, indeed, not a little stream erosion and wave work are at a standstill.

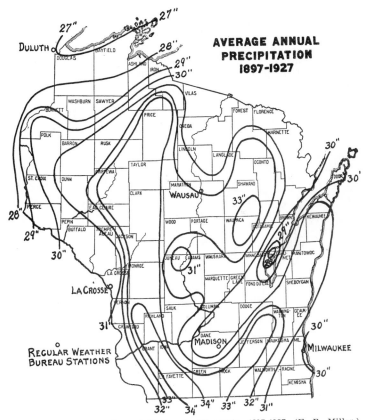

Fig. 5. Average annual precipitation in Wisconsin, 1897-1927. (E. R. Miller.)

Rainfall. The mean annual precipitation, or rainfall and snowfall, in Wisconsin is 31 inches (Fig. 5), ranging from 27 to 34 inches in various parts of the state. About half of the total rainfall comes in May, June, July, and August. The portions of Wisconsin receiving the most rainfall are, in general, the higher northern and southwestern portions. Most of the winter precipitation comes in the form of snow.

Applied Climatology. The important points about the Wisconsin precipitation are (1) that the region is humid and has the topographic development characteristic of a humid region. There is enough precipitation so that all but the smallest streams have water

Conifers, with some mixed hardwoods.

Dwarf oak and pine, including pine barrens.

Oak group, including swamps and prairies.

Maple group.

Fig. 6. Forest map of Wisconsin. (After Wisconsin Geological Survey, 1882.)

the year round. (2) The rainfall supports such vegetation that run-off is well regulated; at least it probably was before the conditions were modified by settlement, the forest land in northern

Wisconsin denuded, many swamps drained, and many of the fields ploughed. (3) The heaviest rainfall comes after the cessation of spring floods due to the melting of the winter's snow, so that stream erosion and transportation are fairly well distributed in spring, summer and autumn.

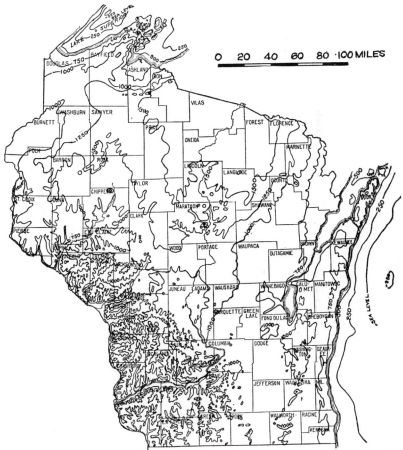

Fig. 7. Map of Wisconsin showing its general topographic form. Contour interval 250 feet. (Based on contour maps of the U. S. Geological Survey, Mississippi River Commission and U. S. Lake Survey, and upon railway profiles. Compiled by F. T. Thwaites.)

It may be assumed that, with the exception of the Glacial Period, Wisconsin has had much the same temperature and rainfall conditions ever since this land was uplifted above the sea during Paleozoic time. Thus it is probable that the land has been sculptured into its present form by processes in every way comparable to those now in operation.

TOPOGRAPHY

General Form. The general topographic form of Wisconsin is indicated in the relief map of the state (Plate I).* The larger part of the state lies between 700 or 800 feet in the southeast and 1600 or 1800 feet above sea level in the north (Fig. 7). The highest point, 1940 feet, is Rib Mountain, Marathon County, near Wausau. The lowest elevation, 581 feet, is the eastern coast at the level of Lake Michigan. The northern coast at the level of Lake Superior, 602 feet, is nearly as low, as is the western border along the Mississippi River, where the elevation ranges from 670 feet at Lake St. Croix, near Prescott and Hudson, to 600 feet at the extreme southwestern corner of the state in Grant County, opposite Dubuque, Iowa. The approximate mean altitude is 1050.

HYDROGRAPHY

Relation to Topography, Geology and Climate. The hydrography, or description of the rivers and lakes and coast, is involved with the features previously discussed as follows. The topography controls the slopes down which water shall flow. This itself, however, is originally determined by the geology, for (a) the dip of the surface of the sedimentary rocks provides original slopes for drainage, (b) the resistance or weakness of rock textures determines where uplands shall be left and lowlands sculptured by stream erosion, (c) the ice invasion, by its erosion and deposition, modifies preglacial drainage, causes waterfalls and rapids, and provides basins for lakes. The humid climate, also, is of importance, determining the steady water supply of the streams and the maintenance of the lakes.

Rivers. The drainage of Wisconsin is treated in detail in the chapters on the several uplands and lowlands. It may be briefly summarized as follows. The state is divided by a major watershed (Fig. 8), which determines (a) that some of the streams shall flow east into the Atlantic Ocean by way of Lake Superior or Lake Michigan, and (b) that the remaining streams—the larger number —shall flow south into the Gulf of Mexico by way of the Mississippi River.

The largest river of the state is the Mississippi, on the western border. Its chief affluents are the St. Croix, Chippewa, Black, and

---

* Excluded from the 1965 reprinting.

Wisconsin rivers, which unite with it in Wisconsin, and the Rock River and several smaller streams which flow through Illinois to the Mississippi. In the St. Lawrence drainage are the Fox-Wolf-Lake Winnebago system, the Menominee on the northeastern boundary between Wisconsin and Michigan and many smaller streams which flow directly into Green Bay, Lake Michigan and Lake Superior.

Fig. 8. The main drainage basins in Wisconsin.

The Lake Michigan streams include the Manitowoc, Sheboygan, and Milwaukee rivers. The chief streams that enter into Lake Superior are the St. Louis, Nemadji, Bois Brule, Bad, and Montreal.

Although smaller than the Mississippi, the Wisconsin River is really the master stream of the state. It rises in Lac Vieux Desert on the Michigan boundary and flows south through the center of the state for four-fifths of its length, before turning westward to the Mississippi.

Lakes. Lake Michigan and Lake Superior are the largest of the Wisconsin lakes, being shared, of course, with the adjacent states. Except in the Driftless Area, there are great numbers of lakes within the boundaries of the state. A recent count made by the conservation commission shows 3834 mapped lakes. Of these the largest is Lake Winnebago. The other lakes fall in four groups.

The first group includes the scattered, moderate-sized lakes in eastern and southeastern Wisconsin. These include the four-lake group of the Yahara River near Madison—Mendota, Monona, Waubesa, and Kegonsa—as well as Lakes Koshkonong, Geneva, Beaver, Puckaway, Poygan, and Shawano, the Oconomowoc and Waupaca groups, and many others.

The second group, including many hundreds of small lakes, lies in the highland lake district of northern Wisconsin, chiefly in Vilas, Oneida and Iron counties. All these lakes are small, but there are few parts of the world where so large a portion of the total area is occupied by lakes.

The third group is in northwestern Wisconsin, especially in Washburn, Burnett, Polk, Barron, and Sawyer counties. These, like the second group, are small lakes, very close together.

Lastly, we have Lake St. Croix and Lake Pepin—long, narrow bodies of water—interrupting respectively the courses of the St. Croix and Mississippi rivers. Allied to them are the hundreds of small floodplain lakes of the Mississippi bottom lands.

Coasts. The coast of Wisconsin is over 500 miles long without counting islands and minor indentations. Between a third and a half of Wisconsin fronts on the water of Lakes Superior and Michigan. Off the coasts are two large promontories, two great bays, and two archipelagos of islands. In Lake Michigan are the Door Peninsula, Green Bay, and Washington and adjacent islands. In Lake Superior the corresponding features are Bayfield Peninsula, Chequamegon Bay, and the Apostle Islands. The remainder of the coast is characterized by great simplicity. There are no large natural harbors on Lake Michigan. The ports of Milwaukee, Racine, Kenosha, Sheboygan, Manitowoc, Kewaunee, Algoma, Two Rivers, and Port Washington, as well as Green Bay and Marinette on Green Bay, are merely the improved mouths of rivers. These have been made into very satisfactory harbors. The harbors of Superior, Ashland, and Port Wing on Lake Superior are larger and better, lying as they do behind sand spits. The lake frontage is, of course, a great resource of Wisconsin in relation to transportation.

The depth of water in Green Bay is from 50 to 120 feet; in the Wisconsin portion of Lake Michigan it is from 280 to 870 feet. The boundary between the states of Wisconsin and Michigan is in the middle of the lake, whose surface is 581 feet above sea level. This means that some of the submerged, eastern portion of the state is 300 feet below the level of the ocean. The same is true of northern Wisconsin, where the deepest part of the bottom of western Lake Superior is about at sea level.

RELATION OF TOPOGRAPHY AND HYDROGRAPHY TO HISTORY

The Fox-Wisconsin Waterway. By all odds the most important topographic feature of Wisconsin in relation to its history is the diagonal valley which extends from Lake Michigan to the Mississippi. This is occupied by the waters of Green Bay, the Fox River and Lake Winnebago, and the lower Wisconsin River.

The traditional site of the landfall of Jean Nicolet, first white man to visit Wisconsin, is at Red Banks, on the southeast coast of Green Bay, where in the early seventeenth century was a Winnebago village. Nicolet, the agent of Champlain, governor of New France, explored this region in 1634. Twenty years or more later two French traders, Radisson and Groseilliers came to this far western land; whether or not they traversed this waterway and saw the Mississippi cannot be proved from the evidence at hand. The first authentic travelers over the Fox-Wisconsin were Louis Jolliet and Father Jacques Marquette, who were sent in 1673 to find the Mississippi. In this they were successful, entering the Great River June 17 at the mouth of the Wisconsin. Thence they turned south and followed the Mississippi as far as the Arkansas. Father Louis Hennepin rescued from captivity by Sieur Duluth was the first to traverse this waterway from the west in 1680.

The Mississippi Waterway. The French had long heard from the Indians of a great central river, which they hoped would lead to the South Sea (Pacific Ocean). After Jolliet had proved that the Mississippi entered the Gulf of Mexico, it was traversed during the entire French regime by traders, soldiers, officers, and missionaries. First of these was Father Hennepin, who came up from the mouth of the Illinois River as far as the Falls of St. Anthony to which he gave their name. Nicolas Perrot, commanding for France on the Mississippi, wintered in 1685-86 near Trempealeau. He later built Fort St. Antoine on Lake Pepin, and there in 1689 he took possession of all the upper Mississippi region for the king of France.

The Bois Brulé-St. Croix Route. Next to the Fox-Wisconsin waterway the most important route from the upper lakes to the Mississippi was via the Bois Brulé (now the Brule) and the St. Croix rivers, with a short portage at upper St. Croix Lake. This was first traversed in 1680 by Duluth, who came from Lake Superior. In 1693 Pierre le Sueur built two forts commanding the ends of this waterway.

Lake Superior, St. Louis River, and Chequamegon Bay. This region was the first to be explored after Nicolet had visited Green Bay. In 1658 Radisson and Groseilliers built the first known dwelling of any white man in Wisconsin on the west shore of Chequamegon Bay. Other traders were here in 1660. From here Father René Ménard started on his journey to the headwaters of Black River, where he was lost in the forest. In 1665 Father Claude Allouez founded the Mission of La Pointe St. Esprit. He built a mission house at the southwest corner of the bay, and from there explored the western and northern shores of Lake Superior. The mission was removed to Mackinac in 1671. Duluth traversed the lake in 1679, ascended St. Louis River and ventured into the lake region of Minnesota.

Thus it was throughout all the early explorations in Wisconsin; the routes of fur traders, missionaries, and soldiers followed the Indian highways. These routes were always directly related to topography and hydrography, controlled ever by the river, the portage, the point, the bay.

La Pointe and La Baye. Two geographical features—a point and a bay—furnished the names most commonly used in the seventeenth century for the leading centers of population in Wisconsin. These names La Pointe—the region near Chequamegon Point and La Baye—the region near Green Bay—were familiar in eastern Canada, and also in France and England.

THE PLACE OF WISCONSIN IN THE PHYSICAL GEOGRAPHY OF THE UNITED STATES

On the basis of various geographic features, especially the geology, topography, climate and vegetation, it is customary to divide the United States in natural regions, physical divisions, or physiographic provinces. The state of Wisconsin lies in two of these provinces—(a) the Laurentian Upland, or Superior Upland, or Lake Superior Highland, or Wisconsin-Michigan Uplands, or Lake Plains, or Lake Region, and (b) the Interior Plains, or Prairie Plains, or Central Low Plains, or Central Lowland. In this book the state will be divided into five provinces, appropriate to this more detailed study in regional geography.

USEFUL GENERAL WORKS ON WISCONSIN PHYSICAL GEOGRAPHY

Case, E. C.   Wisconsin, Its Geology and Physical Geography, Milwaukee, 1907, 190 pp.

Collie, G. L.   Physiography of Wisconsin, Bulletin American Bureau of Geography, Vol. 2, 1901, pp. 270-287.

Davis, W. M.   (On the physiographic features of Wisconsin), Physical Geography, Boston, 1898, pp. 136-137, 197-198, 274; Erklärende Beschreibung der Landformen, Leipzig, 1912, pp. 222-223.

Durand, Loyal, Jr.   The River Systems of Wisconsin, Unpublished thesis, University of Wisconsin, 1925; The Geographic Regions of Wisconsin, Unpublished thesis, University of Wisconsin, 1930.

Hall, James.   Physical Geography (of Wisconsin), Report on the Geological Survey of the State of Wisconsin, Vol. 1, 1862, pp. 1-7.

Lapham, I. A.   A Geographical and Topographical Description of Wisconsin; with Brief Sketches of its History, Geology, Mineralogy, Natural History, Population, Soil, Productions, Government, Antiquities, &c. &c., Milwaukee, 1844, 256 pp; second edition, with the title: Wisconsin, Its Geography and Topography, Milwaukee, 1846, 202 pp.

Martin, Lawrence.   The Physical Geography and Pleistocene of the Lake Superior Region, (a discussion of the topography and glacial geology of the northern two-thirds of Wisconsin), Monograph 52, U. S. Geol. Survey, 1911, pp. 85-117, 427-459; The Physical Geography of Wisconsin, Journal of Geography, Vol. 12, 1914, pp. 226-232.

Mueller, Amy F.   Geology and Physiography of the Wisconsin State Parks, Unpublished thesis, University of Wisconsin, 1927.

Norwood, J. G.   General Observations on the Topography and Climate of Wisconsin,—in Owen's Geological Reconnoissance of the Chippewa Land District of Wisconsin, Senate Ex. Document 57, 30th Congress, 1st Session, Washington, 1848, pp. 110-121.

Penn, Margaret. Physiographic Influences of Population Changes in Wisconsin, Unpublished thesis, University of Wisconsin, 1926.

Timothy, Sister. The Effects of Glaciation on the River Systems of Wisconsin, Unpublished thesis, University of Wisconsin, 1922.

Whitney, J. D. Division of the State of Wisconsin into Districts on Geological and Topographical Grounds, Sketch of their Physical Geography,—In Hall and Whitney's Report on the Geological Survey of the State of Wisconsin, Vol. 1, 1862, chapter on Physical Geography and Surface Geology, pp. 97-139.

GEOLOGICAL HISTORY

Chamberlin, T. C. Historical Geology, Geology of Wisconsin, Vol. 1, 1883, pp. 45-300; see also Geological History of Wisconsin in Snyder, Van Vechten & Co.'s Atlas, Milwaukee, 1878, pp. 148-51.

Hall, James. General Geology (of Wisconsin), Report on the Geological Survey of the State of Wisconsin, Vol. 1, 1862, pp. 8-72.

Hotchkiss, W. O., and Bean, E. F. A Brief Outline of the Geology, Physical Geography, Geography, and Industries of Wisconsin, Bull. 67, Wisconsin Geol. and Nat. Hist. Survey, 1925, 60 pp.

Hotchkiss, W. O. and Thwaites, F. T. Outline of the Geological History of Wisconsin, border printing on Map of Wisconsin Showing Geology and Roads, Wis. Geol. and Nat. Hist. Survey, 1911; see also border printing on Geological Model of Wisconsin, 1910.

Irving, R. D. Sketch of the Geological Structure of Wisconsin, Trans. Amer. Inst. Mining Engineers, Vol. 8, 1880, pp. 479-491.

Lapham, I. A. Geology (of Wisconsin—an account of the historical geology of the State), in Walling's Atlas of the State of Wisconsin, 1876, pp. 16-19.

Thwaites, F. T. The Paleozoic Rocks Found in Deep Wells in Wisconsin and Northern Illinois, Jour. Geology, Vol. 31, 1923, pp. 529-555.

Ulrich, E. O. Notes on New Names in Table of Formations and on Physical Evidences of Breaks Between Paleozoic Systems in Wisconsin, Wisconsin Acad. Sci., Trans., Vol. 21, 1924, pp. 71-107.

ROCKS, MINERALS, AND SOILS

Buckley, E. R. Building and Ornamental Stones of Wisconsin, Bull. 4, Wis. Geol. and Nat. Hist. Survey, 1898, 500 pp; The Clays and Clay Industries of Wisconsin, Ibid., Bull. 7, 1901, 283 pp; Highway Construction in Wisconsin, Ibid., Bull. 10, 1903, 313 pp.

Chamberlin, T. C. Chemical Geology and Lithological Geology, Geology of Wisconsin, Vol. 1, 1883, pp. 3-44.

Chamberlin, T. C. Soils and Subsoils of Wisconsin, Geology of Wisconsin, Vol. 1, 1883, pp. 678-688.

Hotchkiss, W. O., and Steidtmann, Edward. Limestone Road Materials of Wisconsin, Bull. 34, Wis. Geol. and Nat. Hist. Survey, 1914, 137 pp.

Huels, F. W. The Peat Resources of Wisconsin, Bull. 45, Ibid., 1915, 266 pp.

Irving, R. D. Lithology of Wisconsin, Geology of Wisconsin, Vol. 1, 1883, pp. 340-361.

Lawson, P. V. Story of the Rocks and Minerals of Wisconsin, Appleton, 1906, 202 pp.

Ries, Heinrich. The Clays of Wisconsin, Bull. 15, Wis. Geol. and Nat. Hist. Survey, 1906, 247 pp.

Whitney, J. D. Mineralogy (of Wisconsin) in Hall and Whitney's Report on the Geological Survey of the State of Wisconsin, Vol. 1, 1862, pp. 193-220.

Whitson, A. R. Soils of Wisconsin, Bull. 68, Wisconsin Geol. and Nat. Hist. Survey, 1927, 270 pp.

Whitson, A. R., and others. Soils reports: Wisconsin Geol. and Nat. Hist. Survey, Bulls. 11, 23, 24, 28, 29, 30, 31, 32, 37, 38, 39, 40, 43, 47, 48, 49, 52, 53, 54, 55, 56, 59, 60, 61, 62, 72, 77. All except 72A and 77 have maps on scale of either three miles to one inch or one mile to one inch.

CLIMATE

Bormann. Development of Water Power in Wisconsin and Relation of Precipitation to Stream Flow, (map, Average Annual Rainfall of Wisconsin), Monthly Weather Review, Vol. 41, 1913, pp. 1020-1022.

Chief of Engineers. Meteorological Observations by the Lake Survey, Annual Report to the Secretary of War, Cong. Doc. Ser. Nos.—1024, 1079, 1118, 1184, 1285, 1325, 1368, 1413; years,—1859, 1860, 1861, 1862-3, 1866, 1867, 1868, 1869, respectively.

Chief Signal Officer. U. S. Army Annual Reports, 1871—1891 inclusive.

Cox, H. J. Frost and Temperature Conditions in the Cranberry Marshes of Wisconsin, U. S. Weather Bureau, Bull. T, Washington, 1910, pp. 121.

Day, P. C. Precipitation of the Drainage Area of the Great Lakes, Monthly Weather Review, Vol. 54, 1926, pp. 85-106.

Greely, A. W., and Schott, C. A. General Map of Rainfall and Temperature of Wisconsin, Atlas Plate II C, Geology of Wisconsin, 1882.

King, F. H. Physical Features and Climatic Conditions of Northern Wisconsin,—in W. A. Henry's Northern Wisconsin—A Handbook for the Homeseeker, Madison, 1896, pp. 24-40.

Lapham, I. A. (On the climate of Wisconsin), A Geographical and Topographical Description of Wisconsin, Milwaukee, 1844, pp. 84-92; Ibid., 1846, pp. 75-80; Climate of Wisconsin, Trans. Wis. State Agr. Soc., Vol. 1, 1851, pp. 305-324; Ibid., Vol. 2, 1852, pp. 445-455; Meteorology, Northwestern Journal of Education, Science, and General Literature, Vol. 1, 1850, pp. 117-122; On the Climate of the Country Bordering upon the Great North American Lakes, Trans. Chicago Acad. Sci., Vol. 1, 1867, pp. 58-60, and Isothermal Map of Wisconsin; the same map, reproduced earlier, as A Geological and Climatological Map of Wisconsin, Showing the Geological Formations and the Effect of Lake Michigan in Elevating the Mean Temperature of January and Depressing that of July, 36 miles to 1 inch—Insert on G. A. Randall's New Sectional and Township Map of the State of Wisconsin; Rainfall, Geology of Wisconsin, Vol. 2, 1877, pp. 35-44.

Mead, D. W. Mean Annual Rainfall in Wisconsin 1895-1918, in Hydrology, New York, 1919, Fig. 114, p. 207.

Miller, Eric. R.   The Meteorological Influences of Lakes, Proc. 2d Pan.-Am. Sci. Cong., Vol. 2, 1917, p. 189-198; Measurements of Solar Radiation at Madison, Wis., with the Callendar Pyrheliometer, Monthly Weather Review, Vol. 48, 1920, pp. 338-343; Rainfall Maps of Wisconsin and Adjoining States, Trans. Wis. Acad. Sci., Arts, and Letters, Vol. 24, 1929, pp. 501-508; A Century of Temperatures in Wisconsin, Ibid., Vol. 23, 1927, pp. 165-177; Monthly Rainfall Maps of Wisconsin and Adjoining States, Ibid., Vol. 25, 1930, pp. 135-156; Extremes of Temperature in Wisconsin, Ibid., Vol. 26, 1931.

Norwood, J. G.   General Observations on the Topography and Climate of Wisconsin, in Owen's Geological Reconnoissance of the Chippewa Land District of Wisconsin, Senate Ex. Doc. 57, 30th Congress, 1st Session, Washington, 1848, pp. 121-129.

Oldenhage, H. H.   Climatology of Wisconsin, in Snyder, Van Vechten & Co.'s Atlas, Milwaukee, 1878, pp. 18, 151-153, (including a Climatological Map, showing average rainfall and temperature for Summer, Winter, and the Year).

Piippo, Arthur F.   Seventeen Year Record of Sun and Sky Radiation at Madison, Wis., April 1911 to March 1928 incl., Monthly Weather Review, Vol. 56, 1928, pp. 499-504.

Schott, C. A.   Table and Results of the Precipitation in Rain and Snow in the United States and at some Stations in the Adjacent Parts of North America and in Central and South America, Smithsonian Contributions to Knowledge 353, Washington 1872, Reprinted 1881; Tables, Distribution and Variations of the Atmospheric Temperature in the United States and some Adjacent Parts of America, Smithsonian Contributions to Knowledge 277, Washington, 1876.

Smith, L. S.   Climatic Conditions (of Northern Wisconsin), Water Supply Paper 156, U. S. Geol. Survey, 1906, pp. 15-19, including a rainfall map of Wisconsin (see also other Water Supply Papers of the U. S. Geological Survey); Climatic Conditions (in Wisconsin), Bull. 20, Wis. Geol. and Nat. Hist. Survey, 1908, pp. 14-21, et. seq., including 12 separate maps showing rainfall of the state for 1895 to 1905.

Stewart, W. P.   Summary of Climatological Data for the United States by Sections:—Section 58, Northwestern Wisconsin; Section 59, Central Wisconsin; Section 60, Eastern Wisconsin;—U. S. Weather Bureau, Bull. W, 2d ed., Washington, 1926.

U. S. Bureau of Agricultural Economics.   Atlas of American Agriculture, Part 2, Climate:—Section A, Precipitation and Humidity, Washington, 1922; Section B, Temperature, Sunshine and Wind, Washington, 1928; Section I, Frost and the Growing Season, Washington, 1918.

U. S. Patent Office and Smithsonian Institution.   Results of Meteorological Observations Made Under Direction of, 1854-1859, Washington, Vol. 1, 1861, Vol. 2, 1864.

U. S. Surgeon-General's Office.   Meteorological Register for the Years 1822, 1823, 1824, 1825 from Observations made by the Surgeons of the Army at the Military Posts of the United States, Washington, 1826, 63 pp; Meteorological Register for the Years 1826, 1827, 1828, 1829, 1830, (and the preceding as supplement) Philadelphia, 1840, 161 pp.; Meteorological register for Twelve Years from 1831 to 1842 inclusive, Washington, 1851, 324 pp.; Army Meteorological Register for Twelve Years from 1843 to 1852 inclusive, xi, Washington, 1855, 763 pp.

U. S. **Weather Bureau.** Annual Report of the Chief, 1891 to date; Climate and Crop Reports, Monthly Weather Review, and Climatological Data.

Whitson, A. R., and **Baker, O. E.** The Climate of Wisconsin and Its Relation to Agriculture, Agricultural Experiment Station Bulletin 223, University of Wisconsin, 1912, 65 pp; 2nd edition revised 1928; see also climatic summaries in soils bulletins, Wis. Geol. and Nat. Hist. Survey.

Winchell, Alexander. Map of Wisconsin Showing Climate, Walling's Atlas of the State of Wisconsin, Milwaukee, 1876 (with isotherms for Summer, Winter, and the Year).

Williams, F. E. The Climate of Wisconsin, Journ. Geog., Vol. 12, 1914, pp. 232-234.

HYDROGRAPHY

Bird, H. P., and **others.** Report of the committee of the Wisconsin Legislature, 1910, on Water Powers, Forestry, and Drainage, Madison, 1911, 779 pp.

Birge, E. A., and **Juday, C.** The Inland Lakes of Wisconsin,—The Dissolved Gases of the water and Their Biological Significance, Bull. 22, Wis. Geol. and Nat. Hist. Survey, 1911, 259 pp.

Blaisdell, J. J. Forest and Tree Culture in Wisconsin, Madison, 1893, 46 pp., (including a discussion of forests in relation to rainfall, freshets, etc.).

Chamberlin, T. C. Artesian Wells, Geology of Wisconsin, Vol. 1, 1883, pp. 689-701.

Devereaux, W. C. Relation of Deforestation to Precipitation and Run-off in Wisconsin, Monthly Weather Review, Vol. 38, 1910, pp. 720-723.

Griffith, E. M. The Intimate Relation of Forest Cover to Stream Flow, Report of the Committee of the Wisconsin Legislature, 1910, on Water Powers, Forestry, and Drainage, Part 2, 1911, pp. 723-736.

Juday, C. The Inland Lakes of Wisconsin,—The Hydrography and Morphometry of the Lakes, Bull. 27, Wis. Geol. and Nat. Hist. Survey, 1914, 137 pp., (including 29 colored maps).

Kirchoffer, W. G. The Sources of Water Supply in Wisconsin, Bull. 106, University of Wisconsin, 1905, pp. 165-245.

Lapham, I. A. Geological map published in 1869 shows by 100-foot contours the "Probable depth in feet below the level of Lake Michigan at which the azoic or primary rocks may be reached by artesian wells."

Mead, D. W. The Geology of Wisconsin Water Supplies, author's edition Rockford, 1893; The Hydro-Geology of the Upper Mississippi Valley, Journ. Assoc. Eng. Societies, Vol. 13, 1894, 68 pp; The Flow of Streams and the Factors that Modify It, with Special Reference to Wisconsin Conditions, Bull. 425, University of Wisconsin, 1911, 192 pp; see also Exhibit 29 in report by H. P. Bird and others, pp. 737-779.

Smith, L. S. The Water Powers of Wisconsin, Bull. 20, Wis. Geol. and Nat. Hist. Survey, 1908, 354 pp; Wisconsin's Water Power Resources, Wisconsin Engineer, Vol. 13, 1909, pp. 273-285; see also Water Supply Paper 156, U. S. Geol. Survey, 1906, and Water Supply Papers 207, 285, 324, 325, and 354.

Weidman, S., and Schultz, A. R.   Underground and Surface Water Supplies of Wisconsin, Bull. 35, Wis. Geol. and Nat. Hist. Survey, 1915, (accompanied by geological map of Wisconsin, with 250 foot contours on the surface of the pre-Cambrian, and 100 foot contours showing artesian head).

Wisconsin Railroad Commission.   Gazetteer of Streams, Report of the Railroad Commission to the Legislature on Water Powers, Madison, 1915, pp. 489-540; the 1912 edition of the Official Railroad Map of Wisconsin contains a list of rivers and lakes of the state, whose location may be determined by referring to the numbers and letters printed on the border of the map.

APPLIED GEOGRAPHY. A FEW OF THE EARLIEST AND OF THE LATEST CONTRIBUTIONS

Dopp, Mary.   Geographical Influences in the Development of Wisconsin, Bull. Amer. Geog. Soc., Vol. 45, 1913, pp. 401-412, 490-499, 585-609, 653-663, 736-749, 831-846, 902-920.

Gregory, John.   Industrial Resources of Wisconsin, Chicago, 1853, 318 pp; Ibid., 1870, 320 pp.

Hobbins, Joseph.   Health of Wisconsin, in Relation to Physical Features, Geology, Climate, etc., Snyder, Van Vechten & Co.'s Atlas, Milwaukee, 1878, pp. 182-187; also in Descriptive America, Vol. 1, 1884, pp. 110-111.

Hunt, J. W.   Wisconsin Gazetteer, Madison, 1853, 255 pp.

Irving, R. D.   Mineral Resources of Wisconsin, Snyder, Van Vechten & Co.'s Atlas, Milwaukee, 1878, pp. 163-165; The Mineral Resources of Wisconsin, Trans. Amer. Inst. Mining Engineers, Vol. 8, 1880, pp. 478-508.

Kellogg, Louise Phelps.   French Regime in Wisconsin and the Northwest, Madison, 1925, 474 pp.

Lapham, Increase A.   Descriptions of each of the Wisconsin counties and their topography, drainage, resources, industries, population, etc., A Geographical and Topographical Description of Wisconsin, Milwaukee, 1844, pp. 95-252; Wisconsin, Its Geography and Topography, Milwaukee, 1846, pp. 82-202; Report on the Commerce of the Town of Milwaukee, and Navigation of Lake Michigan, pamphlet, 1842; Wisconsin, Her Topographical Features, and General Adaptation to Agriculture, Northwestern Journal of Education, Science, and General Literature, Vol. 1, 1850, pp. 46-49; Statistics Exhibiting the History, Climate, and Productions of the State of Wisconsin, Published by order of the legislature, Madison, 1867, 32 pp. and map; On the Relations of the Wisconsin Geological Survey to Agriculture, Trans. Wis. Agr. Soc., Vol. 12, 1874, pp. 207-210.

Lapham, I. A., Knapp, J. G., and Crocker, H.   Report on the Disastrous Effects of the Destruction of Forest Trees, Madison, 1867, 101 pp.

Martin, Lawrence. The Progressive Development of Resources in the Lake Superior Region, Bull. Amer. Geog. Soc., Vol. 43, 1911, pp. 561-572, 659-669.

Merrill, J. A. Industrial Geography of Wisconsin, Chicago, 1911, 175 pp.

Ritchie, J. S. Wisconsin and Its Resources, with Lake Superior, its Commerce and Navigation, Philadelphia, 1857, 312 pp.

Salisbury, R. D. Geology, Soil, Vegetation, Zoology, Mines, Quarries, and Mineral Springs (of Wisconsin), Descriptive America, Vol. 1, 1884, pp. 109-116.

Schafer, Joseph. Agriculture in Wisconsin, Madison, 1922.

The Wisconsin Number of the Journal of Geography, Vol. 12, 1914, pp. 225-300, by Lawrence Martin, F. E. Williams, C. F. Watson, E. G. Lange, W. S. Welles, Lilla Brabant, R. H. Whitbeck, W. R. McConnell, C. E. Slothower, B. A. Stickle, L. T. Gould, J. A. Merrill, and E. F. Bean.

Turner, Lura A., and J. M. Handbook of Wisconsin, Its History and Geography, Burlington, Wis., 1898, 266 pp.

United States Census. Supplements for Wisconsin on Population, Agriculture, Manufactures, Mines and Quarries, Fifteenth Census of the United States, Taken in the Year 1930.

Whitbeck, R. H. Industries of Wisconsin and their Geographic Basis, Annals Assoc. Amer. Geographers, Vol. 2, 1912, pp. 55-64; Geography and Industries of Wisconsin, Bull. 26, Wis. Geol. and Nat. Hist. Survey, 1913, 90 pp; Geography of the Fox-Winnebago Valley, Bull. 42, Ibid., 1915, 102 pp; Geography of Southeastern Wisconsin, Bull 58, Ibid., 1920, 160 pp.

The most convenient, short history of the state is R. G. Thwaites' "Wisconsin, the Americanization of a French Settlement," Boston, 1908, 432 pp. See also contemporary documents on the French and the British régimes in Wisconsin, Vols. 16-20, Collections Wis. Hist. Soc., 1902-11; earlier volumes of the State Historical Society's Collections and Proceedings; and the Jesuit Relations. For origins of geographical names in Wisconsin see references in Appendix H.

MAPS

General Maps and Index Maps. For general maps of the whole state, geological maps, and glacial maps see Appendix E. Figure 197 and Figures 199 to 205 are index maps. Appendix E explains where and how all these maps may be obtained.

U. S. Geological Survey. Standard quandrangles covering parts of the state and referred to in Appendix E and by regions in the bibliographies of ensuing chapters; special river Maps; for special geological and glacial maps see references in Appendix G; for altitudes at railway stations see Appendix H.

U. S. Lake Survey. (As above).

U. S. Mississippi River Commission. (As above).

Wisconsin Geological and Natural History Survey. Special lake maps; soil maps; lead and zinc district maps; mineral land classification maps; model of Wisconsin, etc. (see Appendices E and G).

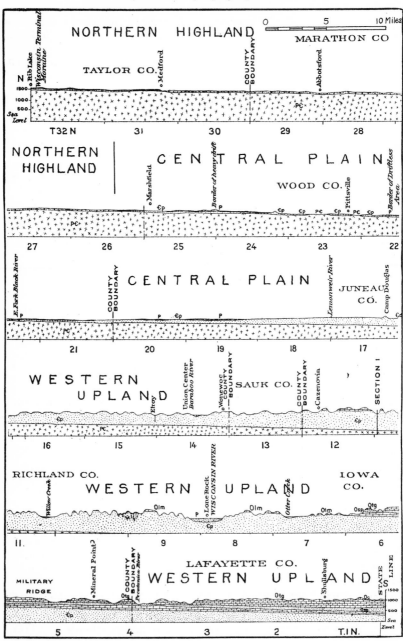

Fig. 9. A north-south section showing the relationships of the Western Upland to the underlying rock. PC—igneous and metamorphic rock of pre-Cambrian age; Cp—Cambrian sandstone; Olm—Lower Magnesian limestone of Ordovician age; Osp, Otg, and Oc—St. Peter sandstone, Black River-Galena limestone, and Richmond shale, all of Ordovician age; P—glacial deposits and river deposits of Pleistocene age.

CHAPTER II

THE GEOGRAPHICAL PROVINCES OF WISCONSIN

THE NATURE OF THE PROVINCES

Driftless Area and Glaciated Region. One simple way to describe the state of Wisconsin is to divide it into two parts,—(a) the Driftless Area and (b) the Glaciated Region. The Glaciated Area is mostly a plain. A large part of the Driftless Area is hilly. The Driftless Area and the Glaciated Region are natural regions, or, as we shall say in this book, geographical provinces.

Plains, Plateaus, and Mountains. From another point of view the state may be said to consist of three natural regions. These are (a) a large area of plains, (b) a smaller area of low plateaus, and (c) a large area of worn-down mountains. The plains are not all of the same level. The plateaus are so cut up by streams as to retain no flat-topped uplands. The former mountains are now worn down so low as to constitute a rather simple plain, although it includes the highest land in the state.

Physiographic Provinces. Another writer, formally resident in our state, in describing the whole United States, has divided Wisconsin into four physical divisions. His "Superior Upland" corresponds essentially with the *Northern Highland* and *Lake Superior Lowland,* described below; his other provinces are called "Eastern Lake Section", "Till Plains", and "Wisconsin Driftless Section." Both this writer and the author of this book agree that there are parts of Wisconsin and the Middle West where details of glacial geology have resulted in distinctions which are more continuous over large areas than bedrock geology is.

The Five Geographical Provinces. It seems best, however, to divide the state into five rather than two or three or four natural regions. These are related in certain ways to the driftless and glaciated areas, and to the plains, plateaus, and worn-down mountains. Three of these geographical provinces are uplands and two are lowlands.

The northernmost of the five geographical provinces is the *Lake Superior Lowland* (Chapter XVII), a part of the Lake Superior basin. The central province, called the *Northern Highland* (Chapter XV), is an upland,—part of the Lake Superior highland. A third division of the state is the large belted plain with curved strips of alternating lowland and upland. This plain is subdivided into three geographical provinces, (a) the *Central Plain* (Chapter XIII), (b) the *Eastern Ridges* and *Lowlands* (Chapter IX), and (c) the *Western Upland* (Chapter III). These geographical regions are not merely physiographic provinces or areas whose surface features have origins of a common nature. The geographical provinces of Wisconsin are also related to the use of the land by plants, by animals, and by man. Each differs from the others in roughness or smoothness of topography, in fertility or sterility of soil, in climate, in adaptation to occupation by wild plants, including forests, by cultivated plants, including crops and orchards, by animals, and by man, as well as in the extent to which white men have developed such resources during the march of Wisconsin history. The Northern Highland partakes of the characteristics of the frontier because it is remote, climatically inferior through altitude and short period between killing frosts, and off the main paths of trans-Wisconsin transportation. The Lake Superior Lowland is a dissected plain with good soil. The Western Upland is rough. The Central Plain, except where glaciated, is smooth but infertile. The Eastern Ridges and Lowlands are smooth, low, fertile, and convenient to through transportation, and hence this is the most densely populated and the richest of the geographical provinces of the state. With time, however, all parts of the state will become more and more alike in use of the land by man.

BOUNDARIES OF THE GEOGRAPHICAL PROVINCES

The boundaries of all five of these geographical provinces are determined largely by the variations of texture and structure in the underlying rocks. The fundamental differences of topography in the five divisions are due to this, and the minor topography in each area shows this relationship clearly. The boundaries of these provinces are shown in Figure 10. They do not follow contours on the topographic map (Fig. 7), they are not determined by climate (Figs. 4, 5), or by vegetation (Fig. 6), or any other single geographical feature. They resemble, but do not exactly correspond to the boundaries on the geological map (Fig. 3). This shows that neither the relative weakness and resistance involved in rock texture, nor the control involved in rock structure, is entirely responsible for the difference in conditions between provinces. Nor is it only the process of wearing down that has determined the con-

Fig. 10. The five geographical provinces of Wisconsin.

trasting topographies of the provinces. It is true that Wisconsin might be divided into two provinces on the basis of process alone,— weathering and stream erosion in one, the Driftless Area, having produced markedly different forms from the glacial erosion and deposition, following weathering and stream erosion, in the glaciated area (Fig. 28). The author has relegated process to a secondary place, however, in drawing the boundaries of the provinces (Fig. 10), making rock characters—texture and structure—a major criterion of classification, with physiographic process and stage for aid in subdivision. That geographical regions so delimited have internal unity and notable contrast with neighboring regions in uses of the land by living things, including man, is a high testimonial to the truism that man adapts himself to his environment and is moulded by it.

THE NORTHERN HIGHLAND

The Highland Is a Peneplain. A plain made by the wearing down of ancient mountains is usually spoken of as a peneplain,— that is, a region worn down nearly to a plain in a place where, formerly, there was rougher topography. The wearing down has been accomplished in a long period of time by the erosive action of streams and the weather. This peneplain is underlain by pre-Cambrian rocks, including igneous, sedimentary, and metamorphic rocks of Archean and Algonkian age.

Throughout the highland of northern Wisconsin, the low plain-like topography truncates highly-folded and complexly-faulted, ancient, sedimentary and metamorphic rocks (Fig. 143). The peneplain surface bevels across granites and associated igneous masses, which must have been originally intruded deep below the surface. This state of affairs is interpreted as indicating that the peneplain was formerly a mountainous region. That the granites and some of the other igneous rocks were intruded deep below the present surface is proved by their character. In order to have cooled slowly enough to have their present coarsely-crystalline character they must have been deeply buried when intruded. The restoration of the folds in the sedimentary rocks, now planed across by erosion, shows that when the folding was completed the region must have risen much higher above sea level than at present, and must have had a mountainous topography. In other words, the topography and rock structure found at present in northern Wisconsin are of the same kinds that would be revealed if the Alps or Rocky Mountains were planed across so that their basement portions were exposed.

Above this peneplain surface rise knobs and ridges of resistant rock. These are residual eminences which stand out by virtue of the superior hardness or resistance of their rocks. They are called monadnocks after a mountain of that type in southwestern New Hampshire.

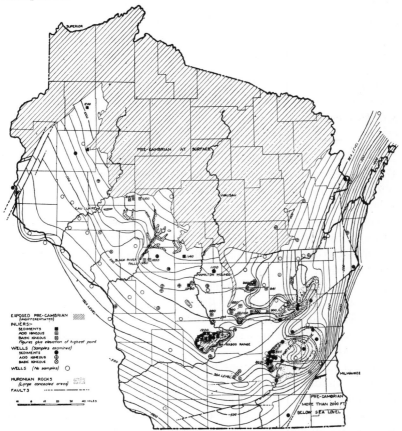

Fig. 11. The peneplain of the Northern Highland of Wisconsin (obliquely-ruled area of exposed pre-Cambrian). Contour lines 100 feet apart show the elevation of the fossil land surfaces of the buried peneplain and its monadnocks. Recent studies of borings suggest that some portions of the buried peneplain have more relief than had previously been supposed. It is also possible that two ancient peneplains are here represented, since the contours in northeastern Wisconsin are closely spaced (see discussion of **The Hidden Peneplain, pp. 390–392**). The contours on this map are based upon data compiled by Thwaites.

It was formerly thought that northern Wisconsin was at one time an island in the sea. The name Isle Wisconsin was applied to it.

The Barron Hills, Baraboo Range, and certain hills near Waterloo, and in the present Fox River valley were thought of as offlying islands. It does not now seem probable, however, that the Northern Highland was ever an island. If there ever was an Isle Wisconsin it must have included the whole state (p. 395), rather than merely the present northern portion where the ancient peneplain is now visible. After the peneplain was formed it seems to have been completely buried beneath the younger sedimentary rocks of the Paleozoic. The northern part of it has only recently been exhumed.

The basin of Lake Superior lies within the Northern Highland. Here the peneplain has been downfaulted, buried, and partly uncovered again.

The Belted Plain of Wisconsin

Relation to Rock Texture. The portion of Wisconsin south of the Lake Superior Lowland and the Northern Highland is a belted plain. It consists of two lowland areas and three upland areas, related directly to the texture and structure of the underlying rocks. These are the Paleozoic rocks which overlie the pre-Cambrian of the Northern Highland. As one goes to the southeast or southwest from the Northern Highland there are found in the following order:

(a) The Cambrian sandstone lowland.

(b) The Lower Magnesian limestone upland.

(c) The Galena-Black River limestone lowland.

(d) or, elsewhere in the state, the upland on the Galena-Black River.

(e) The Niagara limestone upland.

These upland and lowland subdivisions of central and southern Wisconsin, which result in its being a belted plain, are determined by relative weakness or resistance of the underlying rocks. Two rock formations in the state, however, the St. Peter sandstone and the Richmond shale are weak and are usually found only in narrow strips at the bases of the resistant limestones which overlie them. The two chief lowlands (a) on the Cambrian sandstone and (b) on the Galena-Black River limestone were formed because these sedimentary rocks are relatively weak in their resistance to denudation. The three upland areas, cuestas (p. 44), (a) on the Lower Magnesian limestone, (b) on another portion of the Galena-Black River limestone, and (c) on the Niagara limestone, stand up above the adjacent lowlands because of the superior resistance of these limestone formations, just as the durable character of the resistant, igneous and metamorphic rocks of the Lake Superior peneplain makes it stand higher than the adjacent Cambrian sandstone lowland.

Relation to Peneplain of Northern Highland. The present distribution of the uplands and lowlands of the belted plain is not a

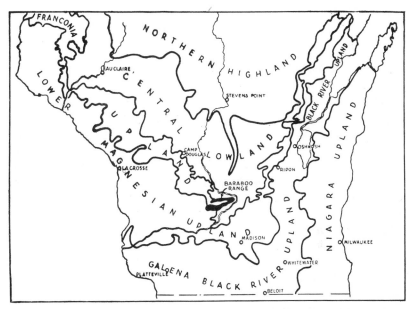

Fig. 12. The cuestas in Wisconsin and the inner lowland which separates them from the exhumed peneplain. The first cuesta is formed by the Lower Magnesian limestone and the Franconia (Cambrian) formation; the second cuesta by the Galena-Black River limestone; the third cuesta by the Niagara limestone; the central plain or central lowland by the lower beds in the Cambrian sandstone; and the Northern Highland by the metamorphic and igneous rocks of the pre-Cambrian. A second lowland, occupied by Green Bay, the lower Fox River, Lake Winnebago, and Rock River, lies on the Galena-Black River limestone just west of the third cuesta. Lake Michigan occupies a third lowland underlain by weak Devonian shales.

permanent but a temporary one, considered from a physiographic standpoint. This is because these sedimentary rocks are being slowly removed from the surface of the pre-Cambrian peneplain of northern Wisconsin. As already stated (p. 36), it seems quite likely that the Cambrian sandstone and some of the overlying Paleozoic rocks formerly covered most, if not all, of the Lake Superior Highland.

That these Paleozoic rocks were more extensive than at present is clearly indicated by the presence of outliers. The outliers are isolated sandstone mounds in the Northern Highland, at some distance from the area of continuous sandstone with which they were formerly connected. These outliers show clearly that the sandstone is wasting away under the attack of stream erosion and weathering. In Wisconsin the highest elevation is now stripped of its sandstone covering and the overlying mantle still lies on the lowland. A similar relationship is found with regard to the edges of the overlying limestone formations. The driftless portion of the escarpment near the edge of the Lower Magnesian limestone is an exceedingly irregular one. Every here and there, an outlier of limestone is isolated upon the sandstone upland, rising above it as a hill or mound. The Niagara limestone bears a similar relationship to the surface of the Galena-Black River limestone.

Another indication that the lowlands of the belted plain are being extended is found in the fact that here and there in central and southern Wisconsin portions of the peneplain of northern Wisconsin have been uncovered. The Baraboo Range of south-central Wisconsin, for example, and similar masses of quartzite and igneous rocks near Waterloo, near Necedah, in the Fox River valley, and at other localities, will be described (p. 387) as former monadnocks upon the pre-Cambrian peneplain. The monadnocks were buried by the Cambrian sandstone and the overlying sedimentary rocks. They are now being exposed as the sandstone and limestone are carried away by erosion. Their superior height results in their being uncovered sooner than the low portion of the peneplain around them. Narrow strips of the peneplain surface have also been exhumed. These are inliers (Fig. 11). They are low and lie along stream courses.

Relation to Geological Structure. The outlines of these zones of lowland and upland in the belted plain do not extend east and west across the state. They curve around the eastern, southern, and western portions of the peneplain of northern Wisconsin, which now projects like a shield. The crescentic form of the several belts of lowland and upland is explained by the relationship of erosion to the structure of the alternating layers of weak and resistant rock. As explained in Chapter I, there is a broad anticlinal fold in Wisconsin. Its main axis extends in a general north and south direction, and it pitches gently southward. The Lake Superior peneplain, originally nearly horizontal, was probably warped into this position after the sedimentary rocks were laid down over it. As a result of the warping, it, therefore, happened that northern Wisconsin became the highest part of the area. It is natural, accordingly, that denudation has first cut through to the older rocks in the northern part of the state. The several sandstone and limestone formations have been worn back in orderly succession, to their present positions, all around the flanks of the pitching fold.

The crescentic form of the Cambrian sandstone lowland is well developed in Wisconsin. The crescentic extent of upland topography in the areas of Lower Magnesian limestone and Galena-Black River limestone is also complete in this state. The parallel curve of the Niagara limestone upland, however, is not wholly developed, part of it being present in the eastern portion of the state, from the Door Peninsula and the upland east of Lake Winnebago southward into Illinois. To the south the Niagara limestone is buried beneath the coal measures and only reappears in northwestern Illinois and to the west in Iowa (Fig. 87), a short distance outside the borders of Wisconsin.

The Belts of Lowland and Upland. Southern Wisconsin may, therefore, be thought of as a belted plain, similar to the belted plains of eastern England and the Paris basin in France. The Northern Highland of Wisconsin was part of the oldland or ancient land surface, upon which the sedimentary rocks of the Paleozoic were deposited. Some of these sediments were derived from the Northern Highland. On this basis Wisconsin has been described as an ancient coastal plain (Fig. 12). The Northern Highland is not known to be the chief or only source of the younger sedimentary rocks, but the analogy is satisfactory in all other respects.

A noteworthy feature of the belted plain of Wisconsin is the different character of the topography on the Galena-Black River limestone (a) in the vicinity of Green Bay, Lake Winnebago, and the Rock River valley, and (b) in southwestern Wisconsin. In the former locality the limestone forms a lowland 580 to 800 feet above sea level, in the latter an upland at an elevation of 900 to 1200 feet. This difference affects so large an area that it has made it desirable to discuss the belted plain, not along the division lines suggested by the geological formations (p. 36), but in the three following geographical provinces:

(a) The Central Plain [the plain of Cambrian sandstone].

(b) The Eastern Ridges and Lowlands [(1) the eastern cuesta of Lower Magnesian limestone, (2) the eastern lowland of Galena-Black River limestone, and (3) the cuesta of Niagara limestone].

(c) The Western Upland [(1) the cuesta of Galena-Black River limestone in southwestern Wisconsin, (2) the cuestas of Lower Magnesian limestone and of the Franconia formation along the western border of the state].

THE ORDER OF DISCUSSION OF THE PROVINCES

In this book the five geographical provinces will be described in order from south to north. The reason for adopting this order is that it seems most logical to begin with the Western Upland because it lies in the Driftless Area. The Driftless Area preserves most of the types of topography that formerly existed throughout Wisconsin. Its topography was made chiefly by weathering, wind work, underground water, and stream erosion. Thus we may study these physiographic processes and the topographic forms they produce before we consider the forms that exist in eastern, central, and northern Wisconsin. Outside the Driftless Area, glacial erosion and glacial deposition, wave work, postglacial stream erosion, and other processes have greatly modified the topography originally made by the weathering and preglacial stream work.

The five geographical provinces of Wisconsin will be discussed in the following chapters, which will explain the origin of the topographic forms and discuss the rivers, lakes, and coasts of Wisconsin, together with a few relationships to man.

BIBLIOGRAPHY

See Chapter 1, pp. 23-29, and lists of articles and maps at the ends of Chapters III, IX, XIII, XV, XVII, and in Appendix G. For a description of physiographic divisions of the United States by N. M. Fenneman, see Annals Assoc. Amer. Geographers, Vol. 6, 1916, 98 pp. and màp (in pocket). For a graphic picture of the physical features of Wisconsin in their nation-wide setting, see A. K. Lobeck's Physiographic Diagram of the United States, 1921.

CHAPTER III

THE WESTERN UPLAND

NATURE OF THE COUNTRY

Scenery of the Western Upland. The landscape in western Wisconsin makes this one of the most attractive parts of the state. Owen, the first geologist to do detailed work in Wisconsin, said in 1847:

"The constant theme of remark, whilst travelling in the regions of the upper Mississippi occupied by the lower magnesian limestone, was the picturesque character of the landscape, and especially the striking similarity which the rock exposure presents to that of ruined structures.

"The scenery on the Rhine, with its castellated heights, has been the frequent theme of remark and admiration by European travellers. Yet it is doubtful whether, in actual beauty of landscape, it is nòt equalled by that of some of the streams that water this region of the far west. It is certain that though the rock formations essentially differ, nature has here fashioned, on an extensive scale, and in advance of all civilization, remarkable and curious counterparts to the artificial landscape which has given celebrity to that part of the European continent.

"The features of the scenery are not, indeed, of the loftiest and most impressive character. There are no elevated peaks, rising in majestic grandeur; no mountain torrents, shrouded in foam and chafing in their rocky channels; no deep and narrow valleys hemmed in on every side and forming, as it were, a little world of their own; no narrow and precipitous passes, winding through circuitous defiles; no cavernous gorges giving exist to pent-up waters; no contorted and twisted strata, affording evidence of gigantic uplift and violent throes. But the features of the scene, though less grand and bold than those of mountainous regions, are yet impressive and strongly marked. We find the luxuriant sward, clothing even down to the water's edge the hill slope. We have the steep cliff shooting up through it, in mural escarpments. We have the stream, clear as crystal, now quiet, and smooth, and glassy, then ruffled by a temporary rapid, or when a terrace of rock abruptly crosses it, broken up into a small romantic cascade. We have clumps of trees, disposed with an effect that might baffle the landscape gardener, now crowning the grassy height, now dotting the green slope with partial and isolated shade. From the hill tops the intervening valleys wear the aspect of cultivated meadows and rich pasture grounds, irrigated by frequent rivulets that wend their way through fields of wild hay, fringed with flourishing willows. Here and there occupying its nook, on the bank of the stream, at some favourable spot, occurs the solitary wigwam, with its scanty appurtenances. On the summit levels spreads the wide prairie, decked with flowers of the gayest hue; its long undulating waves stretching away till sky and meadow mingle in the distant horizon. The whole combination suggests the idea, not of an aboriginal wilderness, inhabited by savage tribes, but of a country lately under a high state of cultivation and suddenly deserted by its inhabitants; their dwellings indeed gone, but the castle-homes of their chieftains only partially destroyed, and showing, in ruins, on the rocky summits around. This latter feature especially aids the delusion; for the peculiar aspect of the exposed limestone and its manner of weathering cause it to assume a resemblance somewhat fantastic indeed, but yet wonderfully close and faithful, to the dilapidated wall, with its crowning parapet and its projecting buttresses and its flanking towers, and even the lesser details that mark the fortress of the olden time."

Location and General Geography. The region with such scenery as Owen describes occupies the western and southwestern portion of the state, comprising part or all of the counties adjacent to the Mississippi River (Fig. 10). The Western Upland contains about 13,250 square miles.

This is a highland region, as the name Western Upland indicates. Most of it is a thoroughly-dissected upland, not a flat-topped or sloping surface as in northern Wisconsin or the region near Lake Michigan. The average elevation of the hilltops above sea level is about 1100 feet in St. Croix and Pierce counties in northwestern Wisconsin, 1280 feet in Vernon County, and 900 to 1200 feet in Grant County in the southwestern part of the state.

The uplands thus stand 100 to 200 feet above the Eastern Ridges and Lowlands to the southeast, and 200 to 350 feet above the Central Plain to the northeast.

In certain ways the Western Upland is similar to the Allegheny and Cumberland Plateaus in the Appalachians. Northwestern Wisconsin near Hudson and River Falls is much like the plateau of western New York. Western and southwestern Wisconsin near La Crosse, Prairie du Chien, and Platteville resemble the rugged plateau of West Virginia or Kentucky.

Aside from the upland itself the strongest topographic features of the region are the great trenches or gorges of the Mississippi and Wisconsin rivers and their numerous branches. The gorge of the Mississippi is incised more than 500 feet below the level of the upland ridges.

State Boundaries in Relation to Topography. The western boundary of Wisconsin is a natural one, following the gorge of the Mississippi. The states of Wisconsin, Iowa, and Minnesota have rarely failed to agree as to their limits, except in a small stretch in Lake Pepin (p. 171), for the main channel of the Mississippi furnishes a definite water boundary. In contrast with this we have the Wisconsin-Illinois boundary, which extends east and west without any natural feature to determine it. The surface features of southwestern and southeastern Wisconsin are just like the adjacent parts of Illinois. Accordingly there was long discussion as to whether the 14 northern counties of what is now Illinois should be in Wisconsin or not. These counties cover 8500 square miles, an area larger than Massachusetts. They include the site of the present city of Chicago and exceedingly valuable coal mines, lead mines, and agricultural lands. The people in several of these counties voted almost unanimously to join Wisconsin; but, rather curiously, the people of Wisconsin voted not to accept them. The Wisconsin-Illinois boundary was finally fixed at the parallel of 42° 30′, or about where it is now; and thus we lost Chicago.

A mistake was made in actually marking the boundary, however, and it is about half a mile too far north on the Mississippi River in Grant County, and a similar distance too far south on Lake Michigan in Kenosha County. It does not extend due east, as provided by the law (p. 487); but it is marked by boundary posts, and is fixed for all time. There is private litigation respecting it, but no suits between Illinois and Wisconsin. A few years ago a problem arose as to its precise location at one point where illicit making of liquor at a small still almost exactly on the State boundary raised the question as to whether the offender should be tried in a federal court in Illinois or in one in Wisconsin.

The river boundary between Wisconsin and Iowa, and between Wisconsin and Minnesota is, in one way, less satisfactory, for it is not actually marked. It follows the middle of the main channel of the Mississippi. Fortunately this main channel has not shifted to any great extent in recent years.

Other state boundary problems related to rivers and to suits in the Supreme Court of the United States between Minnesota and Wisconsin and between Michigan and Wisconsin are discussed on pages 171-172, 422-427, 447-452 and 481-487.

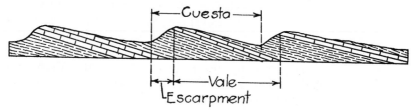

Fig. 13. A series of cuestas and escarpments. (Veatch.)

The Two Cuestas. The upland or plateau region of western Wisconsin consists of two cuestas and one monadnock. A cuesta is an upland belt with a short, steep descent, or escarpment, on one side and a long, gentle slope on the other. The gentle slope usually corresponds to the inclination or dip of slightly-inclined sedimentary rocks. One resistant layer, as of limestone, may determine the whole dip slope. The streams which drain the back slopes of cuestas are called consequent streams, having their courses as a consequence of original topographic surfaces which are roughly parallel to the dip of the rocks. The streams which flow down the escarpments of the cuestas, being opposite in direction to the consequent streams, are called obsequent streams. An escarpment often discloses a basal layer of weak rock, perhaps shale or sandstone. This weak rock usually forms a sloping surface, leading up to a steep cliff, where the resistant overlying stratum comes to the surface in the face of the escarpment.

The cuestas in the Western Upland are similar to the cuestas in eastern Wisconsin (p. 212), but differ from them in three signifi-

A DISSECTED CUESTA IN THE DRIFTLESS AREA NEAR ELROY.
Part of the Western Upland underlain by Cambrian sandstone and mantled by loess.

A. THE MAGNESIAN ESCARPMENT WEST OF THE CHIPPEWA RIVER NEAR KNAPP.

B. CUESTA SURFACE NORTH OF PRESCOTT.

Plain in the foreground is underlain by Lower Magnesian limestone and glacial drift. Flat-topped hill in the background is a mesa of Black River limestone.

cant respects. The first is the dip of the rock, which is somewhat less in western Wisconsin. The second is the presence of large outliers of Galena-Black River on the Lower Magnesian limestone. The third and most important contrast is in the topography, for a large part of the Western Upland is in the never-glaciated or

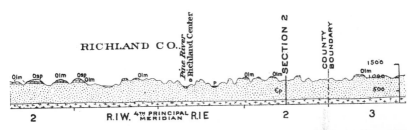

Fig. 14. Cross-section of part of the Western Upland, showing the Lower Magnesian limestone (Olm) and St. Peter sandstone (Osp) resting upon the Cambrian sandstone. The cuesta is thoroughly cut up by streams and all the ridges are narrow.

Driftless Area. These contrasts will be amplified in the subsequent pages. Suffice it here to state that most of the province is not a flat-topped upland or plateau, but a thoroughly-dissected cuesta. With the exception of the area northwest of the Chippewa River, it has no smooth upland areas of notable extent. It is a region of high, narrow ridges and deep, steep-sided valleys.

The northern four-fifths of the Western Upland lies in the belt of Lower Magnesian limestone, and to a smaller extent in the area of the Cambrian sandstone. The southern fifth of the province lies in the belt of Galena-Black River limestone. A small portion of the Western Upland is the Baraboo Range. This is not a cuesta, but an exhumed monadnock made up of pre-Cambrian metamorphic and igneous rocks. The Franconia Upland (Fig. 12) is a part of the Magnesian cuesta (page 50).

THE UPLAND NORTH OF THE WISCONSIN RIVER

Extent and Topography. The part of the Western Upland north of the Wisconsin River, is 180 miles long and 35 to 75 miles wide. The topography is controlled largely by the Lower Magnesian limestone. The Magnesian cuesta in eastern Wisconsin is only 2 to 7 miles wide, while in western Wisconsin it is 10 to 17 times as wide. This is due to the flatter dip and greater thickness of the rocks in the Western Upland.

The general topography of this part of the Western Upland is indicated by the following table, which presents a north-south section, the places listed being 20 to 40 miles apart.

TABLE SHOWING ELEVATIONS IN THE NORTHERN PART OF THE WESTERN UPLAND

Locality	County	Elevation in feet
East Farmington	Polk	1038
Hammond	St. Croix	1103
Ellsworth	Pierce	1070
East of Alma	Buffalo	1240
North of Bangor	La Crosse	1340
Northwest of Norwalk	Monroe	1440
Viroqua	Vernon	1274
Near Richland Center	Richland	1160

The elevations, given above, show the general altitude of the hilltops, which the table shows to have an extreme range of only two or three hundred feet. In the southern counties listed there is considerable relief, the valleys being incised three or four hundred feet below the general level, as is indicated specifically in the following pages.

The Highland Between the St. Croix and Chippewa Rivers. The cuesta-makers are the Lower Magnesian and Franconia formations. The highland is a flat-topped region, whose surface relief is slight. All of it has been glaciated and the relief has no doubt been decreased by the glacial deposition. The rocks in this upland dip southwest and south at a low angle. There are minor folds and occasional faults. The action of erosion upon the nearly-horizontal rock structures has produced a cuesta with a low and irregular escarpment on the east, but no definite escarpment on the north, on account of the thick mantle of glacial deposits.

The Escarpment West of the Chippewa. The eastern escarpment, facing the Central Plain of Cambrian sandstone trends north and south for 20 to 25 miles in St. Croix, Dunn, and Pepin counties. It is a moderately-irregular escarpment with salients extending forward from 3 to 5 miles between stream valley embayments. There are a few detached hills capped by the cuesta-making formation close to the cuesta. One outlying mass of Lower Magnesian limestone, northwest of the city of Menomonie, lies 7 miles east of the escarpment. The height of the escarpment is about 200 feet—221 feet at Knapp and 167 feet west of Downing—and the descent to the sandstone plain is everywhere abrupt.

The Cuesta Surface. The surface of the cuesta slopes rather uniformly westward and southwestward, at the rate of about 9 feet to the mile. The divide is 5 or 6 miles west of the escarpment. The irregularities are due to stream erosion and to glacial deposi-

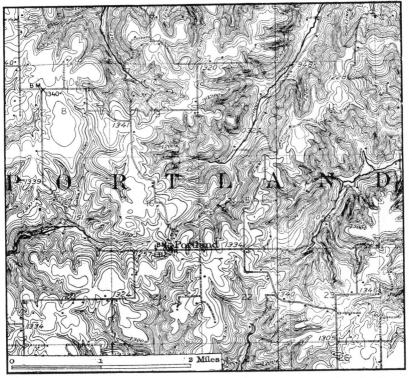

Fig. 15. Topographic map of part of the Western Upland of Wisconsin, showing the deep valleys and rounded ridges. Elevations in feet above sea level. Contour interval 20 feet. For explanation of contours, see Appendix E, p. 497. (From Sparta Quadrangle, U. S. Geol. Survey.)

tion. The former are especially well-marked in the hilly, marginal belt near the St. Croix and Mississippi rivers, and in the east-facing escarpment near the Chippewa valley. A large part of the upland is a smooth plain or low plateau.

Hills on the Cuesta. Rising above the surface of the upland of Lower Magnesian limestone are higher areas capped by the Galena-Black River limestone. They occupy a comparatively small portion of the northwestern upland, but are surrounded by considerable areas of St. Peter sandstone, which has slight relief above the Magnesian limestone upland. The western part of Pierce County, near and to the north of Ellsworth and the southwestern part of St. Croix County, east of Hudson, are made distinctly irregular by the presence of the isolated hills and long ridges of the Galena-Black River limestone. Between these buttes and mesas the Magnesian upland is very smooth, as between Prescott and River Falls. At the borders of the hills of Galena-Black River limestone there is nearly always a steep limestone escarpment, succeeded by a gentle slope of St. Peter sandstone and then by the even Magnesian limestone upland below.

Ridges and Coulees Between Chippewa and La Crosse Rivers. The portion of the Western Upland between the Chippewa and La Crosse rivers is distinctly different from the portion just described. Between the St. Croix and Chippewa rivers is a fairly broad and not very hilly upland, while the higher region between the Chippewa and the La Crosse has been dissected into a system of ridges and valleys, with practically no upland area remaining. The hilltops do not exceed 1100 to 1300 feet above sea level.

The French name coulee is the designation for valley which is prevalent in the region discussed. This region of ridges and coulees occupies all of Buffalo and Trempealeau counties and parts of La Crosse, Monroe, Jackson, Eau Claire, and Pepin counties.

The east-facing escarpment is not at the edge of the Lower Magnesian limestone. It lies 20 to 30 miles farther east. Here erosion has cut down through the calcareous and somewhat resistant upper layers of the Cambrian sandstone. In the weak lower beds this sandstone ceases to be a ridge-maker and the smoother surface of the Central Plain begins. In the upland northwest of the Chippewa River practically all the topography is controlled by the Lower Magnesian limestone, while in the ridge-and-coulee area under discussion this limestone determines the topography of less than a fifth of the region and the Franconia sandstone is the important rock formation.

The reason for this difference seems to be the positions of sandstone and limestone beds in western Wisconsin through folding.

The Mississippi, the master stream of the area, has cut completely through the resistant Lower Magnesian limestone and intrenched itself in the weaker Cambrian sandstone, while in the region north of the Chippewa the Mississippi is still engaged in cutting through the Lower Magnesian limestone.   Its tributaries in the northwestern upland have, therefore, been unable to dissect that upland as deeply as have the tributary streams in the region between the Chippewa and the La Crosse.   This has resulted in an earlier disappearance of the protecting cap of Galena-Black River limestone, the removal of a large proportion of the Lower Magnesian limestone, and the opening out of broad valleys in the weak Cambrian sandstone, particularly in Trempealeau and La Crosse counties.

Moreover, only a narrow strip of the area of ridges and coulees has been glaciated, a belt 3 to 6 miles wide southeast of the Chippewa River.   The lack of simplification by glacial erosion and by glacial deposition, therefore, adds to the contrast of these strikingly-different areas northwest and southeast of the Chippewa.

The Escarpment Southeast of Eau Claire.   The eastern margin of the region of ridges and coulees is an east-facing escarpment. It extends northwest and southeast for 75 miles, between Eau Claire and Tomah.   It is an escarpment underlain by the Cambrian sandstone, which makes an irregular country of marked contrast with the smooth plain of the Cambrian in the Central Plain to the east.   This escarpment is 150 to 300 feet in height. That the sandstone, generally a weak formation, should form an escarpment at all is clear evidence of greater resistance in the upper layers.   It suggests also the very recent removal of the more resistant Lower Magnesian limestone, whose edge has now retreated a score or more of miles to the southwest through weathering and stream erosion.   That the escarpment is retreating is evidenced by its marked irregularity, for it is deeply embayed and has numerous projecting salients and hundreds of detached outlying masses (Figs. 18, 127).

The Dissected Cuesta.   West and southwest of this escarpment we find, not a smoothly-sloping cuesta, but a hilly country, a maze of ridges and coulees.   It is a cuesta, but the cuesta has been thoroughly dissected by stream erosion.   This has been made possible, as already stated, by the position of the master stream and its tributaries in the weak Cambrian sandstone.   The ridges rise 400 feet or more above the valley bottoms.   The sandstone-floored valleys or coulees are a mile or so in width, in contrast to valleys a quarter to a half mile wide in the more resistant Lower Magnesian limestone.   Such ridges as are still capped by the limestone have narrow, craggy, castellated tops, while the lower, sandstone-capped ridges have broad, well-rounded crests.

The extent of removal of the Lower Magnesian limestone from this maturely-dissected cuesta is due in part to the nearness of the Mississippi River and the low baselevel of the streams in the coulees, and also to another factor.   The escarpment and cuesta are crossed by two large stream valleys,—of the Chippewa and Black rivers.   They rise far to the northeast, have great volumes, and well-graded valleys.   These streams and their tributaries have, therefore, dissected the cuesta thoroughly, for they cut through its high eastern part near the escarpment.   If the Niagara cuesta in eastern Wisconsin (p. 229) were crossed by a similar stream, and we could eliminate the later smoothing by glaciation, it would be as greatly dissected as the cuesta in the Driftless Area between the Chippewa and La Crosse rivers.   The latter is now reduced to a hilly area of ridges and coulees.

Dissected Upland Between La Crosse and Wisconsin Rivers

The highland between the Wisconsin and La Crosse rivers is also distinctly different from the area just discussed.   Its topography is intermediate in character between the little-dissected upland north of the Chippewa and the thoroughly-dissected upland between the Chippewa and La Crosse rivers.   It is underlain by the Lower Magnesian limestone, but all the valleys have cut through this formation into the Cambrian sandstone.   The Franconia Upland is a sandstone shelf at the northeastern border of the Magnesian cuesta.

The South-Sloping Cuesta.   The summits of the ridges north of the Wisconsin River slope southward, with the dip of the rocks. The uplands southwest of Sparta, near La Crosse River, stand 1300 to 1440 feet above sea level (Fig. 15), while those in the vicinity of Richland Center, just north of the Wisconsin River, stand at 1100 to 1160 feet (Fig. 59).   This southward descent in a distance of 45 miles is at the rate of $4\frac{1}{2}$ to 6 feet to the mile.

Some of the ridges are continuous for a long distance, as from Sparta to Prairie du Chien by way of Viroqua.   This ridge is over

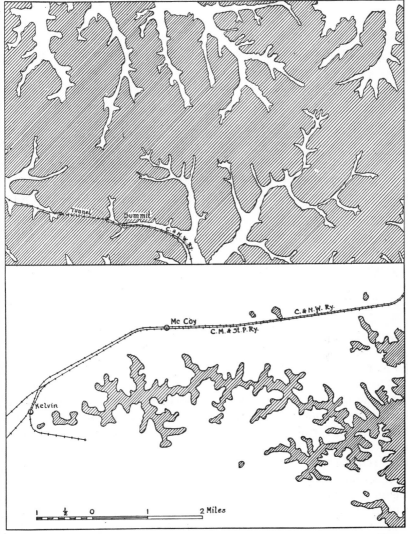

Fig. 16. Two ridges in the Western Upland of Wisconsin south of the La Crosse River. Upper map shows a youthful ridge, capped by resistant limestone. Lower map shows a ridge in old age, made up of weak sandstone. These ridges are only 2 or 3 miles apart. (From Tomah Quadrangle, U. S. Geol. Survey.)

50 miles long. It was probably the route of the St. Paul-Galena Road, a winter highway built in 1850 by way of Black River Falls.

Valleys in the Cuesta. In this highland the valleys are cut 300 to 400 feet below the upland level. The ridge crests, which are distinctly round-topped, are $\frac{1}{10}$ to $\frac{6}{10}$ of a mile wide, while

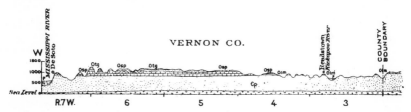

Fig. 17.   East-west section of the Western Upland, showing the Cambrian sandstone overlain by Lower Magnesian, St. Peter, and Galena-Black River formations (Olm, Osp, and Otg.)

the valley bottoms are $\frac{2}{10}$ of a mile to $1\frac{1}{4}$ miles in width. There is, however, a distinct increase in widths of valley bottoms and a diminution in width of ridge crests from the northern to the southern portion of the area. The region is in late youth or very early maturity of the erosion cycle, in contrast with the distinct maturity of the ridge-and-coulee district north of La Crosse.

This is because of the control of stream erosion by the rock structure of the region, and particularly because of the influence of the resistant Lower Magnesian limestone. Between the Chippewa and La Crosse rivers the base of the limestone is 360 to 500 feet above the floodplain of the Mississippi River. From the La Crosse to the Wisconsin River the limestone descends still further. Just above the mouth of the Wisconsin, in the Mississippi valley $3\frac{1}{2}$ miles north of Prairie du Chien, it dips under the present grade of the Mississippi. The Mississippi and Wisconsin have, therefore, been retarded in their down-cutting by this resistant rock, and their tributaries have likewise been retarded in their dissection of the upland. This is why the upland between the La Crosse and Wisconsin rivers has only reached the stage of late youth, while the region north of La Crosse is in the stage of early maturity of stream dissection. Many of the valleys within the cuesta are of a double or benched character. Their sides are marked by rock terraces, but these terraces are due to erosion upon rock layers of unequal resistance to weathering and not to an interruption of the erosion cycle through uplift.

Fossil Stream Courses Within the Cuesta. Ancient stream gravels at Seneca, Crawford County, Windrow Bluff, Monroe County, the Baraboo Bluffs, upwards of a dozen other places in adjacent parts of Wisconsin, and localities nearby in Iowa, Illinois, and Minnesota, preserve evidence of former river courses. The gravels appear to have been laid down 50 or 60 million years ago in creek bottoms and river flats. The streams probably flowed southward and southwestward. They may have had fairly high valley walls. They were long streams, for the gravels in many instances are more than usually well rounded and polished. The gravels are largely made up of quartz and chert, materials composed so largely of silica as to be essentially insoluble. Many of the deposits are associated with clay, sand, and sandstone; they are cemented with iron. To these two circumstances are due the preservation of the ancient gravels. Doubtless they were originally associated with pebbles and fragments of limestone, dolomite, sandstone, shale, and other soluble and perishable rocks, but these are now all destroyed. The gravels are all upon or near ridge tops. They are all that is left to mark the sites of former valleys. The gravels which happen to be preserved through cementation with iron or other binding substances, and those recently released from such cementation, identify bits of extremely ancient stream courses. It is through observation of the points of occurrence of the gravels, through study of the fossils preserved in the chert, and of those in rock ledges which underlie the oldest of the gravels, that we are able to say that the gravels may date back to the Cretaceous, and to assert that there are fragmentary fossil stream courses in the Western Upland of Wisconsin and in portions of the adjacent states. One of these streams lay parallel to and west of the present Kickapoo River. Since the time its course was abandoned, the levels of the ridge tops and valley bottoms have been lowered at least 800 feet by stream work, wind work, underground water and weathering. When streams were flowing in these old courses, the pre-Cambrian peneplain of the Northern Highland of Wisconsin may have lain buried beneath the Paleozoic sediments (p. 369).

Hills Rising Above the Cuesta. Rising above the Western Upland are outlying masses of Galena-Black River limestone, which are separated from the cuesta of southwestern Wisconsin by the lower Wisconsin River. Their area is small, the chief outlier of this sort being in the southwestern part of Crawford County. That the outlying masses of this sort were recently much more extensive is indicated by the presence of considerable areas of weak St. Peter sandstone, as on the ridge west of the Kickapoo River between Viroqua and Prairie du Chien.

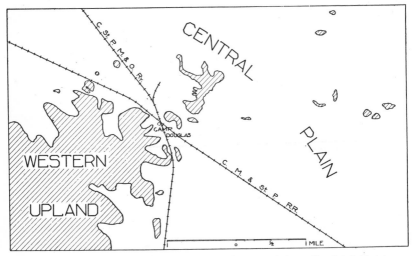

Fig. 18. The Dresbach escarpment near Camp Douglas and the outliers left behind in its recession to the southwest.

The Escarpment Near Camp Douglas. The southern portion of the Magnesian escarpment of western Wisconsin extends southeastward for 50 miles from Tomah to the Baraboo Range in Sauk County. A typical portion of it in the region near Camp Douglas (Pl. XX) is shown in Figure 18, where United States Army officers have mapped the irregular escarpment with its castellated outliers. It has subsequently been mapped more precisely by the U. S. Geological Survey. All of this portion of the escarpment is in the Driftless Area, so that its great irregularity forms a striking contrast with the simpler continuation of the same escarpment northeast of the Baraboo Range in the glaciated region. A remarkable feature of this irregular escarpment is the fact that, as in the area to the northwest, it is not capped at the very front by the Lower Magnesian limestone, which terminates a few miles west of the edge of the escarpment. The escarpment and its outliers are capped by the resistant upper layers of the Cambrian sandstone. These outliers have been thought by some geologists to be monadnocks on a peneplain. For reasons stated later (p. 324) it seems preferable to speak of them, not as monadnocks at all, but outliers of a retreating escarpment. The plain above which they rise will be described in this book as the inner lowland of a

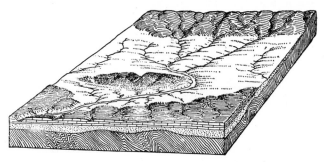

Fig. 19. The Baraboo Range as it will appear after the Magnesian cuesta has retreated a little farther to the west and south. (Davis.)

stripped, belted plain, overlain by level stream deposits. These outliers and the escarpment itself have been carved by weathering, rain-born rills, and wind work into a picturesque, castellated landscape of great irregularity, such as can persist in a never-glaciated, or driftless, area. At the southeast this escarpment joins the upland of the Baraboo Range.

THE BARABOO RANGE

Size and Height. The Baraboo Range extends east and west in Sauk and Columbia counties, having a length of 25 miles and an average width of from 5 to 10 miles. It contains the Devils Lake State Park. The range is a monadnock, similar to those in northern Wisconsin (p. 374). It differs from them in having been completely buried and now being only partially exhumed. As a matter of fact, only the eastern half of this monadnock stands very high above the adjacent country. The western half of the Baraboo Range is still buried in the Western Upland. The range is rather flat-topped, and its surface is at the same level as, or slightly above, the ridges of the dissected cuesta to the west.

The summit altitude of the Baraboo Range varies from 1140 to 1620 feet. The greater part of it is from 1200 to 1400 feet above sea level. The western portion of the range rises only one or two hundred feet above the upland of Lower Magnesian limestone, while the eastern half of the range stands 400 to 800 feet above the plain of Cambrian sandstone.

Geology and Structure. The Baraboo Range is quite different from the rest of the Western Upland in geology and structure. The cuesta is a simple mass of rather weak, Paleozoic sediments,

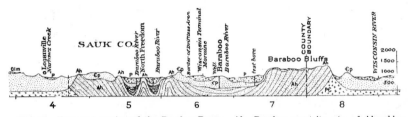

Fig. 20. East-west section of the Baraboo Range. Ah—Baraboo quartzite, etc. of Algonkian (Huronian) age; PC—igneous and metamorphic rocks of pre-Cambrian age; Cp—Potsdam sandstone of Cambrian age; Olm—Lower Magnesian limestone of Ordovician age; P—glacial drift of Pleistocene age.

dipping gently to the southwest. The Baraboo Range is a complex mass of resistant, pre-Cambrian, metamorphic rock, with subordinate amounts of igneous rock. The metamorphic rock is quartzite and slate, with a little iron formation. This is folded into a great trough, or syncline, which trends approximately east and west, its strata standing at all angles from vertical to horizontal (Fig. 20).

Topography and Drainage. Erosion has fashioned this syncline into three topographic features: (1) a broad, flat-topped South Range, sometimes spoken of as the Baraboo Bluffs, (2) a narrow, interrupted North Range, lower than that to the south and connected with it at both the eastern and western ends, and (3) a canoe-shaped intermediate lowland, 400 to 800 feet lower than the North and South ranges. These ranges are made up of the resistant Baraboo quartzite. The intermediate lowland has been excavated in the weak Seeley slate and Freedom iron formation. The lowland is partly floored by the Cambrian sandstone. At one time this sandstone completely filled the lowland, and only part of it has been removed.

The North Range is interrupted by five stream gaps, the South Range by only one. Of these northern gaps, three are now occupied by stream valleys. Two gaps are drift-filled and not used by any river. This is also the case with the one large gap of the South Range, now occupied by Devils Lake (Pls. IX, X, XIV, XXIII).

ELEVATION OF GAPS IN BARABOO RANGE

	In feet above sea level
1. Narrows Creek Gap	880
2. Upper Narrows (Ablemans Gap)	860
3. Broad abandoned gap northwest of Baraboo	960
4. Narrow abandoned gap northeast of Baraboo	960
5. Lower Narrows (Baraboo River)	800
6. Devils Lake Gap	963

The heights given above are on the surface of the glacial deposits and do not represent the elevation of bedrock. In the Devils Lake gap the present surface is at a level of 963 to 1080 feet. A well discloses drift filling of at least 283 feet, so that bedrock is certainly less than 680 feet above sea level. It is probably no more than 500 feet. The Lower Narrows are filled to a depth of 216 feet, so that bedrock is 585 feet above sea level. Each of these gaps is clearly the work of stream erosion. All except the third and fourth in the foregoing table still have precipitous walls.

Besides these six transverse gaps, the North and South ranges are notched on either side by short, deep gorges. These include Fox Glen in the North Range, and Otter, Durwards, and Parfreys Glens, Baxters Hollow—Meyers Mill,—and several other gorges in the South Range.

Preglacial History. The preglacial history of the Baraboo Range may be summarized as follows:

1. The Baraboo Range was a monadnock on a peneplain,
2. It was completely buried by the Paleozoic sandstone and limestone,
3. It has been partly exhumed.

The episode of long-continued erosion which resulted in the production of the Baraboo monadnock on the pre-Cambrian peneplain has already been outlined (pp. 34, 38). The present relief of the range above the Central Plain of Cambrian sandstone is only about half as much as the original relief. The Baraboo monadnock rose to a height of 900 to 1300 feet above the pre-Cambrian peneplain. It seems clear that in pre-Cambrian time some of the glens in the quartzite had been cut, for these stream valleys were filled with the Cambrian sandstone.   This is true for example of Parfreys Glen and Baxters Hollow.   It has not been demonstrated as yet for the large gaps at Devils Lake, Lower Narrows, and Ablemans.

That the range was completely buried beneath the Paleozoic sediments is proved by the sandstone, the preglacial gravel, and the residual limestone soil at the highest points on the quartzite ranges.   The retreating Magnesian cuesta, a few miles to the west and south, has, of course, occupied the site of the whole Baraboo Range, just as it now covers the western extension of the range.

The partial exhumation of the Baraboo Range has proceeded to the point where it is half as high as the original monadnock on the pre-Cambrian peneplain, the present relief being only 400 to 800 feet.   The canoe-shaped interior lowland, floored by the Cambrian sandstone, will decrease in area as the level of the sandstone is lowered.   The occasional hills of sandstone which now rise as high as the quartzite ridges will be removed by erosion, leaving the North and South ranges broader, and standing in greater and greater relief above the surrounding plain and the interior lowland of Cambrian sandstone.   The pre-Cambrian floor of this interior lowland near Baraboo is at an elevation of 310 to 500 feet above sea level.   As a buried and exhumed feature of topography, the Baraboo Range has been called a geographical fossil.

In connection with the preglacial and pre-Cambrian history of the Baraboo Range there has been a complicated series of changes in drainage.   All the details of this are not yet known, but the general sequence is explained in connection with the glaciation and drainage of this region (pp. 125, 190).   For the present it is enough to say that the stream gaps of the North and South ranges are not all due merely to post-Cambrian cutting, through superposition on the buried quartzite monadnock.   Some of them were originally cut by pre-Cambrian rivers, otherwise the Cambrian sandstone and conglomerate could not lie within the stream gaps, as in gaps 3 and 4 (p. 57) north of Baraboo.   Subsequently some river, comparable in size to the Wisconsin, entered the Baraboo Range at either the Lower Narrows or the broad abandoned gap northwest of the city of Baraboo (No. 3, table on p. 57), flowing south through the interior lowland and out through the Devils

Lake gap. It doubtless received tributaries which crossed the North Range by the Ablemans gap, the Narrows Creek gap, and other breaches in the quartzite. It is certain that the gaps were cut by running water. That they have been worn downward from the top of the quartzite ranges is attested by (a) the presence of stream-eroded pot holes near the top of the bluff east of Devils Lake and (b) the rounded chert and quartz pebbles there in ancient river deposits.

Whether this cutting was chiefly before or after the burial and exhuming of the monadnock cannot be stated at present. It is certain that, if it was done before the Cambrian burial, it was all done over again after the burial in Paleozoic time. In the latter case the erosion would be rapid, because the material to be removed was weak sandstone instead of resistant quartzite. It seems improbable, however, that any gap clear through the quartzite monadnock was made in pre-Cambrian time. Gorges were cut in the edges of the North and South ranges, but there is no proof that any gap was cut clear through the South Range in pre-Cambrian time.

The conglomerate and sandstone at the Ablemans gap rest on top of and at both sides of the present gorge in the quartzite. Conclusive evidence that they lie within the quartzite portion of this gorge is not yet available. The sandstone west of Devils Lake may have been deposited in a pre-Cambrian valley which headed thereabouts and extended out to the east and southeast. We have no decisive evidence as yet to show that this valley continued northward past the present side of Devils Lake. The valley west of Devils Lake has gently-sloping walls and a thoroughly-mature aspect. The gorge in which Devils Lake lies is steep-sided and youthful. Both are cut in the same resistant quartzite. If the mature slopes to the west of the lake represent exhumed pre-Cambrian topography, and the youthful gorge walls are post-Cambrian in origin, these contrasting valleys (Pl. XXIII, A) are perfectly in accord.

THE SOUTHWESTERN UPLAND

Topography and Geology. The Southwestern Upland occupies part or all of Grant, Iowa, Lafayette, Crawford, Dane, Green, and Rock counties. It is the southernmost of the two cuestas referred to on page 44. It lies between the Wisconsin and Mississippi rivers in the area west of the outermost terminal moraine of the Wisconsin glaciation. The region is entirely unglaciated, except for a small area of older drift on the southeast. The Southwestern Upland includes the Military Ridge and the slanting upland to the south.

Its topography is suggested by the following table, where the figures show heights above sea level at points on the upland 10 to 25 miles apart. The table is arranged to facilitate the direct comparison of the east-west and north-south sections.

TABLE SHOWING ELEVATIONS IN SOUTHWESTERN WISCONSIN

North of Wyalusing 1180 feet	Fennimore 1220 feet	No. of Dodgeville 1300 feet	Mount Horeb 1226 feet	
Near Glen Haven 1040 feet	Lancaster 1100 feet	Mineral Point 1134 feet	West of New Glarus 1150 ft.	
Near Cassville 980 feet	Platteville 1000 feet	East of Calamine 1020 feet	East of Argyle 1000 feet	West of Janesville 1000 feet
	Hazel Green 960 feet	W. of Gratiot 1060 feet	Near Clarno 950 feet	Near Beloit 960 feet

The predominant rock formation in this area is the Galena-Black River group. Of subordinate importance are the Richmond shale and Niagara limestone, above the Galena-Black River, and below it the St. Peter sandstone, Lower Magnesian limestone and Cambrian sandstone. These rocks all dip southward at a low angle, but there are minor, east-west folds. The latter are topographically unimportant. The rocks are strongly jointed, and thereby may have been somewhat unusually effective in guiding stream erosion.

The Military Ridge. A well known topographic feature in southwestern Wisconsin and eastern Iowa is popularly known as the Military Ridge. In this state it constitutes the divide between the north-flowing tributaries of the Wisconsin River and the south-flowing streams tributary to the Rock and Mississippi. Its crest was followed by the Military Road, built in 1835 from Green Bay to Prairie du Chien, by way of Fond du Lac, Portage, and Blue Mounds. This highway gave the Military Ridge its name. Doubtless the existence of the Military Road was responsible for the location of the territorial road between Milwaukee and the lead and zinc district soon after 1837. It came by way of Madison and joined the Military Road southeast of Mt. Horeb.

Herbert Quick, one of the masters of English prose, has written the most effective description of the Military Ridge and its neighborhood. What he says applies to the heydey of its use as a highway of immigration from Milwaukee and Madison, Wisconsin, to Dubuque, Iowa, over the back slope of the cuesta of the Southwestern Upland.

"It was in the latter part of March. There were snow-drifts in places along the road, and when I reached a place about where Mt. Horeb now is, I had to stop and lie up for three days for a snowstorm. I was ahead of the stream of immigrants that poured over that road in the spring of 1855 in a steady tide . . . .

"As I went on to the westward, I began to see Blue Mound rising like a low mountain . . ., and I stopped at a farm in the foot-hills of the Mound where, because it was rainy, I paid four shillings for putting my horses in the stable . . . .

"I drove out to the highway, and . . . joined again in the stream of people swarming westward. The tide had swollen in the week during which I had laid by . . . The road was rutted, poached deep where wet and beated hard where dry, or pulverized into dust by the stream of emigration. Here we went, oxen, cows, mules, horses; coaches, carriages, blue jeans, corduroys, rags, tatters, silks, satin, caps, tall hats, poverty, riches; speculators, missionaries, land-hunters, merchants; criminals escaping from justice; couples fleeing from the law; families seeking homes; the wrecks of homes seeking secrecy; gold-seekers bearing southwest to the Overland Trail; politicians looking for places in which to win fame and fortune; editors hunting opportunities for founding newspapers; adventurers on their way to everywhere; lawyers with a few books; Abolitionists going to the Border War; innocent-looking outfits carrying fugitive slaves; officers hunting escaped negroes; and most numerous of all, homeseekers 'hunting country'—a nation on wheels, an empire in the commotion and pangs of birth. Down I went with the rest, across ferries, through Dodgeville, Mineral Point and Platteville, past a thousand vacant sites for farms toward my own farm so far from civilization, shot out of civilization by the forces of civilization itself.

"I saw the old mining country from Mineral Point to Dubuque, where lead had been dug for many years, and where the men lived who dug the holes and were called Badgers, thus giving the people of Wisconsin their nickname as distinguished from the Illinois people who came up the rivers to work in the spring, and went back in the fall, and were therefore named after a migratory fish and called Suckers; and at last, I saw from its eastern bank far off to the west, the bluffy shores of Iowa, and down by the river the keen spires and brick and wood buildings of the biggest town I had seen since leaving Milwaukee, the town of Dubuque.

"I camped that night in the northwestern corner of Illinois, in a regular city of movers, all waiting their turns at the ferry which crossed the Mississippi . . ." (from *Vandermark's Folly*, by Herbert Quick, Indianapolis, 1922, pp. 87-88, 92, 105-106; copyright 1921 and 1922; used by special permission of the publishers, The Bobbs-Merrill Company and The Curtis Publishing Company).

The Military Ridge is now traversed by the Chicago and Northwestern Railway from Mount Horeb to Fennimore. The railway was not built until 1881. This is the longest stretch of railway in the state without a bridge over a stream. West of Fennimore the Military Ridge continues to the Mississippi River just south of its junction with the Wisconsin at Nelson Dewey State Park. One reason for the early use of this ridge as a highway was that it was largely treeless.

The Military Ridge is not a symmetrical ridge with even slopes on either side. It is an exceedingly unsymmetrical feature, with a short, steep, northern slope toward the Wisconsin and a long, gentle descent southward to the Illinois line. The position of the Military Ridge is determined entirely by the drainage, which in turn is controlled in part by the geological features. Near the Mississippi River, for example, the crest of the Military Ridge lies 1 to 4 miles south of the Wisconsin River. This is because the north-flowing tributaries of the Wisconsin are short. They are kept from rapid down-cutting and extensive headwater erosion by the resistant character of the Lower Magnesian limestone, in which the bottoms of their valleys are cut. On the other hand, the Military Ridge lies fully 30 miles south of the Wisconsin River near Montfort Junction and Dodgeville. In this region the north-flowing streams are long. They have extended their headwaters a great distance because of the weakness of the Cambrian sandstone in which their lower courses lie.

These facts perhaps make it clear that the Military Ridge is a name applied to a divide on a cuesta. The name is very convenient, for the ridge separates the two parts of the cuesta next to be discussed: (a) the narrow, north-facing escarpment, and (b) the broad, south-sloping upland.

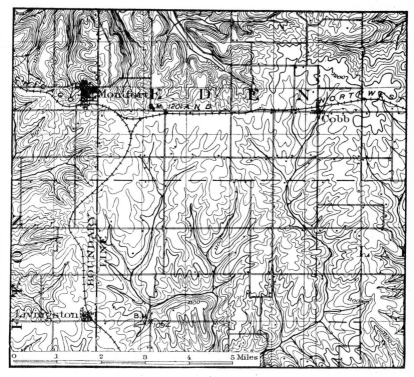

Fig. 21. Topographic map of a portion of the Military Ridge. Contour interval 20 feet. (From Mineral Point Quadrangle, U. S. Geol. Survey.)

The Black River Escarpment. The irregular Black River escarpment in southwestern Wisconsin differs decidedly from the simple feature of the same name in the eastern part of the state (Chapter IX). Between the eastern and western parts, in Columbia and Dane counties, lies a belt where the escarpment is discontinuous, through dissection by stream erosion and burial beneath glacial drift. The Black River escarpment again becomes a conspicuous

feature in the western part of Dane County. It trends east and
west for over 60 miles and then turns northwestward, crossing the
Wisconsin River between its mouth and the Kickapoo, and continuing northwestward into the state of Iowa. The name of the
escarpment is not derived from the Black River of Wisconsin but
from the Black River rock formation, formerly called the Trenton
limestone.

The Wisconsin River flows parallel to this escarpment for a
long distance and then crosses it, so that a portion of the Black
River upland is isolated, standing as a narrow outlier on the
Magnesian cuesta to the north (p. 54). The depth of the gorge or
trench of the Wisconsin River—300 to 400 feet—should be kept
quite distinct from the height of the Black River escarpment, which
is only 100 or 200 feet.

The irregularity of the Black River escarpment is notable. Dendritic valleys indent its front, extending back 4 to 8 miles. Branching ridges between these valleys project forward an equal distance
in front of the general line of the escarpment.   At the ends of
these ridges are isolated limestone outliers. The summits of these
limestone-capped hills rise to the level of the Black River cuesta.

The St. Peter sandstone sometimes makes prominent low cliffs.
It is a weak rock, varies in thickness, and lies upon an irregular
surface of Lower Magnesian limestone, so that its topographic relationships vary greatly.   This is well shown in the valley of Sugar
River which is narrow where the St. Peter sandstone is thick, as
near Rileys, Paoli, and Belleville, and broad where the sandstone
is thin, as at Pine Bluff and between the places listed above.

The Sloping Black River Upland. The upland of Black River
limestone slopes southward at the rate of 6 to 8 feet to the mile.
as is shown in the following table, which should be read from left
to right rather than vertically.   This table also shows that the
western part of the upland is lower than the eastern,—that is, the
slope of the upland is south-southwest rather than south.

This upland is not a smooth plain. It has (a) great valleys,
(b) broad, round-shouldered ridges, and (c) small mounds that
rise above the general level of the ridges. As the chief streams all
flow southward, the ridges trend north and south. The whole upland of southwestern Wisconsin consists of a main east-west ridge
—the Military Ridge—from which half a dozen subordinate ridges

extend southward. These north-south ridges are broad, varying from a mile to 8 and even 12 miles, for the stream valleys are not as wide as in the Magnesian cuesta north of the Wisconsin River.

TABLE SHOWING SOUTHWARD SLOPE OF BLACK RIVER UPLAND

Locality	Elevation above sea level in feet	Locality	Elevation above sea level in feet	Distance in miles	Slope in feet to the mile
Preston...............	1105	Near Fair Play....	900	33	6
Montfort Jct........	1132	Near Buncombe..	880	32	8
Near Dodgeville..	1260	Near Dunbarton..	1060	30	6⅔

They are distinctly not flat-topped ridges, but descend gently from a central divide. They slope eastward and westward toward the stream valleys, and their crests slant southward. The crest of one of these ridges is followed by the Chicago and Northwestern Railway from Montfort, through Platteville, to Cuba.

Relation to Niagara Escarpment. The southern edge of the Black River upland abuts sharply against the Niagara escarpment, which trends northwest-southeast in northwestern Illinois and northeastern Iowa, roughly parallel to the Black River escarpment. This portion of the Niagara escarpment does not enter Wisconsin. It is exceedingly irregular (Fig. 87), being in the Driftless Area, and has numerous salients and many outliers, some of which are in Wisconsin. These outliers include Sinsinawa Mound, White Oak Mound, the Platte Mounds, and the Blue Mounds.

Blue Mounds and Other Niagara Outliers. The Blue Mounds outliers of the Niagara escarpment are 45 to 55 miles northeast of the nearest portions of the Niagara cuesta in Illinois and Iowa, and 69 miles west of the Niagara cuesta of eastern Wisconsin. They may be regarded as having been left behind by the recession of either the escarpment to the east or that to the southwest. Between Blue Mounds and the front of the cuesta near Dubuque are (a) Platte Mounds and (b) Sinsinawa Mound, 25 and 10 miles respectively from the escarpment in Iowa, and (c) the mound near White Oak, Lafayette County, 5 miles from the escarpment in Illinois. Sherrill Mound, northwest of Dubuque, Iowa, and Scales Mound, Charles Mound, and several others, northeast of Galena, Illinois, are all Niagara outliers of the same origin as Blue Mounds.

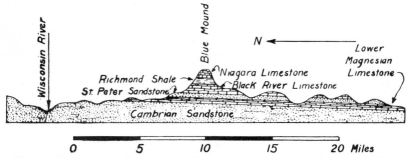

Fig. 22. Cross-section to show the detached mass of Niagara limestone which caps the higher of the Blue Mounds. The underlying shale has been called Hudson River shale as well as Richmond.

The reason for the lack of similar mounds to the east between Blue Mounds and the Niagara escarpment will be discussed in connection with the glaciation of eastern Wisconsin and the destruction of such outliers by glacial erosion.

The general topographic relationships of these mounds is indicated in the table below.

Their summits rise to various levels, standing from 180 to 415 feet above the surrounding upland. Their sizes vary also. Each of these mounds is capped by the Niagara limestone. The sides of each one have an abrupt upper portion, which gives place to a gentle slope on the underlying Richmond shale. The top of Blue Mound is so flat that an oval race track was laid out upon it long

Table Showing Elevations of the Mounds in Southwestern Wisconsin

Name of Mound	Height above sea level	Local relief
West Blue Mound	1716 feet	415 feet
Platte Mound	1420 "	300 "
East Platte Mound	1380 "	180 "
Sinsinawa Mound	1185 "	285 "
Mound near White Oak	1200 "	200 "

ago. In places the rimming wall of Niagara limestone is a genuine precipice, as on portions of the border of Platte Mound. Elsewhere the limestone does not form a precipice, and fragments have rolled and crept a great distance down the shale slope, as on Blue Mound. The mound includes the shale as well as the limestone, and the area of shale at the base of the limestone cap increases southward as the Niagara cuesta is approached. The shale forms a rim about half a mile wide at Blue Mound, while at Sinsinawa Mound it surrounds the limestone-capped summit in a belt 3½ miles wide, except on the sides where there are deep stream valleys.

The summit area of Blue Mound is greater than that of Sinsinawa Mound and some of the other outliers near the Niagara escarpment. This is due to the more resistant character of the rock. In all of the mounds the limestone is somewhat cherty but at Blue Mound nearly all of the limestone has been replaced by quartz. In some ledges over 99% of the rock is silica instead of calcite or dolomite. This renders it exceedingly resistant. This reason for the preservation of Blue Mounds was recognized in 1839 by Owen, who said:

"These isolated and towering mounds, so conspicuous a feature in the landscape of Wisconsin, are evidence of the denuding action to which, under the crumbling hand of time, the surface of our globe is continually subjected, and which the more durable siliceous masses of these hills of flint have been enabled partially to resist."

A gradation form, directly related in origin to the limestone-capped mounds, is the round-topped eminence capped by Richmond shale. Although the shale is exceedingly weak in comparison with the Niagara limestone, the knob or ridge still stands higher than the surrounding area of Black River limestone. Sometimes the shale contains thin beds of limestone, making it a somewhat resistant formation. The following hills (a) the East Blue Mound, (b) the hill south of Platte Mound and east of Platteville, (c) the long, narrow ridge upon which Hazel Green stands, west of Fever River, and (d) the broad, hilly area south of Shullsburg and Gratiot, are all uplands of weak Richmond shale, from which the Niagara limestone has recently been removed. Indeed, the fact that these shale-capped hills have not been worn away but still stand above the sloping Black River upland is proof in itself that the resistant limestone cap has recently been removed from the hilltops.

Valleys in the Southwestern Upland. The valleys in the upland of southwestern Wisconsin are cut rather deeply into the upland.

The grades are steeper than the slope of the upland, so that the valleys increase in depth to the southward. Midway in their courses, and near their mouths, the main valley bottoms are 200 to 300 feet below the ridge tops. These valleys are narrow, com-

TABLE SHOWING FACTS ABOUT THE CHIEF RIVER VALLEYS

Name of River	Elevation at source, in feet above sea level	Elevation at mouth, in feet above sea level
Grant	1100 feet	600 feet
Platte	1100 "	600 "
West Pecatonica	1200 "	800 "
Pecatonica	1200 "	779 "

pared with the ridges between, because the resistant Galena-Black River and Lower Magnesian limestones prevent rapid widening and deepening by stream erosion. The whole region has a different aspect from the Magnesian cuesta north of the Wisconsin. There the valley bottoms are wide and the upland is reduced to skeleton form because the streams are cutting in the weak Cambrian sandstone. The valleys south of the Wisconsin in the Black River cuesta are at a disadvantage, and the upland here is still the dominant feature. The river system is of the regular dendritic pattern, characteristic of stream erosion in homogeneous rocks in a driftless area.

Few geologists or students of physical geography would ascribe valleys in rock to anything but stream erosion. Yet Locke, writing less than a century ago, seems to have been somewhat hesitant about stating the matter. He called the sentences quoted below "a geological speculation."

"It appears evident that the south fork of the Little Makoqueta (a stream barely large enough to turn a mill) has, by abrading its channel for countless ages, worn its bed to the depth of four hundred feet in solid limestone. Is it not probable, then, that the rocks once extended nearly in an uninterrupted level from the heights of the Little Makoqueta to the top of Sinsinawa; and that the mighty Mississippi has rolled its tide long enough to have worn the chasm, the centre of which it is shown to occupy in the section? Is it not probable that the whole surface of the country in that region is now many feet—many hundred feet, indeed—lower than when it first became dry land? Rocks have turned to dust, and the dust been washed away; stones have dissolved, and the solutions have been poured into the sea. The springs of Iowa show that they have levied tribute from the solid rock, and the waters of the Mississippi tell that they are transporting it to ocean depths. The lead ore piled loosely on the top of corroded limestone shows that the matrix of its vein, into which it was originally cast, has abandoned it, to fall down like a ruined wall; a few points, covered by harder materials, remained; gathered the sloping tablets of strata about their shoulders; reared their heads in defiance to a million of storms; and now, in form of conic mountains, point out a few landmarks of earth's olden boundary."

Many of the valleys and slopes of southwestern Wisconsin have been gullied notably in recent years. Some of the stream gullies are 6 to 8 feet deep. There seems to be no reason that this should be interpreted as evidence that anything unusual is taking place in the volume or load of the main streams. It is more probable that gullying has been induced by man's activities in cutting down forests, ploughing fields, excavating mines, or otherwise disturbing nature's balance in the surface drainage or the underground circulation. These gullies lie wholly in unconsolidated materials, never in the rock.

The Upland Is a Cuesta Rather than a Peneplain. There are two ways of explaining the slanting upland south of the Wisconsin River. Either must account for the valleys lying below its level and the mounds rising above it. One is to regard it as a peneplain, with the mounds as monadnocks and the valleys as features produced after the region had been baselevelled and an uplift had taken place. The other is to explain it as a cuesta, or unsymmetrical upland belt, with the mounds as outliers of the retreating escarpment at the front of another cuesta on the Niagara limestone to the southwest and the valleys as features that have always been present since the Niagara limestone was removed from the Galena-Black River surface. On many accounts the latter explanation seems open to least objection. What is to be said applies to the whole Western Upland of Wisconsin and to parts of the surrounding region (Pls. IV, V, XI, B, XIIIA, XXA), as well as to the upland south of the Wisconsin River.

A broad general view in some parts of the upland of western Wisconsin gives the impression of a nearly uniform height of ridge tops. As one looks out over the surrounding country from one of the mounds he sees a rather even sky-line in all directions. When one examines the rock layers in any given locality he may find them dipping a trifle more steeply than the surface inclination of the ridges. When he goes down into the valleys and up to the tops of other ridges, especially if he goes northward or southward, he finds rock layers of different age. He, therefore, may be tempted to conclude that the ridge tops do actually rise to a uniform height, that the apparently-even sky-line is really even, and that a single topographic surface bevels across the several Paleozoic formations.

What he may overlook, however, is that the contour maps show no uniform ridge tops and no even sky-line. They do show steps —escarpments—between the several sets of sloping ridges. Moreover, the features observed are quite as characteristic of cuestas as of peneplains. One may see just such a landscape on the coastal plain cuesta in Alabama, on the Niagara cuesta in eastern Wisconsin, or on the French uplands northeast of Paris where certain battles of the World War were fought.

Other Suggested Peneplains in Wisconsin. It has been suggested at various times and by various geologists, that there are one to four peneplains in western Wisconsin and adjacent portions of Illinois, Iowa, and Minnesota. The author is not convinced that any of these are correctly interpreted as peneplains. He applies the name cuesta (a) to the upland north of the Wisconsin, where Lower Magnesian and Franconia formations cap the ridges, (b) to the upland of Galena-Black River limestone in southwestern Wisconsin, and (c) to the upland of Niagara limestone in Iowa and Illinois, as well as to the continuation of all these features in eastern Wisconsin. The plain at Camp Douglas (pp. 324, 330) is a lowland of weak sandstone, with a veneer of sand which makes the plain very level. There is one plain determined in position by rock structure on the basal layers of the Cambrian sandstone, another on the Lower Magnesian limestone, one on the Franconia, one on the Galena-Black River, and one on the Niagara. The author believes there is no thoroughly-convincing proof that there ever was any peneplain in Wisconsin except the one on the pre-Cambrian in the Northern Highland.

The author has carefully weighed the evidence presented by others and has carried on field studies in Wisconsin throughout a period of ten years. He has concluded that only one of the larger unconformities in Wisconsin and the adjacent states is represented by a peneplain. These unconformities are (1) the pre-Paleozoic in pre-Cambrian time, (2) the pre-St. Peter in Ordovician

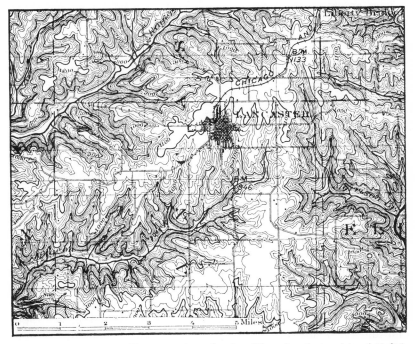

Fig. 23. Part of the Black River cuesta in southwestern Wisconsin. Contour interval 20 feet. (From Lancaster Quadrangle, U. S. Geol. Survey.)

time, (3) the pre-Devonian, (4) the pre-Carboniferous and (5) the Cretaceous, each nearby to the west and south in Minnesota, Iowa, and Illinois, and (6) the preglacial unconformity. The one baselevelled surface related to an unconformity seems to be the exhumed pre-Cambrian peneplain of the Northern Highland and the buried extension of the same surface. The Ordovician unconformity clearly accompanies a hilly topography. The others have had no effect in Wisconsin, so far as the present topography shows. It seems probable that the topography of the Western Upland has all been produced by the action of just such weathering, wind work, and stream erosion as is taking place today. The region need not, necessarily, have ever been less hilly than at the present time.

Arguments Against the Existence of Four Peneplains. Among the reasons for considering that there is no series of four peneplains are the following: (a) Since a dissected peneplain indicates a long stand of the land at one level, followed by an uplift, it would be a remarkable coincidence if four such uplifts happened to have followed one another at regular intervals, and if the upland

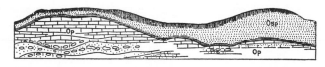

Fig. 24.   The hilly, buried topography of the surface of the Ordovician. Lower Magnesian (Op) limestone at its contact with the St. Peter (Osp) sandstone. (Grant.)

surfaces then chanced to be determined by such rock layers as (1) the Niagara, (2) the Galena-Black River, (3) the Lower Magnesian—all of them resistant limestones—and (4) the more resistant layers in the Cambrian sandstone.

(b) If there really is such a series of four peneplains, then the oldest—the one on the Niagara—should be most cut up by streams, the next in age should not be quite so irregular, the third should be still less dissected, and the fourth should be least hilly of all. Now the fourth—the one on the Cambrian sandstone—is partly masked by sandy deposits, as at Camp Douglas, but the first, second and third upland surfaces are equally hilly, showing no progression or gradation whatever.

(c) If there really are three dissected peneplains and a fourth still being made, then some of the ridges should still be flat-topped for a perfected peneplain contains many areas of unappreciable slope. If such peneplains were cut by erosion, following uplift there should be broader, flat-topped remnants in the area of Lower Magnesian limestone—a more recent peneplain—than in the areas of (a) Galena-Black River and (b) Niagara limestone—the two oldest peneplains. As a matter of fact, every slope in western and southwestern Wisconsin is perfectly drained, except where there are glacial deposits or sink holes. There are absolutely no flat-topped ridges in any one of the three limestone areas. Moreover the Galena-Black River ridges are somewhat broader than the Lower Magnesian limestone ridges, showing that width of ridges is determined by other factors than the ages of the supposed peneplains.

(d) The upland on the Galena-Black River limestone—for which the name Lancaster Peneplain has been proposed—dips down under the Niagara limestone and Richmond shale at the southern border. Its northern border projects out into the air above the St. Peter sandstone and Lower Magnesian limestone. The other

Fig. 25. Cross-section of the Galena-Black River cuesta, showing the escarpment at the northern edge of the so-called peneplain.

supposed peneplains have similar relationships. In other words, each of these surfaces for which the name peneplain has been proposed has border features related to existing escarpments, where a step up or a step down brings one to the next upland surface. To assume warping or faulting exactly at each escarpment seems out of the question, and the dip of the rocks fails to show it. Thus the escarpment borders are hard to explain if a peneplain is suggested, but are perfectly natural accompaniments of a series of upland surfaces, with cuestas determined by rock structure.

The Suggestion of Two Peneplains. Since the first edition of this book was written, another author has suggested that there may be two peneplains upon the cuestas of western Wisconsin and adjacent parts of the Driftless Area. All the objections mentioned above apply to this suggestion also, and none of the supporting arguments seem to be compelling. The author of this suggestion appears to consider that the back slopes of cuestas always must be stripped down to perfect parallelism with the underlying strata; but not all geologists would agree to this premise. He is silent as to the extent and outer limits of the two peneplains which he proposes, but he nowhere implies that the strictly similar cuestas or upland belts of eastern Wisconsin should be called peneplains rather than cuestas.

The Suggestion of One Peneplain Rather than Four. Another possibility is that there was long ago one peneplain upon the Paleozoic sediments of Wisconsin, and that, in its dissection, the four existing steps have been etched out, forming three cuestas and one lowland. This view necessarily involves the existence of the remnants of this peneplain at only such places as (a) the surfaces of Niagara limestone in Illinois and Iowa, (b) the crest of the Military Ridge of Galena-Black River limestone in Wisconsin between Mt. Horeb and the Mississippi River, (c) the crests of certain of the ridges capped by the Lower Magnesian limestone, as in the region south of Sparta and Tomah. Perhaps we should add to this the few, flat-topped areas in the Baraboo Range. There, however, the quartzite folds seem to have been truncated in pre-Cambrian rather than in post-Paleozoic time, and to be still in process of exhumation. These are the only places where it can be argued that remnants of a peneplain are now preserved. The author, however, does not find that the topography and rock structure at these places suggest a peneplain to him. It would be quite proper also to consider the hypothesis that the pre-Cambrian rocks of northern Wisconsin, although baselevelled before the Cambrian, were subsequently bevelled across by a later peneplain. Such a peneplain as this might be thought of as due to either pre-Devonian, pre-Carboniferous, Cretaceous, or Tertiary baselevelling.

Objections to Even a Single Peneplain. The following considerations might be thought to allow or to favor the existence of a single peneplain: there are serious objections to each one.

(a) There is time enough available. The youngest Paleozoic rocks included in the upland are of Silurian age. Surely any surface which was uplifted during the Silurian might long ago have been reduced to baselevel. The history of a region of similar character—the Appalachian Highland—shows that the Allegheny Plateau and the Appalachian ridges, which were uplifted during the late Paleozoic, were certainly reduced to baselevel by Cretaceous time and probably by Jurassic time, being afterwards uplifted, warped, and etched into the present relief. An objection to this argument is that there has also been ample time to destroy such a peneplain.

(b) There may be appropriate sediments in the adjacent regions. The Devonian and Carboniferous rocks of Michigan and

Illinois are made up of just such fine materials—limestone, shale, and sandstone—as might be derived from a region in process of being peneplained. The difficulty is that the sediments do not tell us whether the adjacent lands were nearly-baselevelled or were as hilly as the present Western Upland of Wisconsin, for the material now being carried away by streams is fine sand, silt, and clay.

(c) The Devonian rocks of eastern Wisconsin, northern Illinois, and eastern Iowa are nowhere seen to rest upon a surface of such relief as the present Driftless Area, suggesting that the rocks from the Cambrian to the Silurian were reduced to slight relief before the Devonian sediments were deposited. The same thing applies to the base of the Upper Carboniferous, where there is a marked unconformity. This argument is not known to be valid in the Western Upland of Wisconsin, for there may never have been notable relief in the localities where the Devonian and Carboniferous are now found.

(d) The Cretaceous rocks of Iowa and Minnesota are likewise laid down upon surfaces of less relief than the present Driftless Area, making it possible that the supposed peneplain is of pre-Cretaceous age. No Cretaceous deposits are known to rest upon the upland surfaces within the state of Wisconsin. There are small deposits of siliceous pebbles in such places as the ridge at Seneca, Crawford County, Windrow Bluff, Monroe County, the East Bluff of the Baraboo Range, and other localities in Wisconsin. These gravels are more likely to have been stream bottom deposits within a cuesta of moderate relief than to have been deposits upon a perfected peneplain (page 53).

(e) The ridge tops in southwestern Wisconsin are not known to rise with all the minor anticlines and fall with all the synclines. The ridges are actually higher near certain of these anticlines, however, as at Shullsburg and Gratiot. These rock folds are all low.

(f) A few rivers in southwestern Wisconsin and adjacent states have curves suggestive of intrenched meanders (page 176 and Fig. 63). These are not thought to be inherited from a previous condition of baselevelling, because the larger streams show no such curves, and meandering might be induced by local resistant layers.

The Lack of Preservation of Such a Peneplain. It seems to the author that no remnants of such a Devonian, Carboniferous, Cretaceous, Tertiary, or Quaternary peneplain are now preserved in any part of Wisconsin. No one has ventured to suggest it in eastern Wisconsin, where the upland of Niagara limestone preserves larger smoothly-sloping areas than in the Western Upland. Nor has anyone suggested the baselevelling of the Black River or the Lower Magnesian limestone of eastern Wisconsin. Near the Mississippi and lower Wisconsin rivers there are only slight areas which could possibly be remnants of this ancient peneplain, as on the Military Ridge and Niagara escarpment. These, however, are apparently parts of the etched surfaces which coincide with the resistant limestone layers.

Any previous episodes of fairly complete baselevelling which bevelled across inclined layers should have left bodies of weak sediments overlying each resistant limestone near the point of outcrop of the next overlying resistant limestone. The uncovered portions of these would be wedge-shaped in cross-section and thickest nearest the escarpments. Subsequent etching, by weathering and erosion, might remove absolutely all of these bodies of weak rock. If any of them were preserved they would be close to the escarpments. They would furnish excellent evidence of previous baselevelling, but no such remnants are known to exist in critical areas, although one author considers that he has found them.

Conclusion as to Cuestas. It seems reasonable to hold that the burden of proof lies distinctly with those who believe in the preservation of remnants of earlier peneplains. The chief evidence against them and in favor of the cuesta explanation may be summarized under four heads:

(a) There are no demonstrated remnants of any Paleozoic or later peneplain in this part of Wisconsin. The author has failed to discover any ridges sufficiently flat-topped to even suggest a peneplain to him.

(b) There has been more than ample time since the Cretaceous or Tertiary to destroy all remnants of previous peneplains except in the resistant, crystalline rocks of the pre-Cambrian. Even here the older peneplain seems to have been preserved because it was buried.

(c) If it is argued that the process of etching after baselevelling could produce the existing, slanting surfaces that coincide with resistant limestones, and could remove the weak shale and sandstone, except at the base of escarpments, then the same etching processes could cut out similar surfaces in a series of gently-dipping beds, without peneplanation.

(d) The author, therefore, inclines toward the simpler explanation, that of cuestas, unless some conclusive evidence of peneplanation can be found. All the features of the topography of western and southwestern Wisconsin are easily explained as parts of a system of cuestas in a series of gently-dipping, alternate weak and resistant, sedimentary rocks.

OLD BELMONT STATE PARK

This park is primarily of historic interest, since it is the site of the first state capitol of Wisconsin. It is between Platte and East Platte mounds (p. 66), about six miles northeast of Platteville.

BIBLIOGRAPHY

Alden, W. C. Peneplains and Cycles of Erosion, Professional Paper 106, U. S. Geol. Survey, 1918, pp. 104-105.

Bain, H. F. Zinc and Lead Deposits of the Upper Mississippi Valley, Bulletin 294, U. S. Geol. Survey, 1906, 148 pp; reprinted as Bulletin 19, Wis. Geol. and Nat. Hist. Survey, 1907, (see especially pp. 11-16).

Blanchard, W. O. The Geography of Southwestern Wisconsin, Bull. 65, Wis. Geol. and Nat. Hist. Survey, 1924; The Geography of the Sparta-Tomah Quadrangles, Unpublished thesis, University of Wisconsin, 1917.

Borchers, Irma T. The Geography of the Lead and Zinc Region of the Upper Mississippi Valley, Unpublished thesis, University of Wisconsin, 1929.

Bowmann, F. F. Jr. An Area in the North Baraboo Range, Unpublished thesis, University of Wisconsin, 1925.

Bryan, Kirk, and Powers, Wm. E. Erosion Cycles in the Driftless Area (in press).

Burrows, B. H. Human Adjustments to Natural Environment in St. Croix County, Wisconsin, Unpublished thesis, University of Wisconsin, 1924.

Chamberlin, T. C. Quartzites of Sauk and Columbia Counties, Trans. Wis. Acad. Sci., Vol. 2, 1874, pp. 129-138; Ore Deposits of Southwestern Wisconsin, Geology of Wisconsin, Vol. 4, 1882, pp. 367-571.

Eaton, J. H. Geology of the Region about Devils Lake, Trans. Wis. Acad. Sci., Vol. 1, 1872, pp. 124-128; On the Relation of the Sandstone, Conglomerate, and Limestone of the Baraboo Valley to Each Other and to the Azoic Quartzites, Ibid., Vol. 2, 1874, pp. 123-127.

Folsum, H. W. The Interpretation of the Upland Surfaces in the Driftless Area, Unpublished thesis, University of Wisconsin, 1930.

Grant, U. S.   Lead and Zinc Deposits of Southwestern Wisconsin, Bull. 9, Wis. Geol. and Nat. Hist. Survey, 1903, 103 pp; Ibid., Bull. 14, 1906, 94 pp.

Grant, U. S., and Bain, H. F.   A Pre-glacial Peneplain in the Driftless Area, Science, new series, Vol. 19, 1904, p. 528.

Grant, U. S., and Burchard, E. F.   Lancaster-Mineral Point Folio, Folio 145, U. S. Geol. Survey, 1907.

Hershey, O. H.   Preglacial Erosion Cycles in Northwestern Illinois, Amer. Geol., Vol. 18, 1896, pp. 72-100; The Physiographic Development of the Upper Mississippi Valley, Ibid., Vol. 20, 1897, pp. 246-268.

Hinn, Helen B.   The Geography of the Wisconsin River Valley, Unpublished thesis, University of Wisconsin, 1920.

Hubbard, G. D.   The Blue Mound Quartzite, Amer. Geol. Vol. 25, 1900, pp. 163-168.

Hughes, U. B.   A Correlation of the Peneplains of the Driftless Area; Iowa Acad. Sci., Proc., Vol. 23, 1916, pp. 125-132.

Irving, R. D.   The Age of the Quartzites, Schists, and Conglomerates of Sauk Co., Wis., Trans. Wis. Acad. Sci., Vol. 1, 1872, pp. 129-137; Ibid., Amer. Journ. Sci., 3rd Series, Vol. 3, 1872, pp. 93-99; Geology of Central Wisconsin, Geology of Wisconsin, Vol. 2, 1877, pp. 409-636, (on the Baraboo Range, pp. 504-519).

Jay, E. S.   The Pre-Cambrian Drainage of the Baraboo Region with Special Reference to Devils Lake Gap, Unpublished thesis, University of Wisconsin, 1916.

Johnson, Geneviera C.   The Geography of Green County, Wisconsin, Unpublished thesis, University of Wisconsin, 1918.

Karges, B. E.   Faulting in the Paleozoic Sediments Near Hudson, Wisconsin, Unpublished thesis, University of Wisconsin, 1930.

Kümmel, H. B.   Some Meandering Rivers of Wisconsin, Science, new series, Vol. 1, 1895, pp. 714-716.

Lyon, Lucius, and Messenger.   On the Marking of the Wisconsin-Illinois Boundary, Sen. Doc. 234, 24th Congr., 1st Sess.

MacClintock, Paul.   The Pleistocene History of the Lower Wisconsin River, Journ. Geol., Vol. 30, 1922, pp. 673-689.

Martin, Lawrence.   The Western Uplands, Physical Geography of Wisconsin, Journ. Geog., Vol. 12, 1913, pp. 231-232; Rock Terraces of the Driftless Area of Wisconsin, Bull. Geol. Soc. Amer., Vol. 28, 1917, pp. 148-149.

McConnell, W. R.   Geography of Southwestern Wisconsin, Unpublished thesis, University of Wisconsin, 1917.

Owen, D. D.   Report of a Geological Exploration of Part of Iowa, Wisconsin, and Illinois, House Ex. Doc. 239, 26th Congress, 1st Session, Washington, 1840, 161 pp.,—(see Soils pp. 48-53); Report of a Geological Reconnoissance of the Chippewa Land District of Wisconsin, Senate Ex. Doc. 57, 30th Congress, 1st Session, Washington, 1848, 134 pp; Report of a Geological Survey of Wisconsin, Iowa, and Minnesota, Philadelphia, 1852, 634 pp., and atlas.

Park, E. S.   Geology of an Area in Green County, Unpublished thesis, University of Wisconsin, 1897.

Salisbury, R. D.   On the Northward and Eastward Extension of the Pre-Pleistocene Gravels in the Basin of the Mississippi, Amer. Geol., Vol. 8, 1891, p. 238; Preglacial Gravels on the Quartzite Range near Baraboo, Wis., Journ. Geol., Vol. 3, 1895, pp. 655-667.

Salisbury, R. D., and Atwood, W. W.   The Geography of the Region about Devils Lake and the Dalles of the Wisconsin, Bull. 5, Wis. Geol. and Nat. Hist. Survey, 1900, 151 pp.

Shilling, I. R.   Geography of Vernon County in Wisconsin, Unpublished thesis, University of Wisconsin, 1918.

Smith, Guy-Harold.   The Influence of Rock Structure and Rock Character Upon Topography in the Driftless Area, Unpublished thesis, University of Wisconsin, 1921; Physiography of Baraboo Range of Wisconsin, Pan-American Geologist, Vol. 56, 1931, pp. 123-140.

Smith, W. D.   Geology of the Blue Mounds and the Physiography of the Region Adjacent, Unpublished thesis, University of Wisconsin, 1902.

Strong, Moses.   Geology and Topography of the Lead Region, Geology of Wisconsin, Vol. 2, 1877, pp. 643-752; Geology of the Mississippi Region North of the Wisconsin River, Ibid., Vol. 4, 1882, pp. 3-98; Lead and Zinc Ores, Ibid., Vol. 1, 1883, pp. 637-655.

Swezey, G. D.   On some Points in the Geology of the Region about Beloit, Trans. Wis. Acad. Sci., Vol. 5, 1882, pp. 194-204.

Thwaites, F. T.   Geology of Southern Part of Cross Plains Quadrangle, Dane County, Wisconsin, Unpublished thesis, University of Wisconsin, 1908; Pre-Pleistocene Terraces of the Driftless Area of Wisconsin, Bull. Geol. Soc. Amer., Vol. 39, 1928, pp. 621-641.

Thwaites, F. T., and Twenhofel, W. H.   Windrow Formation, an Upland Gravel Formation of the Driftless and Adjacent Areas of the Upper Mississippi Valley, Bull. Geol. Soc. Amer., Vol. 31, 1920, page 133; Ibid., Vol. 32, 1921, pp. 293-314.

Thwaites, F. T., Twenhofel, W. H., and Martin, Lawrence.   The Sparta-Tomah Folio, U. S. Geol. Survey (in preparation).

Trowbridge, A. C.   Some Partly Dissected Plains in Jo Daviess County, Illinois, Journ. Geol., Vol. 21, 1913, pp. 731-742; Physiographic Studies in the Driftless Area, Bull. Geol. Soc. Amer., Vol. 26, 1915, p. 76; The Erosional History of the Driftless Area, Univ. Iowa Studies, Vol. 9, 1921, pp. 7-127.

Twenhofel, W. H., and Thwaites, F. T.   The Paleozoic Section of the Tomah and Sparta Quadrangles, Wisconsin, Journ. Geology, Vol. 27, 1919, pp. 614-633.

Uber, H. A.   The Terraces of the Wisconsin River Between Prairie du Sac and Prairie du Chien, Wisconsin, Unpublished thesis, University of Wisconsin, 1916.

Van Hise, C. R.   Some Dynamic Phenomena Shown by the Baraboo Quartzite Ranges of Central Wisconsin, Journ. Geol., Vol. 1, 1893, pp. 347-355.

Warren, Anna L.   The Geography of Sauk County, Wisconsin, Unpublished thesis, University of Wisconsin, 1918.

Weidman, S. The Baraboo Iron-Bearing District of Wisconsin, Bull. 13, Wis. Geol. and Nat. Hist. Survey, 1904, 171 pp; Pleistocene Succession in Wisconsin, Bull. Geol. Soc. Amer., Vol. 24, 1913, pp. 697-698.

Whitney, J. D. Report on the Lead Region, in Hall and Whitney's Report on the Geological Survey of Wisconsin, Vol. 1, 1862, pp. 73-420 (including a chapter on "Physical Geography and Surface Geology," pp. 93-139).

Whitson, A. R. and others. Soils Reports, Wisconsin Geol. and Nat. Hist. Survey, Bulls. 13, 30, 38, 40, 53A, 53B, 53C, 54C, 60A, 60B, 60C, 60D, 60E.

Wooster, L. C. Geology of the Lower St. Croix District, Geology of Wisconsin, Vol. 4, 1882, pp. 101-159.

See also bibliographies at ends of Chapters I, VII, and VIII, pp. 23, 181, 207.

MAPS

U. S. Geol. Survey. Quadrangles are now available for a large proportion of the Western Upland. See Fig. 197.

U. S. Mississippi River Commisssion. Sheets 161 (Dubuque) to 185 (Prescott, on the scale of 1: 20,000 (see Fig. 199). Sheets 125 (Dubuque) to 135 (Prescott) on the scale of 1: 63,360; see also atlas of maps accompanying report of G. K. Warren (p. 504).

University of Wisconsin. Model of the Baraboo Range, showing topography and geology. See Plate XIV, facing page 195.

Wisconsin Geological Survey. Atlas Sheets 5 to 9, 1876, scale 1: 63,360, contour interval 50 feet.

Wisconsin Geol. and Nat. Hist. Survey. Nine large scale topographic maps of the lead and zinc district, in atlas accompanying Bulletin 14, 1906; six additional large scale topographic maps of same series, issued in 1909. For areas covered see Fig. 201 in Appendix E. Soil survey maps of Iowa, La Crosse, Juneau, Polk, St. Croix, Pierce, Pepin, Dunn, and Eau Claire counties. For areas covered see Fig. 205 in Appendix E.

CHAPTER IV

THE DRIFTLESS AREA

THE PAINTED STONE

Nearly a century ago Professor Keating, a geologist from the University of Pennsylvania, travelled overland from Chicago to Prairie du Chien, and then went up the Mississippi. In southwestern Wisconsin he noted the absence of the granite bowlders that we now know to have been brought by the continental ice sheet to glaciated areas. He thought of them as erratics of very old, or "primitive" rock, transported during the Flood. He commented upon this change (see p. 103), and he did so at the southern border of the district we now call the Driftless Area.

After crossing the Driftless Area in 1823 Keating observed the resumption of the erratics. He saw them at the first place along the Mississippi where one could possibly do so, unless he were to climb the bluffs. This was at Red Rock, Minnesota, near the northwestern boundary of Wisconsin. Just after he passed Lake St. Croix, Keating tells us that they "landed, for a few minutes, to examine a stone which is held in high veneration by the Indians; on account of the red pigment with which it is bedaubed, it is generally called the painted stone. They remarked that this was the first bowlder of primitive rock, which they had seen to the west of Rock river, and this place corresponds well with that at which these bowlders were first observed by Mr. Colhoun (p. 104) while traveling by land. It is a fragment of sienite, which is about four and a half feet in diameter. It is not surprising that the Indians should have viewed this rock with some curiosity, and deemed it wonderful, considering that its characters differ so materially from those of the rocks which are found in the neighborhood. A man who lives in a country where the highest hills are wholly formed of sandstone and secondary limestone, will necessarily be struck with the peculiar characters of the first specimen of granite that comes under his notice, and it is not to be wondered at, that one who 'sees God in all things', should have made of such a stone an object of worship. The Indians frequently offer presents to the Great Spirit near this stone; among the offerings of their superstition, the party found the feather of an eagle, two roots of the "Pomme de Prairie," (psoralea esculenta, Nuttall,) painted with vermilion; a willow branch whose stem was painted red, had been stuck into the ground on one side, etc. The gentlemen broke off a fragment of this idol, to add to the mineralogical collections, taking care, however, not to leave any chips, the sight of which would wound the feelings of the devotee, by convincing him that the object of his worship had been violated."

Thus we see that Keating, who was almost the first geologist to visit the Driftless Area, clearly recognized its contrast with the surrounding area at both borders. His only geological predecessor, Schoolcraft, also identified its boundary (see p. 106), but at a later time and in another place. Schoolcraft mentions nothing of this sort in his first journey, three years before Keating's. Lieutenant Pike landed at the Painted Stone in 1805, but he did not know its significance. Keating even collected a specimen from the Painted Stone—the first erratic he saw after leaving the Driftless Area. But he was not the real pioneer, for the naked aborigines had painted the stone red, and worshipped it!

NATURE OF THE DRIFTLESS AREA

In thinking of this driftless area it should be recalled that the region is not unusual except in the absence of features of the erosion and deposition brought about by the continental glacier. Much of the United States is also driftless, as in the southern states and a large part of the West, outside the relatively small areas of glaciated mountains. These regions are not spoken of as driftless areas or thought of as exceptional in any essential respect. The Driftless Area of Wisconsin, however, is famous the world over because it is completely surrounded by glaciated territory. It preserves a large sample of what the rest of Wisconsin, as well as northern and eastern United States, were like before the Glacial Period. Within the belts covered by the gigantic continental ice sheets of northeastern North America and northwestern Europe there is no similar region of substantial size which was left bare of glacial ice.

Although it has been suggested that there were tiny glaciers in the Driftless Area, their existence has not yet been agreed to by all geologists.

The Driftless Area is mostly in the Western Upland, but it also extends into the Central Plain and the Northern Highland. It covers an area of nearly 15,000 square miles, roughly 210 miles north and south by 120 miles east and west. This is twice as large as the state of New Jersey, or about as large as Denmark. The

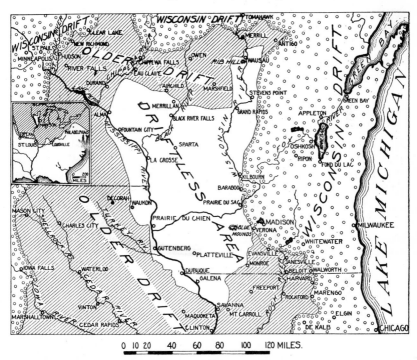

Fig. 26. The Driftless Area of southwestern Wisconsin and adjacent states. For photographs of the Driftless Area see Plates VI, VII, VIII, XIII, XX.

portion of this in Wisconsin is 180 by 120 miles, or 13,360 square miles. The remainder of the Driftless Area is in southeastern Minnesota, northeastern Iowa, and northwestern Illinois.

The Mississippi River flows through the western edge of the Driftless Area, which extends up the Wisconsin River to a point north of Wausau, terminates on the Black River between Neillsville and Black River Falls, and on the Mississippi between the Trempealeau and Chippewa rivers.

TOPOGRAPHY OF THE DRIFTLESS AREA

The Driftless Area is one of the most beautiful parts of the state. Writing in 1854, Edward Daniels, the first state geologist, described the portion in southwestern Wisconsin as follows:

"About one-third of the surface is prairie, dotted and belted with beautiful groves and oak-openings. The scenery combines every element of beauty and grandeur—giving us the sunlit prairie, with its soft swell, waving grass and thousand flowers, the sombre depths of primeval forests; and castellated cliffs, rising hundreds of feet, with beetling crags which a Titan might have piled for his fortress."

This region is not, as a whole, higher than the surrounding region, but the hill tops are distinctly not lower than the adjacent lands. In relation to the encircling glaciated areas it is higher than the Central Plain to the east and northeast; a little lower than the Northern Highland; at about the same level as the Magnesian, Black River, and Niagara cuestas to the west in Minnesota and Iowa; higher than Illinois to the south; and slightly higher than the glaciated portion of the Western Upland to the southeast. Specific altitudes in the Driftless Area are shown in the following table:

TABLE SHOWING ELEVATIONS IN THE DRIFTLESS AREA

Locality	Elevation in feet
Northern Highland, near Wausau	1400
Central Plain, near Necedah	900
Magnesian cuesta, near Alma	1200
Magnesian cuesta, near Richland Center	1160
Baraboo Range, west of Devils Lake	1400
Military Ridge in Black River cuesta near Mt. Horeb	1200

CAUSE OF THE DRIFTLESS AREA

The Driftless Area Is Not a Nunatak. The cause of the Driftless Area is not the simple one which might at first suggest itself. On the borders of the Greenland and Antarctic ice sheets of the present time there are small driftless areas. It was so at the maximum of past glaciation. Similar areas never overridden by ice are found in all glaciated mountains, as in Alaska, the Alps, the

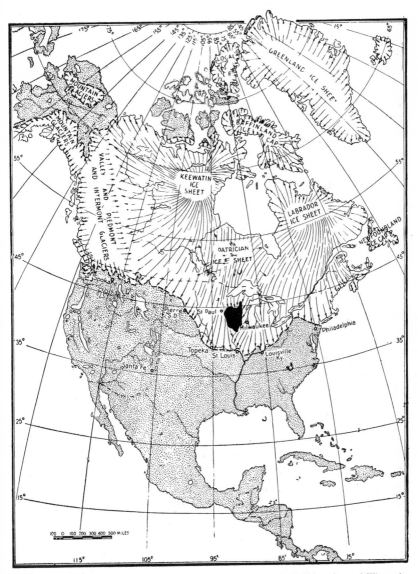

Fig. 27. North America during the Glacial Period, showing the Driftless Area of Wisconsin and adjacent states—in black; other areas never glaciated are dotted; islands peripherally driftless, have dotted borders; the Labrador, Patrician and Keewatin centers of glaciation are located approximately. This map is drawn as if the maxima of glaciation were known to be synchronous.

Himalayas, New Zealand, and Patagonia. In Greenland peaks which rise up like islands through a sea of ice are called *nunataks*. They are free from glaciation because of their height. The Drift- less Area of Wisconsin is not a nunatak area.

The Driftless Area Is Neither an Island Nor a Lake Bed. About the middle of the last century, when the drift deposits were thought of as laid down in the ocean, the Driftless Area was considered to have been an island. Its lack of height above the surrounding areas, and our modern knowledge of drift deposits as glacial and not marine, render this explanation untenable.

It was also once thought to represent a lake bed. The deposits of fine silt or loess in Wisconsin, then thought to be lacustrine, are now known to be chiefly wind-laid (p. 135).

The Elements of Topography and Time. Instead, it is driftless because of three factors:

(a) The highland to the north furnished temporary protection from ice invasion;

(b) The more rapid movement of glacial lobes in the lowland to the east and the region to the west resulted in the final joining of these ice lobes south of the Driftless Area, so that it was completely surrounded by the continental gla- cier;

(c) The termination of the forward movement and the begin- ning of retreat came before there was time for the ice from the north, east, and west to cover the driftless rem- nant.

Such was the relation of the Driftless Area to (a) the topography of the adjacent region and (b) the element of time. It is assumed that the advances of the glacial lobes east, north, and west of the Driftless Area took place at the same time. There is no direct evidence of this, but no facts are known that disprove it.

Topographic Influences Outside. The upland which lent tem- porary protection to the Driftless Area was the Northern High- land. The lowland which rapidly led the eastern lobe southward past the Driftless Area was the river valley now occupied by Lake Michigan. The similar feature to the west was not so low, being the valley of the Red and Minnesota rivers in Minnesota and the valley of the Des Moines River in central Iowa.

Glacial Lobes. The ice which covered nearly all of northeastern and eastern Wisconsin during the Glacial Period is thought to have come chiefly from the Labrador ice sheet, east of Hudson Bay (Fig. 27). At least three advances in northwestern Wisconsin came from the north or northwest, as a part of either the Keewatin, or the Patrician, ice sheets west of Hudson Bay. The ice which covered northwestern Wisconsin and Minnesota and Iowa came chiefly from the Keewatin center (Fig. 27).

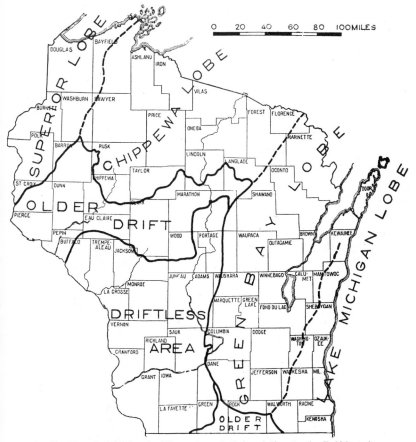

Fig. 28. The glacial lobes in Wisconsin and their relation to the Driftless Area.

In advancing over Wisconsin the ice sheet or continental glacier, was divided into several lobes (Fig. 28), determined in position by the preglacial configuration of the land. These were the Lake

Michigan and Green Bay lobes on the east and the Lake Superior and Minnesota lobes on the west. On the north two minor branches of the Lake Superior lobe, one from the bay at Ashland, the other from the bay east of Keweenaw Point in upper Michigan united to form the Chippewa lobe. The movement of these ice lobes is known by the glacial scratches, or striae, upon the rock ledges and by the transported rocks, or erratics, which have been traced to their sources, often in Michigan, Minnesota, or Canada. These lobes advanced until they completely coalesced.

There were several stages of glacial advance, the latest being called the Wisconsin stage of glaciation. The earlier ones are known as the Iowan, Illinoian, Kansan, and pre-Kansan or Nebraskan. Their deposits are spoken of collectively as older drift. It is not yet decisively established that we do not now live in an interglacial period or that the ice may not advance again. The most recent glacial deposits in the United States are called Wisconsin drift, whether found in the state of Wisconsin or elsewhere.

The Ice East of the Driftless Area. To the east of the Driftless Area, the lobe from the Labrador center advanced down the preglacial river valleys now occupied by Lake Michigan, Green Bay, and Lake Winnebago. By the time it had advanced farthest in southern Illinois, about 300 miles south of Wisconsin, it had spread westward in this state nearly as far as Wausau and Madison. It covered the rather-low southeastern part of the Northern Highland, the eastern part of the Central Plain, and the low eastern region of Galena-Black River limestone. Thus it had expanded where ice movement was relatively easy. No doubt its later rate of broadening and westward expansion over those parts of the Driftless Area which are higher would have been slower. We know that such was the case with the expansion across the high eastern part of the Baraboo Range (p. 121).

The Ice North of the Driftless Area. In the meantime the lobe of the Labrador glacier in the lowland now occupied by Lake Superior, was advancing more slowly. It was retarded on the highland between western Lake Superior and the northern portion of the Driftless Area. Therefore the time necessary for the rapid southward movement of the Lake Michigan ice past practically the whole length of Wisconsin and Illinois, was only long enough for the Lake Superior ice to ascend to the crest of the Northern High-

THE DRIFTLESS AREA AT THE WISCONSIN STAGE OF GLACIATION

The relations of the border of the continental glacier to topography; the extent of Glacial Lake Wisconsin; the positions of some of the glacial streams which laid down deposits of gravel and sand outwash plains and valley trains many of which are now dissected into terraces.

land and advance down its southern slope and across part of the Central Plain to the vicinity of Eau Claire, Neillsville, Wisconsin Rapids, and Merrill. The later expansion of this ice across the Driftless Area would have been more rapid. Except in the high Western Upland, it would have been almost as fast as that of the ice of the Green Bay-Lake Michigan lobe on the east.

The Ice West of the Driftless Area. During the same period of time it seems likely that the ice from the Keewatin center of glaciation was advancing rapidly southward, down what are now the valleys of the Red River of the North, the western part of the Minnesota River, the Des Moines River in Iowa, and part of the Mississippi valley. It finally reached the vicinity of St. Louis, nearly as far south as the Lake Michigan ice in southern Illinois. This lobe also expanded laterally with southward extension. At its maximum it was probably confluent with the Lake Michigan lobe in northwestern Illinois.

These ice tongues on either side of the Driftless Area, therefore, joined each other at a point only 50 miles south of the Wisconsin line. The Keewatin ice in Minnesota and Iowa was reinforced by a little ice from the Lake Superior glacier. This resulted in the overriding of the upland in Wisconsin northwest of the Chippewa River. The later expansion of this ice would doubtless have been as rapid as that for some time before.

Results of a Longer Time of Glaciation. Topography has played the major rôle in the formation of such a Driftless Area as we have, but now the element of time comes in. If the Glacial Period had lasted a little longer and the snowfall had been sufficiently great or the melting sufficiently slow, there should have been the following consequences:

(a) the Labrador, Patrician and Keewatin ice sheets would have advanced still farther south across Missouri and from Illinois into Kentucky;

(b) there would have been further lateral expansion of these glaciers, covering more of the eastern and western edges of the Driftless Area and encroaching more on the south;

(c) the ice from the north would have moved southward from the Northern Highland, whose retarding influence was then decidedly lessened;

(d) these lobes acting together would have completely covered the Driftless Area.

However, time did not permit this. Forward movement decreased in rate and the ice borders melted back. This was all in one of the periods of early glaciation. We do not know that the Driftless Area was completely surrounded more than once. Indeed it was never completely encircled by ice unless, as already stated, the Labrador and Keewatin lobes advanced at the same time. In the latest, or Wisconsin, stage of glaciation the ice fell far short of the borders of the Driftless Area, except where the Green Bay lobe spread a little farther westward than any of its predecessors. Thus we have all the borders of the Driftless Area made up of older drift (Fig. 26), except in eastern Wisconsin between Wausau and Madison.

The Work of Weathering and Underground Water in the Driftless Area

Weathering and the Residual Soils. Throughout the Driftless Area the work of weathering has continued since long before the Glacial Period and has produced a deep mantle of residual soil or geest. This forms a notable contrast with the remainder of the state, where the continental glacier scraped away nearly all the residual soil and left a sheet of transported soil. The latter is essentially unaltered, except in the areas of older drift.

The Dodgeville and Baxter silt loams of the Wisconsin Soil Survey represent the residual soil of limestone. The Boone fine sandy loam is that of sandstone. The Marathon and Mosinee loams are residual from granite, greenstone, and crystalline rocks.

The residual material in the limestone belts is chiefly a fine brown or reddish clay, representing the more or less insoluble residue from the decay of the limestone. It also contains numerous fragments of flint or chert, made of silica and therefore relatively insoluble and not much weathered. It grades downward into solid rock, passing through successive zones of larger and larger limestone fragments (Fig. 32).

The thickness of the residual soil in the Driftless Area varies considerably. The average thickness may be a little over 7 feet. This is based upon about 1800 measurements. The average thickness of residual soil on slopes is about 4⅗ feet, on ridges 8 feet, on broad uplands 13½ feet, in valleys 7 to 18 feet, and the maximum recorded thickness is 70 feet or more. The latter is on an upland. It has been computed that the removal of about 100 feet

A CRAG IN THE DRIFTLESS AREA OF WISCONSIN.

Five-column rock near Readstown, Kickapoo valley, due to weathering and wind work. The columns are sandstone; the cap is limestone. Such fragile forms as this are absent in the glaciated northern and eastern parts of the state. Crags of this type are more common in the arid western part of the United States than in the humid Middle West.

A. MONUMENT ROCK NEAR VIROQUA.

B. DRESBACH CLIFF NEAR SPARTA.

of limestone by weathering leaves 10 feet of residual soil. The limestone regions have thicker residual soil than the sandstone areas, but the latter, including rock fragments as well as soil, may have a thicker residual covering than the limestone.

Crags and Pinnacles. As a result of the process of weathering and of erosion by the wind the Western Upland abounds in picturesque rocky crags and pinnacles. There are also crags of this sort east of the Western Upland near the Wisconsin River. Such rocks and crags are present in the Driftless Area, but generally absent in glaciated regions. Some of these are columns recently separated from adjacent precipices, as at Stand Rock in the Dells of the Wisconsin (p. 351), the Turks Head and the Devils Door at Devils Lake, and many others.

Fig. 29. Stand Rock, a Driftless Area crag at the Dells of the Wisconsin. (Hobbs, after Salisbury and Atwood.)

Other crags rise above the level of the surrounding country as solitary rocky towers. They are especially abundant in the St. Peter sandstone. Monument Rock, in Vernon County, 7 miles south of Viroqua, is a conspicuous crag of this sort (Pl. VIII, A). It is 35 or 40 feet high and twice as wide at the top as at the base. The Devils Chimney and Picture Rock in Dane County are similar, precipitous towers of sandstone.

There are also large hills with crags, chimneys, and towers at their borders. Among such hills is Roche á Cris in Adams County. It rises to a height of 225 feet. Friendship Mound is 85 feet higher. Mosquito Mound in Portage County, Pilot Knob, Rattlesnake Rock, and Petenwell Peak, the latter 230 feet high, all in Adams and Juneau counties, are picturesque hills and crags of the Cambrian sandstone. Necedah Mound is a quartzite hill. Gibraltar Rock in Columbia County is a glaciated crag of the same type as Roche á Cris, but capped by St. Peter sandstone, and somewhat rounded on the eastern side by the ice.

Natural Bridges. Another result of the operation of weathering has been to make natural bridges. In the Driftless Area two of these are of fair size and there are numerous smaller arches of the same nature.

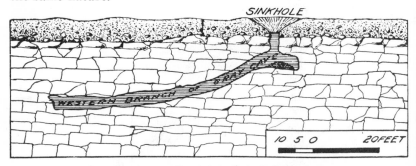

Fig. 30. A cave and sink hole in the Driftless Area.

The arch at Rockbridge spans a stream which has cut under a rock bridge. Nearby the abandoned lower valley joins that of a larger stream. As the bridge is in sandstone it is not due to solution. It is due to weathering along a vertical joint plane. The main stream seems to have swung against the spur on one side and the tributary on the other, undercutting the cliff at this point. Thus the tributary was diverted and passes through an arch about 10 or 12 feet high, with a span of 15 or 20 feet. It is 8 miles north of Richland Center and is well shown on the Richland Center Quadrangle.

A second natural bridge is 1¼ miles northeast of Leland in Sauk County. The bridge is an arch with a 35 foot span. It is 25 to 35 feet high. The rock ledge which spans the opening is less than four feet wide at the top. This bridge is also in sandstone. It is not related to any stream, but apparently results from weathering and the removal of grains of sand by the wind and of blocks of sandstone by gravity. Below the arch of the bridge is a cave $7\frac{1}{2}$ feet high and 25 feet long.

There is a small natural bridge southwest of Madison near the Devils Chimney in the Town of Primrose. The arch is about 6 feet high and has a span of 8 feet.

The shallow caves in the Mississippi bluffs near La Crosse seem also to be due to weathering.

Solution. While the chemical action of underground water goes on near the surface, there is much dissolving of soluble portions of the rock deep below the surface. The water which soaks into the ground penetrates to considerable depths. It dissolves little of the quartz in the sandstone belts of the Driftless Area, but in the limestone areas it takes away much of the lime.

Sink Holes. Parts of the Driftless Area abound in sink holes and caves and these have been produced chiefly by the solvent action of the underground water, aided by the abundant joints in the rock.

The sink holes are sometimes at the entrances of caves. They are circular or elliptical depressions, some dry, others containing ponds. Some of them are 60 feet in diameter and 20 feet deep. One sink hole near Blue Mound is 100 by 200 feet, and 20 feet deep. They average over 5 feet in depth and the greatest number are not over 10 or 15 feet in diameter. That they are still being formed by solution and by falling in of cavern roofs and entrances is shown by the recent killing of trees adjacent to some sink holes, as near Blue Mound. There are at least 70 sink holes in Vernon, Crawford, and Richland counties alone.

Caves. Some of the caves of Wisconsin are listed below.

LIST OF A FEW OF THE WISCONSIN CAVES

Name or location of cave	County
Cave near Blue Mound[a]	Dane
John Gray Cave near Rockbridge[b]	Richland
Eagle Cave northeast of Blue River[c]	Richland
Bear Cave near Boscobel[d]	Crawford
Richardson Cave near Verona[e]	Dane
Cave near Wauzeka[f]	Crawford
" " Highland	Iowa
" " Shullsburg	Lafayette

Name or location of cave	County
" " Mazomanie[g]	Iowa
" " Viroqua	Vernon
" " Wilson	St. Croix
" " Castle Rock	Grant
" " Platte Mound	Grant
" north of Eagle	Richland

a One mile northwest of Blue Mounds railway station, on the N. W. ¼ of section 6, T. 6 N., R. 6 E., Township Blue Mounds.
b Seven miles northwest of Richland Center, on the S. E. ¼ of the N. W. ¼ of Section 25, T. 11 N., R. 1 E., Township Rockbridge.
c Six miles northeast of Blue River, on the N. W. ¼ of Section 19, T. 9 N., R. 1 W., Township Eagle.
d Six and one-half miles northeast of Boscobel, on the N. W. ¼ of the N. E. ¼ of Section 28, T. 9 N., R. 3 W., Township Scott.
e Three and one-half miles north of Verona, on the N. E. ¼ of the N. E. ¼ of Section 5, T. 6. N., R. 8 E., Township Verona.
f Northwest of Wauzeka near Little Kickapoo lead mine in the N. W. ¼ of Section 10, T. 7 N., R. 5 W.
g Five miles north of Blue Mound.   Entrance blocked.

There are doubtless scores of other caves and they are practically all within the Driftless Area. The lead and zinc district abounds in natural caves. Indeed the mining in early days consisted in some cases only of the removal of lead deposits that lined caverns and cavities. Many of these were small, but certain of them contained chambers as much as 35 feet long and 6 or 8 feet wide. The cavern just cited was 60 feet below the surface.

Some facts about the dimensions of caves in the Western Upland are summarized in the table below. Their forms are indicated by the maps reproduced in Figure 31.

TABLE SHOWING FACTS ABOUT CERTAIN CAVES IN WISCONSIN

Name	Location	Total length in feet	Width in feet	Height in feet	Depth below surface	Geological formation
Blue Mound Cave	Near Blue Mound, Dane County	250	5½	5	25	Galena limestone
John Gray Cave	Rockbridge, Richland County	710	2–14	3–5	40	Lower Magnesian limestone
Bear Cave	Near Boscobel, Crawford County	800	7–60	5–40	75	" " "
Eagle Cave	Near Eagle, Richland County	960	70	7½	50	" " "
Richardson Cave	North of Verona, Dane County	Over 500	3–25	3–10	40	" " "

The table shows that these Wisconsin caves are relatively small affairs, compared with the mammoth caverns of Kentucky, Tennessee, Indiana, Virginia, and some other parts of United States. They are not insignificant features, however, and there may be still larger ones within the Driftless Area. The Bear Cave has

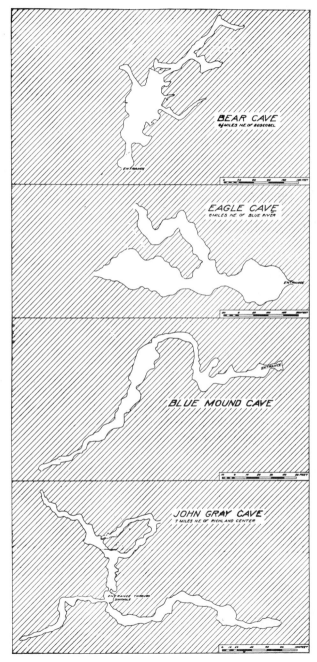

Fig. 31. Outline maps of four caves in southwestern Wisconsin. (Lange.)

two levels, the upper one containing one room 60 feet wide, 65 feet long, and 40 to 60 feet high, while the Eagle Cave has one room 100 feet wide and 30 or 40 feet high.   There are many other caves, whose entrances have recently been blocked by falling in of rock, as is well known to farmers on the limestone uplands of the Driftless Area.   There are probably still other hidden caves in the localities where well drillers report that their bits suddenly fell 4 or 5 feet, after penetrating through solid rock.

Several of these caves contain stalactites and all are more or less filled with mud.   Enlarged fissures extend below the bottoms of some caves.

On the whole there seem to be more caves in the Lower Magnesian than in the Galena-Black River or the Niagara limestone of the Driftless Area.   Few caves are known in the Black River or Lower Magnesian limestone in the glaciated, eastern part of the state.

Caves Limited to Driftless Area.   So far as now known, only one or two of the caves in eastern Wisconsin due to the solvent action of underground water are in the area of Wisconsin drift. One of these is near Maribel in Manitowoc County.   Another is the Richardson Cave in Dane County.   Its entrance lies within a quarter of a mile of the outermost terminal moraine.   There are enlarged fissures and tiny caves near Wilson (p. 251) in northwestern Wisconsin.   These are in the region of older drift.   It is improbable that there are great numbers of caves in the eastern part of the state but that every one happens to be covered by glacial drift, and that only a tiny proportion of them accidentally uncovered by streams or by man in connection with all the work of railway grading, highway building, and artificial excavation.

The absence of numerous caves in eastern Wisconsin seems to be due to sculpture by ice during the Glacial Period.   The continental glacier nearly everywhere eroded down to a level below the bottoms of the preglacial caves, as will be explained in a later chapter (p. 251).

Since the end of the Glacial Period there has not been time enough for underground water to make new caves in the glaciated territory, for solution and weathering perform their work so slowly that the delicate glacial scratches are not yet removed from the rock ledges.   This suggests that the caves of the Driftless Area

are chiefly the work of underground solution in preglacial time
and that they have been exceedingly long in process of formation.
As with numerous other phenomena, we should not know of this
were it not for the preservation of the physiographic features of
preglacial time in the Driftless Area.

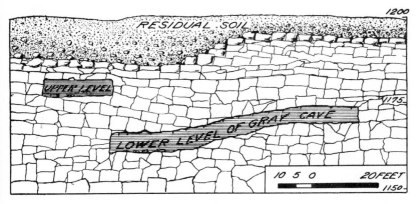

Fig. 32. Two levels of the same cave in the Driftless Area. Varying thickness of
residual soil.

BIBLIOGRAPHY

Aldrich, Mildred. A Comparison of Agricultural Conditions in the Driftless
and Glaciated Portions of Wisconsin, Unpublished thesis, University of
Wisconsin, 1912.

Alden, W. C. Criteria for the Discrimination of the Age of Glacial Drift
Sheets, Journ. Geol., Vol. 17, 1909, pp. 694-709; Quaternary Geology of
Southeastern Wisconsin, U. S. Geol. Survey, Prof. Paper 106, 1918, pp.
103-111, 171-172.

Bowman, Isaiah. (On plants in the Driftless Area), Forest Physiography,
New York, 1911, p. 497.

Chamberlin, T. C. (*An early explanation of the Driftless Area*, see also R.
D. Irving, and N. H. Winchell), in Snyder, Van Vechten & Co.'s Atlas of
Wisconsin, Milwaukee, 1878, p. 151; Annual Report, Wis. Geol. Survey
for the Year 1878, Madison, 1879, pp. 21-32; Trans. Wis. Acad. Sci.,
Vol. 5, 1882, pp. 268-270; Geology of Wisconsin, Vol. 1, 1883, pp. 269-
270; Maps of Driftless Area, Atlas Plate II, Geol. of Wis., 1881; 3rd
Ann. Rept., U. S. Geol. Survey, 1883, Plates 28, 29, 31, and 35; 7th Ann.
Rept., Ibid. 1888, Plate 8; (on soils) Ibid., pp. 678-688.

Chamberlin, T. C. and **Salisbury, R. D.** The Driftless Area of the Upper
Mississippi Valley, 6th Annual Report, U. S. Geol. Survey, 1885, pp.
199-322, with maps, (*the most complete discussion of the Driftless Area,
and of the residual soil and loess within it*).

Dana, J. D.   (On the Driftless Area), Amer. Journ. Sci., 3rd Series, Vol. 15, 1878, pp. 62-64, 254-255.

Davis, W. M.   (On the Driftless Area), Guidebook for the Transcontinental Excursion of 1912, American Geographical Society of New York, Boston, 1912, pp. 16, 87-89.

Daniels, Edward.   (On the Driftless Area), First Annual Report on the Geological Survey of the State of Wisconsin, Madison, 1854, pp. 11-12; (on caves) Ibid., pp. 15, 26; Proc. Boston Soc. Nat. Hist., Vol. 4, 1854, p. 387.

Folsom, W. H. C.   (On Driftless Area phenomena), Fifty Years in the Northwest, St. Paul, 1888, pp. 383-384.

Gilbert, G. K.   (On the Driftless Area), Geological Guidebook of the Rocky Mountain Excursion, Compte Rendu de la 5me. Session, Washington, 1891, Congrès Géologique International, Washington, 1893, pp. 289-290; see also *S. F. Emmons*, Ibid., pp. 298-301.

Grant, U. S.   (On the Driftless Area), Bull. 14, Wis. Geol. and Nat. Hist. Survey, 1906, pp. 12-14.

Grant, U. S. and Burchard, E. F.   Geological Atlas of the United States, Lancaster-Mineral Point Folio, No. 145, 1907.

Hall, James.   (On the Driftless Area), Report on the Geological Survey of the State of Wisconsin, Vol. 1, 1862, pp. 6-7.

Hodge, J. T.   (On the Driftless Area), On the Wisconsin and Missouri Lead Region, Amer. Journ. Sci., Vol. 43, 1842, p. 36.

Irving, R. D.   (*The first correct explanation of the Driftless Area*, see also N. H. Winchell), The Quaternary Deposits of Central Wisconsin, Geology of Wisconsin, Vol. 2, 1877, pp. 608-611, 632-635, and Plate XXV A; Annual Report, Wis. Geol. Survey for 1876, Madison, 1877, p. 15; Origin of the Driftless Area of the Northwest, Amer. Journ. Sci., 3rd Series, Vol. 15, 1878, pp. 313-314; Driftless Region of Wisconsin, op. cit., pp. 406-407; Trans. Am. Inst. Mining Engineers, Vol. 8, 1880, pp. 491, 504; see also boundaries of Driftless Area on map facing p. 506; see boundaries of Driftless Area on Map of Quaternary Formations of Wisconsin, Atlas Plate II, Geology of Wisconsin, 1881; (on crags and pinnacles), Geology of Wisconsin, Vol. 2, 1877, pp. 428, 523, 564, 567-568, 572-574, 606-607.

Kay, G. F., and Apfel, E. T.   The Pre-Illionian Pleistocene Geology of Iowa, Iowa Geol. Survey, Vol. 34, 1929, pp. 24-32, 45-48.

Keating, W. H.   (*The first observation of the Driftless Area phenomena*), Narrative of an Expedition to the Source of St. Peter's River, Lake Winnepeek, Lake of the Woods, etc. etc., Performed in the Year 1823, Philadelphia, 1824, Vol. 1, pp. 200, 263, 287-288.

Lange, E. G.   Caves of the Driftless Area of Southwestern Wisconsin, Unpublished thesis, University of Wisconsin, 1909.

Lapham, I. A.   (On the Driftless Area), A Geographical and Topographical Description of Wisconsin, Milwaukee, 1844, p. 59; Wisconsin, Its Geography and Topography, Milwaukee, 1846, p. 57; Geological Formations of Wisconsin, Trans. Wis. State Agr. Soc., Vol. 1, 1851, pp. 125-126; Geology, in Walling's Atlas of the State of Wisconsin, Milwaukee, 1876,

p. 19; also some time before 1862, see Lapham quoted by Whitney in Geology of Wisconsin, Vol. 1, 1862, p. 119, and Geol. Survey of Illinois, Vol. 1, 1866, p. 161; Report for 1874, Geology of Wisconsin, Vol. 2, 1877, p. 57.

Leverett, Frank, and **Sardeson, F. W.** Surface Formations and Agricultural Conditions of the South Half of Minnesota, Minnesota Geol. Survey, Bull. 14, 1914.

Long, S. H. (On Driftless Area phenomena), Voyage in a Six-Oared Skiff to the Falls of Saint Anthony in 1817, Collections Historical Society of Minnesota, Vol. 2, Part 1, 1860, pp. 27, 29, 50.

MacClintock, Paul. The Pleistocene History of the Lower Wisconsin River, Journ. Geology, Vol. 30, 1922, pp. 673-689.

Martin, Lawrence. Rock Terraces of the Driftless Area of Wisconsin (abstract), Geol. Soc. America, Bull., Vol. 28, 1917, pp. 148-149; A Pennsylvanian's Discovery of the Driftless Area, Bull. Geogr. Soc. Philadelphia, Vol. 31, 1923, pp. 32-39 (140-147); (On the Driftless Area), Monograph 52, U. S. Geol. Survey, 1911, pp. 429, 438, 454; The Discovery of the Painted Stone—An Early Observation of the Driftless Area, Journ. Geog., Vol. 14, 1915, pp. 58-59.

McGee, W J (On the Driftless Area), On the Relative Position of the Forest Bed and Associated Drift Formations in Northeastern Iowa, Amer. Journ. Sci., 3rd Series, Vol. 15, 1878, p. 341; On the Complete Series of Superficial Deposits in Northeastern Iowa, Proc. Amer. Assoc. Adv. Sci., 27th Meeting, 1878, Salem, 1879, pp. 198-201; The Drainage System and the Distribution of the Loess in Eastern Iowa, Bull. Philos. Soc. Wash., Vol. 6, 1884, pp. 93-97; The Pleistocene History of Northeastern Iowa, 11th Annual Rept., U. S. Geol. Survey, Part 1, 1891, pp. 202-206, 277, 295-303, 353-354, 357, 435-450, 548, 566, 571-572. Map showing Lake Hennepin, facing p. 577. Plate 44 shows the border of Driftless Area, distribution of residual soil, of loess, etc., in northeastern Iowa.

Mansfield, G. R. The Baraboo Region of Wisconsin, Journ. Geog., Vol. 6, 1908, pp. 286-292; Glacial and Normal Erosion in Montana and Wisconsin, Ibid., pp. 309-312.

Murrish, John. (On the Driftless Area), State of Wisconsin, Report on the Geological Survey of the Lead Regions, 1871, pp. 13-14, 16; (on sink holes), pp. 21-22.

Norwood, J. G. (On the Driftless Area),—in Owen's Geological Reconnoissance of the Chippewa Land District of Wisconsin, Senate Ex. Document 57, 30th Congress, 1st Session, Washington, 1848, p. 105.

Orr, E. Exposures of Iowan and Kansan (?) Drift, East of the Usually Accepted West Boundary Line of the Driftless Area, Proceedings Iowa Acad. of Science, Vol. 14, 1907, pp. 231-236.

Owen, D. D. (*The third mention of the Driftless Area phenomena*)—in Geological Exploration of Part of Iowa, Wisconsin, and Illinois, House Ex. Document 239, 26th Congress, 1st Session, Washington, 1840, second footnote on page 46; (on occurrence of erratics at borders of Driftless Area),—in Wisconsin, pp. 65, 109, 111, 112,—In Iowa, pp. 70, 71, 72, 75, etc; (on soils) Ibid., pp. 48-53; (on prairies) Ibid., see township descriptions, pp. 70-115; (on the loess) in Geological Reconnoissance of the Chippewa Land District of Wisconsin, Senate Ex. Document 57, 30th Congress, 1st Session, Washington, 1848, p. 57; (on drift deposits near the Driftless Area), Ibid., pp. 68-70.

Park, E. S.   Map showing part of border of Driftless Area and location of boulder finds,—Geology of an Area in Green County, Wis., Pl. 9, p. 80, Unpublished thesis, University of Wisconsin, 1897.

Percival, J. G.   (On the Driftless Area), Annual Rept. Geol. Surv. Wis., 1855, pp. 30-31; Ibid., 1856, pp. 17-18.

Pike, Z. M.   (On the Driftless Area phenomena), An Account of Expeditions to the Sources of the Mississippi, etc., During the Years 1805, 1806, and 1807, Philadelphia, 1810, p. 23.

Salisbury, R. D.   (On the Driftless Area), Descriptive America, Vol. 1, 1884, p. 109; Trans. Wis. Acad. Sci., Vol. 6, 1885, p. 48; Loess in Wisconsin Drift Formation, Journ. Geol., Vol. 4, 1896, pp. 929-937.

Salibury, R. D., and Atwood, W. W.   Drift Phenomena in the Vicinity of Devils Lake and Baraboo, Wisconsin, Journ. Geol., Vol. 5, 1897, pp. 131-147; Devils Lake and the Dalles, Bull. 5, Wis. Geol. and Nat. Hist. Survey, 1900, pp. 78-79, 142-146.

Sardeson, F. W.   On Glacial Deposits in the Driftless Area, Amer. Geol., Vol. 20, 1897, pp. 392-403.

Shaw, James.   (On the Driftless Area), Geol. Survey of Illinois, Vol. 5, 1873, pp. 4-5, 30-33; Ibid., Economical Geology of Illinois, Vol. 3, 1882, pp. 25-28.

Schoolcraft, H. R.   (*The second mention of the Driftless Area phenomena*), Remarks on the Lead Mine Country on the Upper Mississippi (Addressed to the Editors of the New York Mirror) pp. 294-307, especially p. 306, in Schoolcraft's Narrative of an Expedidtion through the Upper Mississippi to Itasca Lake, etc., in 1832, New York, 1834; not signed or dated in original report, but reprinted verbatim in Schoolcraft's "Summary Narrative" (1855, pp. 560-572) as "Brief Notes of a Tour in 1831, from Galena, in Illinois, to Fort Winnebago, on the Source of the Fox River, Wisconsin. By Henry R. Schoolcraft," addressed to George P. Morris, Esq., New York.

Squier, G. H.   Depth of the Glacial Submergence on the Upper Mississippi, Science, Vol. 4, 1884, p. 160; Studies in the Driftless Region of Wisconsin, Journ. Geol., Vol. 5, 1897, pp. 825-836; Ibid., Vol. 6, 1898, pp. 182-192; Ibid., Vol. 7, 1899, pp. 79-82; Peculiar Local Deposits on Bluffs Adjacent to the Mississippi, Trans. Wis. Acad. Sci., Vol. 16, 1908, pp. 258-274.

Strong, Moses.   (On the Driftless Area), Geology of Wisconsin, Vol. 2, 1877, pp. 665-667; Vol. 4, 1882, pp. 88, 92-94; Annual Rept., Wis. Geol. Survey for 1876, Madison, 1877, p. 10; Atlas Plate II, Geology of Wisconsin, 1881; (on caves and sink holes), Geology of Wisconsin, Vol. 2, 1877, pp. 661-662; Ibid., Vol. 4, 1882, pp. 75-76, 96-98; (on prairies), Ibid., Vol. 2, 1877, pp. 660-661, and Pl. 27; map showing prairies, Ibid., Atlas Plate IIa.

Tenney, H. A. (On the Driftless Area) in Daniels' First Annual Report on the Geological Survey of the State of Wisconsin, Madison, 1854, pp. 69-74.

Thwaites, F. T. Mysteries of Devils Lake, Madison Democrat, Feb. 18, 1908; Pre-Wisconsin Terraces of the Driftless Area of Wisconsin, Bull. Geol. Soc. America, Vol. 39, 1928, pp. 621-642; Outline of Glacial Geology, 1927, (on the Driftless Area), pp. 148-150.

Thwaites, F. T. and **Twenhofel, W. H.** Windrow Formation; An Upland Gravel Formation of the Driftless and Adjacent Areas of the Upper Mississippi Valley, Bull. Geol. Soc. America, Vol. 32, 1921, pp. 293-314.

Trowbridge, A. C. The Erosional History of the Driftless Area, Iowa University Studies, Studies in Natural History, 1921, Vol. 9, No. 3; The History of Devils Lake, Wisconsin, Journ. Geology, Vol. 25, 1917, pp. 344-372.

Weidman, S. (On the Driftless Area), Geology of North Central Wisconsin, Bull. 16, Wis. Geol. and Nat. Hist. Survey, 1907, pp. 548-565; (on residual soil), Ibid., pp. 552-554, 560-562, 680; see also other pages in Bull. 16 and discussion of soils in northwestern Wisconsin in Bulls. 11 and 23; (on Driftless Area deposits), Ibid., Bull. 13, 1904, pp. 101-102; Pleistocene Succession in Wisconsin, Bull. Geol. Soc. Amer., Vol. 24, 1913, pp. 697-698.

Whitbeck, R. H. Contrasts between the Glaciated and Driftless Portions of Wisconsin, Bull. Philadelphia Geog. Soc., Vol. 9, 1911, pp. 114-123; Economic Aspects of the Glaciation of Wisconsin, Annals Assoc. Amer. Geographers, Vol. 3, 1913, pp. 62-87.

White, C. A. (On the Driftless Area), Geological Survey of the State of Iowa, Vol. 1, 1870, p. 87.

Whitney, J. D. (*The first description of the Driftless Area*), Surface Geology of the Lead Region, in Hall and Whitney's Report on the Geological Survey of the State of Wisconsin, Vol. 1, 1862, pp. 114-139, with a map of the Driftless Area, Fig. 2, facing p. 118; Geol. Survey of Illinois, Vol. 1, 1866, pp. 160-162; Economical Geology of Illinois, Vol. 1, 1882, pp. 123-124.

Whitson, A. R., Weidman, S., and **others.** (On loess and on residual soil in the Driftless Area), Soil Surveys of Iowa, Juneau and La Crosse counties, North Central Wisconsin, and the South Part of Northwestern Wisconsin, Bulls. 11, 23, 30, 38, and 40, Wis. Geol. and Nat. Hist. Survey, 1903 to 1914. See also Soil Survey of Viroqua Area, Bureau of Soils, U. S. Dept. of Agriculture; and D. D. Owen's soil report of 1840, with analyses and township descriptions.

Whittlesey, Charles. (On the Driftless Area), Smithsonian Contributions to Knowledge, Vol. 15, 1867, p. 20 and map facing title page; Proc. Amer. Assoc. Adv. Sci., Vol. 15, 1867, p. 49.

Winchell, N. H. (*The first correct explanation of the Driftless Area*, see also R. D. Irving), Fifth Annual Report, Minn. Geol. and Nat. Hist. Survey, 1877, pp. 9, 34-38; see also First Annual Rept., Ibid., 1873, pp. 46, 61; Fourth Annual Rept., Ibid., 1870, pp. 5, 21, 59-62; Geology of Minnesota, Vol. 1, 1884, pp. 117, 213, 227, 245, 260, 275, 278, 311-

312, 317, 406; Ibid., Vol. 2, 1888, pp. 2, 14-15; (on Red Rock), Ibid., p. 398.

Worthern, A. H.   (On the Driftless Area), Geol. Survey of Illinois, Vol. 1, 1866, pp. 27-33; Ibid., Economical Geology of Illinois, Vol. 1, 1882, pp. 22-26.

See also the bibliographies at the ends of Chapters I and III and in Appendix G.

MAPS

For maps showing the borders of the Driftless Area in Wisconsin, see the publications listed above, and especially:

J. D. Whitney, 1862.
R. D. Irving, 1877, 1879, 1881.
Moses Strong, 1881.
T. C. Chamberlin, 1881, 1883, 1888.
T. C. Chamberlin and R. D. Salisbury, 1885.

W J McGee, 1891.
E. S. Park, 1897.
R. D. Salisbury and W. W. Atwood, 1900.
U. S. Grant and E. F. Burchard, 1906.
Samuel Weidman, 1907.

For topographic maps within the Driftless Area, see end of Chapter III, p. 80.

CHAPTER V

THE DISCOVERY AND EXPLANATION OF THE DRIFTLESS AREA

OBSERVATIONS AND THEIR INTERPRETATION

Discovery by the Indians. We do not know the time and place of the first recognition of the absence of erratic bowlders in what we now call the Driftless Area. It was known to the Indians: they actually worshipped an erratic bowlder near its border (p. 105). It was probably observed by some of the first white men who came to the Upper Mississippi. The following geologists are among those who have discussed it from first-hand knowledge and speculated upon its origin.

1823—W. H. Keating. The first recognition and understanding of the Driftless Area phenomena, of which we know, was made by the geologist, Keating, whose observations were published in 1824. He traveled overland from Chicago in 1823, as already stated (p. 81), accompanying Major Long's Second Expedition. He entered what is now the state of Wisconsin about where the border of the Driftless Area crosses from Wisconsin into Illinois, southwest of Monroe. He then crossed the Military Ridge some distance west of Blue Mound and went to Prairie du Chien. The party travelled up the Mississippi valley to Minneapolis, some of the members going overland through Iowa and Minnesota, and others, including Keating, by boat.

They observed the absence of erratic bowlders almost at once on entering the Driftless Area. They also noticed when they left the bowlderless region, though not until they had passed through the loess-covered region of thin, older drift and reached the latest drift, with abundant erratics, west of Lake St. Croix.

After describing the abundant bowlders of hornblendic granite between the Rock and Pecatonica rivers in northern Illinois, Keating speaks of crossing the latter stream and coming within sight of the Platte Mounds. He then says:

"No granitic blocks are to be seen; this is accounted for by the fact that we are no longer upon the alluvial formation, but upon the magnesian limestone which rises to a greater height, constituting the dividing ridge between the Mississippi, Rock River, and the Wisconsan, and perhaps connecting itself with what have been termed the Wisconsan hills."

Later Keating records:

"The first boulders which had been seen from Rock river, were observed by Mr. Colhoun at about seven miles from Fort St. Anthony; they consisted of granite."

This is clearly the border of the Wisconsin drift sheet near Minneapolis and St. Paul. Though not observing any erratics in the older drift, as indeed is none too easy today, Colhoun and the land party did see the difference in topography close to the very border of the Driftless Area which lies a short distance south of Lake Pepin, for Keating says:

"A very great change in the country above Lake Pepin was visible. The bluffs were not so high, they were more frequently interrupted, and gave a new character to the scenery of the river."

In the meantime the river party, with Keating himself, had been on the Mississippi, where the border of the Driftless Area cannot be easily detected. Within a short distance of the mouth of the St. Croix, Keating and his party landed and examined the Painted Stone, as described at the beginning of Chapter IV, identifying one of the first erratics west of the Driftless Area that can be seen without leaving the Mississippi River.

This Painted Stone was likewise visited in 1805 by Lieut. Z. M. Pike. From his description we may determine just where it was situated. He was ascending the Mississippi and "encamped on a prairie on the east side, on which is a large painted stone, about 8 miles below the Sioux village." It appears from Pike's map that Sioux village was on the site of the present city of St. Paul.

The Painted Stone seems to have given the name to the modern village of Red Rock, Minnesota. Prof. N. H. Winchell says:

"At Red Rock, the rock itself is a bowlder of granite, originally light-colored, but stained with the Indian's 'red paint,' and more recently girdled by successive belts of bright vermillion with oil and lead. On the end lying away from the river is a representation of an Indian's head surmounted by eagle's feathers.

W. H. C. Folsom says: "The peculiarity of the painted boulder from which Red Rock took its name is that it was a shrine, to which from generation to generation pilgrimages were made, and offerings and sacrifices presented. Its Indian name was 'Eyah Shah,' or 'Red Rock.' The stone is not naturally red, but painted with vermillion, or, as some say, with the blood of slaughtered victims. The Indians call the stone also 'Waukan,' or 'mystery.' It lies on a weathered stratum of limestone, and seems to be a fragment from some distant granite ledge. The Dakotahs say it walked or rolled to its present position, and they point to the path over which it traveled. They visited it occasionally every year until 1862, each time painting it and bringing offerings. It is painted in stripes, twelve in number, two inches wide and from two to six inches apart. The north end has a rudely drawn picture of the sun, and a rude face with fifteen rays."

After passing Lake St. Croix in 1817 S. H. Long says:

"The bluffs are more regular both in their height and direction than they are below Lake Pepin, and the valley of the river more uniform in its width. The stratifications of the bluffs are almost entirely sandstone, containing clay or lime in greater or less proportions. The pebbles are a mixture of primitive and secondary stones of various kinds. Blue clay or chalk is frequently to be found."

At a place called the Crevasse, between Lake Pepin and Prescott, Long found what he considered "agates of various hues, calcedony, flint, serpentine, ruby, and rock crystal, etc." These minerals were probably agate and calcedony, garnet, and other fragments of igneous rock from the trap and granite rocks to the northwest. On a high hill on the Iowa side between Trempealeau and La Crosse he found only "crystals of iron ore, silicious crystallizations, beautifully tinged with iron, some of them purple others reddish, yellow, white, etc., crusts of sandstone strongly cemented with iron, and I think set with solid crystals of quartz, etc." These are all minerals such as occur in the local sandstone and limestone.

The "primitive stones" found west of Lake St. Croix and the "serpentine," "rubies" and agates, north of Lake Pepin, as well as the change in the character of the bluffs, would have told a geologist that he was in a drift-covered rather than a driftless area, especially in contrast with the presence of nothing but local materials on the bluff south of Trempealeau, which is in the Driftless Area. It is no injustice to this army officer, however, to give priority to Keating who saw such things and knew their significance.

1831—H. R. Schoolcraft. During a journey overland from Galena to Portage via Blue Mound and Madison, Schoolcraft notes:

"And the occurrence of *lost rocks* (primitive boulders) as Mr. B. happily termed them, which are first observed after passing the Blue Mound, becomes more frequent in this portion of the country, denoting our approach to the boulders of the northwestern primitive formation." Published in 1834.

1839—D. D. Owen. In connection with the possible use of "erratic boulders" of "granite, greenstone, porphyry, and other primitive rocks" for millstones, Owen says:

"They are much more frequent towards the heads of the streams, than they are near the Mississippi river. In crossing the line between ranges seven and eight of the fourth principal meridian, they commence very abruptly, and are found in great numbers, and sometimes of very large dimensions." Published in 1840.

This is interpreted as a location of the eastern border of what we now call the Driftless Area somewhere between Cross Plains and Verona, and probably on the Mineral Point Road west of Madison.

In his descriptions of individual townships Owen speaks in many places of the occurrence of "boulders, detached and worn masses of transported granite, and other crystalline rocks." He mentions all the erratics that he and his parties saw; and they examined every quarter section of every township. He describes the rock ledges and soil of all the townships in which no erratic bowlders were observed. These bowlderless townships are all in what we now call the Driftless Area. Among the occurrences mentioned in Wisconsin are the following:

> Northwest of Madison, R. 8 E., Towns 8 and 9,—"numerous boulders of hornblende rock, some of them very large (say 10 or 12 feet high)."
>
> West of Madison, R. 8. E., Town 7,—"in the north are numerous ponds; the middle and south not well watered."
>
> Southeast of Monroe, R. 8 E., Town 1,—"a few boulders are to be seen now and then."

South of Monroe, R. 7 E., Town 1,—"some boulders," and reference to granite specimen in his collections.

There are similar descriptions for the glaciated territory just west of the Driftless Area in Iowa. Thus we see that, just as Keating determined the southern and approximate northern borders of the Driftless Area in 1823, so Owen made observations from which he could have determined its eastern and western boundaries even more accurately in 1839. So far as known, however, he made no specific mention of the lack of erratics in the Driftless Area.

1841—J. T. Hodge. While studying the lead and zinc district of southwestern Wisconsin Hodge made the first attempted explanation of the Driftless Area, as far as known. It is the fourth known mention of the phenomena. He says:

"In the western part of Wisconsin, there are no primary bowlders, no loose rocks but those which once evidently formed a part of the formations on which they now repose; in the eastern part of the territory, however, and to the west in Iowa, such bowlders are not wanting. Whether this region may have been in part protected by the high lands to the north of it, and the progress of the bowlders been thus intercepted and turned aside, must be determined by more extended observations. This supposition is rendered more plausible by the unusual course of the Wisconsin river, it suddenly turning from a south to a west direction. In its valley, however, where it flows towards the west, no bowlders are found except the small pebbles brought down by the river itself." Published in 1842.

1844—I. A. Lapham. In a description of the "Mineral Country" (south of the Wisconsin River and west of Madison) Lapham says:

"The theoretical geologist will find a hard problem to solve in his endeavor to account for the almost total absence of those boulders of primitive rock in the mineral district, which are so abundant elsewhere in the Territory."

Lapham himself evidently kept on trying to solve this problem, though he published little upon it. In 1861 he said:

"A portion of the limestone district of Wisconsin, lying west of Sugar River, and south of the great dividing ridge running parallel with, and a few miles south of the Wisconsin River, is known as 'the mineral region', and is destitute of drift materials."

In a discussion of the Driftless Area published in 1862, Whitney says that the boundary of the bowlderless district has been continued to the north of the Wisconsin River "chiefly on the authority of Mr. Lapham, who has not, however, made any minute exploration for the purpose of defining the boundary of the drift, so that it must be accepted as only approximately correct."

Whitney also quotes Lapham on the Driftless Area in his Illinois report published in 1866.

In 1874 Lapham wrote:

"It still remains to be accounted for that in the lead region in this state there is almost a total absence of the Drift phenomena."

In his annual report written in 1874, and published in 1877, Lapham says:

"Though the lead region is supposed to have been exempt from the influence of the glaciers which have distributed so much drift material over adjacent districts, there are some facts still requiring explanation, particularly the one first noticed by Prof. Whitney, of the occurrence of blocks of St. Peters sandstone resting upon formations of later age. The boundary of the glacial drift through Green county has now been accurately traced. The occurrence of drift material in the valleys of the Mississippi and lower Wisconsin is rightly attributed to river transportation from above."

1847—J. G. Norwood. On his way from Lake Superior to the Mississippi, by way of the Montreal and Wisconsin Rivers, Norwood observed that:

"Not a boulder, nor scarcely a pebble, is to be seen after passing the first ten miles below Whitney's rapids; showing, conclusively, that the forces which transported the immense numbers of erratic blocks, met with in other sections of the territory, did not tend in this direction."

The place of observation is probably not far south of Nekoosa on the Wisconsin River.

1854—Edward Daniels. In a description of the Geology of the Lead Mines, Daniels says:

"A remarkable fact in the superficial deposits of this region is the entire absence of the drift so abundantly represented over the north-west generally, by boulders, gravel, sands, and clay. So far as my observation extends, not a single boulder or gravel stone can be found over the whole district. The surface material bears no evidence of distant origin, and unless some of the clays shall be proven diluvial we have here no traces of transported drift. Whatever then may have been the agency which dispersed the huge masses of rock, fragments of native copper, beds of sand and gravel, so lavishly over the surrounding country, we know, that by some peculiarity of position the lead region was above its reach. Widely removed as this circumstance may seem from practical matters, it has nevertheless a most important bearing upon the economic value of the district to which it relates. For had it been otherwise the whole surface would have been covered with loose deposits, often of great thickness, burying all indications of the presence of lead veins, rendering discovery exceedingly doubtful, and profitable mining a practical impossibility. The precise boundary of the district thus destitute of drift, is not yet ascertained."

1855—**J. G. Percival.** The most serious early attempt to explain the Driftless Area was made by Percival, who was also one of the first to describe the loess. "The mineral district does not appear to have been invaded to any extent by the gravel and bowlder drift, which has covered so extensively other parts of the surface in this and the adjoining states. Apparently the bold escarpment, backed by the high ridges and prairies, along the south side of the Wisconsin river from a point not far east of the Blue Mounds, has obstructed the course of the drift current, and turned it east and south around the east point of the ridge at those mounds. An opening near the source of Sugar river seems to have given passage to that current, by which large accumulations of gravel drift have been formed along the west side of the valley of that river, near Exeter, and of bowlder and gravel drift farther east, while scattered bowlders, usually of no great size, are found in the side valleys, and on the slopes of the adjoining ridges and prairies, towards the west, as far south at least as the vicinity of Monroe. In the tract of country occupied by the blue limestone and upper sandstone, between the high prairie, west of Janesville, and the ridge of the lower magnesian, south of Madison, accumulations of such diluvial drift are comparatively small and unfrequent, but with occasional exceptions, while on the north of that ridge they are large and extensive; that ridge having also acted apparently as an obstruction to their progress. My observations in that part of the country, covered more or less by this diluvial drift, have been very limited, and a farther consideration of its extent must be deferred to a future occasion. The bowlders and smaller rock fragments, composing this drift, are chiefly derived from primary and trap rocks, though partly from the flints (hornstones and quartz) accompanying the limestones, particularly the lower magnesian. Small nodules of hematite, and of iron pyrites partly converted into hematite, such as occur at the junction of the blue limestone and upper sandstone, are frequently found in this drift and scattered on the adjoining surface.

In the immediate vicinity of the Mississippi, on the surface of the higher ridges and prairies adjacent, accumulations of drift are occasionally found, in some instances quite extensive, composed of a fine sand, usually yellow or light brown, as if formed from the sandstone adjoining that river towards the north. These are generally arranged in hillocks, with intervening round hollows or basins, such as are common in drift districts. This sand, on the surface, is mixed more or less with mould, forming a light soil but at a small depth is sufficiently pure for mortar. A tract of 2-3 square miles, covered with such drift, and remarkable for its hillocks and hollows, extends from the bluffs of the Mississippi to the valley of the Great Menominee, S. W. of Jamestown village, and similar accumulations are met with on the high lands, adjoining the Mississippi, between Potosi and Cassville. On the summits of the river bluffs, particularly in the vicinity of Cassville, small rolled fragments of the same materials as those composing the gravel drift, above noticed, are often profusely scattered. These facts indicate the passage of a peculiar drift current along the course of the Mississippi, and it is worthy of remark, that the points where those accumulations are most remarkable are a little below two large bends in that river, namely, that from south to southeast just above Cassville, and that to the south between Dubuque and Potosi. Such a deflection would naturally cause an eddy, and thus lead to those accumulations."

1862—J. D. Whitney. In this year Whitney published a rather full discussion of the Driftless Area, accompanied by the first Driftless Area map ("Diagram of the Region Destitute of Drift and Boulders in Wisconsin, Iowa, and Minnesota"). He had worked in parts of the Driftless Area since 1858 or earlier. He is quoted below only in part:

"If we consider the magnitude and universality of the drift deposits in the northern United States, and especially in northern Wisconsin, we shall be the more astonished to learn that throughout nearly the whole Lead Region, and over a considerable extent of territory to the north of it, no trace of transported materials, boulders, or drift can be found; and, what is more curious, to·the east, south, and west, the limit of the productive Lead Region is almost exactly the limit of the area thus marked by the absence of the drift.

The conclusions to which we have been led by the study of this driftless region are as follows:

1st. That there has existed, ever since the period of the deposition of the Upper Silurian, a considerable area, chiefly in Wisconsin and near the Mississippi river, which has never been sunk below the level of the ocean, or covered by any extensive and permanent body of water, and which, consequently, has not only not received any newer deposit than the Upper Silurian, but has also entirely escaped the invasion of the drift, which took place over so vast an extent of the northern hemisphere.

2nd. That the extensive denudation which can be shown to have taken place in this region, as witnessed by the outliers of rock still remaining, and the general outline of the surface, has not been occasioned by any currents of water sweeping over the surface, under some great general cause, but that it has all been quietly and silently effected by the simple agency of rain and frost, acting uninterruptedly through a vast period of time.

3rd. That during a long period this island, as it may be called, remained uninhabited by animals, but that at the close of the drift epoch, when the surrounding region itself became peopled with numerous animals and plants, it was the residence of a variety of species, some of which are now extinct, while others still exist.

4th. That a large portion of the superficial detritus of the West, even in those regions where drift boulders are met with, must have had its origin in the subaerial destruction of the rocks, the soluble portion of them having been gradually removed by the percolating water, while that which remains represents the insoluble residuum, the sand and clay, which was originally present in smaller quantity in the strata thus acted on."

1864—Charles Whittlesey. It seems probable that Whittlesey was unconvinced as to the existence of an isolated Driftless Area and thought that erratic bowlders might sometime be found within it, for he seems to have never mentioned the phenomena. He was the ablest observer and interpreter of the drift phenomena in Wisconsin in early days. He worked in various parts of the state for 20 years, beginning about 1844. He visited Dubuque as early as 1848 or 1851. In 1864 he stated:

"Between the northern limits of the Mississippi tertiary and the southern edge of the glacial drift, there is in Ohio, Kentucky, Indiana, and Illinois a belt of debatable ground, the outlines of which are not easily defined." In another paper, written at the same time, he says: "Judging by the boulders, the southern edge of the icefield on the west, extended from near Dayton, Ohio, in a northwesterly course, to and past Dubuque in Iowa."

Thus he recognized the absence of erratics in parts of Illinois and Iowa close to southwestern Wisconsin, but did not mention the glaciated country to the south which isolates the Driftless Area. His map, published in 1867, agreed with his description, drawing the glacial boundary between Dubuque and Prairie du Chien. It is odd that he does not mention the erratics to the south in Illinois, for as early as 1848 he knew of the correct southern boundary of the drift near the junction of the Kaskaskia and Mississippi rivers.

1874—Moses Strong. The tracing of the boundary of the Driftless Area in Green County, as alluded to above by Lapham in 1874, was the work of Moses Strong. In 1877 Strong said:

"The northern boundary of the driftless region lies far to the north of the Lead region. The eastern line was found in Green county, and traced out with all possible accuracy. For a particular description of it, reference is made to the geological maps; in brief, however, it is as follows: It commences on the west side of the Pecatonica river, crossing the state line at the southwest corner of the town of Cadiz. From here it proceeds almost in a straight line to the city of Monroe. Thence north, it runs along the divide between the Pecatonica and Sugar rivers, until about two miles south of New Glarus, where it takes a northeasterly course, and passes out of the county about a mile west of Belleville. The course thus indicated is its present line as shown by erratic bowlders lying upon the surface. If the drift deposits originally extended farther westward, no trace thereof now remains. East of the line described, bowlders are found in all parts of the county, with more or less frequency. The boundary line, where boulders are now found, does not appear to conform at all to the surface features, but crosses the valleys of the streams, and the ridges between them, with equal impartiality."

Strong also mapped the northern border of the Driftless Area near the Mississippi River, his descriptions being published in 1877 and 1882. The boundary of the Driftless Area on the map of Quaternary Formations of Wisconsin, 1881, is apparently based chiefly upon the detailed mapping of Moses Strong and R. D. Irving. Irving's map, 1879, and the present geological maps of the state place the boundary on the Mississippi near Alma rather than farther south at Fountain City, where Strong drew it some time before 1881.

1877—R. D. Irving. The first correct explanation of the origin of the Driftless Area seems to have been made by R. D. Irving, upon the basis of field work which began in 1874. N. H. Winchell and T. C. Chamberlin independently reached similar conclusions at about the same time (see below). Irving spoke of the Driftless Area in 1876 and published his explanation in 1877. How much earlier he reached this conclusion is not known. He said, in part:

"The first and most striking fact that presents itself to the investigator of the drift phenomena of Wisconsin is the existence of an extensive *driftless region,* the remainder of the state at the same time displaying an altogether extraordinary development of the drift materials. In the driftless region, which occupies 12,000 square miles of the southwestern part of Wisconsin, or nearly one-fourth the entire area of the state, the drift is not merely insignificant, but absolutely wanting. Except in the valleys of the largest streams, like the Wisconsin and Mississippi, not a single erratic bowlder, nor even a rounded stone, is to be seen throughout the district; whilst the exception named is not really an exception, the small gravel deposits that occur on these streams having evidently been brought by the rivers themselves, during their former greatly expanded condition, from those portions of their courses that lie within the drift-bearing regions."

He then discussed the outline of the driftless area, its topography and altitude, and gives an excellent map of its borders in central and southern Wisconsin. His explanation was as follows

(the italics are Irving's) : *"The Driftless Region of Wisconsin owes its existence, not to superior altitude, but to the fact that the glaciers were deflected from it by the influence of the valleys of Green Bay and Lake Superior*. Some writers have thrown out the idea that the driftless area is one of present great altitude compared with the regions around it, and that by virtue of this altitude during the Glacial period it caused a splitting of the general ice sheet, itself escaping glaciation. This idea may have arisen from the fact that in the southern part of the area the district known as the 'Lead Region' has a considerable elevation; but the facts heretofore given have shown that in reality the driftless area is for the most part, *lower* than the drift-covered country immediately around; the greatest development, for instance, of the western lateral moraine of the glacier of the Green Bay valley, having been on the very crown of the watershed between the Lake Michigan and Mississippi river slopes, whilst the driftless region is altogether on the last named slope. Moreover, to the north, towards Lake Superior, and to the west in Minnesota, the whole country covered with drift materials lies at a much greater altitude."
* * *

"That it was not invaded from the north is evidently due to the fact that the glacier or glaciers of that region were deflected to the westward by the influence of the valley of Lake Superior."
* * *

"Future investigations will undoubtedly bring out a close connection between the structure of the Lake Superior valley and the glacial movements south of it." * * * "The main ice sheet coming from the north met, in the great trough of Lake Superior, over 2,000 feet in depth, an obstacle which it was never able to entirely overcome, and so reached further southward in small tongues composed perhaps of only the upper portions of the ice. These tongues being deflected westward by the rock structure of the country, and having their force mainly spent on climbing over the watershed, left the region further south untouched. The eastern part of the Lake Superior trough is not nearly so deep as the western, and the divide between Lake Superior and the two lakes south of it never attains any great altitude, so that here the ice mass, having at the same time perhaps a greater force on account of its nearness to the head of the ice movement on the Laurentian highlands of Canada, was able to extend southward on a large scale, producing the glaciers of the Green Bay valley and of Lake Michigan.

"Although quite crude in its details, I am convinced that the main points of the explanation thus offered for the existence of the driftless region in the northwest will prove to be correct. To obtain a full elucidation of the subject, much must yet be done in the way of investigation, not only in Wisconsin, but over all of Minnesota and the state south, in order that the details of the ice movement for the whole northwest may be fully understood."

Irving shows the extent of the Driftless Area in his geological map of Wisconsin made in 1879. See also map of Quaternary Formations of Wisconsin, 1881.

1877—N. H. Winchell. In a discussion of Houston County, Minnesota, published in 1877, Winchell says:

"There being no foreign drift in this county, these streams run in their ancient channels and several hundred feet below the general upland level."

"The true northern drift is not spread over this county. It contains no drift clay, nor boulders of foreign origin. There is a thin deposit of foreign gravel at Riceford, in the extreme southwestern part of the county, and there is a terrace along the Mississippi river that is made up of gravel and sand of northern origin, but this county wholly escaped the operation of those forces which spread the well-known drift clay and boulders over the most of the state. Whether any former glacial era caused it to be covered with the ice of the northern glaciers cannot be determined, since the materials left by that era, if any there were, may have been decomposed, and may have entered into the stratified clays and the soils of the Mississippi valley further south under the combined influence of time, and the intense activity of the destructive forces of the latest glacial era."

"As to the cause of this exemption of a part of southwestern Minnesota, and portions of Wisconsin, Iowa and Illinois adjacent, from the forces of the northern drift epoch, there has been but one opinion advanced so far as the writer is aware. It is that of Prof. J. D. Whitney, who attributes it to the *non-submergence of this region since the deposit of the Silurian rocks and their elevation above the ocean.* If it were demonstrated or generally believed that the prevalence of the drift in other parts of the Northwest, in the same latitude, is due to the submergence of the continent beneath the ocean since the Tertiary age, this assumed cause would be apropos. But on the contrary it is pretty generally agreed by geologists, both in America and Europe, that the drift is due to the former existence of glaciers that covered the surface of the country, and, moving generally southward, not only brought from the northern regions the foreign substances that constitute the drift, but required, for their existence, that the land surface should be raised several hundred feet at least above the ocean during their prevalence. Again there is every reason to suppose this region *has been submerged* since the age of the Silurian. It is difficult to conceive what could have produced the horizontal lamination of the loess loam, unless it be attributed to the action of standing, or but slightly agitated water. This loam not only exists along the immediate river valley, but is spread widely over the highlands of the whole district. It is true there is no evidence of its having been the product of marine depositions, on the contrary it is evidently of fresh water origin, but that the country has been deeply submerged and remained so for a long period within recent geological time can hardly be questioned. There is also reason to believe that some portions of it were buried beneath the waters of the Cretaceous ocean."

"In examining the topography and the geological structure of the country lying to the north of this so-called driftless tract, it is evident that the great valley of the Lake Superior region, once occupied by glacial ice, would overflow, both first and last, along the lines of the lowest outlet, and that perhaps the higher and less passable parts along its southern barrier-shore would never be entirely surmounted. The continental glacier, in this region, would flow toward the southwest or south, guided by the main topographical features. In north-central Wisconsin is an isolated area of granitic and metamorphic rock, which not only extends to the shore of Lake Superior, but wedges out northeastwardly in the form of a long, high and persistent point or spur, in the southern part of Lake Superior, known as Kewenaw Point, in the State of Michigan. It is plain to see that this point would act on a crowding but somewhat flexible mass of ice as an entering wedge to split it into two main masses, and that the widening of the wedge, in the granitic region of northern Wisconsin, would perpetuate the division so as to cause, if other topography were favorable, a constant flow along the northwest side, and another in a more southerly direction, that would spread over northern Michigan and find its easiest exit through the valleys of lakes Michigan and Huron. According to Prof. R. Irving, and Messrs. Foster and Whitney, the western end of Lake Superior lies in an Archaean synclinal trough running southwesterly. This again would divert the flowing ice over the northeastern portions of Minnesota to the expense of northern Wisconsin. Glacial scratches on the rocks at Duluth, at the western extremity of the lake, have a west-south-westerly direction.

"Now it is a striking coincidence that this driftless tract lies nearly south and in the lee of this wedge-like area of metamorphic rock, and would be protected from the ice-flow by it. It is hence reasonable to infer that the absence of the drift in this region is due to the existence of this protecting barrier lying to the north of it in Wisconsin, while further to the south the two main branches of the ice-flow again united and spread, before their final retirement, a continuous sheet of drift over central Illinois, and southern Iowa."

Winchell previously mentioned the Driftless Area in 1873 and 1876, but without explanation.

1878—T. C. Chamberlin. In 1878 Chamberlin said:

"One prodigious tongue of ice ploughed along the bed of Lake Michigan, and a smaller one pushed through the valley of Green Bay and Rock River, while another immense ice stream flowed southwestward through the trough of Lake Superior and onward into Minnesota. The diversion of the glacier through these great channels seems to have left the southwestern portion of the state intact, and over it we find no drift accumulations."

In 1879 Chamberlin reviews the explanations by Irving and by Winchell, and then says:

"My own view, entertained some two years previous to the publication of those sketched above, involves a combination of these views, and some supplementary elements that seem essential to anything like adequacy; for when we have combined the above views, and given full emphasis to the agency of the highlands in crowding the ice aside, and to that of the great lake troughs in leading it away, we still have a troublesome residuum to explain; for, as previously stated, the ice, nevertheless, mounted the heights in sufficient massiveness to maintain its onward flow for 100 miles. It cannot be said to have spent its force, for the momentum of a glacier is insignificant, on account of the slowness of its motion.

"The disappearance of this stream on the southern slope, I have attributed to the wasting to which it was subjected.*"

See also the boundaries of the Driftless Area in the map of Quaternary Formations of Wisconsin, published in 1881, and probably compiled by Chamberlin from the detailed mapping by Irving and by Strong.

1878—W J McGee. The first mention of the Driftless Area by McGee is in a paper published in 1878 in which he calls attention to erratics in an Iowa area previously thought to be driftless. In 1884 he described the drainage of the driftless portion of Iowa. In that great classic of American physiography—McGee's "Pleistocene History of Northeastern Iowa,"—the Driftless Area is discussed in as great detail for Iowa as in that other classic which covers the whole area—Chamberlin and Salisbury's "Driftless Area of the Upper Mississippi Valley." The latter was published in 1885, the former in 1891; but McGee's field work was chiefly between 1876 and 1881 and most of the report was written in the latter year. McGee did additional field work in 1888 and published his report three years later.

His descriptions of the Driftless Area topography, of the residual soil or geest, and of the loess are excellent. All may well be read today for application to the corresponding features in Wisconsin.

McGee believed, however, that the deposits of fine silt, or loess, of the Driftless Area "represent the mud of a Pleistocene Glacier, and that they were accumulated either within ice-bound lakes or in cañons excavated in the ice sheet."

This conclusion he published in 1879 and again in 1882 and 1891. In the latter year he published his map of Lake Hennepin, which he thought to have covered the whole Driftless Area of Iowa, Wisconsin, Illinois, and Minnesota. He correlated it with one of the pre-Wisconsin glacial advances. Because of the topography of the Driftless Area and because of lack of evidence of subsequent earth movements such as would be necessary if there were a glacial lake here, it is now thought that no such lake ever existed and that the loess in Wisconsin is wind-blown.

---

*Original footnote. Discussion attending the reading of Prof. Irving's paper before Wisconsin Academy of Science, December, 1877.

Except for this one error of interpretation, McGee's great monograph on a representative part of the Driftless Area still stands as when written,—a notable contribution by one of the masters of American physiography, thoroughly observed, well reasoned, and delightfully described.

1885—T. C. Chamberlin and R. D. Salisbury. The most complete discussion of the Driftless Area is Chamberlin and Salisbury's "Driftless Area of the Upper Mississippi Valley," published in 1885. It was based upon much additional field work, beginning in 1883 and chiefly by the junior author. There are numerous excellent maps. The gist of the explanation is contained in the closing, happily-phrased paragraph, as follows:

"Diverted by highlands, led away by valleys, consumed by wastage where weak, self-perpetuated where strong, the fingers of the mer de glace closed around the ancient Jardin of the Upper Mississippi Valley, but failed to close upon it."

In the maps accompanying this report the border of the Driftless Area is drawn much farther to the northwest and west than is now the practice.

1907—Samuel Weidman. One of the chief additions to our knowledge of the Driftless Area since 1885 is based upon the work of Weidman, who has mapped a large northeastern extension of the area in the Wisconsin valley near Wausau.

Other geologists have relocated parts of the borders of the Driftless Area in Wisconsin, Minnesota, Iowa, and Illinois in recent years, with slight modifications in detail.

BIBLIOGRAPHY

See Chapters III and IV, pp. 77-80, 97-102, and Appendix G.

CHAPTER VI

THE GLACIAL PERIOD IN THE WESTERN UPLAND

THREE SCENES AT DEVILS LAKE PARK

Those who visit the state park at Devils Lake (p. 55) may picture three strikingly-different scenes. The first is the region as a stream valley. The river is the Wisconsin, flowing into the Devils Lake gap from the north, turning abruptly eastward near the present site of the Kirkland Hotel and its cottages, and then flowing around the Devils Nose and southwestward toward Prairie du Sac. If you had climbed the bluffs at that time you might have found the Devils Door, or Turks Head, or some of the other picturesque crags and pinnacles of today. If they were not there it is certain there were others equally bizarre. You would have looked down, however, on a different scene.

The surface of the present lake lies 500 feet below the East Bluff, but the waters of the preglacial Wisconsin River flowed along at a level at least 280 feet lower. The gorge was then 800 or 900 feet deep. The scene must have been even more picturesque than that in the present gorge below Niagara Falls. The river had less volume than the Niagara, but the gorge was deeper and more beautiful. At that time there was no lake. There was no hill at the railway cut east of the station. There was no hill where the wooded ridge now extends across the valley north of Devils Lake. There was no level land where the various groups of cottages now stand. The tumbled blocks in the talus slopes looked as they do now. The bluffs were much as today, except that they overlooked a deeper valley and, therefore, appeared much higher.

The second scene is that of Devils Lake during the Glacial Period. East of the site of Devils Lake railway station was an ice tongue, a lobe of the continental glacier. Since its terminus lay in a narrow valley it looked much like the larger glaciers in Alaska or in the Alps. No present-day ice tongue of Glacier National Park, of Mt. Rainier, or of the Canadian Selkirks, equals it in

A. THE SHEER ICE CLIFF AT THE TERMINUS OF NUNATUK
GLACIER, ALASKA.

There was once a similar cliff at each end of Devils Lake, Wisconsin.

B. THE TERMINAL MORAINE AT THE SOUTHEASTERN END
OF THE DEVILS LAKE GAP.

It occupies the site of a former ice cliff much like that shown in the upper picture.

A. TALUS OF WEATHERED QUARTZITE BLOCKS AT DEVIL'S LAKE, IN THE
DRIFTLESS AREA.

Steep slope with numerous crags and pinnacles.

B. THE DEVILS NOSE

A moderate slope of quartzite from which all the talus and all crags and pinnacles were
removed by the continental glacier. This is about a mile from the bluff shown in the upper
picture.

size or impressiveness. North of the lake was another ice lobe of equal beauty. They ended in sheer ice cliffs one or two hundred feet high, like the glacier in Plate IX, A. Between them was the glacial lake, dotted with icebergs, and at a much higher level than the present lake. This was at least 20,000 years ago, perhaps 80,-000 years, perhaps much more.

The third scene is that of today, with the lake, the moraines, the bluffs, the fields, and the forests, a gem of true mountain scenery, such as cannot be seen elsewhere east of the Rockies.

ICE INVASIONS IN THE WESTERN UPLAND

Devils Lake is at one of the three parts of the Western Upland ever invaded by the continental glacier. These three areas are (a) the eastern part of the Baraboo Range, (b) the northwestern highland, near the St. Croix and Chippewa rivers, and (c) the southeastern part of the upland near Beloit and Monroe. Indirectly glaciation wrought minor changes in the uplands of the Driftless Area and in most of its valleys.

GLACIATION OF THE BARABOO RANGE

The Advancing Glacier. The glaciation of the eastern half of the Baraboo Range was accomplished by the Green Bay lobe of the Lake Michigan glacier. The ice was moving westward across the Central Plain of Cambrian sandstone. On reaching the Baraboo Range, its forward movement was retarded by the high quartzite hills. The ice north and south of the range accordingly moved forward more rapidly than the ice which had to rise up over this obstacle. Lobes therefore developed, one north and one south of the Baraboo Range.

The lobe to the north quickly over-topped the low North Range and filled the interior valley east of the city of Baraboo, while the lobe to the south was advancing to about the position of the present city of Merrimac, in the lowland between the Magnesian escarpment and the South Range. In the meantime the ice between these lobes was slowly ascending the high South Range and advancing westward across its summit. By the time this ice on the upland had reached a point a little over 4 miles east of where Devils Lake is now (Fig. 35) the northern and southern lobes had moved into the Devils Lake gorge, from the north and south respectively, and were less than two miles apart.

Just then the strong forward movement ceased and all the ice fronts began to melt back. If the forward movement had lasted a very little longer these valley lobes would have come together and the unglaciated portion of the upland east of Devils Lake

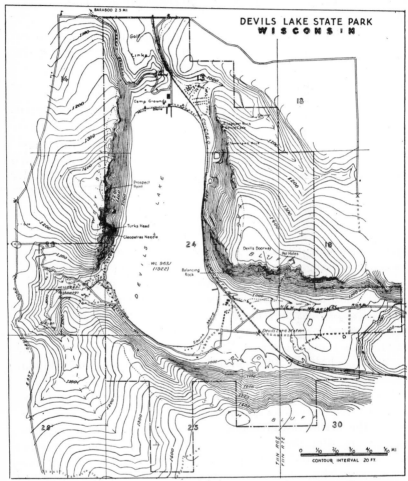

Fig. 33. Map of Devils Lake State Park. The line of dashes and dots is the boundary of the park.

would have been a detached driftless area. If it had lasted still longer the upland ice would have covered these few square miles and there would have been no unglaciated territory east of Devils Lake. As it is, the cessation of rapid forward movement, at the time when it did end, resulted in the formation of an eastward-projecting portion of the main Driftless Area, which includes the former stream valley at Devils Lake and the upland east of it for a little over 4 miles. Thus the peculiar shape of the glacial boundary at the Baraboo Range (Fig. 35) was determined by (a) the topography of this quartzite upland and (b) by the time of cessation of strong forward movement at the margin of the continental glacier.

Removal of Talus. The extent to which glacial erosion modified the Baraboo Range is not known exactly, though one fact is clear. The ice carried away the huge blocks of quartzite in the extensive talus slopes which bordered the steeper slopes of the Baraboo Bluffs. We know this from the talus slopes of quartzite now present in the never-glaciated Devils Lake gap, in contrast with the absence of such talus at the glaciated Lower Narrows gap northeast of Baraboo, and on the Devils Nose (Pl. X, B), and in the region to the east of Devils Lake.

Terminal Moraines. At the ice fronts, strong terminal moraines were formed, especially at the ends of the valley lobes north and south of Devils Lake. These are ridges of glacial drift, built at a former terminus of the glacier. They are made up partly of unassorted bowlders, sand, and clay, a deposit called till, and partly stratified gravel, sand, and clay. From the ice fronts, streams flowed away, leaving a sloping deposit called an outwash plain or valley train and made up of rounded gravel and fine sand. Where there was a continuous southward-sloping valley before the Glacial Period, there is now a much shallower valley, interrupted by two terminal moraines, outwash and lake deposits, and the present lake. Devils Lake is only 43 feet deep, but the thickness of the lake clay and glacial outwash between the lake bottom and the rock floor of the valley may be estimated at not less than 300 or 350 feet. A well at the north end of Devils Lake passed through 283 feet of glacial drift without reaching bedrock.

Devils Lake With Icebergs. Between the two ice fronts in the Devils Lake water gap there eventually came to be a temporary glacial lake, at a level much higher than the present body of water. The lake extended nearly a third of the way up the bluffs. It covered all the low ground east and west of the present lake. Icebergs floated about in this larger lake, and rafted erratics of granite, greenstone, and other foreign bowlders out into the small valley of the Driftless Area to the west, where they may be found today. The outlet of this body of water may have been in or at the edge of the ice northwest of the present lake. A thick mantle of clay was spread over the till which had been deposited directly by the glacial lobes and over the outwash gravels laid down by the glacial streams.

Glacial Lake Baraboo. There was a much larger glacial lake west of the site of the city of Baraboo in the lowland or valley between the quartzite ranges. This has been called Glacial Lake Baraboo (Fig. 34). It was a bay on the western side of Glacial Lake Wisconsin (Fig. 133), with which it was connected by narrow straits at Ablemans and a broader strait east of Reedsburg. Beach deposits and ice-rafted erratics are found within the limits of this lake between Baraboo and Ablemans. It was at least 120

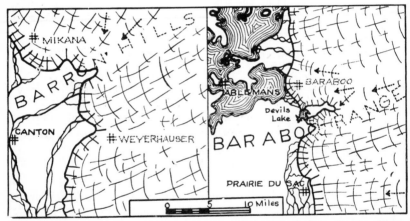

Fig. 34. The continental glacier covering the eastern portions of the Baraboo Range and the Barron Hills, with lobate margin due to topography. Glacial streams flowing away from the melting ice and depositing outwash gravels. Glacial Lake Baraboo and Glacial Devils Lake in right-hand map.

feet deep. There is lake clay below the outwash gravels at the border of the terminal moraine. In the extension of this lake basin up the Baraboo valley, northwest of Ablemans, erratic bowlders are found at Reedsburg.

There were similar streams and temporary lakes in front of various parts of the continental glacier, as in the valleys northeast of Devils Lake. Because of these lakes and glacial rivers the interior valley of the Baraboo Range and the region north and south of it were built up by deposits of glacial outwash and lake clay, giving the Baraboo Range a little less relief above the surrounding region than before the glacial invasion.

Changes in Drainage. The drainage of the Baraboo Range was modified tremendously by glaciation. The Wisconsin River was diverted during the Glacial Period. Before the Ice Age the river crossed the Baraboo Range by way of the Lower Narrows and the Devils Lake watergap. Now it makes a great bend to the east, passing near Portage and returning to its former valley a short distance west of Merrimac. Even here it is not exactly in its old channel. The precise time and mechanism of this diversion is not yet known.

The drainage at the Lower Narrows was reversed in direction (Fig. 35). Skillet Creek was diverted by the building of a termi-

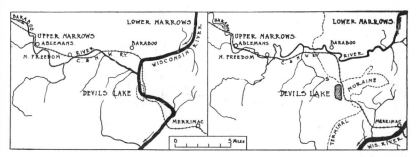

Fig. 35.   Preglacial and present drainage in the Baraboo Range near Devils Lake State Park.   Skillet Creek is at S.

nal moraine and outwash plain across its lower course. It now turns at right angles, and flows northwestward, having an extremely-picturesque, postglacial, rock gorge at Pewits Nest. In the Lower Narrows the glacial deposits fill the watergap to a depth of more than 216 feet, and at Baraboo in the interior valley to 276 feet.

WISCONSIN DRIFT NEAR THE ST. CROIX RIVER

The Thick Glacial Deposits. A small portion of the northwestern upland was glaciated during the latest advance of the continental ice sheet into Wisconsin, when all of the eastern and northern parts of the state were ice-covered. At this time glacial drift was deposited in the northwestern part of St. Croix County and most of Polk County. This drift is very thick. It is not weathered, and has a rougher topography than that of the region to the southeast.

A Drift-Mantled Escarpment. The escarpment of the northern border of the Magnesian cuesta extends east and west for 35 miles in Polk and Barron counties. Nearly all of it is so mantled in the glacial deposits that it is inconspicuous. We do not know how much it was modified by glacial sculpture.

OLDER DRIFT IN THE NORTHWESTERN UPLAND

The earlier glaciation of northwestern Wisconsin was by ice which advanced into the state from the northwest and covered the area north and northwest of Alma on the Mississippi River. The proportions of this ice which came from (a) the Lake Superior lobe of the Labrador ice sheet, (b) the Lake Superior lobe of the Patrician ice sheet, and (c) the Minnesota lobe of the Keewatin ice sheet, has not yet been worked out in detail. The drift is weathered (see p. 128), has few lakes or other undrained depressions, scarcity of hilly morainic topography, and differs in other respects from the younger or Wisconsin drift. In the parts of Buffalo, Pepin, Pierce and St. Croix counties where it is found, this drift has probably levelled up preglacial inequalities and made the region less rugged than it was before the Glacial Period.

GLACIAL EROSION IN NORTHWESTERN WISCONSIN

The Topography of the Upland. The exact extent to which the northwestern upland was sculptured by the continental glacier is not yet known. It has been stated that the cuesta surface northwest of Chippewa River (p. 46) is much less irregular than the region of ridges and coulees between the Chippewa and the La Crosse rivers (p. 48). One region is glaciated, the other driftless. It, therefore, seems probable that the difference is due to glaciation, for the rocks in the two regions are similar. The difference is akin to that between the glaciated cuesta of west central New York in the Allegheny Plateau and the driftless continuation of the same region to the south in Pennsylvania and West Virginia. In northwestern Wisconsin, however, we are not ready to state as yet whether the change was brought about chiefly by glacial erosion or by glacial deposition. Moreover the preglacial topography of the regions north and south of the Chippewa River may not have been exactly the same, because of certain relations of the limestone (p. 49). It seems likely that glacial deposition may have been more important than glacial erosion in producing the rather even upland near the Chippewa and St. Croix. This does not deny that glacial sculpture may also have been effective.

An Ice-Sculptured Escarpment. In one respect we have the suggestion of considerable glacial erosion in the northwestern upland. The Magnesian escarpment, which extends north and south in St. Croix, Pierce, and Pepin counties, is much less irregular than the continuation of the same escarpment to the southeast in the portion of the Driftless Area between the Trempealeau River in Buffalo County and the Baraboo Range in Sauk County. That this north-south escarpment was not simplified quite as much by glacial abrasion as the Niagara escarpment (p. 246) seems to be due to the fact that the ice advanced eastward over it from the back slope of the Magnesian cuesta, instead of moving parallel to it, as was the case in eastern Wisconsin. There are small caves near the edge of the escarpment in the region of older drift near Wilson, so that it is clear that glacial erosion was less effective than in eastern Wisconsin (p. 251). This was because the ice was thin and weak, since the escarpment lies not far from the extreme limit of glaciation and may not have been subject to glacial abrasion as long as those in eastern Wisconsin. It should not be thought, however, that the glacier did no work. A great amount of erosion was necessary in removing all the outliers from the escarpment and making its outline as simple as it is.

OLDER DRIFT WEST OF THE ROCK RIVER

Extent and Character. An area of older drift extends into the Western Upland from the Rock River at Janesville and Beloit to a line running west of Martintown, Monroe, Paoli, and Verona. The ice which overrode this area was the Lake Michigan-Green Bay Glacier. Near Beloit glacial striae point east and west. In Green County they even trend northwestward.

The characteristics of this area are, on the whole, more like those of the driftless than of the glaciated part of Wisconsin. The drift border is very difficult to locate, being determined in many localities upon the basis of rare, single erratics. Lakes and upland swamps are lacking. The drainage upon the surface of glacial deposits is mature. Ground moraine and terminal moraine are usually indistinguishable, but there are a few, weak, discontinuous moraines.

The Weathered Drift. The drift mantle is thin and deeply weathered. Originally this drift was probably thick, and not weathered at all. The limestone pebbles and limy clay are leached away from the upper 2 or 3 feet of drift, which are often red. Below is brownish drift, with limestone pebbles so weathered that they crumble in the hand, or pitted and etched by solution till they have lost all glacial form and striation. The body of the drift

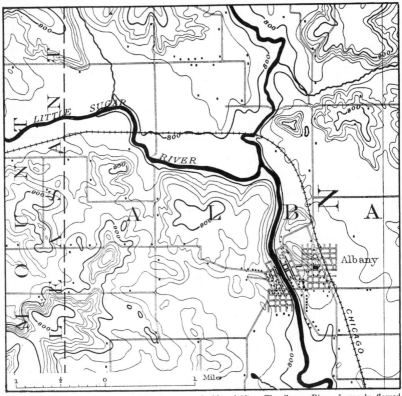

Fig. 36. Stream diversion within the area of older drifts. The Sugar River formerly flowed in the broad abandoned valley northeast of Albany. It now has a narrow rock gorge west of the city where it was, as it were, let down or superimposed by the glacial drift. An explanation of the word superimposed is given on page 192. (From Brodhead Quadrangle, U. S. Geol. Survey.)

below this weathered material is blue, pink, gray, or buff. It is not weathered, and 80 or 90% of the pebbles are subangular, striated limestones. It has been estimated that 15 or 20 feet of the surface of the older drift may have been carried away by solution in order to leave the 2 or 3 feet of residual red clay at the surface.

In this clay are insoluble pebbles of chert, quartz, quartzite, and fine-grained crystalline rocks. There are some till deposits and some gravel deposits that are not weathered so much, either because of (a) position, where erosion has recently removed the overlying weathered drift, or (b) nature of material, as in a sandy deposit through which the underground water moves rapidly and has sometimes been thought to dissolve relatively little. As already explained, there was much more time between the glacial advances than has elapsed since the last ice sheet melted away.

Stream Diversions. Within this belt of older drift most of the streams are in their preglacial rock valleys. There are exceptions, where part of the course of a stream has been diverted by the glacier. At Albany, Green County, the Sugar River has a narrow rock gorge west of the broad abandoned valley followed by the railway. There are also diversions due to drift dams across the mouths of valleys. An illustration of this is the small creek northwest of Dayton, Green County. It abruptly leaves a broad, upper valley and flows through a narrow, rock-walled valley near Belleville, because of a drift dam across the mouth of its lower valley. Other examples are found near Verona, Brodhead, and Monroe. In these drift dams the glacial deposits are still thick.

The valley of the Sugar River, which crosses this area of older drift, is floored with a valley train of gravel and sand, chiefly and perhaps entirely of Wisconsin age.

Transitional Topography. This region of older drift is included in the Western Upland rather than in the Eastern Ridges and Lowlands because of its transitional topography. It is more like the Western Upland, even though once glaciated. The bedrock and the glacial features result in making the topography partake of the characteristics of both provinces, with broader valleys and gentler slopes than the Driftless Area to the west, but, on the whole, systematic erosion topography rather than aimless glacial features.

DEPOSITS OF GLACIAL AGE WITHIN THE DRIFTLESS AREA

Nature of Deposits. The boundary of the Driftless Area is taken as the outermost of the terminal moraines, or the line of outermost erratic material carried to its present resting place directly by the ice. Thus the boundary marks the farthest place at which the glacier actually stood and, when melting, deposited its unstratified till or bowlder clay. In front of the glacial lobes, however, deposits were made during the Glacial Period, which lie within the territory over which the ice never advanced and which is otherwise driftless. Not all these deposits are of foreign glacial material. There are three kinds:—(a) lake deposits, (b) river deposits, and (c) wind deposits. For the sake of convenience reference will be made here to some of the Driftless Area deposits in the Central Plain and Northern Highland, as well as in the Western Upland.

GLACIAL LAKE DEPOSITS

Relation to Slopes. The topography of the land at the border of an ice sheet results in the deposition of (a) river deposits, if a slope leads away from the glacier, and (b) lake deposits, if the slope is toward the ice.

Glacial Lake Wisconsin. An extensive deposit in a temporary body of water at the glacier margin was laid down in Glacial Lake Wisconsin (see p. 337), which lay in the Driftless Area north of the Baraboo Range, between the Wisconsin and Black rivers. Lake clay and sand, and erratic bowlders, rafted out into the Driftless Area from the glacier front by floating icebergs, are the chief deposits of this glacial lake. It probably covered at least 1825 square miles, being larger than Green Bay or three-fourths the size of Great Salt Lake. It rose to a height of 960 to 1000 feet above sea level, being 70 to 150 feet deep. It existed so short a time that the shorelines and deltas at its borders are in most places too faint for recognition. Some of the lake deposit was subsequently covered by sandy, glacial stream deposits, and by dunes and loess. The waters of this lake probably escaped to the west over the Black River divide in Wood County.

The Mississippi Lake. The presence of rare erratic bowlders on the Mississippi bluffs in the Driftless Area may be explained by the possibility that the glacial Mississippi was temporarily dammed at the south. Such a damming would form a long, narrow, valley-lake. Erratics might be rafted out by floating ice to positions high above the present floodplain, and into the mouths of tributary, non-glacial valleys, as in Grant River where such erratics of granite, diorite, porphyry, and quartzite have been found.

On the other hand, the small pebbles of granite, trap, porphyry, jasper, quartzite, and quartz, found near the Mississippi at several localities east of Trempealeau, La Crosse, and De Soto, at elevations of 380 to 480 feet above the river, may very well be older drift, rather than berg-rafted lake deposits, since the boundary of the Driftless Area a short distance to the northwest at Alma is by no means established. Such isolated patches of quartz pebbles as those near Seneca, Crawford County, are of preglacial origin (p. 53).

It has been considered by one author that certain deposits at Bridgeport and Wauzeka near the mouth of the Wisconsin River represent older drift from Iowa. Upon this hypothesis the Mississippi would have been dammed south of Prairie du Chien, a glacial lake would have occupied its valley above this point, and its outlet would have drained eastward up the valley of the present lower Wisconsin River.

Deposits of Glacial Streams

Outwash Plains and Valley Trains. The streams from the melting glacier also flowed out into and across the Driftless Area, where they made great deposits of erratic bowlders, gravel, and sand. The parts of the Driftless Area covered by these stream deposits is shown in Plate VI. Another map (Fig. 163) represents the same thing in the region near Stevens Point and Wausau.

These deposits are of two sorts:—(a) broad outwash gravel plains, made up of coalescing alluvial fans and deltas, as in Adams, Wood, and Portage counties, between the Wisconsin River and the terminal moraine of the Green Bay lobe; (b) long, narrow, valley trains, as in the Wisconsin River between Sauk City and Prairie du Chien, in the Mississippi valley between Lake Pepin and Dubuque, and in the Black River between Hatfield and Trempealeau. Subsequent down-cutting of the streams has separated these valley trains into detached terraces (p. 156).

The broad outwash plains are frequently of great area. The glacial gravel and sand in Adams County covers nearly 400 square miles of the Driftless Area. The narrow valley trains in the Driftless Area are often of great thickness, the outwash gravel deposits in the Mississippi and Wisconsin valleys being 100 to 200 feet thick. Often these outwash gravels are partly or wholly covered by later deposits, such as the postglacial river bottom silts of the Mississippi, or by wind-blown sand and loess.

Most of these outwash gravel deposits were probably built by glacial streams during the latest advance of the ice sheet. Thus they are of Wisconsin age.

Older Outwash in the Wisconsin Valley.   Glacial waters also flowed down the Mississippi, Wisconsin, Black, and other stream

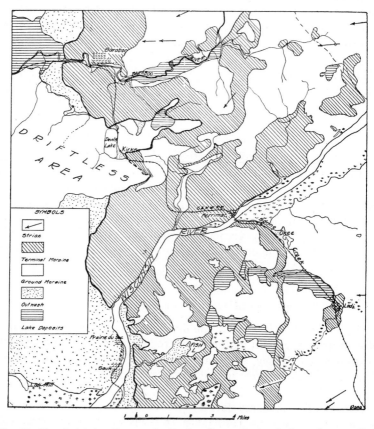

Fig. 37.   The glacial deposits near the Baraboo Range, the terminal moraine near Prairie du Sac and the head of the valley train of outwash gravels which extend down the Wisconsin River.   (Alden).

valleys of the Western Upland, during the earlier periods of glaciation.   They undoubtedly deposited outwash gravels.   Most of these accumulations of older outwash were subsequently buried or eroded away.

There are thin, scattered deposits of thoroughly-weathered drift east and west of Wauzeka and on the rock terrace at Bridgeport, near the mouth of the Wisconsin River. These are reddish, insoluble clays with rounded pebbles of quartzite, quartz, chert, and resistant igneous rocks. The clay is leached of all its soluble material, but there are rare limestone pebbles in the deposit. The nature of this drift seems to indicate that it represents outwash rather than till. Its relationships favor the view that it is not western drift from Iowa, but an ancient outwash deposit from an ice front somewhere east of Prairie du Sac.

There are also weathered gravels near Sauk City. The lime is entirely leached from the upper layers, though limestone pebbles are abundant below. As the same thing occurs in some deposits of undoubted Wisconsin age, as in Waushara County, it does not appear necessary that the outwash near Sauk City is of Illinoian age. It seems more probable that this is Wisconsin outwash.

Valley Fill in the Driftless Area. The most recent outwash deposits of the Western Upland are discussed more fully in the following chapters (pp. 156, 188, 202). There are valley train gravels of glacial origin, now represented by terraces. Many stream valleys are deeply filled with deposits of glacial time but not of glacial material. In general the valleys of the Driftless Area in the Western Upland are flat-floored. Hardly any of these streams are now cutting in rock.

Deposits of Driftless Area Streams

Within the Driftless Area the deposits of glacial time which do not contain any glacial material are found in the Kickapoo, Grant, Pecatonica, La Crosse, and many other rivers. Nearly all streams of any size, built up, or aggraded their courses so that they kept pace with the glacier-fed streams which crossed the Driftless Area. Certain smaller streams failed to keep up and were dammed at the mouth, where temporary lakes (p. 172) or small swamps were formed. Extensive alluvial fans were built.

One reason for up-building or aggradation by these non-glacial streams during glacial time was that the precipitation over the whole Driftless Area was undoubtedly increased by the presence of the surrounding continental glacier and the cold that it brought. The preglacial vegetation of the Driftless Area may have all been killed, for botanists tell us that the plant forms there now are not significantly different from those of the surrounding glaciated territory.   One author states, however, that the Driftless Area is interesting ecologically because it contains many plants, such as mosses and other low forms, peculiar to it alone.   Frost action in the Driftless Area was probably deeper and more effective during the Glacial Period than at present.   These three factors, (a) increased water supply, (b) barren surface, and (c) increased weathering through frost action, all tended to allow the Driftless Area streams and their tributaries to erode rapidly at their headwaters and, therefore, to transport more material.   The deposition by the heavily-laden glacier-fed streams, to which these non-glacial, Driftless Area streams were tributary, caused the latter to deposit also.

Wind Work During and Since the Glacial Period

Conditions Favoring Transportation by the Wind.   The work of the wind during and immediately after the Glacial Period was much different from the present and the preglacial wind work. The outwash gravel plains and the valley trains contained great quantities of finely-comminuted material.   This included (a) the milky sediment ground fine by erosion of the rock beneath the glacier and carried by ice-born streams, (b) fragments of rock due to erosion of the gravel and sand of the glacial streams, and (c) finely-weathered material from the hills of the Driftless Area. The first two kinds of material were continually being deposited by the glacial streams, which constantly shifted over the outwash plains and valley trains as they built up their beds.   As soon as such deposits dried in the sun, the dust and finer sand were picked up and transported by the wind, a process frequently observed near glacial streams in Alaska.

Dunes.   The sandstone ledges of the Driftless Area were barren of vegetation during the Glacial Period.   This gave the wind an opportunity to carry away the finer portions of the weathered sandstone.   The barrenness of driftless ridges and valley trains gave the wind free opportunity to transport fine sandy material, both glacial and weathered, and to accumulate sand in dunes.   Dunes are found in the Wisconsin valley east of Necedah, and between Lone Rock and Sauk City, in the Mississippi valley near Trempealeau and Onalaska, and at various other points on the terraces. They are apt to be covered with grass and trees, so that only a few of them are still in motion.   Man's activities in cutting timber and ploughing has set the dunes to moving again here and there.

During the Glacial Period, then, (a) the supply of fine, transported material carried and deposited by the glacial streams and (b) the release of fine, weathered material previously held down by vegetation, gave the winds an opportunity to transport and deposit very fine sand and dust to which it could not have had access before the Glacial Period. This condition lasted during the melting of the glaciers and subsequently up to the time when vegetation encroached upon and once more covered the Driftless Area. The glacial deposits near the Driftless Area also supplied much material to the wind before vegetation began to cover them.

The Loess. The finer deposits resulting from this increased wind work are called loess, a silt intermediate in fineness between clay and sand. There was a long period during which the origin of the loess was in dispute. Some held that it was a wind deposit; others, that it was laid down in lakes and by rivers. There are surely some water-laid deposits of loess, chiefly to the south of Wisconsin, but it is now practically agreed that most, if not all, of the Wisconsin loess of the uplands is wind-deposited.

Nature of Loess. The loess deposits are commonly:

    (a) fine, homogeneous, without stratification;

    (b) friable, soft enough to be easily removed with a spade;

    (c) in cliffs, with vertical faces or with steeper erosion slopes than any other unconsolidated material;

    (d) made up largely of quartz, but so full of lime as to effervesce freely in acid;

    (e) light brown, buff, or yellow in color.

Loess may have nodules or tubular masses of lime; it may contain freshwater shells.

A glacial deposit of similar appearance has sometimes been mistaken for loess. This is a lake clay without conspicuous layers. It is found, among other places, in the upper parts of railway cuts in the terminal moraines east of Devils Lake and east of Wisconsin Dells (Kilbourn), as well as the region near Madison. The presence of large ice-rafted bowlders in this lake clay is often the only thing distinguishing it from loess, especially as it is similar in color and often stands up with vertical cliffs.

Distribution of Loess. The chief loess deposits in this state are located in the Driftless Area and in the portion of it just east of the Mississippi River. The thicker loess deposits are usually within 10 to 20 miles of the river, and rarely more than 30 or 40 miles. Thinner deposits of loess are found 50 to 80 miles from the Mississippi. There is loess in the region of older drift in northwestern Wisconsin, and doubtless in many other areas as yet unmapped. It is not confined to the region of older drift or to the Driftless Area, but overlies the latest drift as well. The loess deposits extend high upon the upland above the bottomlands of the Mississippi, being found more than 300 feet above the present floodplain near Cassville and Dubuque, and over 600 feet above the Mississippi near Viroqua.

The Knox silt loam and Marshall silt loam, as classified by the Division of Soils of the Wisconsin Geological Survey, are made up largely of loess. This wind-blown soil covers 350 square miles of Iowa County and probably even more of Grant County nearer the Mississippi. In La Crosse County the loess covers about 200 square miles. It is known to occupy 870 square miles near the Mississippi River in the region of older drift and the driftless area of Pierce, Pepin, Dunn, Clark, Jackson, and Eau Claire counties. Thus it includes over 1400 square miles of 6 counties in western Wisconsin where the soils are mapped.

Thickness of Loess. The thickness of loess averages from 10 to 15 feet, running up to 60 feet near the Mississippi and down to a foot or less at a considerable distance. The undecayed, angular, rock fragments, which make up this deposit, are .05 to .005 millimeters or less than $\frac{1}{500}$ of an inch in diameter. These fragments grade from fine sand near the river to impalpable dust at a distance of a few miles.

Reason for Distribution of Loess. The general character of this material, capable of being carried in suspension by the wind, its coarseness and thickness near the Mississippi River, with gradation eastward, and its existence at variable heights above river level, all seem to point to its transportation by the wind. Much of that in Wisconsin originated, in times of low water, on the great valley train of the Mississippi and the valley trains of the tributary rivers which flowed from the ice sheet to the west across the portion of the Driftless Area in Minnesota and Iowa.

The absence of thick loess deposits adjacent to the great outwash deposits of the middle Wisconsin River, in Adams and adjacent counties, may be due to the following:

(a) The glacial material here is coarse, the fine silt generally being carried much farther from the glacier before deposition.

(b) The local glacial climate may have been wetter, resulting in less dry material for wind transportation.

(c) The Cambrian sandstone may have yielded less fine material by preglacial weathering than the Ordovician limestones to the west near the Mississippi.

(d) There is less upland area on which wind-blown loess could be deposited and exempted from later burial by stream and lake deposits.

(e) The wind-blown loess, from the weathered limestone and sandstone of central Wisconsin, carried eastward by the prevailing westerly winds, would fall on the slumping border of the stagnant, retreating glacier. Here it might become mixed with the till, sand and gravel, so that it made no thick deposits, as it did on the hills of the Driftless Area near the Mississippi.

Accordingly it is not surprising to find the loess absent in Marathon and adjacent counties of the Northern Highland and Central Plain. There is a little in Juneau County, but none to the east in Waushara County. Of course, the loess was deposited here and there in central Wisconsin, but it is thinner than that near the Mississippi. Nevertheless at least a part of the finer glacial material near the surface and much of the clayey soil in eastern Wisconsin is due to wind work during and just after the Glacial Period.

PRAIRIES IN THE DRIFTLESS AREA

The extent of country which was treeless when white men first came to Wisconsin is indicated in Figure 38. The proportion of prairie land is much less than in Iowa and Illinois. It is larger, however, in western and southwestern Wisconsin than in the northern and eastern parts of the state (Fig. 114). Most of the prairies were in the oak forest. The maple forest and the woodland of conifers, or mixed conifers and hardwoods, had few open spaces.

The location of certain of these prairies is indicated by the names still current, as Prairie du Chien, Prairie du Sac, Prairie La Crosse—the last now shortened to La Crosse—and Muscoda.

Fig. 38. The distribution of prairies—shown in white—in the Western Upland of Wisconsin.

Muscoda is an Indian name, meaning "land destitute of trees," the present form being a corruption of Maskoute or Mashkodeng.

The cause of the treeless condition has not been settled. Among the suggestions are peculiarities of soil, dampness, dryness, excessive evaporation, fires, underlying rock, and local topography. In the Western Upland the prairies are in both driftless and glaciated territory, on ridges, and in valleys. Many of them are in loess-covered areas where the soil is exceedingly fine-grained, but only a small proportion of the area of loess is treeless. In Iowa and Illinois, however, where there is more loess, there are larger areas of prairie.

RELATIVE VALUE OF DRIFTLESS AND GLACIATED LAND

A study of the percentage of improved land and the value of crops in Wisconsin shows that, except in the pre-Cambrian area, the Driftless Area is inferior to the glaciated portion of the state. Through the addition of the drift materials to the soil and, even more, through the smoothing of topography, it appears likely that the glaciated area has benefited to such an extent that the increased valuation to agriculture in the southern part of Wisconsin may be estimated at about thirty million dollars a year. Whitbeck's summary follows:

"Comparison of Per cent of Improved Lands. — Twelve typical counties, 6 sandstone and 6 limestone, excess in favor of drift, 10½%.

Comparison of Crop Values per Square Mile of Total Area.—Excess in favor of drift, in sandstone belt, 10%; in limestone belt, 31%.

Comparison in Per cent of Uncleared Land (southern half of state.)—In driftless area, 27%; in drift area, 11%.

Comparison of Crop Values per Square Mile in Four Counties Crossed by the Terminal Moraine.—Sandstone, average value all crops per square mile, drift, $2,776; driftless, $1,968. Limestone, average value all crops per square mile, drift, $3,828; driftless, $2,690. Excess in favor of drift of 40%.

Comparison of Crop Values in Two Chains of Townships, one on each side of the terminal moraine.—Excess in favor of drift, in sandstone belt, 11%; in limestone belt, 36%.

Comparison of Crop Values per Square Mile in Forty Townships Chosen at Random.—Excess in favor of drift, in sandstone belt, 23% ; in limestone belt, 38%.

Productivity of the Soil.—Average of four principal crops, (corn, oats, barley, potatoes) :

	BUSHELS PER ACRE	
	Sandstone	*Limestone*
Drift	47.8 bu.	52.0 bu.
Driftless	35.5 bu.	51.1 bu.
Excess	12.3 bu. =33%	.9 bu. =2—%

Comparisons in Dairying and Stock Raising.—Of the leading ten counties in the production of cheese, Nos. 1, 2, 4, 7, 9, 10 in rank are drift-covered; Nos. 5, 6, 8 are driftless; No. 3 is half drift-covered.

Average production of cheese per county, drift, 10,000,000 lbs; driftless, 7,000,000 lbs.

Of the leading ten counties in the production of butter, Nos. 1, 2, 3, 4, 5, 9 in rank are drift-covered; Nos. 6, 7, 8, 10 are driftless.

Average production of butter per county, drift, 4⅓ million lbs; driftless, 3½ million lbs.

Excess in favor of drift, cheese, 40% ; butter, 25%.

In the number of all farm animals, excess in favor of drift counties, 14%".

BIBLIOGRAPHY AND MAPS

See Chapters III and IV, pp. 77-80, 97-102, and Appendix G.

CHAPTER VII

THE MISSISSIPPI IN WISCONSIN

PERE MARQUETTE AND THE FATHER OF WATERS

In the year 1673 Père Marquette and the Sieur Jolliet first saw the Father of Waters. They reached the Mississippi at its junction with the Wisconsin, between Prairie du Chien and the Nelson Dewey State Park of today. Marquette says that he "entered Mississippi on The 17th day of June, with a Joy that I cannot Express."

Marquette described the river as follows:

"The Mississippi takes its rise in various lakes in the country of the Northern nations. It is narrow at the place where Miskous [Wisconsin River] empties; its Current, which flows southward, is slow and gentle. To the right is a large Chain of very high Mountains, and to the left are beautiful lands; in various Places the stream is Divided by Islands. On sounding, we found ten brasses of water. Its Width is very unequal; sometimes it is three-quarters of a league, and sometimes it narrows to three arpents."

Thus the early observers tell us of the source and mouth of the Mississippi, of the high bluffs on either side, the islands, and the gentle current. The width of the channel, 3 arpents to three-quarters of a league, or 576 to 9,500 feet, was merely an estimate. The depth of 10 brasses, or about 53 feet, seems much greater than that of today. Now, as then, the uplands adjacent to the Mississippi are "beautiful lands."

THE SCENERY OF THE MISSISSIPPI

The gorge or trench of the Mississippi along the western border of the state furnishes the most rugged topography and picturesque scenery to be found in Wisconsin. Indeed, it seems certain that, for travellers, it is destined to be one of the most attractive regions between the Appalachians and the Rocky Mountains. It may be seen comfortably from a river steamer, or from the Chicago,

Burlington and Quincy Railway on the eastern bank, or the Chicago, Milwaukee and St. Paul on the west.   In many respects it is similar and not inferior to the gorge of the Rhine in Germany.

Even the ruined castles of the Rhine are simulated in the rocky cliffs along the Mississippi.   The geologist Owen described this effectively in 1839, when he spoke of the limestone cliffs of Wisconsin and Iowa as follows:

"Sometimes they may be seen in the distance, rising from out the rolling hills of the prairie, like ruined castles, moss-grown under the hand of time.

"Sometimes they present even when more closely inspected, a curious resemblance to turrets, and bastions, and battlements, and even to the loopholes and embrasures of a regular fortification. Sometimes single blocks are seen jutting forth, not unlike dormer windows rising through the turf-clad roof of an old cottage; and again, at times, especially along the descending spurs of the hills, isolated masses emerge in a thousand fanciful shapes, in which the imagination readily recognizes the appearance of giants, sphinxes, lions, and innumerable fantastic resemblances."

Even the historical associations of the Rhine are not entirely wanting, for the Mississippi had a long history of Indian occupation, even before the coming of white explorers.   The effigy mounds of the aborigines are very abundant along the Mississippi bluffs, and the two centuries and a half of Caucasian travel along the Father of Waters are replete with interesting events.

Many of the early travellers looked upon the Mississippi with keen appreciation.   In 1821 the geologist Schoolcraft, viewing the Mississippi at its confluence with the St. Croix near Prescott, said:

"There is an island in the Mississippi opposite its junction.   At this place, the river bluffs assume an increased height, and more imposing aspect, and in the course of the succeeding fifty miles, we are presented with some of the most majestic and pleasing scenery which adorns the banks of the upper Mississippi.   In many places the calcareous bluffs terminate in pyramids of naked rocks, which resemble the crumbling ruins of antique towers, and aspire to such a giddy height above the level of the water, that the scattered oaks which cling around their rugged summits seem dwindled to the most diminutive size; at others, the river is contracted between two perpendicular walls of opposing rock, which appear to

have been sundered to allow it an undisturbed passage to the
ocean, and not unfrequently, these walls are half buried in their
own ruins, and present a striking example of the wasting effects
of time upon the calcareous strata of our planet. Sometimes, there
is a rock bluff on one bank, and an extensive plain of alluvion on
the other, contrasting with the finest effect, the barrenness of the
mineral, with the luxuriant herbage, and the rural beauty, of the
vegetable kingdom. Again, the hills recede from either shore, and
are veiled in the azure tint of the distant landscape, while the river
assumes an amazing width, and is beautified with innumerable
islands, and we find ourselves at once bewildered between the in-
finity of its channels, and the attractive imagery of its banks. Nor
is the presence of animated nature wanting, to enrich and beautify
the scene. The deer is frequently seen standing in the cool current of the stream, gathering the moss from the hidden rocks below, or surveying our approach from the grassy summit of the im-
pending cliff, with an unconcern, which tells us how little it is ac-
quainted with the sight of man. The whole tribe of water-fowl
are found upon the river, and by the variety of their plumage,
and their shapes—the wildness of their notes—and the flapping
of their wings, serve to diversify the scene, while the well known
notes of the robin, and other singing birds upon the shores, which
are the same that we have listened to in childhood, recall a train
of the most pleasing reflections. Nor is the red man, the lord of
the forest, wanting. His cottage is disclosed by the curling smoke
upon the distant hills, where he surveys with a satisfied eye the
varied creation upon the plains below;—the deer—the elk—the
water fowl—the river which floats his canoe—the trees which
overshadow the grassy hills upon which he reposes during the
heats of noon—the thickets, where he arouses the sleeping bear—
the prairie, which gives vigour to his constitution."

STREAM EROSION FOR COMPARISON WITH GLACIAL EROSION

We, who live in Wisconsin, are favored with an especial opportunity to compare and contrast the work of glaciers with the work
of rivers.

The features of the Western Upland due to sub-aerial processes,
especially to river erosion and river deposits, are not less striking
than the glacial features of this province (Chapter VI) and of the
Eastern Ridges and Lowlands (Chapter X). Most of the Western

Upland lies in the Driftless Area. Therefore the phenomena of
stream work are shown clearly and without significant complica-
tion. As this is river erosion and river deposition by the main
stream and tributaries of the largest river in the United States—
the Mississippi—it displays the features of stream work on an un-

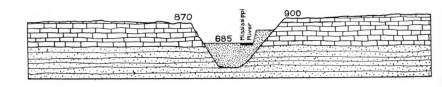

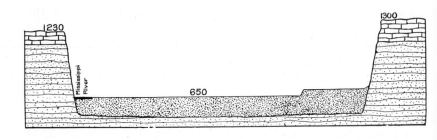

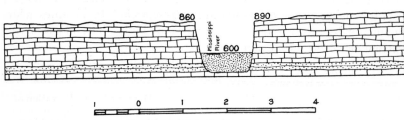

Fig. 39. Cross-sections to show the variations in the width of the Mississippi gorge and the
height of the bluffs, from Prescott (upper), to Trempealeau (middle), and Grant County near
Dubuque (lower). Brick pattern, limestone; dots between horizontal lines, sandstone; dots
without lines, glacial gravel and sand and floodplain alluvium.

usually effective scale. These features may very well be compared
and contrasted with the topographic forms and physiographic
phenomena to be described in eastern Wisconsin, especially as that
more populous portion of the state is underlain by the same rocks
and, before the Glacial Period doubtless was much like the Western
Upland.

BASIN, GORGE, AND CHANNEL

The Mississippi has a basin, a gorge, and a channel. The Mississippi basin, popularly called the Mississippi Valley, includes the basins of its tributaries. Its limits are the divides which separate it from the neighboring streams. Thus the basin of the Mississippi extends from the watershed of the Lake Michigan-St. Lawrence drainage in eastern Wisconsin to the water parting of the Red River-Hudson Bay system in Minnesota and the Columbia River in western Montana. The gorge of the Mississippi is the youthful, steep-sided trench which follows the western boundary of Wisconsin. Between its bluffs this gorge contains the still narrower channel actually occupied by the waters of the Mississippi.

The Mississippi gorge is deeper than that of Niagara but not so narrow. It is not so deep as many mountain gorges, such as the Royal Gorge of the Arkansas River in Colorado or the Grand Canyon of the Colorado River in Arizona. The name trench perhaps describes it better than the name gorge, though the latter is preferred because it helps emphasize the youthful character of the upper Mississippi.

THE GORGE OR TRENCH OF THE MISSISSIPPI

General Description. This gorge has a length of over 200 miles in Wisconsin, extending from Prescott, at the mouth of the St. Croix River, to southwestern Grant County opposite Dubuque, Iowa. In the gorge or trench are bottomlands 1 to 6½ miles wide, bordered by steep bluffs 230 to 650 feet high. The places listed below are 20 to 50 miles apart.

FACTS ABOUT THE GORGE OF THE MISSISSIPPI IN WISCONSIN

Locality	Width of gorge in miles	Height of bluffs above floodplain in feet	Elevation of floodplain in feet above sea level
Prescott	1¼	230	677
Pepin	3	400	668
Buffalo City	4⅓	500	657
Trempealeau	6½	611	649
De Soto	3	500	617
Prairie du Chien	2½	500	612
Cassville	1½	300	602
Dubuque	1	260	592

The Mississippi Bluffs. These bluffs are exceedingly steep, descending more than 500 feet near Trempealeau, for example, within a horizontal distance of 800 feet. In fact a portion of each bluff is likely to be a precipice. It is this retention of steep cliffs and precipices on the walls of the gorge that lead us to call it a gorge and to classify it as young. In most places the gorge wall

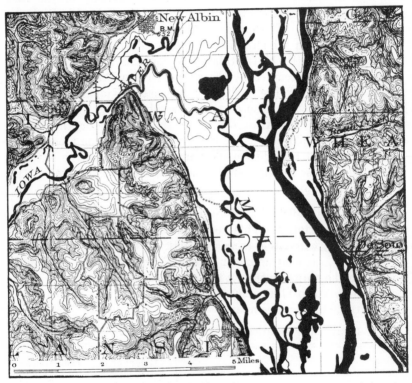

Fig 40. The bluffs of the Mississippi where the rock is chiefly weak sandstone and the gorge is broad. Contour interval 20 feet. (From Waukon Quadrangle, U. S. Geol. Survey.)

consists of steep slopes on the weaker rocks and precipices where the more resistant layers of sandstone or limestone come to the surface.

The bluffs in the Iowa-Wisconsin portion of the river were well described in 1891 by McGee, who said:

"The most prominent geographic feature of the driftless area is the Mississippi cañon—a steep-sided, flat-bottomed gorge ranging from a mile to 7 or 8 miles in width, and gradually diminishing in depth southward from nearly 500 feet at the north line of the State to less than 200 feet at the tip of 'Cromwell's Nose.' Yet, although a veritable rock-bound cañon, similar in genesis to that of the Colorado and in depth approaching that of the Hudson, this gorge is not confined between continuous palisades, but rather guarded by lines of isolated or nearly isolated bluffs stretching

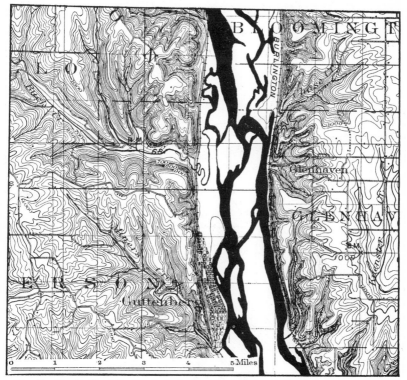

Fig. 41. The bluffs of the Mississippi where the rock is chiefly resistant limestone and the gorge is narrow. This is 50 miles downstream from the area shown in Figure 40. Contour interval 20 feet. (From Elkader Quadrangle, U. S. Geol. Survey.)

along either side of the great river like lines of giant sentinels. Sometimes, indeed, the bluffs are closely crowded, and for a score of miles the cliff wall may be broken only by narrow ravines or the constricted gorges of petty streams; but again the interspaces widen and the bluffs contract until the cañon wall, as seen from the river, becomes but a line of isolated buttes, now round-topped, again crowded with a fillet of precipitous rock, and always forest clad on the north slopes but grassed toward the sun. Yet viewed from one of their own summits, the sentry-like bluffs are seen not as buttes but as salients—the extremities of divides stretching and rising far into the interior of the strongly undulating plain forming the driftless area. At the same time, too, the interspaces are

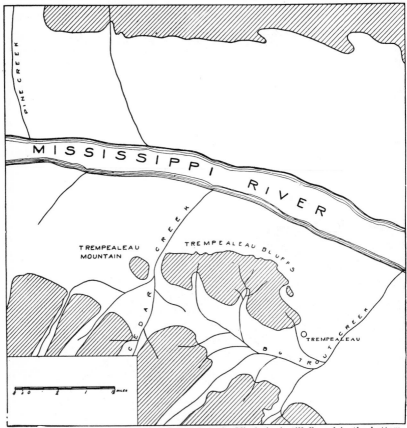

Fig. 42. The rock hill at Trempealeau when the Mississippi still flowed in the bottom-lands to the northeast.

found to be but broad reentrants or amphitheaters, bounded by the converging sides of the salients, themselves sculptured into lines of bluffs as high and steep as those overlooking the river channel. Yet now and then there may be seen mural precipices nearly as steep as the Palisades of the Hudson, and continuous perhaps for two, three, or even five miles."

PERROT STATE PARK

One exceptional feature of the Mississippi gorge is related to the bluffs. This is the high rock hill south of the middle of the gorge at the village of Trempealeau. This main hill, usually called Trempealeau Bluffs, terminates at the north in an isolated knob,

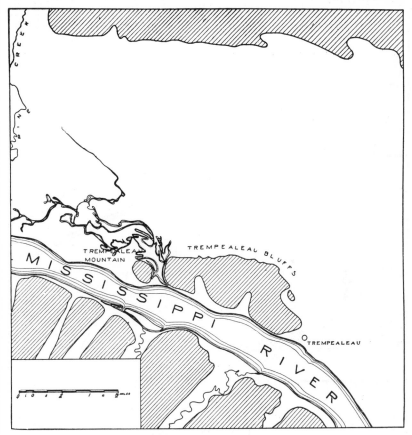

Fig. 43. The rock hill at Trempealeau with the Mississippi in its present channel on the southwest of La Montagne qui trempe á l'eau, or the hill in the river.

known as Trempealeau Mountain. The Winnebago Indians are said to have called this whole rocky eminence *Hay-nee-ah-chah,* or soaking mountain, the Dakota Indians *Min-nay-chon-ka-hah,* or bluff in the water, and the Sioux *Pah-hah-dah,* or mountain separated by water. Accordingly the French continued the same name in the form *La Montagne qui trempe á l'eau,* or the hill which soaks in the water.   Our modern name Trempealeau is made up of the last four words of the French phrase.   It has also been translated Mountain Island, the mountain that is steeped in the water, the bluff rising out of the water, the mountain which sinks, or inclines, or dips in the water.   It might well be called the hill with its base in the water, or, simply, the hill in the river.

Perrot State Park is a place of great geological and historical interest.   Within its boundaries are 910 acres of land on Trempealeau Bluffs and Trempealeau Mountain (Fig. 44).   It was named

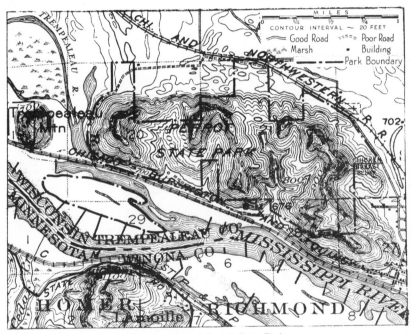

Fig. 44. Map of Perrot State Park.

in honor of Nicholas Perrot, who wintered near Trempealeau in 1685-86.   The whole rock island at Trempealeau is a precipitous, isolated series of knobs, the highest of which rises 500 feet above the bottomlands.   Between the Minnesota upland and Trempealeau Bluffs is a channel less than a mile wide, occupied by the Mississippi River.   A broad expanse of bottomland stretches eastward 3½ miles from the Trempealeau Bluffs to the cliffs of the Wisconsin upland.   It is not now occupied by any river.   In one sense

A.   THE MISSISSIPPI RIVER AT PRAIRIE DU CHIEN.

B.   THE MISSISSIPPI RIVER AT TREMPEALEAU.

A. VALLEY OF THE WISCONSIN RIVER FILLED WITH GLACIAL OUTWASH.
Looking southward from the Baraboo Range.

B. POSTGLACIAL GORGE OF ST. CROIX RIVER AT INTERSTATE PARK.

Trempealeau Bluffs do not now constitute a real island, laved by waters on all sides, but a hill on the eastern border of the river.

The rocks exposed are Lower Magnesian limestone, and the following Cambrian formations: Madison, Jordan, Trempealeau, Franconia, Dresbach (see p. 4). The Lower Magnesian limestone persists only as isolated remnants which cap the higher knobs. The explanation of this isolated rock hill in the Mississippi bottomland is this. The Trempealeau Bluffs were originally a part of the Minnesota upland. The Mississippi River used to flow in the broad, abandoned trench east of Trempealeau Mountain. It received tributaries from the west (Fig. 42), including Cedar Creek, which seems to have isolated the northern end of Trempealeau Mountain from the main hill, or Trempealeau Bluffs. Another tributary, Big Trout Creek, flowed southward where the Mississippi is now. The glacial floods in the Mississippi deposited sand and gravel up to a higher level than at present. In flowing upon this filling, the Mississippi got into the valleys of Cedar and Big Trout Creeks (Fig. 43) where it has flowed ever since. As the Mississippi River is the boundary between Wisconsin and Minnesota, a bit of the latter has been given this state by this prehistoric stream diversion.

There is a similar group of three large, detached, rock hills in Minnesota (Fig. 53) at the northwestern end of Lake Pepin, but here the main river has returned to its valley, and the lateral gorges are abandoned. They are followed by the Chicago, Milwaukee, St. Paul and Pacific railway between Red Wing and a point west of Lake City, Minnesota, opposite Stockholm, Wisconsin.

THE MISSISSIPPI FLOODPLAIN

The gorge walls along the Mississippi are youthful, but the bottomland and the river itself have a more mature aspect. The floor of the gorge shows two conspicuous features: (a) the floodplain of the river, which occupies most of the bottomland, and (b) the terraces, which are narrow and discontinuous (Figs. 47, 49, 51).

The floodplain slopes southward from an elevation of about 677 feet at Prescott to 592 feet at Dubuque. Thus it descends only 85 feet in Wisconsin. The distance between these points is 213 miles, but the river curves back and forth from one side of the floodplain to the other, so that the water actually flows 259 miles. The grade of the river, therefore, is a little less than 4 inches to the mile.

The floodplain material is clay, silt and loam, sometimes sandy and often dark with organic matter. It may be 10 to 30 inches thick and is underlain by several feet of sand, which often grades into coarse gravel 3 to 6 feet below the surface. There are sometimes low knolls, rising 5 to 10 feet above the adjacent basins. Many of these knolls are made up of $1\frac{1}{2}$ to 3 feet of fine sand, beneath which is coarser sand. The basins contain pools or lakes or swamps, where the fine silt and decayed vegetation constitute the floodplain material. There are also bayous, and abandoned channels upon the floodplain (p. 174). In places the peat deposits are 10 feet or more in thickness. Always, however, the floodplain material is a surface film compared with the great thickness of glacial outwash below.

It has sometimes been remarked that the precipitous bluffs of the Mississippi are out of harmony with the gentler slopes of some of its tributary valleys. Another observation of apparent incongruity has to do with the relative sizes of stream channel and gorge. The explanation of the two points is a common one. It does not appear that the Mississippi is too small for its gorge. The gorge was occupied by larger streams during the Glacial Period. They may have deepened it somewhat. They probably eroded the ends of the spurs. They certainly deposited a great thickness of gravel and sand (p. 156). This deposit has hoisted the river up to a level where the bluffs are farther apart than if the stream was flowing on the rock floor below.

The Mississippi. One of the most impressive features of a great river is the volume of water, flowing away to the sea. This was well described by the geologist, Keating, in 1823, when he went up the Mississippi from Prairie du Chien:

"The first day's voyage on the Mississippi was delightful to those who had never been on that river before; the magnificence of the scenery is such, its characters differ so widely from those of the landscapes which we are accustomed to behold in our tame regions, its features are so bold, so wild, so majestic, that they impart new sensations to the mind; the very rapidity of the stream, although it opposes our ascent, delights us: it conveys such an idea of the extensive volume of water which this river ceaselessly rolls toward the ocean. The immense number of islands which it imbosoms, also contributes to the variety of the scenery by presenting it constantly under a new aspect."

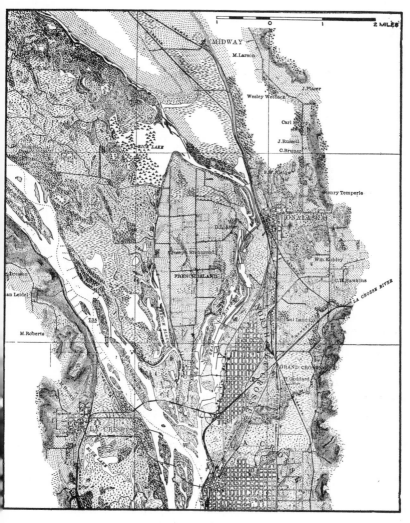

Fig. 45. The gorge of the Mississippi at La Crosse and Onalaska, showing the bluffs, several terraces, the floodplain with sloughs and lakes, and the channels of the Mississippi, Black and La Crosse rivers with their sand bars. (Mississippi River Commission.)

A number of facts about the river itself are summarized in the following table. It is based upon the Mississippi River Commission charts of 1893 to 1895 and data furnished by the U. S. Engineer Office at St. Paul, Minnesota. They do not represent exactly the same stage of water, though they are roughly comparable. The stream-flow figures represent the extremes between 1838 and 1930.

FACTS ABOUT THE MISSISSIPPI IN WISCONSIN

Locality	Depth of river in feet	Width of river in feet	Proportion of flood plain occupied	Flow in cubic feet per second	
				Maximum	Minimum
Near Prescott below Lake St. Croix	5 to 12	1550	26%	134,000	3,000
At outlet of Lake Pepin, above mouth of Chippewa River	4 1/2 to 27	1500	9%		
At East Winona, Wis	4 1/2 to 20	1500	10%	142,000	6,000
Near La Crosse, above mouth of Black River	4 1/2 to 11	2400	9%		
At Prairie du Chien, above mouth of Wisconsin River	5 to 26	1400	11%		
Opposite Bagley, Wis., at Clayton, Iowa	17 to 37	725	8%	190,000	16,000
North of Dubuque, Ia., at southern boundary of Wisconsin	4 1/2 to 22	1800	27%		

The table shows that the depth of water varies from $4\frac{1}{2}$ feet to 37 feet. The width of the river is from 725 feet to 2400 feet. The channel occupies from one-twelfth to one-quarter of the bottom-land. The flow of water ranges from 3,000 to 190,000 cubic feet per second, being less than one-eighth as great as that on the Mississippi delta below New Orleans. The current averages about 3 miles an hour. The river is usually frozen over from the middle of November to the end of March.

High water in Wisconsin is likely to be in April or May, although sometimes there are floods in October. Low water is in December, January, or February, with a secondary low water stage in the summer. The vertical range from high to low water is 18 to 22 feet. At extreme high water a large proportion of the flood plain is covered. At extreme low water the channel averages approximately 8 feet in depth but because of sand bars the controlling depth is $4\frac{1}{2}$ feet. The continual shifting and building of these bars entail considerable maintenance to keep the channel open for navigation. The original project adopted by the Government was to provide a 6-foot depth at low water. It was attempted to do this by open river improvements, i. e., wing dams, closing dams, etc. Between 90 and 95 per cent of the channel had been improved to provide for 6 feet of water when the existing project of a 6-foot depth was increased to 9 feet and the method of improvement changed from open river regulation to canalization.

The continual shifting of sand bars is shown by the changes in depth of the main channel from year to year in the longitudinal profile of the Mississippi River (Fig. 46) near Beef Slough and Alma, Buffalo County. Here the river was deepened 3 feet at one point from 1912 to 1913. Less than a mile down stream it was built up 3 feet in the same period. Nearby the bed of the river was filled over 4 feet from 1913 to 1914.

In Wisconsin the Mississippi is not a meandering stream, as the river is between Illinois and Louisiana. There are no long tiresome detours, no cutting off of oxbows, and little destruction of farm land by river erosion. The process of shifting in position and of rising floods, however, is what has made the flood plain.

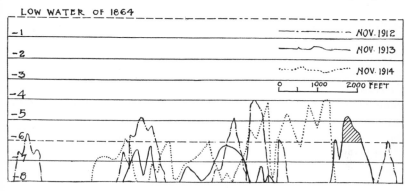

Fig. 46.   Variations in depth of the main channel of the Mississippi near Alma from 1912 to 1914.   (U. S. Engineer Office.)

Some of the smaller tributaries of the Mississippi, however, meander across the flood plain of the master stream.

At the borders of the main channel and some of the larger sloughs are low embankments, called natural levees. Accordingly the flood plain often slopes away from the river rather than toward it.

The commerce of this great waterway grew slowly from 1823 to 1846, a little more rapidly in 1851, and increased by leaps and bounds from 1851 to 1858. The first steamboat appeared on the Wisconsin portion of the Mississippi in 1823. There were about 300 steamboats in 1860. With railway building from 1857 to 1862, and in subsequent years, river transportation rapidly shrank to small proportions. It reached its low point about 1920 when the only traffic was local. In 1925 a common carrier started a towing service with small barges and in the fall of 1926 the "Inland Waterways" established barge service for local and through freight. Traffic has since then increased rapidly and it is expected that such traffic will soon far surpass the maximum of the period 1851 to 1858.

MISSISSIPPI TERRACES

General Description. The bottomland of the Mississippi contains not only a floodplain; it has also a conspicuous series of terraces. These furnish the sites of all the important cities and villages along the Mississippi. These terraces do not usually occupy the whole width of the bottomland outside the actual channel of the river, as they do at Trempealeau. In some places the terrace land occupies only half the bottom of the gorge. This is the case at Prairie du Chien and La Crosse. Still more frequently the terraces are narrow and close to the bluffs, as on the eastern side of Lake Pepin, and at Cassville. Often they are entirely cut away and the floodplain occupies the whole distance between the bluffs. In no case do they extend continuously more than a few miles up and down the river. These terraces are sometimes spoken of as the second bottoms.

La Crosse Terrace. A few of the larger terraces are described below. The dimensions of all the terraces are summarized in the table on pages 165-166. All the terraces are faced by rather steep scarps. For example at La Crosse (Fig. 45) the floodplain occupies the western side of the bottomland, and there is no terrace at the base of the Minnesota bluff, except in the valley mouth at La Crescent. The present channel of the river is about half way between the bluffs. At the city of La Crosse the terrace is bordered by a steep west-facing scarp 30 feet high. The surface of the terrace rises less than 10 feet more in the city. It ends in a low, east-facing scarp, where the abandoned valley of La Crosse River lies within the terrace near the base of the Wisconsin bluff. The northern end of this terrace is cut into several parts between (a) the La Crosse River, (b) the Black River, and (c) the French Slough of the Mississippi.

Onalaska Terraces. While there is only one terrace preserved at La Crosse, one of the lower of the Mississippi River series, there

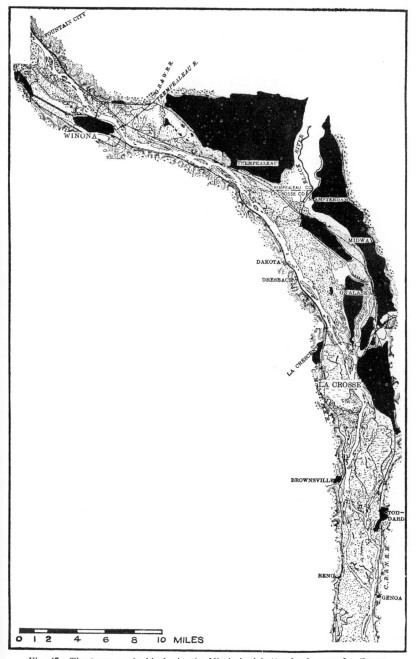

Fig. 47. The terraces—in black—in the Mississippi bottomlands near La Crosse.

are three or four nearby to the north of Onalaska and North La Crosse, separated by steep terrace scarps. The levels represented are (a) the 25 to 35 foot terrace of (1) Brices Prairie, (2) French Island, (3) North La Crosse, (4) Onalaska, and (5) a smaller remnant, (b) the 50 foot terrace east of North La Crosse, (c) the 80 foot terrace upon which Onalaska is situated, and (d) perhaps a 100 foot terrace represented by isolated hills between Onalaska and the bluff. Upon the Onalaska terraces are irregularities due to sand dune accumulation. The 100 foot level may be entirely due to dunes and not a terrace accumulation.

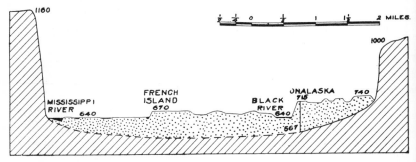

Fig. 48. The terraces near Onalaska.

Trempealeau Terraces. Extending northward from La Crosse and Onalaska is a prominent group of terraces. At Trempealeau, 15 miles above La Crosse, they occupy the whole width of the bottomland, due to the diversion of the Mississippi into the narrow, post-glacial channel west of Trempealeau Mountain (p. 151). The main terrace lies about 60 feet above the river. At the north this terrace is separated from the floodplain of the Mississippi and of Pine and Trempealeau creeks by a conspicuous 55 foot scarp. South of the rock hill at Trempealeau there are lower terrace remnants and also hills 90 or 100 feet above the river. Their tops may represent either an older and higher terrace or a group of dunes.

Prairie du Chien Terrace. There is a single, broad terrace at Prairie du Chien, 55 miles south of La Crosse. It ascends from the river to the east bluff, and slopes northward from the mouth of the Wisconsin River. Thus it seems to be part of a great alluvial fan, built by the glacial Wisconsin in the Mississippi bottomlands. At the western or Iowa bluff there is no terrace, except in the valley mouths where McGregor and North McGregor are located. The main river is now near the western edge of the flood-plain. The latter includes the island west of the channel used by steamboats which land at Prairie du Chien. At the steamer landing, near the Chicago, Milwaukee, St. Paul and Pacific station and the Dousman House, there is a steep ascent to a 15 or 20 foot level. This part of the terrace is separated from that to the east by the slough called the Marais de St. Friol. In and about the city of Prairie du Chien the terrace is 25 feet above the Mississippi. Southeast of the city, the terrace is 25 feet higher; but this is all one terrace, not two levels. This terrace is terminated on the south by a scarp 35 to 40 feet high, cut by the Wisconsin River.

Bagley and Cassville Terraces. At Bagley and Cassville the modern floodplain occupies nearly the whole width of the bottom-land. There is a narrow terrace about 15 to 20 feet above the floodplain at each of these villages, and a smaller 55 foot terrace at Cassville, perhaps due to a rock ledge.

Cochrane-Waumandee Terraces. The Cochrane-Buffalo City-Waumandee terraces constitute a slightly-dissected group 20 feet or so above the river. Near Waumandee Lake there are dunes on the 20 foot surface, and a higher terrace at about 80 feet.

Teepeeota Point Terrace. Opposite Alma is the Teepeeota Point terrace. It lies midway in the bottomlands, with floodplain on either side. This terrace is more than 9 miles long. Its southern portion is rendered irregular by dunes. It is entirely in Minnesota.

West of Teepeeota Point is the Waubasha terrace in Minnesota. The Iowa and Minnesota terraces are not discussed in this book, though their distribution is summarized in the table on pages 165, 166.

Terraces Near Lake Pepin. East of Lake Pepin occurs one of the best series of terraces and scarps in Wisconsin. West of the mouth of Chippewa River a high terrace breaks down to the river in a single, magnificent 90 to 95 foot scarp (Fig. 50). East of the Chippewa near Nelson is a 40 foot scarp separating a lower terrace from the floodplain. East of Lake Pepin, however, the 90 to 95 foot scarp separates into three smaller bluffs, so that the terraces 30, 45, and 105 feet above the Mississippi are still preserved. At the village of Pepin, a few miles to the north, there are only

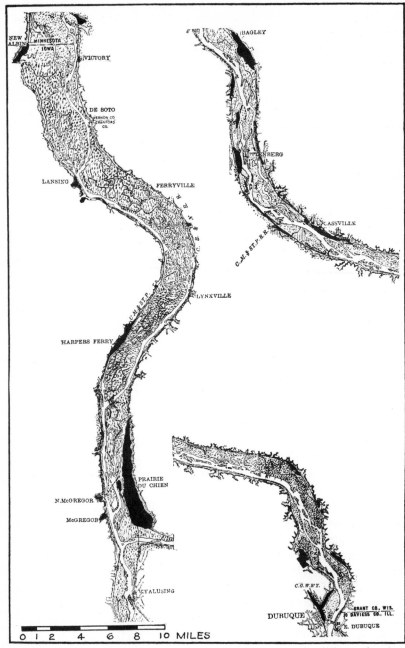

Fig. 49. The terraces—in black—in the Mississippi bottomlands of the southwestern portion of Wisconsin.

two terraces. The lower of these, which furnishes the village site, is broadest and stands about 45 feet above the river. The higher one is at an elevation of 105 feet.

Bay City-Diamond Bluff Terraces. At the northern end of Lake Pepin a series of terraces extends from Bay City to Diamond

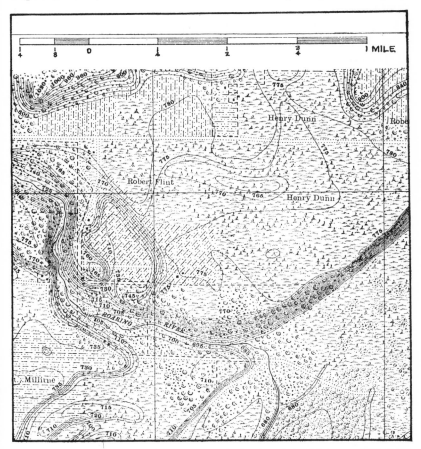

Fig. 50. Topographic map of three terraces at the junction of the Chippewa and Mississippi rivers. The terrace scarps rise 30, 50, and 95 feet respectively above the floodplain at this point. Contour interval 5 feet. (Mississippi River Commission.)

Bluff. These are at elevations of about 20, 40, 60 and 90 feet above the river. Of these the lowest terrace, upon isolated remnants of which the villages of Diamond Bluff and Bay City are located, is smallest. The highest terraces cover the larger area. An intermediate terrace, containing the village of Trenton, is nearly cut in two near its northern extremity, where Trimbell River comes very close to the Trenton Slough of the Mississippi River. Several of these terraces, and certain others along the Mississippi in Wisconsin, are highest next the river. Some of them slope away toward the bluffs, some descend toward the base of the next terrace scarp, and some descend toward the river. These variations may be due in cases to irregularity of original deposition, in other cases to subsequent stream dissection.

Terraces Near Lake St. Croix. Prescott, south of Lake St. Croix, is built on the northernmost of the terrace groups of the Mississippi River in Wisconsin. The city is on the lowest terrace of a series of three, being about 95 feet above the river. There are other terraces at 155 and 235 feet.

The upper levels seem to be made up of the coarsest sand and gravel, as if the high valley trains, subsequently dissected into terraces, were deposited when the ice was nearby, and the lower ones after the glacier had receded some distance to the north and northwest. This is also true of the terraces of the Chippewa River near Eau Claire.

The Prescott terraces appear to be built upon rock benches, so that the sand and gravel is more or less a veneer. This does not seem to be true of any of the terraces farther south in Wisconsin, except possibly the higher of the two terraces at Cassville.

Terrace Distribution. Figures 47, 49 and 51 show the distribution of these terraces. The table (pp. 165, 166) summarizes their elevations and dimensions. It is based upon an intensive study of the large-scale charts of the Mississippi River Commission and upon brief field studies in selected areas. There may be a 5 to 10 foot error in the level listed at some points, as all the terraces show variations due to (a) original deposition, (b) subsequent dissection, (c) deposition of sand dunes. The elevations in the table represent the height of the larger part of the terrace above the present floodplain.

The Problem of Correlating Terrace Groups. The terraces listed in the table might possibly be grouped into three systems—a high group, one or more intermediate groups, and a low group. It is clear that certain of the terraces in the mouths of tributary streams correspond to slightly-lower terrace remnants farther from the bluffs of the main Mississippi trench. Thus the terrace at French

Island west of Onalaska is quite a little higher than the small, isolated terrace still farther west. The 40 foot terrace in the middle of the valley west of Smiths Landing and Diamond Bluff might conceivably be the equivalent of the 65 foot terrace at Diamond Bluff rather than the correlative of the 40 foot terrace at the same place. This condition is repeated at many points along the Mississippi in Wisconsin. It seems clear, therefore, that additional field work will be necessary before the several systems of Mississippi River terraces can be definitely correlated. Moreover they must be carefully correlated before the rates of southward inclination of the terrace systems can be worked out.

Northward Increase in Number and Height of Terraces. At least one tentative conclusion may now be stated, even without further field work. (1) There are more and higher terraces in northern than in southern Wisconsin; (2) the gradients of the individual terraces increase in steepness toward the north; (3) each single terrace in southwestern Wisconsin seems to split into two or more terraces in northwestern Wisconsin.

The Mississippi terraces in Wisconsin south of La Crosse appear to be in two series. These are (a) the 20 foot terraces and (b) the 40 to 50 foot terraces, as may be seen by inspecting the table on page 166. North of La Crosse, however, there are three or more persistent terraces. They rapidly increase in height to more than 100 feet above the Mississippi floodplain. The intermediate terraces appear at such levels as to suggest (a) that the 40 foot terrace south of La Crosse may split and form both the 60 and the 105 foot levels as it increases in height to the north, (b) that the 20 foot terrace south of La Crosse may be similarly related to both the 30 and the 45 foot levels farther north, and (c) that the 20 foot level north of La Crosse is entirely independent of the 20 foot level to the south. All this must be regarded as tentative rather than final.

Likeness of Terrace History to That of Glacial Lake Beaches. The hypothesis is advanced that the Mississippi terraces are similar, in number and in degree of inclination to the abandoned beaches of the Glacial Great Lakes (Chapters XII, XVIII). The beaches in southern Lake Michigan are essentially horizontal, while those to the north (a) increase in altitude and (b) split into separate strand lines. Of course the beaches were originally horizontal, while the river terraces slanted southward. In the table

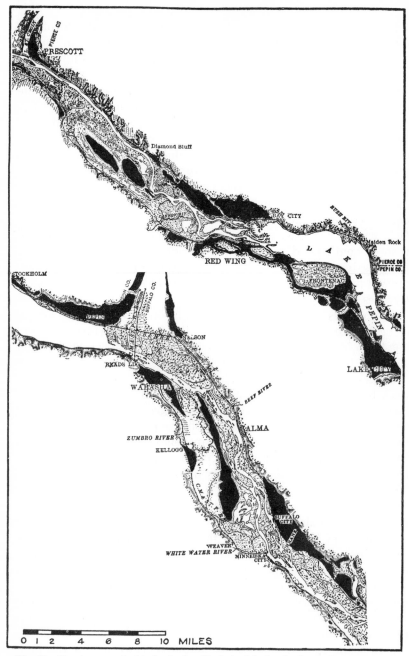

Fig. 51. The terraces—in black—in the Mississippi bottomlands of the northwestern portion of Wisconsin.

TABLE SHOWING DIMENSIONS OF THE MISSISSIPPI TERRACES

Locality	Average elevation of surface above flood-plain, in feet	Maximum length and width of terrace, in miles	Remarks
Prescott, Wis.	95	4½ by ⅜	Gravel terrace, possibly on rock bench.
Prescott, Wis.	155	3¾ by ⅜	Perhaps chiefly a rock bench.
Prescott, Wis.	235	-----------	Small remnants of coarse gravel, probably on a rock bench.
North of Etter, Minn.	80	-----------	Perhaps rock benches.
West of Smith's Landing and Diamond Bluff, Wis.	45	7½ by ⅝	Gravel terrace in middle of bottomlands.
Diamond Bluff, Wis.	40	2¼ by ⅛	
Diamond Bluff–Trenton, Wis.	65	6½ by ½	
Red Wing, Minn.	40	5 by ⅜	Extends south to Frontenac behind rock hills.
Red Wing, Minn.	120	⅝ by ¼	
Hager-Bay City, Wis.	90	4 by 1⅜	
Bay City, Wis.	75	2 by ⅝	
Bay City, Wis.	60	⅜ by 1⁄16	
Bay City, Wis.	20	⅝ by ⅛	
Wacouta, Minn.	70	1¼ by ⅜	
Frontenac village, Minn.	105	1¼ by ½	
Frontenac station–Lake City, Minn.	105	4⅜ by ⅞	
Lake City, Minn.	105	1⅜ by ⅜	
Florence-Lake City, Minn.	85	3 by 5⁄16	
Frontenac-Lake City, Minn.	65	7⅜ by ⅜	
Lake City, Minn.	40	4¾ by 1⅜	
Stockholm, Wis.	45	1¼ by ¼	
Bogus Creek, Wis.	105	-----------	Two small remnants.
Pepin, Wis.	45	6 by 1	
Pepin–Chippewa River	105	5 by ¾	
Near Chippewa River	225	⅜ by 3⁄16	Possibly a drift remnant or rock hill.
Near Chippewa River	30	1⅛ by ⅜	
Nelson, Wis.	45	3⅛ by ¾	
Wabasha, Minn.	30	4⅞ by ¾	
Kellogg, Minn.	30	1¾ by ¾	
Teepeeota Point terrace, opposite Alma, Wis.	30	9⅛ by ¾	Large terrace in middle of bottomlands.
Cochrane-Waumandee Lake, Wis.	20	11 by 1¼	
Waumandee River, Wis.	80	1⅜ by ⅜	
Minnesota City, Minn.	35	2 by ¼	
Winona, Minn.	20	3¾ by ⅝	

TABLE SHOWING DIMENSIONS OF THE MISSISSIPPI TERRACES–Continued

Locality	Average elevation of surface above flood-plain, in feet	Maximum length and width of terrace, in miles	Remarks
Bluff Siding, Wis.	45	⅝ by 1⁄16	
Northwest of Trempealeau Mountain, Wis.	30	-----------	Several remnants.
East of Trempealeau Bluffs, Wis.	60	10¼ by 3¾	Partly alluvial fan of Black River
East of Trempealeau Bluffs, Wis.	90	-----------	Several remnants.
East of Trempealeau, Wis.	35	2 by ¾	
Lytle-Brices Prairie, Wis.	35	5 by 1	
New Amsterdam-Onalaska, Wis.	80	15 by 3	
French Island-West La Crosse, Wis.	25	4¾ by 1½	
West of French Island, Wis.	25	¾ by ⅛	
Onalaska, Wis.	100	-----------	Small remnants, perhaps not a terrace.
South of Onalaska, Wis.	25	2 by ½	
North La Crosse, Wis.	25	1¾ by ⅝	
East of North La Crosse	50	⅞ by ¼	
La Crescent, Minn.	50	1⅝ by ⅝	
La Crosse, Wis.	40	5¾ by 2¼	
Brownsville, Minn.	65	-----------	Small remnant, perhaps rock terrace.
Stoddard, Wis.	30	1¾ by ½	
Near Reno, Minn.	30	-----------	Two small remnants.
New Albin, Iowa	30	-----------	Remnant in tributary.
Columbus-Lansing, Iowa	35	-----------	Three remnants.
North of Harpers Ferry, Iowa	40	-----------	Two small remnants
Harpers Ferry, Iowa	40	3½ by ½	
McGregor and North McGregor, Iowa	40	-----------	Two small remnants in side valleys.
Prairie du Chien, Wis.	25	7½ by 1⅝	A large alluvial fan of Wisconsin River.
Bagley, Wis.	20	2½ by ⅝	
Jung, Iowa	20	1¼ by ⅛	
Buck Creek, Iowa	20	1½ by ⅛	
Guttenberg, Iowa	20	2¼ by ¼	
Cassville, Wis.	55	1⅝ by ⅛	Possibly rock bench in Galena-Black River limestone.
Cassville, Wis.	20	4 by 3⁄16	
South of Potosi, Wis.	20	1½ by ¼	
Leisure Creek, Iowa	40	½ by ⅛	
Platte River, Wis.	40	-----------	Two small remnants
Edmore, Iowa	40	1⅛ by ⅝	
Rutledge Siding, Wis.	40	1¼ by ¼	
Dubuque, Iowa	40	1½ by ¼	

(pp. 165, 166) the terrace levels have been stated in relation to the gently-sloping surface of the present Mississippi floodplain.

The reason for the northward inclination of the Lake Michigan beaches and of the abandoned beaches of Glacial Lake Agassiz is that the land has been rising toward the north during and since the Glacial Period (pp. 300, 461). The beaches in southern Lake Michigan are essentially horizontal because they are south of the Whittlesey hinge line (Fig. 118), which is thought to cross the Mississippi at or near La Crosse. If this is the case, it furnishes a simple and rational explanation of the few and low terraces to the south and the many and high terraces to the north.

It is, of course, recognized that there are other controls for terraces, besides the tilting of the land. They may be related to rock ledges, as seems to be the case on the St. Croix River at the Interstate Park. They may be related to the volume of water and the load of gravel, sand, and mud which was carried by the Mississippi. It may turn out that a few of the terrace remnants—all of which now seem to be of latest-glacial or Wisconsin age—are

Fig. 52. The channel and bottomlands within the gorge of the Mississippi, showing the buried portion of the gorge. Op, Lower Magnesian limestone or Prairie du Chien formation; Osp, St. Peter sandstone; Opy, Platteville limestone; Og, Galena limestone; Om, Richmond shale; Sn, Niagara limestone; Ql, loess; Qot, outwash terrace deposits and alluvium. (Grant.)

related to one or another of the earlier drift sheets (p. 88). Certainly a system of similar terraces was formed during each of the earlier stages of glaciation; but all or nearly all of these, like the pre-Wisconsin glacial lake beaches, were subsequently destroyed.

The Mississippi terraces were made by glacial streams. These streams originally deposited the sand and gravel. Subsequently they cut away the larger part of it. What remains constitutes the terraces. The effect of tilting, during the existence of the glacial streams which made the terraces, would be to form just such a terrace system as we have. If there were more tilting north of a hinge line at La Crosse, then the terraces should rise to higher and higher levels, as is the case. There should be more terrace levels north of La Crosse than to the south. This also appears to be the case. Finally, the tributaries of the Mississippi north of the hinge line should have more terraces than the tributaries to the south.   This also appears to be the case, as witnessed by the Chippewa River with five or more terraces in contrast with the Wisconsin which has only two or three.

Thickness of Terrace Material.   The thickness of the terrace material is remarkable.   Close to the bluffs it may be (a) 75 to 80 feet, as at Fountain City and Maiden Rock, or (b) 103 feet, as at Alma, or (c) 148 feet, as at Onalaska, or (d) 172 feet, as at Cassville.   Farther out in the middle of the bottomland it may be (e) 170 feet, as at La Crosse, or (f) only 147 feet, as at Prairie du Chien.

Origin and Age of Terraces.   The terrace material is sand and gravel, sometimes with a little silty material, and frequently with dune sand on the surface.   Where the terrace material is pebbly or gravelly it contains a variety of igneous and metamorphic crystalline rocks, as well as the local sandstone, and flint, and sometimes limestone.   The crystalline rocks show that the terrace material is glacial outwash.   The terraces are of no great age within the Glacial Period since stream erosion has not dissected them greatly, except for the trimming of their edges by the Mississippi and its glacial ancestors.   The presence of limestone and the lack of weathering in most of the crystalline pebbles indicate that most of the terraces represent valley train outwash of the Wisconsin stage of glaciation.

Buried Portion of the Mississippi Gorge

The gorge of the Mississippi was formerly cut down to a lower level, as is shown by the following table, based upon well records in the Mississippi gorge of Wisconsin, Iowa, and Minnesota.

The figures in the right hand column, printed in heavy face type, represent wells near the middle of the valley, the others are at or near the blúffs and do not show the maximum depth of the gorge.

ROCK FLOOR OF MISSISSIPPI RIVER IN WISCONSIN

Locality	Depth to rock in feet	Elevation of surface, in feet above sea level	Elevation of rock floor, i feet above sea level
Opposite Prescott, Pierce Co., at Hastings Prairie, Minn..........	200	about 720	520a
Opposite Hager, Pierce Co., at Red Wing, Minn..............	160	708	548
Opposite Stockholm, Pepin Co., at Lake City, Minn.	207	710	503
Opposite Nelson, Buffalo Co., at Wabasha, Minn..............	165	700	535
Alma, Buffalo Co..	103	662	559
Opposite East Winona, Trempealeau Co., at Winona, Minn.	150	662	512a
La Crosse, La Crosse Co.................................	170	674	504
Opposite Victory, Vernon Co., at New Albin, Iowa............	130	650	520
Prairie du Chien, Crawford Co...........................	147	639	492
Cassville, Grant Co.....................................	172	630	458
Opposite Rutledge Siding, Grant Co., near Eagle Point, Iowa....	160	600	440
South of above, at Julian Hotel, Dubuque, Iowa..............	210	615	405

a Or less. The level near St. Paul may be slightly below 500 feet.

The table shows that the gorge has been buried to a depth of 100 to 200 feet. The rock floor of the Mississippi slopes southward, indicating that the river flowed southward before the Glacial Period as it does now. Its grade is a little steeper than that of the present river, and may be as steep as six inches to the mile.

LAKES OF THE MISSISSIPPI

Lake Pepin. In the gorge of the Mississippi, just west of the Chippewa River in Pierce and Pepin counties, is Lake Pepin. It is 1 to 2½ miles wide and nearly 22 miles long, covering 38½ square miles. The maximum depth of the lake is 56 feet, but most of it is 20 to 35 feet deep. Father Hennepin observed in 1680 that "its Waters are almost standing, the Stream being hardly perceptible in the middle." Hennepin called Lake Pepin "the Lake of Tears."

The cause of this lake is the delta of the Chippewa River, which lies in the gorge of the Mississippi at the southeastern end of the lake. This delta is covered by modern floodplain deposits. The reason that the small Chippewa River was able to bring more material than the larger Mississippi could carry away is that the grade of the Chippewa is much steeper than that of the Mississippi. The tributary stream carried more and coarser debris than the master stream could remove. Accordingly a lake was dammed back in the gorge of the Mississippi. The gorge has been filled by the main stream to a depth of at least 30 to 50 feet since the dam was built, as we know from the depth of Lake Pepin. The Mississippi has built out a delta in the northern end of the lake and this is still growing. The lake must originally have extended much farther upstream toward Prescott. Certainly the head of the lake was not long ago at least five miles farther upstream at Hager, Wis., and Red Wing, Minn. The Sioux Indians are said to have traditions regarding this. Below Red Wing three large and several

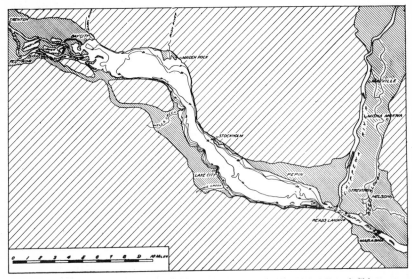

Fig. 53 Lake Pepin in the Mississippi bottomlands, showing the delta of Chippewa River which causes the lake. Contour interval 10 feet.

small lakes lie between the distributary channels of the Mississippi. The water in the northwestern end of Lake Pepin has been shoaled to less than half the depth it must have had originally, indicating the process by which the lake will eventually be destroyed.

It has been stated that Lake Pepin has varied notably in level within historic times. This is upon the basis of stumps of trees in the channel of the Mississippi at Red Wing, Minn. These stumps are possibly to be interpreted as evidence that the channels are shifting and being built up. It is thought that the French explorers found the main Chippewa flowing in what is now called the Beef River Slough below Wabasha, Minn. The change from this to the present channel, opposite Reads Landing, Minn., would cause the level of Lake Pepin to rise slightly.

The shores of Lake Pepin are partly the high rock bluffs of the Mississippi gorge, partly the Mississippi terraces, and partly the very low modern deposits made by streams and waves. The larger stream deposits are the deltas of the Mississippi and the Chippewa at the head and foot of the lake respectively, and the smaller deltas of Rush River near Maiden Rock and of Isabel Creek at Bay City. Other notable features of the low shorelines are the spits, made by waves and currents. Pairs of these spits converge in V-shaped points, or cusps, enclosing triangular swampy areas. There are cusps at Stockholm and Maiden Rock on the Wisconsin shore, and even better ones at the Point Au Sable, Central Point, and Lake City cusps on the Minnesota shore.

Conditions at Prairie du Chien. The question naturally arises as to why the Wisconsin River has not made an enlargement of the Mississippi at Prairie du Chien. The fact is that the waters of the Mississippi are backed up somewhat at Prairie du Chien, though not enough to make a second Lake Pepin. There may have been such a lake there in the past, however; but, if so, it has been completely filled by the deposits of the Mississippi. The terrace at Prairie du Chien constricts the Mississippi somewhat, so that it should be able to carry away the Wisconsin River detritus more effectively than it can that of the Chippewa.

Lake St. Croix (p. 202) is of the same type as Lake Pepin.

State Boundary in Lake Pepin. An interesting matter of political geography has to do with the boundary between Wisconsin and Minnesota in Lake Pepin. The law provides that the boundary shall follow the main channel of the Mississippi River along the western border of this state (p. 485). In Lake Pepin there is no accepted channel. The steamboats may go almost anywhere, except at the inlet and outlet. There has been difficulty in administering the fish and game laws because, until recently, the open season differed in the two states. Accordingly there has been controversy, the state of Minnesota claiming the boundary should be in the middle of the lake half way between the shores, while the state of Wisconsin contended that it should follow the usual route traversed by steamboats. This route happens to be much nearer the Minnesota shore, because the larger boats call only at Lake

City on the Minnesota side. Accordingly Wisconsin claims nearly three-fourths of Lake Pepin, while Minnesota contends that we are entitled to only half.

From the point of view of geography there certainly is no main channel. Moreover, the line of deepest water—if it were held that that constituted a main channel—does not coincide with the route usually followed by steamboats. Figure 53 shows that the line of deepest water is in many places on a broad, flat-floored portion of the lake bottom, much of it occupying half or a third the width of the lake. Where it is narrower it is even closer to the Minnesota shore than the steamboat route. The whole controversy really turns on the question as to whether the body of water at Lake Pepin is river or lake. Geographers can have no hesitancy in calling it lake, just as is the case in Lakes Ontario, Erie, or Huron, which are broad stretches of water in the St. Lawrence River system.

Lakes and Swamps in Side Valleys. Small bodies of water of the same type as Lake St. Croix occupy valley mouths farther south.

Many valley mouths along the Mississippi have contained lakes of this type, but most of them have now been destroyed by stream fillings or by plant growth. Others have been modified by the work of man in building dams or highway fills. Waumandee Lake, northwest of Fountain City in Buffalo County, is an illustration. There are two small lakes in the mouth of the Platte River in Grant County. The swamp representatives of similar lakes now extinct lie at the mouth of Copper Creek south of Ferryville in Crawford County, and at the mouth of Wind Creek north of Diamond Bluff in Pierce County. There are swamps in the mouths of other tributaries, where the water has never been deep enough to form a lake.

Small Lakes of the Mississippi Floodplain. One of the regions of most abundant lakes in Wisconsin is the present floodplain of the Mississippi. These lakes are all small and all shallow. Few of them are more than 5 or 6 feet deep. They are relatively inaccessible because of the swampy bottomlands surrounding them. Only a few of them have names. Yet their total number is very great. This may be judged from the fact that in an area of about 20 square miles of the floodplain there are over 200 lakes and ponds. This was in the Wisconsin portion of the Mississippi bottomland in Crawford County between Lynxville and De Soto. The number was counted on one of the detailed maps of the Mississippi River Commission, and sloughs and bayous still connected with the river were not included. The lakes range in size from bodies of water half a mile long and an eighth of a mile wide to pools only

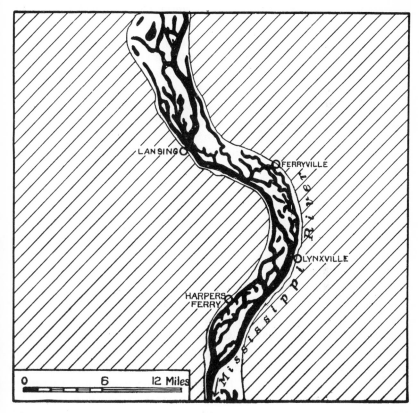

Fig. 54. Channels and backwater sloughs of the Mississippi. The curvature of the trench walls suggests erosion by a much larger river than the present Mississippi. Such streams, too small for their valleys are technically referred to as underfit rivers.

one or two hundred feet in diameter. There may be as many as a thousand of these floodplain lakes in the Mississippi floodplain of Wisconsin.

Not all of these lakes are as small as those just mentioned. Rice Lake north of La Crosse and the upper and lower lakes north of the head of Lake Pepin are each a mile and a half long and a mile wide. McGregor Lake at Prairie du Chien is four-fifths of a mile long and nearly half a mile wide. There are still larger ones west of the main channel of the Mississippi in the Minnesota portion of the floodplain. These large bodies of water are exceptional, however, and the small lakes are more common.

The floodplain contains not only a main broad channel of the Mississippi, sometimes divided into two or more channels, but also numerous narrow sloughs or bayous. Some of these sloughs are the outlets of tributary streams in which the water flows rapidly. Others are cross sloughs which leave the main channel and join it again. Still others are abandoned channels of the Mississippi, and contain long, narrow, crooked, bodies of stagnant water without perceptible current. These backwater sloughs are more closely allied to the floodplain lakes than to the channels of the river and its tributaries. At times of high water the Mississippi spreads over a large part of the bottomland, covering sloughs, lakes, and floodplain.

Besides Lake Pepin and the St. Croix-Waumandee type there are lake basins (a) in portions of abandoned channels now cut off from the main stream by deposits of detritus or accumulations due to plants, (b) in low portions of the floodplain between the deposits at the borders of existing channels or bayous and the rock bluffs or the edges of terraces, and (c) in hollows formed by the growth of a delta into a lake.

There are hundreds of illustrations of the first or abandoned-channel type of floodplain lake, for example Long Lake near Trempealeau, Round Lake near La Crosse and the Marais de St. Friol at Prairie du Chien. The last is partly due to an artificial dam. McGregor Lake is not of exactly the same type, being held in between two main channels of the river on the island at Prairie du Chien. The lower Mississippi flows in great ox-bow curves, but as there are none of these curves, or meanders, in the main channel of the Mississippi in Wisconsin there are no ox-bow lakes among the abandoned channels.

The lakes between floodplain deposits and the rock bluffs or terrace scarps are illustrated by Rice Lake (Fig. 45), which lies close to Black River north of Onalaska. It is a good-sized lake, but much of it is only one or two feet deep and the western two-thirds is filled with growing plants. This lake is fast being converted into a swamp. There is a similar body of water in the main flood-plain north of Waumandee Creek, and another is being made at Goose Bay near the head of Lake Pepin.

The lakes between distributaries of a delta are exemplified by Upper Lake and Lower Lake at the head of Lake Pepin. A third lake of this type is in process of formation near Bay City between Lower Lake and Lake Pepin. It is now nearly enclosed and is already partly filled with marsh plants. A fourth lake will soon be formed between the Main and South channels at the head of Lake Pepin. Upper Lake and Lower Lake were 5 to 7 feet deep in

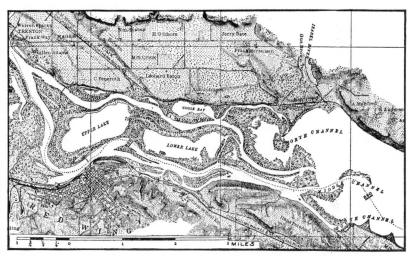

Fig. 55. The small lakes in the Mississippi floodplain at the head of Lake Pepin. (Mississippi River Commission.)

1897. West of the former in Minnesota is another lake of the same origin which has been almost entirely filled by vegetable accumulation. Only a little open water remains at the eastern edge of the lake basin, the remainder being a level marsh.

Swamps of the Mississippi Floodplain. It has already been shown that many lakes of the Mississippi floodplain are in process of extinction by filling and that others have been completely filled and converted into swamps. There are also vast areas of swamp land which have never been in lakes or abandoned channels. These marshes cover hundreds of square miles. Some of them are too wet for utilization. Others are dry enough during portions of the year so that they form meadow land of good quality. Many mea-

dows and areas underlain by peat support marsh grass but cannot be utilized because they are so swampy. Because they are inundated several times a year, they are too soft to support horses
and wagons, so that the wild marsh grass remains uncut. The
detailed maps of the Mississippi River Commission distinguish the
marshes from the meadows. The forested areas of the floodplain,
where elm, birch, maple, oak, and basswood trees grow profusely,
are often on the higher lands. Willows grow on lower portions of
the floodplain. As the natural levees, which border parts of the
main channel, are higher, drier, and more favorable to tree growth
than the swamps and meadows away from the river, the traveller
who merely sees the floodplain from a steamboat in the main channel is likely to get an impression that there is less swamp land
than is really the case.

HISTORY OF THE MISSISSIPPI

Age of the Gorge. The recent history of the Mississippi may be
summarized as follows. We do not know just when the river came
into existence; but at some time before the Wisconsin stage of the
Glacial Period the existing gorge or trench was cut. We know
this from the presence of glacial till within the gorge at points
west of Lake St. Croix and others south of Dubuque. It appears
likely that the gorge is not only pre-Wisconsin, but completely preglacial, although complete proof is not yet at hand.

We do not know whether much of the deepening of the gorge is
due to glacial streams or not. In fact the early history of the
gorge is not at all well understood. The gorge may be antecedent
to the warping and folding of the Paleozoic rocks. It might even
be superimposed from overlying strata, now removed.

Lack of Halts in Down-Cutting. If there had been one or more
peneplain stages at the elevation of the Niagara, Galena-Black River or Lower Magnesian limestones, the Mississippi could not fail to
record it by possessing intrenched meanders or by having left
old river gravels at the upper limits of the gorge. But there are
no such features preserved in Wisconsin or the adjacent states, so
that we assume that there was no halt in the down-cutting which
formed the gorge.

Persistence of Direction. The river never flowed northward, as
has been thought by some authors on the basis of the widening of
the gorge upstream from Dubuque to Trempealeau. If it had, the

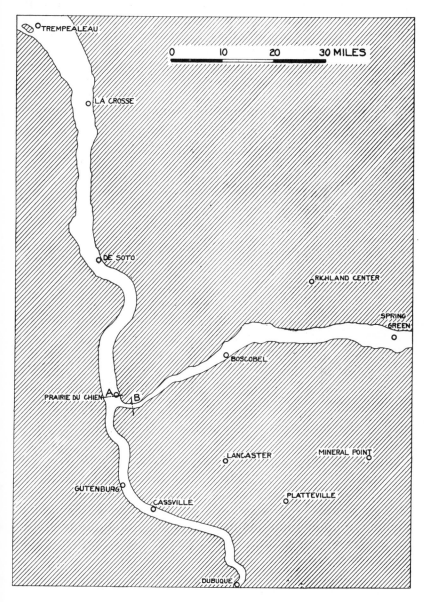

Fig. 56.   The Mississippi and Wisconsin gorges narrowing downstream in relation to resistant limestone at A and B, Broad portions of gorge in weak sandstone.

place of beginning of the greatest constriction should not so exactly coincide with the dipping of the resistant Lower Magnesian limestone beneath the grade of the river at Prairie du Chien (see A, Fig. 56). The Wisconsin River also widens upstream, though rivers usually widen downstream. It would be necessary to assume that the Wisconsin formerly flowed eastward, were it not for the

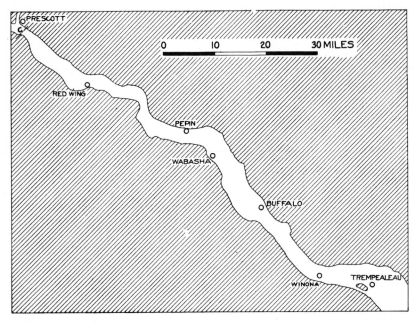

Fig. 57. The Mississippi in its broad gorge at Trempealeau in sandstone, and its narrow gorge at Prescott in limestone.

fact that the narrowest place is exactly where the same resistant limestone dips beneath the river grade (B, Fig. 56). Again, the theory of former northward flow meets an insurmountable obstacle in the narrowing upstream in the Mississippi from Trempealeau, where the broad gorge lies entirely in the weak Cambrian sandstone, to Prescott, where the narrow gorge again has the resistant Lower Magnesian limestone below river grade (C, Fig. 57). Finally, the rock floor of the Mississippi gorge slopes southward. The suggestion that the Mississippi is an antecedent stream, that is, one that was there before a regional uplift of earlier geological history, does not appear to be irresistibly compelling.

Period of Filling With Outwash. During the Glacial Period the Mississippi gorge was partly filled with outwash gravel and sand, the highest terraces being over 100 feet above the present flood-plain and the total filling being more than 250 feet in some places. It was during this aggradation by the enlarged Glacial Mississippi that the major portion of the dust and fine sand for loess was exported eastward by the wind.

Period of Terrace Cutting. An episode of erosion and terrace cutting followed this enormous filling. It is evident that the terraces formerly extended clear across the gorge bottom, for one terrace still does so at Trempealeau. Moreover there are terrace remnants not only near the bluffs but also in the very middle of the bottomland, as in the great terrace that extends southward from Teepeeota Point past Alma and Buffalo City. The tributary streams as well as the Mississippi itself have participated in the gigantic task of removing this terrace material, as is shown by numerous terrace remnants now cut off from the base of the bluffs by small creeks from side valleys.

Reason for Terrace Cutting. The change from deposition to erosion may have been related to the large volume of water from the southward outflow of Glacial Lake Agassiz. The Glacial River Warren, as this stream was called, was large, and flowed for a long time. It flowed out of a lake, and, therefore, had relatively little sediment compared with the earlier glacial streams in the Mississippi valley which came directly from the ice. The outlets of other glacial lakes, as the Kettle-St. Croix outlet of Glacial Lake Nemadji (p. 452), the Upper St. Croix Lake outlet of Glacial Lake Duluth (p. 458), the Black River outlet of Glacial Lake Wisconsin (p. 340), and the Portage outlet of Glacial Lake Jean Nicolet (Glacial Lake Oshkosh) (p. 302) likewise delivered great volumes of water to various parts of the Mississippi valley. As these waters were less heavily laden with sediment than their predecessors which came directly from the melting ice, we might perhaps assume that the change from deposition to erosion came as a result of the floods from the glacial lakes.

Another possibility is that the change came as a result of uplift. If the hinge lines which cross Lake Michigan north of Milwaukee cross the Mississippi valley in Wisconsin, then we have terraces on the Mississippi responding to the same conditions as the beaches along Lake Michigan (p. 298). The terraces appear to be split into several levels north of the hinge line, for there are fewer terraces to the south (p. 167).

A third possible cause of the change from deposition to erosion is merely the recession of the ice. This resulted in the Mississippi's laying down its coarser load farther north and having only fine detritus at this time, and eventually no coarse sand and gravel at all. This decrease of load would, likewise, enable it to erode terraces in the valley train gravels previously deposited. Its tributaries would keep pace with it in the down-cutting, especially such tributaries as the Wisconsin and Chippewa which came directly from the glacier. In the Wisconsin and Chippewa valleys there has likewise been a change from deposition to terrace cutting. As the Wisconsin never had very much water from the small and short-lived Glacial Lake Jean Nicolet, and the Chippewa carried no water from a glacial lake, it seems probable that this third explanation is the true one, or that it is some combination of the second and third explanations here outlined. It appears as if the tilting as well as the decreased stream load were necessary to explain the change from deposition to terrace cutting.

Period of Floodplain Deposition. There has been a third change in the history of the Mississippi. The present floodplain is floored chiefly with clay and silt, beneath which is sand and then gravel. This fine floodplain material is still being deposited. The change from the terrace-cutting back to the modern era of floodplain deposition, seems to have taken place because the volume of the Mississippi was suddenly decreased. It had had the water from the melting ice and from glacial lakes. Its volume decreased to that dependent upon the rainfall of its own drainage basin. It could not carry all the load delivered by its headwaters and tributaries, so it resumed deposition, as is shown by the formation of Lake Pepin and the other lakes.

Summary. Such is the known history of the Mississippi. We may say that there are four episodes: (1) the preglacial and possibly glacial gorge-cutting, (2) the period of deposition, (a) by glaciers, (b) by glacial streams, and (c) by non-glacial tributary streams of the Driftless Area during the Glacial Period, (3) the period of terrace-cutting during the late stages of the Glacial Period, and (4) the period of floodplain deposition and lake formation still in progress. Each of the Mississippi tributaries in the Western Upland, even those lying wholly in the Driftless Area, has had essentially the same episodes in its history.

BIBLIOGRAPHY

Calvin, Samuel. Some Features of the Channels of the Mississippi River between Lansing and Dubuque and their Probable History, Proc. Iowa Acad. Sci., Vol. 14, 1907, pp. 213-220; Geology of Allamakee County, Iowa Geol. Survey, Vol. 4, 1895, pp. 44-54; Geology of Dubuque County, Ibid., Vol. 10, 1900, pp. 473-475.

Chamberlin, T. C., and Salisbury, R. D. Driftless Area of the Upper Mississippi River, 6th Annual Rept., U. S. Geol. Survey, 1885, (Erosion and the Results), pp. 221-239; (Terraces of the Glacial Flood Deposits), pp. 308-311.

Featherstonhaugh, G. W. Report of a Geological Reconnaissance Made in 1835 from the Seat of Government by the way of Green Bay and the Wisconsin Territory to the Coteau de Prairie, Senate Document 333, Washington, 1836, (on the origin of Lake Pepin), pp. 132-133; A Canoe Voyage up the Minnay Sotor, 2 volumes, London, 1847, (on Wisconsin and Mississippi rivers), Vol. 1, pp. 191-258, 270-273; Ibid., Vol. 2, pp. 15-22, 28-33, 113.

Gingnass, Michael. (Description of Fox, Wisconsin, and Mississippi rivers and Lake Pepin),—in Collections State Historical Society of Wisconsin, Vol. 17, 1906, pp. 24-25.

Hennepin, Louis. A New Discovery of a Vast Country in America, 1679-1682, two parts, London, 1698, 355, 178 pp., (on Mississippi River and Lake Pepin, Vol. 1, pp. 180-182).

Hershey, O. H. The Physiographic Development of the Upper Mississippi Valley, Amer. Geol., Vol. 20, 1897, pp. 246-268.

Kay, G. F., and Apfel, E. T. The Pre-Illinoian Pleistocene Geology of Iowa, Iowa Geol. Survey, Vol. 34, 1929, pp. 24-32, 45-48, 154-157.

Keating, W. H. Narrative of an Expedition to the Source of St. Peters River, etc., Philadelphia, 1824, (on Mississippi River features), pp. 236-237, 265-266, 271-272, 278-280, 293-294.

Lees, J. H. Earth Movements and Drainage Lines in Iowa, Proc. Iowa Acad. Sci., Vol. 21, 1914, pp. 173-180.

Leonard, A. G. (On terraces) Geology of Clayton County, Iowa Geol. Survey, Vol. 16, 1906, pp. 287-289.

Leverett, Frank. The Preglacial Valleys of the Mississippi and its Tributaries, Journ. Geol., Vol. 3, 1895, pp. 740-763; The Lower Rapids of the Mississippi River, Ibid., Vol. 7, 1899, pp. 1-22.

Leverett, Frank, and Sardeson, F. W. Surface Formations and Agricultural Conditions of the South Half of Minnesota, Minnesota Geol. Survey, Bull. 14, 1914.

MacClintock, Paul. The Pleistocene History of the Lower Wisconsin River, Journ. Geol., Vol. 30, 1922, pp. 673-689.

McGee, W J   (On the gorge of the Mississippi), 11th Annual Rept., U. S. Geol. Survey, Part 1, 1891, pp. 367-372; (on Mississippi River terraces), Ibid, pp. 425-426.

Marquette, Jacques.   (On the Mississippi River), Jesuit Relations, 1673-77, Thwaites' edition, Vol. 59, Cleveland, 1900, p. 109; Hennepin's translation in "A New Discovery of a Vast Country in America," Part 1, London, 1698, pp. 325-327; see also Joliet's map, 1674, reproduced in Revue de Géographie, February, 1880, and in the Jesuit Relations, op. cit., Vol. 59, 1900, facing p. 86.

Martin, Lawrence.   Gravel Terraces of the Mississippi River in Wisconsin (abstract), Annals, Assoc. Amer. Geogr., Vol. 7, 1918, p. 79; The Gorge of the Upper Mississippi as a Rival of the Rhine Gorge, Bull., Geogr. Soc. Philadelphia, Vol. 14, 1916, pp. 127-147; Valley Lakes Due to Variation in Stream Load, Monograph 52, U. S. Geol. Survey, 1911, p. 438.

Merrick, G. B.   Old Times on the Upper Mississippi, Cleveland, 1909, 303 pp.

Nicollet, J. N.   Report Intended to Illustrate A Map of the Hydrographical Basin of the Upper Mississippi River, Senate Doc. 237, 26th Congress, 2nd session, Washington, 1843, 170 pp.

Pike, Z. M.   An Account of Expeditions to the Sources of the Mississippi, etc., during the Years 1805, 1806, and 1807, Philadelphia, 1810, 277 pp., (on Mississippi River in Wisconsin), pp. 10-23, 94-102, Appendix to Part I, pp. 2-4, 43-50.

Schoolcraft, H. R.   Narrative Journal of Travels, etc., Albany, 1821, (on Trempealeau Mountain), pp. 334-335; (on Lake Pepin), pp. 324, 327-331; (on gravel deposits), pp. 169, 170, 179.

Thwaites, F. T.   Pre-Wisconsin Terraces of the Driftless Area of Wisconsin, Bull. Geol. Soc. Amer., Vol. 39, 1928, pp. 621-641.

Trowbridge, A. C.   The Erosional History of the Driftless Area, Iowa Univ. Studies, Studies in Nat. Hist., Vol. 9, No. 3, 1921.

Warren, G. K.   Bridging the Mississippi River between St. Paul, Minn., and St. Louis, Mo., Senate Ex. Document 69, 45th Congress, 2nd Session, Washington, 1878; also published in Appendix X3 of the Report of the Chief of Engineers, 1878; Valley of the Minnesota River and of the Mississippi River to the Junction of the Ohio,—Its Origin Considered, Amer. Journ. Sci., 3rd series, Vol. 16, 1878, pp. 417-431.

Whitney, J. D.   (On Prairie du Chien terrace), Geol. Survey of Iowa, Vol. 1, 1858, pp. 15-16.

Winchell, N. H.   Alluvial Terraces of Houston Co., Fifth Annual Rept. Minn. Geol. and Nat. Hist. Survey, 1877, pp. 38-41; Geology of Minnesota, Vol. 1, 1884, pp. 227-230 (see also other Minnesota county reports, on Winona, Wabasha, Goodhue, Dakota, and Washington counties); Changes of Level in Lake Pepin, Geology of Minnesota, Vol. 2, 1888, pp. 3-6, 25-26.

For other literature see Chapter III and Appendix G.

MAPS

See end of Chapter III, p. 80.

CHAPTER VIII

THE RIVERS WITHIN THE WESTERN UPLAND

A BARRIER AND ITS PASSES

The Western Upland is a topographic barrier between the low plains of eastern Wisconsin and the Mississippi waterway. This barrier may be crossed with ease:

(1) in southwestern Wisconsin and northern Illinois, where it is low;

(2) along the lower Wisconsin River;

(3) along the Black River;

(4) along the Chippewa River;

(5) in St. Croix and Polk counties, where it is low.

The valley of the Lower Wisconsin River furnishes the best of these passes. It is the only natural highway in the 100 miles between the Black River and northern Illinois. Thus it is similar in position to the Cumberland Gap of the Appalachian Highland, though more like the gap of the Great Kanawha River in furnishing a water grade across the whole plateau. The latter is a narrow, steep-sided gorge, and hence the Indians used Cumberland Gap when they established the trail which came to be known as the Warriors Path. Daniel Boone laid out the famous Wilderness Road to Kentucky in 1775, crossing the Appalachian Barrier at Cumberland Gap. Nearly a century earlier, however, the Wisconsin River route across the Western Upland was utilized by Radisson and Groseilliers, followed by Marquette and Joliet, and many others.

The heavy travel of today does not follow the Wisconsin River pass, for it does not form a direct part of the route between Chicago and St. Paul. No railway with heavy traffic uses the second, third, or fourth passes listed above. The most important railway routes across the Western Upland barrier are (a) the Omaha Line through St. Croix County, and (b) the Chicago, Milwaukee, St. Paul and Pacific, and the Chicago and Northwestern along a sixth route which follows the La Crosse River after tunneling from the headwaters of the Baraboo River and the Lemonweir branch of the Upper Wisconsin. The route of the Northwestern Railway along the Military Ridge, the Green Bay and Western Railway along Trempealeau River, and the Chicago, Milwaukee, St. Paul and Pacific along the Wisconsin River and the Chippewa River also cross the Western Upland, but not with important trunk lines.

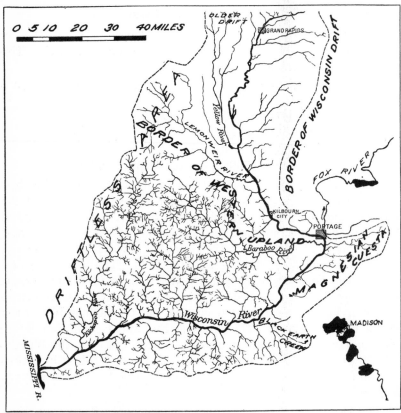

Fig. 58. The Wisconsin River and its tributaries in the Western Upland. In the Driftless Area the drainage pattern is highly dendritic, in contrast with the simpler drainage in the Central Plain between Wisconsin Dells (Kilbourn) and Wisconsin Rapids (Grand Rapids) near the border of the Wisconsin drift.

These passes across the Western Upland are of decided importance in our early history, and control all local travel today. All but one are directly determined by rivers within the Western Upland. The lower Wisconsin is the largest and most important of these rivers.

THE LOWER WISCONSIN RIVER

The Gorge of the Wisconsin. The valley of the Wisconsin River in the Western Upland is strikingly different from the middle and upper sections of the same river valley in the Central Plain (p. 353) and the Northern Highland (p. 418). It is a great gorge or trench, extending westward from the terminal moraine at Prairie du Sac to the Mississippi at Prairie du Chien.

This gorge is over 4 miles wide at Prairie du Sac, narrowing to 2 miles at Muscoda, 40 miles downstream to the west, and to half a mile at Bridgeport, 35 miles farther west near Prairie du Chien. The walls of the gorge rise abruptly 300 to 400 feet.

The river descends from 740 feet at Prairie du Sac to 615 feet near Prairie du Chien, or at the average rate of about $1\frac{3}{5}$ feet to the mile. Accordingly the current of the river is gentle. The water surface has flat reaches and steeper pitches, however, the detailed surveys showing grades as gentle as $\frac{9}{100}$ of a foot per mile, interrupted by descents of as much as $3\frac{7}{10}$ feet per mile. There are no genuine rapids in this part of the stream, for it lies in the Driftless Area. It furnished an ideal canoe route in early days. The river is shallow, however, so that canalization would have been necessary if heavy transportation had ever come this way. The valley is broad enough so that roads and railways were easily built; the roads are sandy and the extensive swamps rendered construction difficult in some places. Nevertheless the Wisconsin River route is not an important highway today. It resembles the Wilderness Road through Cumberland Gap as a geographical feature which was taken advantage of in early times and has subsequently suffered a period of decline.

A Neglected Highway. The lower Wisconsin was relatively as important in the seventeenth and eighteenth centuries as any valley route in the world. It was part of the diagonal highway from the Great Lakes to the Mississippi by way of Green Bay and the Fox River. The only rival canoe route to the Mississippi was that by the Desplaines and Illinois rivers. If one had been asked to pick out sites for great cities he might well have chosen: (a) Green Bay, on the Green Bay of Lake Michigan at the mouth of the Fox; (b) Portage, at the Fox-Wisconsin divide; (c) Prairie du Chien, on the Mississippi at the mouth of the Wisconsin; or (d) Chicago, on Lake Michigan, close to the Desplaines-Illinois divide; and (e) St. Louis, close to the junction of the Illinois, Mississippi, and Missouri. The Fox-Wisconsin route would seem to have had the advantage, for it is as natural a highway as the Mohawk Valley in New York. A Chicago might have grown up at Green Bay or a St. Louis at Prairie du Chien. The canal at Portage was completed at about the same time as the canal at Chicago

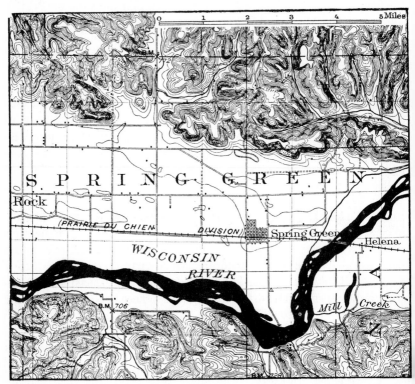

Fig. 59. The broad portion of the Wisconsin gorge cut chiefly in weak sandstone. Compare with Figure 60, which represents an area sixty-five miles downstream. Contour interval 20 feet. The Tower Hill State Park is located southeast of Spring Green on the south bank of the Wisconsin River. (From Richland Center Quadrangle. U. S. Geol. Survey.)

and the federal government spent half a million dollars in improving the lower Wisconsin River. The invention of the railway put an end to canal and river traffic. Neither Chicago nor St. Louis owes its present preeminence exclusively to the Great Lakes and the Mississippi. Nor could the natural highway of the Wisconsin, although traversed by the Milwaukee-Madison-Prairie du Chien railway route, save the lower Wisconsin from becoming a neglected highway. The Western Upland can be crossed farther north by the railways from Chicago to St. Paul-Minneapolis. In recent years, however, the Chicago, Milwaukee, St. Paul and Pacific Railway has developed and used its Madison-Prairie du Chien line much more than previously, for passenger traffic as well as for freight. Nevertheless the greatest and most promising natural

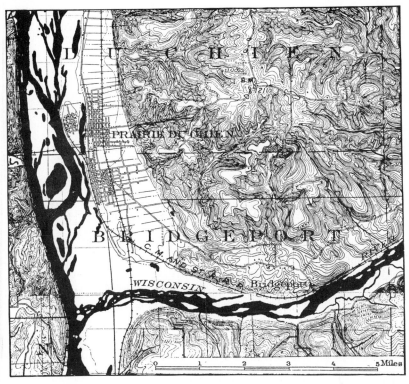

Fig. 60. The narrow portion of the Wisconsin gorge, cut chiefly in resistant limestone. The map also shows the rock terrace at Bridgeport. Contour interval 20 feet. (From Waukon and Elkader Quadrangles, U. S. Geol. Survey.)

highway in Wisconsin two centuries ago is by no means the busiest and most popular route of communication of today.

Wisconsin River Bottomland. The broad floor of the Wisconsin gorge has a floodplain, in the midst of which the shallow modern river flows over shifting sand bars. Some of these bars have been observed to move down stream as much as 800 feet in a year. Father Marquette said in 1673 that the "Meskousing" River "is very wide; it has a sandy bottom, which forms various shoals that render its navigation very difficult. It is full of Islands covered with vines. On the banks one sees fertile land, diversified with woods, prairies, and hills."

Above this floodplain rises a series of terraces made by the erosion of the valley train of outwash sand and gravel which the streams from the ice front at and east of Prairie du Sac built up during the Glacial Period. Upon those terraces are sand dunes, heaped up by the wind. Wells show that the filling of glacial sand and gravel has a depth of 125 to 150 feet.

MOUTH OF WISCONSIN RIVER
VIEWED FROM TOP OF BLUFFS ON THE MISSISSIPPI RIVER

Fig. 61. Sand bars in the Wisconsin River (Warren.)

Wisconsin River Bluffs. The steeply-rising bluffs are much like those of the Mississippi. The cliffs are likely to be near the top of the bluff where a resistant sandstone stratum outcrops beneath a capping of limestone. The lower slopes are gentler because they are masked by talus. They often contain great, angular blocks of limestone which have slid down from above, and deposits of residual clay which have crept down from the limestone or the loess above it. As on the Mississippi, the south-facing bluff is likely to be grassy. This was commented upon by Owen in 1847, before much timber had been cut away by man. Owen said:

"Bold exposures of rock, with a grassy bank beneath * * * are, for the most part, only on the south and western sides of the hills; the north and eastern declivities are more rounded and most generally overgrown with trees and shrubbery * * * It seems as if the alternate thawing and freezing on the sunny side has caused a more rapid decay of the rock, which scaling and slipping off, sometimes in large masses, slips down the side of the hill; this together with the rapid transition from heat to cold on the southern exposure, probably prevents trees from coming to maturity on that side."

Another important factor, in making south-facing bluffs grassy rather than covered with trees, is the retention of moisture on the north-facing slope, which does not receive the direct rays of the sun.

Rock Terrace at Bridgeport. Near the mouth of the Wisconsin there is a rock terrace on the northern side of the river. It is 100 to 160 feet above the stream channel north of Bridgeport. It has a width of half a mile to a mile and a length of about 6 miles. Thus the Wisconsin valley is here a double feature, a narrow gorge within a broader trench (Fig. 60). This terrace is not interpreted as indicating a halt in the down-cutting by the stream, but merely a difference in the resistance of the rocks in which the valley is cut. The broad upper portion of the valley near the top of the bluffs is in the weak St. Peter sandstone, in the less resistant portion of the lower Magnesian limestone, and in the Galena-Black River limestone, while the narrow inner gorge lies in the more resistant portion of the Lower Magnesian limestone. Farther to the north and east the dip of the rocks carries the Lower Magnesian to a higher level than that of the present stream and so the valley has been widened out in the underlying Cambrian sandstone which is weak. It grows wider upstream in a manner abnormal to streams in homogeneous rocks, but this is perfectly natural in view of the weakness of the Cambrian sandstone to the northeast. A thin limestone member of the Cambrian series sometimes forms narrow rock terraces at certain points east of Bridgeport.

The Wisconsin as a Superposed River Upon the Baraboo Range. The Wisconsin River formerly flowed across the Baraboo Range, as is shown by the abandoned water gaps at Devils Lake and the

Lower Narrows (p. 125). These gaps show, by their steepness and freshness, the recency of occupation by a large river. The way in which it crossed the high Baraboo ridges seems to indicate that it had this course at a time when the Baraboo quartzite was completely buried beneath the Cambrian sandstone and overlying formations up to the Niagara limestone. As the stream cut down and discovered the quartzite in its bed, it eroded into the top of the Baraboo Range and gradually produced the water gaps. Such a stream, let down upon underlying resistant beds, is termed a superposed river.

While the river was slowly cutting the water gaps in the quartzite, which are narrow because the quartzite is exceedingly resistant, the tributary streams were making wider valleys in the weak Cambrian sandstone. The whole process of opening out the interior valley of the Baraboo Range and making the adjacent portion of the lowland of the Central Plain to the north, south, and east took little if any longer than the cutting of the water gaps. Thus the exhuming of the buried quartzite and the restoration of the Baraboo Range to its former topographic prominence came, not before, but during the time that the Wisconsin was carving the Devils Lake water gap.

The Wisconsin as a Diverted Stream. One interesting possibility as to the still earlier history of the Wisconsin River is that it once flowed through Madison and Janesville (Fig. 62), though at a far higher level than the present stream. The Yahara, now a branch of the Rock River, may be the beheaded remnant of the Wisconsin. When the original Wisconsin first began to flow, after the region had been uplifted from the ocean, it acquired its course as a consequence of the southward dip of the sedimentary rocks. It would have been distinctly abnormal for the river to have turned abruptly to the west near Merrimac and flowed to the Mississippi nearly at right angles to its present course. The dip of the rocks made a topographic surface upon which a southward course past Madison was much more natural, especially as the axis of the arch lay west of Merrimac.

The diversion of the Wisconsin to its present westward course may, then, have been accomplished by headwater erosion on the part of the short stream along the lower Wisconsin valley, which we may call the Kickapoo-Wisconsin since it was originally a short eastern branch of the Kickapoo River, heading near Muscoda. The Kickapoo-Wisconsin extended its headwaters eastward, beheading the several streams which now rise on the Military Ridge. It finally tapped the south-flowing Yahara-Wisconsin, beheading it at a point just north of the Magnesian cuesta. This west-flowing Kickapoo-Wisconsin had an advantage over the south-flowing Yahara-Wisconsin (a) because the former was at a lower level on account of being tributary to the large Mississippi, and (b) because the Yahara-Wisconsin had to cut through the resistant Lower Magnesian limestone. Of course such a hypothetical diversion as this was long ago. Direct evidence of it, as for example quartzite pebbles in the Yahara valley, has been largely removed or concealed by the subsequent glaciation. The philosophical basis of this stream diversion is sound. The processes assumed are familiar ones. No one has suggested a more plausible explanation of the course of the Wisconsin, from Merrimac to Prairie du Chien, which is highly abnormal. Complete verification of this hypothetical explanation will doubtless be found in years to come.

TOWER HILL STATE PARK

This park is located in the Western Upland, on the south bank of the Wisconsin River, about three miles southeast of Spring Green. State Trunk Highway 11 parallels the park for a short distance. The rocks exposed in the park include 6 feet of Dresbach sandstone overlaid by 140 feet of Mazomanie sandstone. These soft sandstones facilitated the digging of the old shot tower which consists of a vertical shaft from the top of the bluff. The bottom of the shaft is connected by tunnel to the foot of the bluffs.

The park is of historic interest because of the part the shot tower played in the early development of southwestern Wisconsin.

THE BARABOO RIVER

Of the numerous tributaries of the Wisconsin in the Western Upland may be mentioned the Baraboo River, Dell Creek, west of Wisconsin Dells, Pine River near Richland Center, the Kickapoo River near Wauzeka, and Black Earth Creek.

The Baraboo River rises northwest of Elroy near Kendall and flows parallel to and inside the east-facing escarpment of the Western Upland to Reedsburg. At Ablemans, a short distance south of Reedsburg, it enters the Baraboo Range by a water gap in the

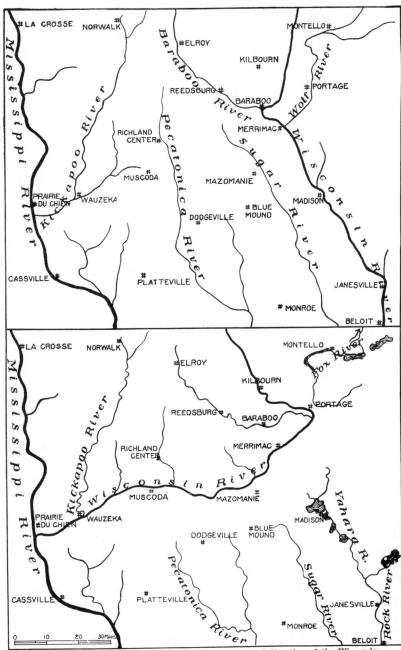

Fig. 62. Hypothetical representation of the capture and diversion of the Wisconsin
River by a branch of the Mississippi in preglacial time.

quartzite, flowing eastward through the interior valley past the city of Baraboo. It then turns northward and leaves the Baraboo Range by the Lower Narrows water gap, where the Wisconsin River formerly flowed in the opposite direction. It joins the present Wisconsin River in the Central Plain near Portage.

The present grade of the stream descends from over 1000 feet near Kendall to 960 feet at Elroy, 860 feet at Reedsburg, 840 feet at Baraboo, and 780 feet at the mouth near Portage, or at the rate of about 3½ feet to the mile.

Formerly the Baraboo was tributary to the Wisconsin near the city of Baraboo where both streams flowed at a level about 250 feet lower than at present (Fig. 35). The glacial diversion and filling have also resulted in the filling of the Baraboo valley upstream from the terminal moraine, as at Reedsburg in the Driftless Area where the river deposits are about 125 feet thick.

All of this course west of the city of Baraboo is in the Driftless Area and the pattern of the stream and its tributaries is tree-like or dendritic.

The interlocking ridges and valleys very well illustrate the character of the eastern portion of the upland. There is little limestone left on the ridges, and the skeleton pattern of topography in this geographical province is well developed.

The Chicago and Northwestern Railway has taken advantage of the Baraboo valley; and, as the valley trends in the right direction, one of the trunk lines across Wisconsin was built through the border of the Western Upland. Some of the grades between Reedsburg and Camp Douglas, and between Merrimac and Madison, are heavy. Much of the line is crooked, adding to the expense of double-tracking in a narrow valley. It would be expensive to double-track through the ridges, as in the tunnels north of Elroy and west of Kendall. All these factors have recently made it necessary to divert the faster trains and most of the freight traffic from this route through the Western Upland to a new line which traverses the Central Plain north of Portage and Camp Douglas, thus avoiding the rough highland.

DELL CREEK

East of the Baraboo River is a smaller stream called Dell Creek. It rises in the Western Upland and flows southeastward across the Central Plain, finally turning at right angles west of Mirror Lake to enter the Wisconsin in the Lower Dalles near Wisconsin

Dells. Thus it consists of a dendritic, preglacial, headwater portion in the Western Upland, where the valley slopes are gentle, and a more youthful, lower course, which is postglacial and contains rock-walled gorges (p. 351).

Although wholly in the Driftless Area, Dell Creek has been subject to glacial diversion. It appears probable that before the Glacial Period it entered the Baraboo Range by a water gap north of Baraboo, joining the Wisconsin River near the city of Baraboo. Thus the diversion at the right-angled turn in Dell Creek is due to the terminal moraine and the deposits of glacial outwash and lake sediment northwest of Baraboo.

CONTRAST OF PINE RIVER AND OTTER CREEK

The tributaries of the Wisconsin on the north are long and have gentle grades, while those on the south are short and slope more steeply. Thus Otter Creek of Iowa County, Blue Mound Creek, Fennimore Creek, and other streams flowing northward from the Military Ridge form a striking contrast with Pine River, Kickapoo River, Honey Creek, and other south-flowing tributaries of the Wisconsin. Otter Creek rises west of Dodgeville at an elevation of 1200 feet and descends 500 feet to the Wisconsin in 7 miles. Nearly opposite it is Pine River which rises at about the same elevation but is 33 miles long and, therefore, has not nearly so steep a grade. The south-flowing streams follow the dip of the rocks. The north-flowing streams flow against the dip on the Black River escarpment. All these tributaries of the lower Wisconsin have the dendritic, drainage pattern, lack of lakes, and accordant relationship of main and side streams that is typical of the Driftless Area.

BLACK EARTH CREEK

One of the southern tributaries of the Wisconsin River is notable because it furnishes an important part of the low-grade railway route across the state. This is Black Earth Creek, which rises a short distance west of Madison and enters the Wisconsin valley at Mazomanie. The divide is only 80 feet above Madison and 150 feet above Mazomanie. This low pass is crossed by the Chicago, Milwaukee, St. Paul and Pacific Railway which follows the Wisconsin valley to Prairie du Chien. It was the first railway built across Wisconsin from Lake Michigan to the Mississippi River, being completed from Madison to Prairie du Chien in 1857.

A. DRIFTLESS AREA TOPOGRAPHY WITHIN ONE OF THE CUESTA AREAS IN WESTERN WISCONSIN. NEAR BLACK EARTH, DANE COUNTY.

B. AERIAL VIEW OF THE CONFLUENCE OF THE WISCONSIN WITH THE MISSISSIPPI.

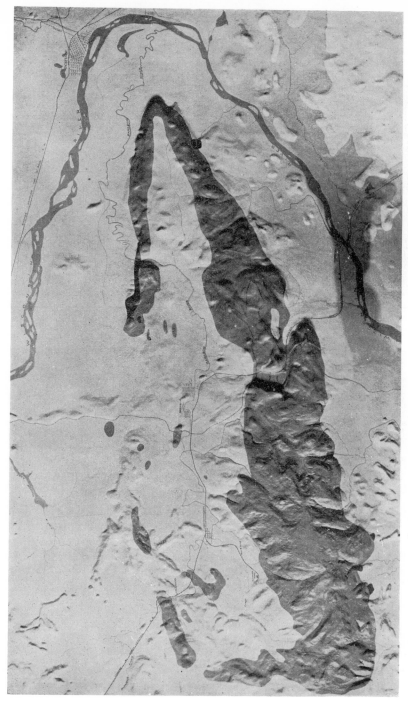

MODEL OF THE BARABOO RANGE, SHOWING THE PRESENT COURSE OF THE WISCONSIN RIVER AND ITS ABANDONED

Two isolated hills stand in the middle of the Black Earth valley near its head. They seem to be related to the topography developed before the St. Peter sandstone was deposited as well as to that of the present. The present divide is several miles east of the pre-glacial divide, and the low col has been produced by the coming together of streams which were eating back by headwater erosion. The outlet of extinct Lake Middleton, just west of Lake Mendota, was to the westward down the Black Earth valley, for the Yahara valley at Madison was blocked by the ice sheet during the existence of this lake.

THE KICKAPOO RIVER

The largest and simplest Driftless Area stream in Wisconsin is the Kickapoo River. It rises in Monroe County near Summit and Wilton, entering the Wisconsin about 65 miles to the south at Wauzeka. Its grade from 992 feet at Wilton to 644 feet at Wauzeka is about 5 feet to the mile, but at the headwaters the grade is steeper. The dendritic pattern of this river valley is characteristic of a valley in flat-lying rocks in a region never glaciated. The cross-section of its headwaters along the line of the Chicago and Northwestern Railway shows the general character of the Western Upland, which is here a hilly region with narrow ridges rather than a sloping highland as in the Niagara cuesta of eastern Wisconsin. The railway, which follows the Baraboo valley from Reedsburg to Elroy (p. 193), plunges into a tunnel just west of Kendall, emerging in the eastern headwaters of the Kickapoo River. It crosses a spur west of Wilton by a second tunnel, traverses part of the western headwaters of the Kickapoo, and leaves the valley by a third tunnel west of Summit (Fig. 16). The valley of the Kickapoo, near Steuben, contains the most perfect intrenched meanders in Wisconsin. These meanders, like the ones near Benton (Fig. 63), are thought to be related to causes other than ancient peneplanation. The necks of the oxbows are cut through so that isolated hills stand within semicircular valleys.

VALLEYS SOUTH OF THE MILITARY RIDGE

The valleys of the Western Upland in southwestern Wisconsin are narrower than those north of the Wisconsin because they are mainly cut in resistant limestone rather than weak sandstone (p. 52). The general courses of the Platte and Grant rivers have already been described (pp. 68, 72). It has been suggested that the courses of some of these streams, as well as those north of the Wisconsin River, have been notably influenced by joint planes. Some waterpower was developed in southwestern Wisconsin in the days of grist mills, though less than would be possible if the region were outside the Driftless Area and had rapids and waterfalls. One of the most interesting, modern developments is that near Benton, where a fall is developed by diverting the Fever River through a tunnel in the rocky neck between two great curves in the river at Horseshoe Bend (Fig. 63).

The Pecatonica and Sugar rivers will be described fully in Chapter XI, where they are compared and contrasted with the streams in the glaciated area of eastern Wisconsin.

THE LA CROSSE RIVER

The tributaries of the Mississippi north of the Wisconsin include the La Crosse, Black, Trempealeau, Buffalo, Chippewa, St. Croix, and numerous smaller streams. As the La Crosse River lies entirely in the Driftless Area and as its headwater region has been mapped in detail (Sparta and Wilton Quadrangles), it will be described somewhat more fully than the size and importance of the stream would otherwise warrant. The valley of the La Crosse is important, however, in furnishing the route of the main line of the Chicago, Milwaukee, St. Paul and Pacific Railway across the Western Upland from the Central Plain to the Mississippi. It is also followed by the main line of the Chicago and Northwestern Railway between Chicago and the Black Hills of South Dakota.

The La Crosse River rises close to the escarpment at the edge of the Western Upland, the divide being narrow enough to be pierced by railway tunnels only a quarter mile long. It has a wide valley opened out in the weak Cambrian sandstone, and the stream may be said to be in a stage of late youth or early maturity of its erosion cycle. It rises at an elevation of about 1000 feet, having a steeper grade in the headwaters than near the mouth. In the headwaters, above Sparta, it descends at the rate of $13\frac{1}{3}$ feet to the mile. From Sparta to the Mississippi at La Crosse it descends at the rate of only $5\frac{1}{2}$ feet to the mile. Having passed the stage of early youth and being in the Driftless Area, it has no rapids or waterfalls or lakes in its course. Following the period of valley filling, the streams meandered, and in some cases erosion has exposed rock ledges. Trout Falls north of Sparta is due to this cause.

The valley of the La Crosse River has been partly filled, the depth of the alluvial material at West Salem and Sparta being 112 and 60 feet respectively.

The valley is rather wide in the headwater region, where the Western Upland attains less height than to the west near the Mississippi. In the middle course, near Sparta and Bangor, the valley bottom is one to two miles wide. The walls of the valley rise to heights of 200 to 500 feet within short distances of the river. The tributaries of La Crosse River come in at the grade of this main stream, as should be expected in a never-glaciated region. These tributaries have cut up the portion of the Western Upland traversed by the La Crosse till little of the upland remains. It flows through a region of rugged hills rather than a plateau. The ruggedness is systematic in relation to ridges and spurs that interlock with valleys and their tributaries. The tops of some of the hills, capped by the Lower Magnesian limestone and by resistant layers of the Cambrian sandstone, are croned by castellated peaks and battlemented ridges.

THE BLACK RIVER

The youthful, steep-sided valley of the Black River in the Western Upland is strikingly different from that of the same stream in the Central Plain (p. 360). In the upland its valley is similar to that of the La Crosse in most respects. It differs from it chiefly because the Black rises outside the Western Upland and flows completely across it. This results in a flatter gradient in the Black River than in the La Crosse. It also results in a more thorough dissection of the portion of the upland traversed by the river. Moreover, the Black River rises in the glaciated portion of Wisconsin. It carried vast floods from the melting ice and transported great quantities of sediment across this part of the Driftless Area during the Glacial Period. There are terraces along the lower Black River.

The Black River does not enter the Mississippi River as soon as it reaches the Mississippi valley, but flows southward nearly 15 miles on the Mississippi floodplain (Figs. 45, 64). At the city of La Crosse the Black and La Crosse rivers enter the master stream independently within a few yards of each other.

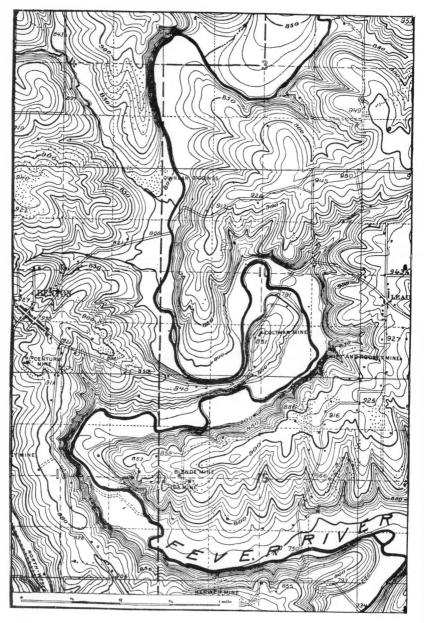

Fig. 63. The great bend of the Fever River near Benton. It may be an intrenched meander.

THE TREMPEALEAU RIVER

This stream is like the La Crosse in having its drainage basin entirely within the Western Upland. Its headwaters are close to the Central Plain, however, and the divide is low, so that a trans-Wisconsin railway, the Green Bay and Western, on its way from the Mississippi to Lake Michigan follows the easy grades of Trem-

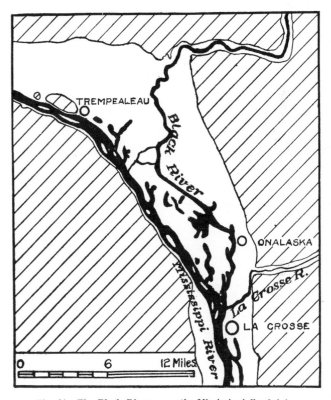

Fig. 64. The Black River upon the Mississippi floodplain.

pealeau River. As in all the valleys of the Western Upland, especially in the Driftless Area, this leads it through a hilly country of charming scenery.

THE BUFFALO RIVER

This stream belongs entirely to the Western Upland. It differs from the La Crosse and Trempealeau rivers, however, in two important respects. Its headwaters lie in the region of older drift. Buffalo River therefore carried glacial floods across the Driftless

Area during one of the earlier stages of the Glacial Period, as witnessed by glacial gravels. Secondly, it has been involved in a process of stream diversion. A large stream, formerly tributary to the Chippewa, now forms the headwaters of the Buffalo. By

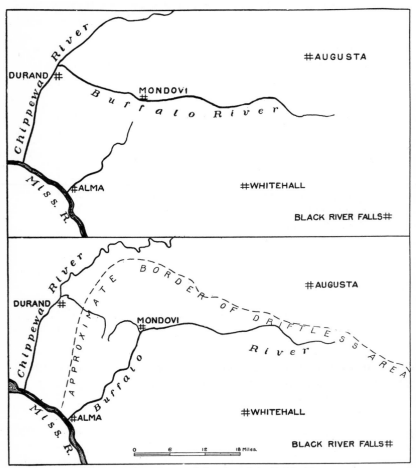

Fig. 65. The capture and diversion of the Buffalo River.

this diversion the Buffalo River more than doubled its length and greatly increased its volume.

The diversion took place at Mondovi in Buffalo County where Buffalo River now turns abruptly from a westward course in a broad valley and flows south and southwest in a narrower valley to the Mississippi. The beheaded stream to the west of Mondovi is called Bear Creek. It enters the Chippewa near Durand.

Whether this stream diversion was a result of blocking of the mouth of Bear Creek with glacial drift and outwash, or is due to headwater erosion in preglacial time, has not been established. The point of capture lies in the Driftless Area, and the older drift crosses the beheaded valley a short distance to the west, making the glacial diversion theory quite plausible. On the other hand, the mouth of the Buffalo River is lower than that of Bear Creek and the Chippewa, so that the headwater erosion theory might equally well furnish the explanation, especially as the Buffalo River valley at and just below the point of capture is rather broad for a postglacial feature. The Kickapoo River seems to have long ago made just such a diversion of the upper Wisconsin (p. 190). It is interesting to note that the first geological map of the United States, published in 1809, shows the Buffalo River as a tributary of the Chippewa. This involves an early recognition of the unity of this valley, but should not be interpreted as evidence that the diversion took place only a century ago.

The Chippewa River

What has already been said of the Black River in contrast with the La Crosse (p. 197) applies with equal force to the Chippewa. The course of the Chippewa River across the Western Upland differs from that of the Black in one salient respect. It lies wholly in glaciated territory. Accordingly, the valley is wider and the valley walls are less precipitous. It lies in a region of the older drift, however, and there are no rapids, waterfalls, or lakes in this portion of the stream course. The grade of 2.3 feet per mile is nearly as gentle as that of the Black River.

The most striking feature of the lower Chippewa River is the great accumulation at its mouth. This detritus, supplied by glacial floods during the latest as well as the earlier glaciations, was greater in amount than could be carried away by the Mississippi. Ordinarily a tributary is unable to bring more material than can be carried along by the master stream, but the Chippewa is exceptional in this respect among all the affluents of the Mississippi. Accordingly it has been able to force the Mississippi over to the western edge of its floodplain, damming back the master stream and forming Lake Pepin (p. 169).

The Chippewa also has a striking system of gravel terraces (p. 159). Its course in the Central Plain (p. 360) and in the Northern Highland (p. 413) is much steeper than the lower Chippewa.

THE ST. CROIX RIVER

The stream which forms the northwestern boundary of the state has most of its course in the Northern Highland and Central Plain (pp. 361, 413). Within the Western Upland it is similar in general features to the Black (p. 197) and Wisconsin (p. 185), but it more nearly resembles the Chippewa. The valley of the St. Croix is less rugged than that of the Black because of glacial erosion and glacial deposition. Part of its present course lies in the older drift and part in the latest or Wisconsin drift.

Before the Glacial Period the upper St. Croix had a course to the west in Minnesota. Its middle course in the St. Croix Dalles (p. 364) is postglacial. Before the Glacial Period its lower course was occupied by a short stream, whose headwaters were the Apple River tributary of today. The St. Croix River was the outlet of two of the glacial lakes in the Lake Superior basin (Fig. 185). North of Stillwater, Minn., the valley is fairly wide, with gently-sloping, terraced sides. South of Stillwater the valley is more deeply set in the upland, as to the southwest of River Falls and at Hudson, where the Chicago, St. Paul, Minneapolis, and Omaha Railway—the Omaha—crosses it on the main line from Chicago to St. Paul. It has a system of gravel terraces, increasing in number upstream (p. 366).

One tributary of the St. Croix, the Kinnikinnic, is notable because it rises on the Magnesian cuesta and flows completely across the higher Black River cuesta in a postglacial rock gorge.

The St. Croix River and the Bois Brule River furnish the route for the proposed Lake Superior and Mississippi Canal, which as government engineers have estimated, would cost between seven and eighteen million dollars.

LAKE ST. CROIX

The lower St. Croix River is dammed back by the Mississippi River deposits, forming a long narrow body of water similar to Lake Pepin, except that it is in a tributary instead of the main valley.

Lake St. Croix extends northward from Prescott to Stillwater, Minn. It is 23 miles long, and varies from a quarter mile to a mile and a half in width. The depth of water is 30 to 50 feet, shoaling to $3\frac{1}{2}$ feet on certain sand bars. The range of level in the water surface is 16 to 22 feet, depending upon floods in the Mississippi and upper St. Croix rivers. It is navigable by steamboats from Prescott to Stillwater, and even to Osceola, though there is little steamboat transportation now.

As is shown in Figure 66 the southern end of the lake at Prescott is separated from the Mississippi by a sand bar. The shores of the lake are steep, rock bluffs rising 100 to 135 feet within a short distance of the water. The slopes below lake level are even steeper, and the lake basin has a broad, flat floor with water over 40 feet deep.

Kinnikinnic River has built a delta into Lake St. Croix. Opposite Afton, Minn., the Catfish Bar extends nearly across the lake, as does the Willow River Bar at Hudson. The depth of water in the lake shows that the valley train of outwash gravel and sand in the Mississippi was built up at least 40 feet after Lake St. Croix was formed.

The scenery at Lake St. Croix was well described by Schoolcraft in 1832.

"The River St. Croix has one peculiarity, to distinguish it from all other American rivers. It has its source and its termination in a lake, and each of these bears the same name with itself. The lake at its mouth is not less than thirty miles in length, and is, probably, nowhere, much over a mile wide. Its banks are high and afford a series of picturesque views, which keep the eye constantly on the stretch. The country is an upland prairie interspersed with groves and majestic eminences. The waters are beautifully transparent, and the margin exhibits a pebbly beach, so cleanly washed that it would scarcely afford earth enough to stain the fairest shoe. If "Loch Katrine" presents a more attractive outline of sylvan coast, it must be beautiful indeed. We went up it, turning point after point, with the pleasure that novelty imparts, aided by the chanting of our canoemen. We were in hourly expectation of reaching its head for our night encampment; but we saw the sun set, casting its golden hues and its deep shadows over the water, and going down in a gorgeous amphitheatre of fleecy

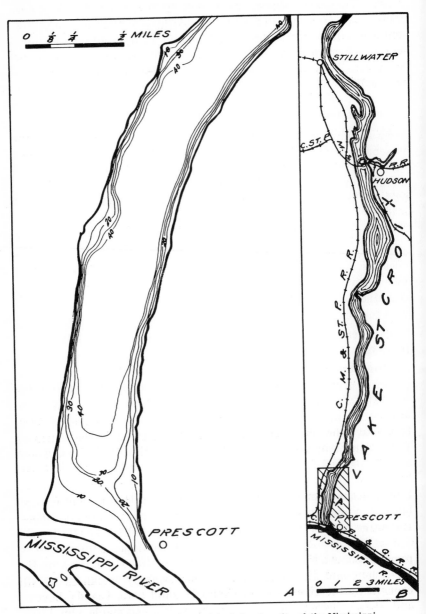

Fig. 66. Lake St. Croix, dammed back by deposits of the Mississippi.

clouds. The moon almost imperceptibly shone out, to supply its place, creating a scene of moonlight stillness. Nothing could present a greater contrast to the noisy scene of horses and horsemen, war and bloodshed, which, we were then unconscious, was about being acted, so near to us. We allude to the pursuit and destruction of the Black Hawk's army."

Coulees Along the Mississippi

The smaller tributaries of the Mississippi are similar to the La Crosse River. Many of the valleys of these streams are called coulees (p. 48). They are youthful valleys, especially well seen to the south of Alma and Fountain City, where there has been no glaciation. The grade of one of the largest of these streams is gentle enough so that it is followed by the La Crosse and Southeastern Railway between Stoddard and Viroqua.

Another coulee has had its lower course diverted by the Mississippi. This is near Wyalusing, south of the Wisconsin River at one of the new state parks.

Nelson Dewey State Park

The Nelson Dewey State Park is situated at the junction of the Mississippi and Wisconsin rivers, just south of Prairie du Chien. Its area is 1650 acres. The establishment of this park was one of the wisest acts in the history of the state.

The physiographic features of the park are notable. It contains the termination of the Military Ridge in Wisconsin. This is a hill more than 500 feet above the adjacent bottomlands, or 1180 feet above the sea level. The park has $2\frac{1}{2}$ miles of the bluffs of Wisconsin River and 3 miles of the bluffs of the Mississippi. These are 400 to 565 feet high. They expose the Galena-Black River limestone above, the St. Peter sandstone in the middle, and the Lower Magnesian limestone at the base. Weathering and erosion have fashioned these bluffs into a pleasing variety of cliffs, slopes, and crags. These are especially well preserved because this is part of the Driftless Area (p. 81). Some of the upper cliffs are sheer precipices. At the base of the bluffs are the floodplains of the Wisconsin and the Mississippi,—flat, swampy expanses of silt brought by the rivers,—with channels, sand bars, islands, bayous, and lakes. There is oak forest on the bluffs, and maple, elm, birch, and willow on the drier parts of the floodplain.

The coulee which forms the southeastern boundary of the park, as constituted in 1915, has had its stream diverted from a former course by the Mississippi River. It now ends nearly half a mile north of the village of Wyalusing. Formerly it terminated half a mile south of the village. The rock hill at Wyalusing is one side of its abandoned valley and the high bluff is the other. The

Fig. 67. Map of Nelson Dewey State Park.

coulee has been shortened nearly a mile by the process called stream diversion, for the Mississippi has diverted the creek by cutting into the side of its valley. The abandoned valley furnishes the site for Wyalusing.

Nelson Dewey Park is conveniently reached from Prairie du Chien, Wyalusing, or Bridgeport. Visitors and campers will add to their enjoyment if they provide themselves in advance with copies of Charts 165 and 166 of the Mississippi River Commission map, or with the Elkader Quadrangle of the U. S. Geological Survey map (for prices see Appendix E).

Nothing adequate can be said here of the historical associations and scenic delights of this state park, with its broad views over the Wisconsin and Mississippi rivers, where Marquette and Jolliet came more than 250 years ago.

BIBLIOGRAPHY

Allen, C. J. Survey of Mississippi, St. Croix, Chippewa, and Wisconsin Rivers, House Ex. Document 39, 46th Congress, 2nd Session, Washington, 1880, 108 pp; The Improvement of the St. Croix River, House Ex. Doc. 40, Ibid., 1880, 8 pp.

Blanchard, W. O. The Geography of Southwestern Wisconsin, Bull. 65, Wis. Geol. and Nat. Hist. Survey, 1924, 105 pp.

Harder, E. C. The Joint System in the Rocks of Southwestern Wisconsin and its Relation to the Drainage Network, Bull. 138, University of Wisconsin, 1906, pp. 207-246.

Hennepin, Louis. A New Discovery of a Vast Country in America, 1679-1682, London, 1698, (on Wisconsin and Fox Rivers, Part 1, pp. 255-260).

Hobbs, W. H. Examples of Joint Controlled Drainage from Wisconsin and New York, Journ. Geol., Vol. 13, 1905, pp. 363-374.

Irving, R. D. (On Devils Lake Drainage), Geology of Wisconsin, Vol. 2, 1878, pp. 507-509; (on lower Wisconsin River), Ibid., pp. 420-421.

Johns, R. B. The Physiography and Geology of the La Crosse River Valley, Unpublished thesis, University of Wisconsin, 1900.

Kümmel, H. B. Some Meandering Rivers of Wisconsin, Science, new series, Vol. 1, 1895, pp. 714-716.

La Salle, Robert Cavelier, sieur de. Description of Wisconsin Rivers, 1682, —in Collections Wis. Hist. Soc., Vol. 16, 1902, pp. 105-107.

MacClintock, Paul. The Pleistocene History of the Lower Wisconsin River, Journ. Geology, Vol. 30, 1922, pp. 673-689.

Marquette, Jacques. Jesuit Relations, 1673-77, Thwaites edition, Vol. 59, Cleveland, 1900, (on Wisconsin River, p. 107); Hennepin's translation in "A New Discovery of a Vast Country in America," Part I, London, 1698, pp. 325-327.

Martin, Lawrence, Williams, F. E., and Bean, E. F. Baraboo Range and Devils Lake, A Manual of Physical Geography Excursions, Madison, 1913, pp. 149-169.

Salisbury, R. D. Preglacial Gravels on the Quartzite Range near Baraboo, Wis., Journ. Geol., Vol. 3, 1895, pp. 655-667.

Salisbury, R. D., and Atwood, W. W.   The Geography of the Region about Devils Lake and the Dalles of the Wisconsin, Bull. 5, Wis. Geol. and Nat. Hist. Survey, 1900, pp. 61-72, 128-142.

Sears, C. B.   Lake Superior and Mississippi Canal, House Document 330, 54th Congress, 1st Session, Washington, 1896, 65 pp.

Strong, Moses.   (On streams in the lead region), Geology of Wisconsin, Vol. 2, 1877, pp. 652-658; (on the history of the Wisconsin and Mississippi Rivers), Ibid., Vol. 4, 1882, pp. 92-95.

Thwaites, F. T.   Geology of the Southeast Quarter of the Cross Plains Quadrangle, Dane County, Wisconsin, Unpublished thesis, University of Wisconsin, 1908; Pre-Wisconsin Terraces of the Driftless Area of Wisconsin, Bull., Geol. Soc. America, Vol. 39, 1928, pp. 621-642.

Thwaites, R. G.   Down Historic Waterways, Chicago, 1890, 1902, pp. 237-293.

Trowbridge, A. C.   The Erosional History of the Driftless Area, Iowa Univ. Studies, Studies in Nat. Hist., Vol. 9, No. 3, 1921; The History of Devils Lake, Wisconsin, Journ. Geology, Vol. 25, 1917, pp. 344-372.

Upham, Warren.   The St. Croix River Before, During, and After the Ice Age, Report of the Commissioner of the State Park of the Dalles, 1896, pp. 45-58; Time of Erosion of the Upper Mississippi, Minnesota, and St. Croix Valleys, Science, new series, Vol. 8, 1898, p. 470; Pleistocene Ice and River Erosion in the Saint Croix Valley of Minnesota and Wisconsin, Bull. Geol. Soc. Amer., Vol. 12, 1901, pp. 13-24.

Warren, G. K.   Report on the Transportation Route along the Wisconsin and Fox Rivers in the State of Wisconsin, Senate Ex. Document 28, 44th Congress, 1st Session, Washington, 1876, 114 pp., also published as Appendix T, Part 2, Report of Chief of Engineers, U. S. Army for 1876, with atlas of maps.

Weidman, S.   The Baraboo Iron-Bearing District of Wisconsin, Bull. 13, Wis. Geol. and Nat. Hist. Survey, 1904, pp. 101-102.

Wooster, L. C.   (On the St. Croix drainage), Geology of Wisconsin, Vol. 4, 1882, pp. 131-138.

MAPS

See end of Chapter III, p. 80.

CHAPTER IX

THE EASTERN RIDGES AND LOWLANDS

The Most Densely Settled Portion of the State

A map showing density of population in Wisconsin (Fig. 68) makes it clear that eastern Wisconsin contains a large proportion of the people of the state. Only two counties outside this area have a density of population exceeding 50 to the square mile. Six of our eight largest cities are in the Eastern Ridges and Lowlands. These are Milwaukee, Racine, Oshkosh, Sheboygan, Madison and Green Bay. The reasons for this are not simple. There is the length of time since settlement. It is nearly 300 years since the first white man came, and he established his earliest villages and farms in eastern Wisconsin. There is the proximity to Lake Michigan and to Illinois. There is the intelligence and industry of the German population in this district. Their prosperity may have induced others to settle here rather than farther west or north. Finally, there are (a) level topography, (b) fertile soil, and (c) favorable climate. These last three are factors of prime importance. Thus the physical geography of eastern Wisconsin assumes an importance that makes it of interest to study and explain the origin of the surface features in this geographical province.

The topographic features to be described are distinct, but they are low. The escarpments are the only hilly places. The dominant thing in eastern Wisconsin is the plain.

The General Geography of the Province

The geographical province of The Eastern Ridges and Lowlands (Fig. 10) occupies all of the state between Lake Michigan and the Western Upland and Central Plain. On the west the boundary of this district follows the contact of the Cambrian sandstone with the Lower Magnesian limestone from the Menominee River in Marinette County to the Wisconsin River in Sauk and Columbia

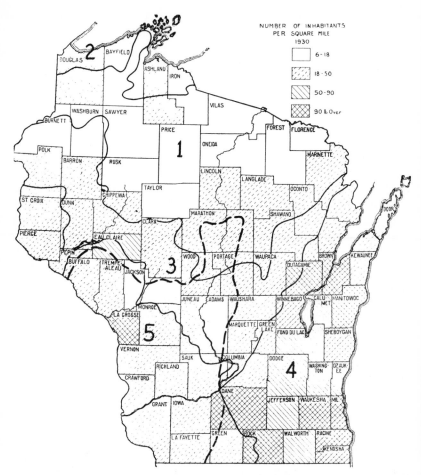

Fig. 68. The densely-settled eastern portion of Wisconsin, where nearly every county has more than 50 people to the square mile and eleven counties exceed 90 to the square mile. The boundaries of the geographical provinces and of the Driftless Area are also shown; 1—Northern Highland; 2—Lake Superior Lowland; 3—Central Plain; 4—Eastern Ridges and Lowlands; 5—Western Upland; Driftless Area delimited by dashed line. (After U. S. Census.)

counties near Prairie du Sac. From this point southward to Illinois the boundary selected is (a) the terminal moraine at the edge of the latest drift sheet and (b) the Rock River below Janesville. The province includes about 13,500 square miles, exclusive of the portions submerged beneath the waters of Green Bay and Lake Michigan in Wisconsin, which cover 7500 square miles in addition.

The general topographic features of the province are

(a) the ridges of Lower Magnesian and Black River limestone, on the west;

(b) the lowland of Galena-Black River limestone, containing Green Bay, Lake Winnebago, and the Rock River;

(c) the upland of Niagara limestone, adjacent to Lake Michigan;

(d) the lowland of Devonian shale in the valley now occupied by Lake Michigan.

Fig. 69. Cross-section of the cuestas of the Eastern Ridges and Lowlands and the rocks which underlie them. The rock formations named "Trenton limestone" and "Cincinnati or Hudson River shale" in this cross-section are now called Black River limestone and Richmond shale respectively.

An east-west section of these topographic features is shown in Figure 69. The ridges and lowlands extend north-northeast and south-southwest in four parallel belts.

The crest of the western limestone ridge is 700 to 1100 feet above sea level. The axis of the Green Bay-Lake Winnebago-Rock River lowland lies 50 to 250 feet lower. The crest of the Niagara upland is 740 to 1200 feet above the sea. The axis of the lowland occupied by Lake Michigan is over 1400 feet lower than the crest of the Niagara upland, or from 80 feet above to 323 feet below sea level. Both the Niagara upland and the Lower Magnesian limestone ridge have a crest some distance west of their middle lines, with a steep, west-facing escarpment and a long, gentle, eastward slope.

The Topography is Controlled by Cuestas

It has already been explained (p. 36) that alternate weak and resistant rock layers having a moderate inclination will be carved by streams and the weather into a belted plain. This plain will have parallel strips of upland and lowland corresponding respectively to the more important resistant and weak strata, as shown in Figure 69. The uplands are called cuestas and the lowlands have sometimes been called vales. A cuesta is a ridge which has a steep escarpment on one side and a long gentle slope on the other.

It is obvious that the topography of the eastern ridges and lowlands of Wisconsin is controlled by cuestas. The westernmost ridge is the rather low, narrow cuesta formed by the resistant Lower Magnesian limestone. It will be alluded to hereafter as the Magnesian cuesta. The eastern upland is the higher and broader cuesta of Niagara limestone. The intermediate Green Bay-Lake Winnebago-Rock River lowland lies upon the belt of Black River and Galena limestone, with the gentle back slope of the Magnesian cuesta for one wall and the steep escarpment of the Niagara cuesta for the other. This is the second lowland in order outward from the oldland of the exhumed Northern Highland peneplain (p. 36). The first is the inner lowland of the Central Plain of Cambrian sandstone. The third lowland, east of the Niagara cuesta, forming the basin of Lake Michigan, lies in the belt of weak Devonian shales above the Niagara limestone. The basin of Lake Michigan seems to show a series of minor cuestas along the cross-section from Port Washington, Wis., to Muskegon, Mich., but this is not revealed in other cross-sections of the basin. East of Lake Michigan the drift obscures the underlying rock completely. The Lake Michigan lowland, half of which lies in the state of Wisconsin, owes its abnormal depth, as compared with the other lowlands, chiefly to glacial erosion rather than weathering and stream work, while the two cuestas and their intermediate lowland in eastern Wisconsin, though also modified by glaciation, are normal products of weathering and stream work in a belt of cuesta-making rocks.

The Richmond shale and the St. Peter sandstone, whose outcrops form narrow strips at the bases of the Niagara and Black River limestones, respectively, have been much less important factors in the topographic situation, except as they help the resistant limestones above to stand out in escarpments.

THE MAGNESIAN CUESTA

Topography. The cuesta of Lower Magnesian limestone varies in elevation above sea level. The following table shows the elevation of its crest in various parts of eastern Wisconsin.

A small part of this elevation may be due to the covering of glacial drift. In all cases it is thought to represent the altitude of the crest of this limestone within 20 to 40 feet but not all of it has been mapped topographically. Some of the elevations listed below may be lower than normal because of being located in stream valleys.

TABLE SHOWING ELEVATION OF MAGNESIAN CUESTA

Locality	County	Elevation in feet above sea level
Pound	Marinette	724
Lena	Oconto	716
Hortonville Junction	Outagamie	800
Winneconne	Winnebago	754
Rush Lake Junction	Winnebago	846
Green Lake	Green Lake	805
Markesan	Green Lake	853
Rio	Columbia	928
Poynette-Arlington	Columbia	1100
Dane	Dane	1160
Lutheran Hill	Dane	1240

Variations from North to South. This limestone cuesta varies greatly in altitude and in width. The altitudes given in the preceding table show local variations in altitude and also a general increase in height from northeast to southwest. The Columbia County portion is about 500 feet higher than that in Marinette County. This seems to be related chiefly to the warping of the peneplain upon which the limestone and underlying sandstone rest, but perhaps also in part to greater glacial erosion at the northeast.

In parts of Marinette, Shawano, Outagamie, Winnebago, Green Lake, and Columbia counties, the width of the cuesta is only two to seven miles, in contrast with ten to twenty miles north of Madison. The reason for these variations in width is, in some cases, the relation of the erosion surface to the dip. The broader portion of the cuesta is in the area of flatter structure to the south. The

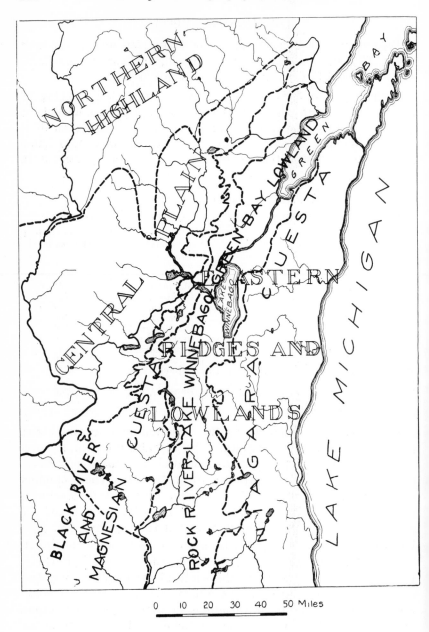

Fig. 70. The Eastern Ridges and Lowlands, made up of Magnesian cuesta, Green Bay-Lake Winnebago—Rock River lowland, Niagara cuesta, and Lake Michigan lowland.

maximum variation in thickness of the limestone from north to south is only a moderate number of feet. This is quite inadequate to account for the variations in width of the cuesta without the difference in dip, which amounts to several degrees. Another factor entering into the width of the Magnesian limestone cuesta is the resistant character of portions of the overlying Galena-Black River limestone. Its hardness causes it to resist weathering and erosion. This enables it to retain an elevated position in a subordinate cuesta on the back slope of the Magnesian limestone cuesta. This reduces the width of the belt of Lower Magnesian limestone.

The narrowness of the Magnesian cuesta in comparison with the inner lowland of Cambrian sandstone, which is two to five times as wide, is determined partly by the weakness of the latter, but chiefly by the thickness. The Cambrian sandstone is 500 to 800 feet in thickness, and the Lower Magnesian limestone 250 feet or less.

The Magnesian Escarpment. A west and northwest-facing escarpment terminates the Magnesian cuesta. From its crest one overlooks the lowland of the Central Plain. The escarpment in eastern Wisconsin is 175 miles long. It is 300 feet high in Dane and Columbia counties. A good place to see the high portion of the escarpment is along the line of the Chicago and Northwestern Railway between Dane and Lodi, and along the Chicago, Milwaukee, St. Paul, and Pacific Railway between Arlington and Poynette. The face of the escarpment descends abruptly. The easiest possible grade for the latter railway, even in a stream valley cutting the escarpment face, is over 80 feet to the mile.

The larger part of the escarpment, however, is much lower, especially where the rocks dip steeply in northeastern Wisconsin. In Marinette County, for example, the escarpment is only a little over fifty feet high. Here the Chicago, Milwaukee, St. Paul and Pacific Railway descends from 724 feet at Pound on the cuesta to 672 feet at Beaver on the inner lowland, with a grade of less than 14 feet to the mile. The lowness of the escarpment here, however, is due to (a) glacial erosion of its crest, (b) glacial filling of the lowland it overlooks. Both of these subtract from its height. This is certainly the case in Outagamie County, between Black Creek and Seymour on the Green Bay and Western Railway. There the whole Magnesian cuesta is represented by a low rock cut at the crest of the escarpment, west of which the inner lowland is a great swampy flat, with deep glacial filling.

The escarpment is unusually simple in outline, although here and there its front projects in great salients. In Shawano County a triangular area projects seven miles. There are similar salients in southwestern Outagamie County and in Winnebago County south of Lake Poygan. The few reentrants are complementary to these, as in Columbia County near Cambria. As has been seen in western Wisconsin (p. 55), it is normal for a retreating escarpment to have enormous numbers of these salients and reéntrants,

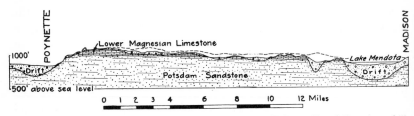

Fig. 71. Cross-section of the Magnesian cuesta north of Madison. Dotted line shows hill crests. Potsdam sandstone is now called Cambrian sandstone.

so the small number gives the Magnesian escarpment abnormal simplicity. Imperfect mapping, due to lack of well records, and failure to find ledges in an unsettled country, may also explain a little of the apparent simplicity of the escarpment, but not all of it.

The escarpment is likewise abnormal in the absence of great numbers of salients, cut off and converted into isolated, flat-topped buttes or mesas of sandstone with a limestone capping. There is a large mass of this sort in Green Lake County (Fig. 86) surrounded by narrow valleys near Princeton and Green Lake. There are a very few small ones. The northern 160 miles of the escarpment is almost entirely without such outliers, though there should be hundreds or thousands of them. In a small area in Columbia County south of the Baraboo Range, however, there are a few of these outliers in front of the escarpment, especially in the region south of the Wisconsin River between Prairie du Sac and Portage. Gibraltar Rock, 1240 feet high and capped by the St. Peter sandstone, is the highest of these outliers. There are similar hills capped by the Lower Magnesian limestone in the country east of Okee. These outliers have been isolated by the erosion and retreat of the escarpment of the Magnesian cuesta in preglacial time.

The Back Slope of the Magnesian Cuesta. The dip slope of the Magnesian cuesta is a gentle one. From Rio to Columbus in Columbia County the upland descends only about 96 feet in 16¼ miles, or not quite six feet to the mile. The slope is about the same in Marinette and Oconto counties near Pound and Stiles Junction. At that point the cuesta is 10 miles wide but the hilltops have an eastward descent of only about 62 feet. In its general eastward slope the surface of the cuesta is exactly that of one made by weathering and stream erosion, acting upon a gently-dipping limestone bed in a region never glaciated. The drainage, however, has been greatly modified, and the northern part of the cuesta seems much less hilly than before the Glacial Period (p. 252). Toward the south, the streams which drain the back slope of the cuesta have grades steeper than the dip of the limestone. They have, therefore, cut through it and have opened out wide lowlands in the weak Cambrian sandstone underneath. A good illustration of this is found north of Lake Mendota in Dane County, (see Fig. 75 and page 223). Such erosion gives the back slope of the cuesta a relief of as much as 200 to 300 feet in places, making it a rather hilly country.

THE GREEN BAY-LAKE WINNEBAGO-ROCK RIVER LOWLAND

Topography. The lowland between the Magnesian and Niagara cuesta, in the belt of Galena-Black River limestone and St. Peter sandstone, slopes eastward with a descent of 200 to 350 feet. The northern end of its axis is 250 to 300 feet lower than the southern end. The following table shows the elevations of various portions of the Green Bay-Lake Winnebago-Rock River lowland. At the north, part of it lies 100 to 120 feet beneath the waters of Green Bay; at the south, near the Illinois line, it is nearly 1000 feet above sea level. The elevations are given in feet above sea level and the localities listed are 40 to 45 miles apart from north to south and at the western and eastern edges of the lowland respectively. They represent the approximate elevation of the surface of the Galena-Black River limestone ledges beneath the glacial drift, which has an average thickness of over 100 feet and a maximum thickness of over 400 feet in a few valleys.

TABLE SHOWING ELEVATION OF THE GREEN BAY-LAKE WINNEBAGO-
ROCK RIVER LOWLAND

WESTERN EDGE OF LOWLAND		EASTERN EDGE OF LOWLAND	
Locality	Elevation	Locality	Elevation
Porterfield, Marinette Co.	672	In Green Bay (bottom)........	460
Pulaski, Shawano Co.........	797	Green Bay, Brown Co..........	480
Ripon, Fond du Lac Co.....	1000	Fond du Lac, Fond du Lac	
Sun Prairie, Dane Co..........	1000	Co.....................................	650
Near Clinton, Rock Co......	950	Oconomowoc, Waukesha Co.	700
East of Stoughton, Dane		Sharon, Walworth Co............	700
Co.....................................	1000	Palmyra, Jefferson Co.........	750

Variations of Altitude and Width. The altitude and width of
this limestone belt vary less from north to south than the Lower
Magnesian limestone. Its general elevation shows the same control by warping. The steeper dip seems to be responsible for the
width of only 12 to 20 miles in the northern 140 miles, in contrast
with a width of 40 to 50 miles in the southwestern part of the low-

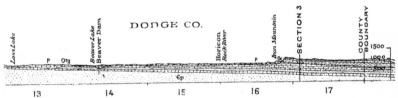

Fig. 72. The Rock River lowland, with the Niagara cuesta on the east.

land. The Black River and Galena limestones have a combined
thickness of about 250 feet, but erosion has given them an outcrop
width of 12 to 15 miles, while the adjacent Lower Magnesian limestone, with about the same thickness, forms a belt of country less
than half as wide.

Reasons for Description as a Lowland. That the Galena-Black
River limestone group should form an upland in one part of the
state and a lowland in another, is abnormal. It may possibly be
explained in part by the shaly character of the Galena-Black River
limestone in the north. As a matter of fact most of this eastern
belt of Galena-Black River limestone would not be regarded as a
lowland at all, were it not for the Niagara upland on the east. If

the shoreline of the ocean or of Lake Michigan were where the eastern edge of the Black River dips under the Richmond shale and Niagara limestone, we should, doubtless, describe this limestone belt as a cuesta. It would be much like the Alabama and New Jersey cuestas, having a low west-facing escarpment overlooking a diminutive lowland on the back of the Magnesian cuesta, and an eastward slope to the shore. But since the Niagara upland is present, it seems better to speak of this portion of the Galena-Black River limestone as a lowland. The northern part of it, in Green Bay, is unquestionably a lowland. The lowland is really a plain of Richmond shale, but practically all the shale seems to have been stripped off by glacial erosion.

The Black River Escarpment. The western edge of the Galena-Black River limestone is so resistant in places as to form a low escarpment. This escarpment is, variably, (a) an inconspicuous ledge or (b) a higher cliff, in several cases exceeding the crest of the Magnesian escarpment in height, or (c) entirely wanting, or (d) buried beneath the glacial drift. Near Seymour, in Outagamie County on the Green Bay and Western Railway, the escarpment is a more conspicuous feature than that of the Lower Magnesian limestone. Its crest is much higher than that of the cuesta to the west. Forty miles south of this, however, in Winnebago County, the Black River escarpment is an inconspicuous feature, seen near the quarries between Oshkosh and Omro. Fifteen miles south of this at Ripon, the Black River escarpment is again a conspicuous feature. It continues southward with considerable strength and the Black River limestone stands considerably higher than the back slope of the Magnesian cuesta in places. Brandon, Fond du Lac County, some distance back from the edge of the Black River escarpment, is 144 feet higher than Markesan, Green Lake County, on the slope of the Magnesian cuesta.

The outline of the Black River escarpment is similar to that of the Magnesian limestone, having salients and embayments. Its border displays an unusual lack of irregularity in plan. Near the southwest it is most irregular, and there the isolated outlying hills capped by the Black River limestone are even more plentiful than in the corresponding part of the Magnesian escarpment.

Two kinds of valleys indent the edge of the escarpment. One is narrow and occupied by torrential streams, such as Mitchell's Glen and Arcade Glen southwest of Ripon. These are postglacial gorges in which the streams descend by rapids and waterfalls from the crest of the Black River escarpment to the back slope of the Magnesian cuesta. Mitchell's Glen, nearly 100 feet deep, is very narrow and steep-sided. The other sort of stream valley which crosses the Black River escarpment slopes in the opposite direction across a low part of the escarpment. These are larger streams, like the Menominee River in Marinette County, the upper Fox River in Winnebago County, and the several headwaters of the Rock River in Columbia and Dane counties. These valleys are broader and make wide gaps in the escarpment. Some of these gaps are of preglacial origin. They are of special interest in view of the scarcity of such stream gaps in the Niagara cuesta to the east.

The Floor of the Lowland. The floor of the Green Bay-Lake Winnebago-Rock River lowland is the back slope of the Galena-Black River limestone. It extends from the crest of its escarpment eastward to the base of the Niagara escarpment. It has already been stated that this floor is inclined, so that the eastern edge is lower than the western, and slopes northward so that the Green Bay end is nearly 300 feet lower than the part near the Illinois line. The floor of the lowland may be divided into three parts (a) the submerged part, north of the city of Green Bay, (b) the middle area of rather smooth plain, and (c) the southern, hilly area.

Most of the northern division of the lowland lies beneath the waters of Green Bay. West of the southern half of Green Bay, which has a length of 90 miles in Wisconsin, is a narrow strip of monotonously-smooth plain, sloping eastward toward the bay. Here glacial lake deposits, described later (p. 265), have made a plain which is smoother than the preglacial surface of the eastward-dipping limestone. This plain is only 20 to 80 feet above the lake and the streams have not been able to cut deeply into it.

The middle division of the lowland, from Green Bay to Dodge County, is also a plain, in this case covered by glacial deposits (p. 255). Its preglacial rock surface (p. 252) may have been even more hilly than the present drift surface (Fig. 72). The drift gives it a local relief of 50 to 150 feet, especially where long, narrow moraines and oval drumlins are found. Elsewhere it is a prairie plain of slight relief.

The southern, hilly region is also one of glacial deposits; but the topography in much of this area is determined by the underlying rock, rather than the overmantle of drift. The eastern part, in Jefferson, Walworth, and part of Rock County, is less hilly than the western, in Dane County. The eastern area is so much like the middle division of the lowland that it will not be described further, except in connection with the glacial topography (Fig. 91). The rougher western half may be typified by a description of the region near the capital of the state.

The Region Near Madison. In Dane county near Madison is a hilly region whose surface topography (lower map), and whose bedrock and approximate preglacial topography are shown in Figure 73. It extends over into the equally high, Magnesian cuesta to the north. The streams have cut down through the Lower Magnesian limestone into the underlying Cambrian sandstone. The region, therefore, has topography related to four rock formations, the Cambrian or Potsdam sandstone, Lower Magnesian limestone, St. Peter sandstone, and Galena-Black River limestone, in addition to the unconsolidated, overlying, glacial drift. The four rock formations all dip southward at angles of less than half a degree, so that for purposes of topographic description the structure may be spoken of as essentially horizontal. The texture of the four rock formations is summarized by saying that two resistant formations, the Galena-Black River and Lower Magnesian, alternate with two weak formations, the St. Peter and Potsdam.

The processes that have acted upon these formations are (a) normal erosion agencies—weathering, stream work, and, in very much smaller measure, the work of underground water and of the wind —and (b) glacial agencies—the rasping ice itself, and the melting ice, which deposited its load on disappearing and also supplied the glacial-stream and lake deposits.

The stage of advancement of this region in its cycle of erosion under the action of normal agencies was what is generally known as late youth. This cycle was then halted by the accident of glaciation, which restored the region to extreme youth, except in so far as the topography inherited from the stage of the normal cycle before glaciation is revealed through the mantle of unconsolidated glacial drift.

Fig. 73.   Maps showing the topography near Madison on the surface of the rock (upper map) and on the surface of the glacial drift (lower map).   The preglacial topography was probably similar in the main to that in the upper map, upon which drainage has been drawn for completeness.   A few features like the Capitol, University of Wisconsin, Maple Bluff, and Picnic Point have been added to the upper map for the sake of location.   Contour interval 50 feet.   (Upper map by Thwaites, lower map based upon Madison Quadrangle, U. S. Geol. Survey.)

The region is an irregular one. The hilltops rise to elevations of 1000 to 1100 feet (Fig. 73). The visible valley bottom of the master stream, the Yahara, a western branch of the Rock River, is 844 to 849 feet above sea level at the surfaces of Lakes Waubesa and Mendota. Thus the visible relief, or difference in altitude between hilltops and valley bottoms, is about 250 to 300 feet. If we could remove the glacial deposits we should find that the highest

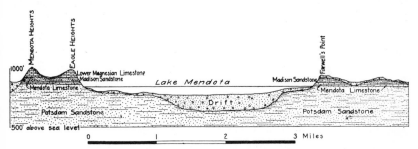

Fig. 74. Cross-section of the Yahara valley at Madison where the Cambrian sandstone (Potsdam) forms a wide valley.

hills rise 900 to 1100 feet above sea level and the Yahara valley bottom lies 550 to 650 feet above sea level, giving a relief of 500 feet. This is probably a close approximation to the preglacial relief.

The Yahara valley is 2½ to 5 miles wide and a geological map (Fig. 75) and cross-section of it shows the relationships of the four geological formations to the topography produced by the action of the normal erosion agencies. The resistant Black River limestone is seen to cap the highest hills, the weak St. Peter sandstone forms steep, short slopes, the resistant Lower Magnesian limestone caps lower hills (Fig. 74), and the weak Cambrian sandstone forms the broad valley bottoms.

The geological map (Fig. 75) shows that the valley is exceptionally wide, because the Cambrian sandstone is so weak that the normal erosion agencies were able to open out a great lowland after they had cut through the more resistant Lower Magnesian limestone. The remnants of Black River limestone are far apart and cap only the highest hills. The region is at the very edge of this formation. In the northern part of the area the Black River is entirely lacking on the back slope of the Magnesian cuesta.

The ice, which came into this region from the northeast, did not modify the amount of relief of the Madison region to a revolutionary extent by sculpture, for it probably eroded somewhat equally from the hilltops and the valley bottoms. It did erode many feet of rock from the hilltops, however, and it probably widened the valley a great deal and deepened it in the lake basins. By deposition, however, it altered the topography tremendously, making it much less hilly than it was before being glaciated. It also greatly altered the drainage. The topography in late youth of a region of alternate, weak and resistant, horizontal formations was doubtless characterized by well-developed dendritic drainage. The present extreme youth of the postglacial drainage, however, has the Four-Lake system of the Yahara headwaters, the many swamps and smaller lakes, and the aimless pattern of streams and lakes and swamps on the glacial drift (Fig. 73), although the main streams seldom diverge far from their former courses.

This type of topography is characteristic of the southwestern portion of the Green Bay-Lake Winnebago lowland. Farther east, in eastern Dane and western Jefferson counties, there are moderate-sized lowlands of St. Peter sandstone on the eastern headwaters of Rock River. This sandstone determines the topography in larger areas here than anywhere else in the state. Isolated masses of Black River limestone rise above it, but the region is much less hilly than near Madison, being covered deeply by glacial deposits.

In hilliness this portion of the Galena-Black River limestone belongs quite as much with the upland of southwestern Wisconsin as with the lowland near Green Bay and Lake Winnebago. Since it forms a convenient gradation zone, the recently glaciated part of it has been discussed with the lowland. The boundary between two geographical provinces grading into each other has to be an arbitrary one, and the border of the Wisconsin drift seems to be as satisfactory a boundary as can be chosen.

THE NIAGARA CUESTA

Topography. The upland between Lake Michigan and the Green Bay-Lake Winnebago-Rock River lowland is underlain by the Niagara limestone, with an exceedingly narrow strip of Richmond shale at the western border, and a still smaller area of Devonian shale and shaly limestone near Lake Michigan. This upland is

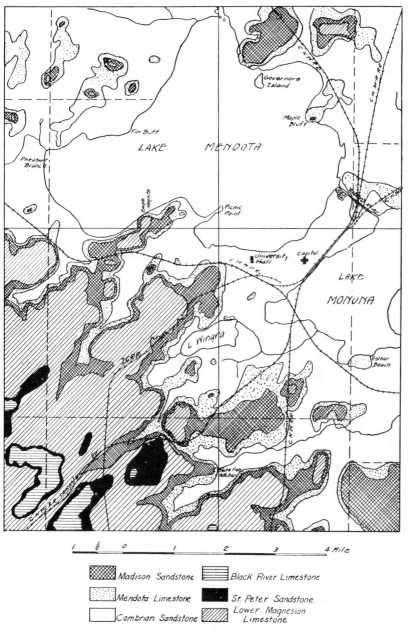

Fig. 75. Geological map of the region near Madison. (Thwaites.)

unsymmetrical. The eastern border is everywhere lower than the western. The middle portion is over 300 feet higher than the northern and southern portions. The significant altitudes are shown in the table on this page, all altitudes being given in feet above sea level. The figures are arranged in order from north to south, the several east-west sections being 40 to 45 miles apart The figures in the table represent the approximate height of the rock beneath the glacial drift, which is over 100 feet thick near the Illinois boundary. The drift covering in northeastern Wisconsin is everywhere thin.

TABLE SHOWING THE APPROXIMATE ALTITUDE OF THE ROCK SURFACE
IN THE NIAGARA UPLAND

WESTERN PORTION		EASTERN PORTION	
Locality	Altitude	Locality	Altitude
Rock Island	890		
Washington Island	740		
Eagle Bluff near Peninsula Park	780		
Near Sturgeon Bay, Door Co.	820		
Near New Franken, Brown Co.	809	Algoma, Kewaunee Co.	590
Near Quinney, Calumet Co.	1120	Manitowoc, Manitowoc Co.	500
Near Iron Mountain, Dodge Co.	1200	Port Washington, Ozaukee Co.	600
Waukesha, Waukesha Co.	820	Cudahy, Milwaukee Co.	565
West of Genoa Junction, Walworth Co.	870	Near Pleasant Prairie, Kenosha Co.	530

Geological Relationships. The Niagara cuesta is an upland 7 to 20 miles wide at the north on Washington Island and the Door Peninsula, and 25 to 45 miles wide at the south between Milwaukee and the Illinois line. The Niagara limestone is 450 to 800 feet thick and the Richmond shale at its base has a thickness of 200 to 500 feet. The dip is steepest in the northern part of the state, but nowhere exceeds 2 to 5 degrees. The formation makes a topographic feature only a little wider than the adjacent lowland of Galena-Black River limestone, although the Niagara limestone is between twice and three times as thick.

The Niagara limestone upland appears in its characteristic topographic expression in the upland of eastern Wisconsin. This thick, continuous, and hard formation is everywhere a cuesta-maker. Over 500 miles to the east, in northern New York, the Niagara limestone also appears as a cuesta-maker. It forms an upland or ridge in practically all of the 900 miles of its circuitous course from Niagara Falls to Wisconsin. It trends westward and northward through the Province of Ontario, to the peninsulas and islands between Georgian Bay and Lake Huron, westward through the upper peninsula of Michigan, and southward through eastern Wisconsin. After passing out of Wisconsin it is again a cuesta-maker in northwestern Illinois and eastern Iowa. It is not a topographic feature in northeastern Illinois where buried by drift and by the Coal Measures. It is again a cuesta-maker part of the way northwestward through Canada to the Arctic Circle, except where buried by drift or by younger rocks.

The Niagara Escarpment. The west-facing escarpment of the Niagara upland overlooks the Green Bay-Winnebago-Rock River lowland and is characteristically developed east of Lake Winnebago, where it is known locally as *The Ledge.* There the escarpment descends abruptly to the lake. At High Cliff, south of Clifton in Calumet County, it falls 223 feet, or from 970 to 747 feet above sea level, in less than 700 feet horizontally. South of Stockbridge the crest of the escarpment is at an elevation of 1060 feet above sea level and is 313 feet higher than the base. It continues southward into Fond du Lac County with about the same altitude.

This escarpment extends across the state of Wisconsin for over 230 miles, but is nowhere so conspicuous a topographic feature as it is east of Lake Winnebago. To the north its height diminishes slightly. In Door Peninsula and Washington Island it rises only 160 to 220 feet above Green Bay, which, however, is 100 to 144 feet deep. It also decreases in height to the southward, mainly because of the covering of glacial drift, but perhaps also because of local folds in the rock, or thinning of the limestone, or greater glacial erosion in places. Well records show that the escarpment stands over 400 feet above the adjacent lowland at Horicon. The escarpment near Waukesha and Oconomowoc is inconspicuous as a present-day topographic feature, but well records show that it is fully 120 feet high.

In outline the Niagara escarpment is even simpler than the Magnesian escarpment (p. 215). Its salients and embayments, so far as we know the geology beneath the glacial drift, break the line of the escarpment relatively little. There are no spurs, pinnacles, or outlying masses along 99% of its front. One exception occurs at the small islands in Green Bay, including Chambers Island, which lies seven miles in front of the escarpment. It is recognized, of course, that there are breaks in the escarpment at several straits in Green Bay, like Death's Door and the Rock Island Channel, and

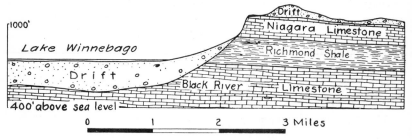

Fig. 76.   Cross-section of the Niagara escarpment near Fond du Lac.

the valleys at Sturgeon Bay, Ellison Bay, and, perhaps, Ephraim. This general simplicity of the Niagara escarpment in eastern Wisconsin (Fig. 84) is best appreciated by comparing it with the Niagara escarpment in Iowa a few miles west of the Wisconsin boundary, or in illinois (Fig. 85) just south of the Wisconsin boundary, where there is a maze of projecting spurs and detached outliers up to a distance of 40 miles from the escarpment. Yet the topography in the two areas was initially determined by (a) identical processes, acting upon (b) the same texture and structure of rock formation, for (c) the same length of time. The operation of the process of glaciation (p. 246) in eastern Wisconsin but not in Illinois and Iowa has produced fundamentally different forms. The extent of irregularity depends upon a number of the factors, as is explained in the next chapter.

The influence of the Richmond shale upon the Niagara escarpment may be illustrated by comparison with its influence at the cataract of Niagara. There the removal of the weak shale by the river in the plunge pool at the base of the falls undermines the resistant Niagara limestone, causing the falls to recede. There is no similarly powerful and rapid-working agency to remove the Richmond shale at the base of the Niagara escarpment in eastern Wisconsin, but the shale is so weak that it has been worn back nearly as fast as the limestone and forms only a narrow strip in the gentler slope at the base of the limestone (Fig. 76), which almost everywhere descends abruptly, and, often, in nearly vertical precipices.

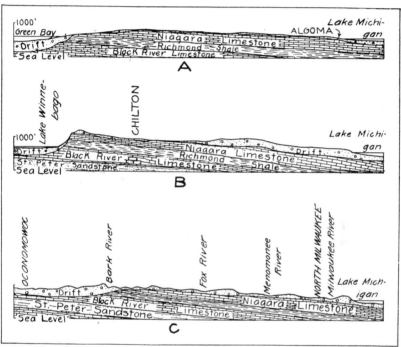

Fig. 77. Three east-west cross-sections of the Niagara cuesta, showing variations in height and in relation to the covering of glacial drift.

Gaps in the Niagara Cuesta. In contrast with the Magnesian and Black River escarpments, the Niagara escarpment and cuesta are remarkable for the absence of transverse gaps. The southern two-thirds of the cuesta is crossed by only one stream flowing from the Galena-Black River lowland to Lake Michigan. The single exception is the Manitowoc River, northeast of Lake Winnebago. Two railways take advantage of the gap it follows.

Quite in contrast, the Magnesian cuesta in the northern 100 miles is crossed by the Menominee, Peshtigo, Oconto, and Fox rivers, and several smaller streams flowing from the inner lowland of Cambrian sandstone. The Black River escarpment is crossed in 180 miles by the four larger streams mentioned above and by the Pensaukee and the western headwaters of the Rock River.

Although the Niagara escarpment is unbroken by stream gaps for the southern 170 miles, the northern portion of the Niagara cuesta is breached by several gaps. The widest of these lies between the end of Door Peninsula and the Garden Peninsula of upper Michigan. It is 30 miles wide and is interrupted by Washington Island, 8 miles long, another smaller island in Wisconsin, and three large and several small islands farther north (Fig. 83). Thus the broadest actual gap in this thirty miles is only about five miles in width.

Another gap is found 65 miles to the south at Sturgeon Bay (Fig. 120). This gap is less than a mile wide at the east, increasing to two miles at the western side of the cuesta. All these northern gaps are now occupied by the waters of Green Bay and Lake Michigan.

Whether these gaps are preglacial stream courses or due to the work of glacial erosion (p. 304) is not quite settled, but it seems likely that they are due to a combination of the two.

The gap occupied by the Manitowoc River, is filled with glacial drift. It is not certain that this was occupied by a stream which went through in preglacial time, for its headwaters are at a higher level than the floor of the Green Bay-Lake Winnebago lowland.

The Niagara Upland. The upland on the back slope of the Niagara cuesta is a region of very moderate relief, with glacial deposits forming the greatest irregularities. This is partly because of glacial erosion (pp. 249, 252) and partly because a dip slope of homogeneous limestone does not have as rough topography if no streams cross it as it would have if it were crossed by several streams from an adjacent lowland. The short streams of the cuesta in the high western part of the upland are of small volume. They have, therefore, been unable to cut deeply and to make the region hilly, as large streams in a water gap would do. In the eastern part, the upland itself has descended to a height of 700 feet, or only 120 feet above the level of Lake Michigan. The erosion by the

THE BACK SLOPE OF THE NIAGARA CUESTA NEAR PEWAUKEE.

A. LONGITUDINAL VIEW OF ONE OF THE OVAL HILLS OF GLACIAL DRIFT, OR DRUMLINS, IN EASTERN WISCONSIN.

Drumlin two miles northeast of McFarland, Dane County. It is steepest at the north end.

B. TRANSVERSE VIEW OF DRUMLIN TWO MILES NORTH OF SULLIVAN, JEFFERSON COUNTY.

largest streams, like the Milwaukee River near its mouth, results in a maximum relief of only 100 to 120 feet by cutting into the glacial drift and the rock. The greatest relief resulting from the glacial deposits lying upon the rock surface is 100 to 200 feet.

The slope of the drift-covered upland from the crest to the wave-cut cliffs of Lake Michigan is at an average rate of about 12 feet to the mile. The upland descends from 1050 feet at the crest east of Lake Winnebago to 638 feet at Mosel near Lake Michigan. This is 32½ miles. It descends from 1100 feet at the escarpment near Hartland, Waukesha County, to 700 feet near Lake Michigan.

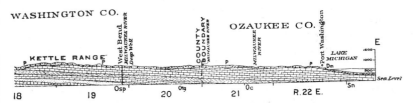

Fig. 78. Cross-section to show the back slope of the Niagara cuesta.

The Pishtaka or Fox River of Waukesha, Racine, and Kenosha counties and several smaller streams have a longitudinal trend. The Milwaukee River flows eastward down the dip slope to within 7 miles of Lake Michigan. At Fredonia, Ozaukee County, it turns southward at right angles and flows parallel to the coast for 32 miles before entering the lake at Milwaukee. There are several similar cases. Whether these are due to minor cuestas on the back slope of the Niagara upland in the alternating weak and resistant portions of the Niagara limestone formation, or, as seems more likely, to the lateral moraines of the Lake Michigan glacier, has not yet been established. The drainage is discussed more fully in Chapter XI.

The eastern termination of the Niagara upland is masked by the waters of Lake Michigan, the origin of whose basin is so related to the events of the Glacial Period that it is postponed for later treatment (p. 237).

BIBLIOGRAPHY

Alden, W. C.   Milwaukee Special Folio, Geologic Atlas of the United States, No. 140, U. S. Geol. Survey, 1906; Quaternary Geology of Southeastern Wisconsin, Prof. Paper 106, U. S. Geol. Survey, 1918, pp. 1-102.

Buell, I. M.   Geology of the Waterloo Quartzite Area, Transactions Wis. Acad. Sci., Vol. 9, 1893, pp. 255-274.

Bussewitz, W. R.   Dodge County, privately printed, 1926.

Case, E. C.   Description of Models Illustrating the Physical Geography of Wisconsin, Bull. 3, Milwaukee State Normal School, 1907, pp. 1-19.

Chamberlin, T. C.   Geology of Eastern Wisconsin, Geology of Wisconsin, Vol. 2, 1877, pp. 95-405; Historical Geology,—Lower Magnesian limestone, St. Peter sandstone, Trenton limestone, Galena limestone, Hudson River or Cincinnati shale, Niagara limestone—Geology of Wisconsin, Vol. 4, 1883, pp. 138-260.

Cramer, Frank.   On a Recent Rock-Flexure, Amer. Journ. Sci., 3rd Series, Vol. 39, 1890, pp. 220-225.

Daniels, E.   The Mines of Wisconsin, Trans. Wis. State Agr. Soc., Vol. 4, 1857, pp. 356-362; (on the Niagara and the Potsdam), Proc. Bost. Soc. Nat. Hist., Vol. 6, 1859, pp. 309-310.

Fischer, C. W.   Geography of Manitowoc County, Wisconsin, Unpublished thesis, University of Wisconsin, 1921.

Hollister, D. E.   The Geology of the Cuestas of the Great Lakes Region, Unpublished thesis, University of Wisconsin, 1926.

Irving, R. D.   Note on the Age of the Metamorphic Rocks of Portland, Dodge County, Wis., Amer. Journ. Sci., 3rd Series, Vol. 5, 1873, pp. 282-286; Geology of Central Wisconsin, Geology of Wisconsin, Vol. 2, 1877, pp. 409-636.

Jackson, C. T.   (On the Clinton iron ore), Proc. Bost. Soc. Nat. Hist., Vol. 6, 1859, p. 341.

Lapham, I. A.   On the Geology of the South-eastern Portion of the State of Wisconsin,—in Foster and Whitney's Geology of the Lake Superior Land District, Senate Ex. Document 4, Special Session, 1851, Washington, 1851, pp. 167-173; Geological Formation of Wisconsin, Trans. Wis. State Agr. Soc., Vol. 1, 1851, pp. 122-128; Discovery of Devonian Rocks and Fossils in Wisconsin, Amer. Journ. Sci., 2nd Series, Vol. 29, 1860, p. 145.

Lee, F. K.   The Geography of Juneau County, Wisconsin, Unpublished thesis, University of Wisconsin, 1916.

Martin, Lawrence.   The Eastern Ridges and Lowlands, Physical Geography of Wisconsin, Journ. Geog., Vol. 12, 1914, pp. 228-230.

Percival, J. G.   General Reconnoissance (of Wisconsin) and the Quartz Rock of the Baraboo and of Portland, Annual Rept. of the Geol. Survey of the State of Wisconsin, Madison, 1856, pp. 64-103; Report on the Iron of Dodge and Washington Counties, State of Wisconsin, Milwaukee, 1855, 13 pp.

Platt, R. S. A Detail of Regional Geography (Door Peninsula), Annals, Assoc. Amer. Geographers, Vol. 18, 1928, pp. 81-126.

Scheuber, F. A. Abandoned Shorelines of Lake Mendota (Wisconsin), Unpublished thesis, University of Wisconsin, 1916.

Stromme, O. U. Geology of Madison and Parts of Adjacent Townships, Dane County, Wisconsin, Unpublished thesis, University of Wisconsin, 1907.

Suydam, V. A. Geography of Ripon and Vicinity, 9 pp.

Thwaites, F. T. Map of Preglacial Topography near Madison, In Bull. 8, Wis. Geol. and Nat. Hist. Survey, 2nd edition, 1910; Geology of the Vicinity of Lakes Waubesa and Kegonsa, Dane County, Wisconsin, Unpublished thesis, University of Wisconsin, 1906; Geology of the Southeast Quarter of the Cross Plains Quadrangle, Dane County, Wisconsin, Unpublished thesis, University of Wisconsin, 1908.

Warner, J. H. The Waterloo Quartzite Area of Wisconsin, Unpublished thesis, University of Wisconsin, 1904.

Whitbeck, R. H. Geography of the Fox-Winnebago Valley, Bull. 42, Wis. Geol. and Nat. Hist. Survey, 1915, 105 pp; Geography of Southeastern Wisconsin, Bull. 58, Wis. Geol. and Nat. Hist. Survey, 1920, 160 pp.

Whitson, A. R., and others. Soils Reports, Wis. Geol. and Nat. Hist. Survey, Bulls. 24, 29, 37, 39, 47, 48, 49, 52D, 53A, 53B, 54D, 55, 56A, 56B, 56C, 59A, 59B, 59C, 61A, 61B, 61C, 62A.

Note: The publications dealing with the glacial geology, soils, and hydrography of eastern Wisconsin are listed at the ends of Chapters X, XI, and XII. Some of these also contain geological and physiographic data.

MAPS

U. S. Geol. Survey. Topographic maps as follows: Neenah, Fond du Lac, Hartford, West Bend, Port Washington, Baraboo, Poynette, Cross Plains, Madison, Sun Prairie, Waterloo, Watertown, Oconomowoc, Waukesha, Milwaukee, Milwaukee Special, Evansville, Stoughton, Koshkonong, Whitewater, Eagle, Muskego, Bay View, Brodhead, Janesville, Shopiere, Delavan, Lake Geneva, Silver Lake, Racine, Geneva, Waukegan, Ripon, and Neshkoro Quadrangles, (see Fig. 197).

See also generalized topographic map of eastern Wisconsin (Pl. II A) in the atlas accompanying the Geology of Wisconsin.

U. S. Lake Survey. Corps of Engineers, Survey of Northern and Northwestern Lakes: Lake Survey Charts 7, 70, 71, 72, 73, 74, 715, 723, 725, 728, 734, 737, 743, 745, 747, 1475, various scales from 1: 8,000 to 1: 5,000,000, (see Fig. 199).

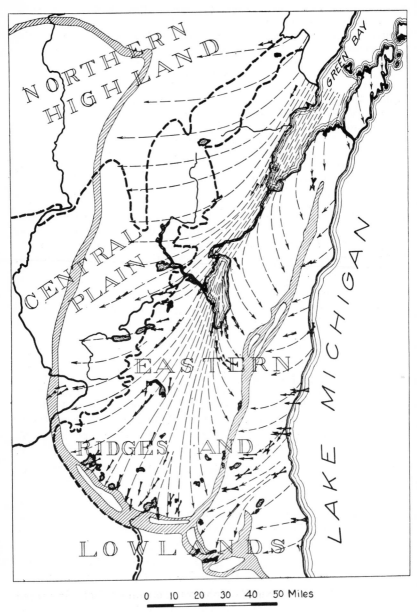

Fig. 79. The Green Bay lobe and Lake Michigan lobe of the continental glacier, with arrows indicating the direction of glacial striae. (Modified from map by Alden.)

CHAPTER X

THE GLACIATION OF EASTERN WISCONSIN

The Oval Hills of Glacial Drift

In southeastern Wisconsin there are more than 1400 oval hills of glacial drift in an area of 4200 square miles. There are fully as many of these oval hills in the northeastern part of the state. They are called drumlins and they were made by the continental glacier. Wisconsin is famous in the world outside for two of its geographical features. One of these is the Driftless Area (Chapter V), the other is the drumlins (p. 257).

These oval drumlins have one peculiarity. Their longer axes are always parallel to the direction of ice movement. Therefore they tell us the directions in which various parts of the continental glacier moved in eastern Wisconsin.

Lobes of the Advancing Ice Sheet

The Eastern Ridges and Lowlands were entirely covered by the Labrador ice sheet of the continental glacier (Fig. 27). This ice sheet was lobate near the borders, the lobes being determined by broad valleys and lowlands. In these lowlands the ice moved faster than on the intervening ridges and uplands. The Lake Michigan and Green Bay lobes had the extent shown in Figure 79. The arrows on this map indicate the directions of ice movement, revealed by glacial striae on the rock ledges, as well as by the drumlins. From this it is seen that the Lake Michigan lobe advanced southward down the shallow river valley which occupied the site of the present deep lake and westward across the Niagara upland. The Green Bay lobe was a branch of the Lake Michigan glacier. Its size and position suggest a thumb on a mitten (Fig. 80). This smaller lobe advanced down the Green Bay-Lake Winnebago-Rock River lowland. The Green Bay lobe was unsymmetrical. It spread further to the west than to the east of the axis of the lowland, because it was opposed on the east by the Niagara upland and the

Lake Michigan glacier. It had a free opportunity to expand to the west over the Magnesian cuesta and into the portions of the Northern Highland and Central Plain which now lie next the Driftless Area.

Fig. 80. Lobes and moraines of the continental glacier in the Great Lakes region. (Taylor and Leverett.)

GLACIAL EROSION

The sculpturing of eastern Wisconsin by this ice greatly modified the preglacial topography. The ice moved across Wisconsin for long ages. The continental glacier advanced not once, but several times. Each glacial epoch was probably of greater duration than the time since the last ice sheet melted away. If it is granted that moving ice, carrying a load of rock in the basal layers, can abrade the surfaces over which it moves, then the assumption of great glacial erosion rests only upon the question of time. The scratched, striated, and polished rock surfaces prove that continental glaciers can erode, and of time there is more than sufficient. The proofs of unusually-effective glacial erosion in eastern Wisconsin are to be found in the following:

(a) the rock basin character of Lake Michigan;

(b) the similar form of Green Bay;

(c) the submerged hanging valley relationship of Green Bay and Lake Michigan;

(d) the absence of the Richmond shale from the floor of the Green Bay-Lake Winnebago-Rock River lowland;

(e) the amount of quartzite derived from certain small ledges;

(f) the simple outlines of the limestone escarpments;

(g) the absence of residual soil on the surfaces of the cuestas;

(h) the paucity of caves and sink holes;

(i) the absence of marked ridges and valleys upon the cuesta surfaces;

(j) the topographic contrast between the glaciated and drift-less portions of Wisconsin, and the gradation from one to the other in the border region.

TOPOGRAPHIC FEATURES DUE TO GLACIAL EROSION

Rock Basin Character of Lake Michigan. One part of eastern Wisconsin where a great amount of glacial erosion took place was the basin of Lake Michigan. Years ago the basins of the Great Lakes were correctly ascribed to glacial erosion. Later there was a long time when the erosive power of glaciers was questioned. Indeed it was said that the basins of the Great Lakes were due to the joint action of "preglacial erosion, glacial corrosion, glacial accumulation blocking up outlets, depression due to ice occupancy, and general crust movements, together with possible unascertained agencies." Now that even greater, glacial sculpture is known to have taken place in Alaska, Greenland, Norway, the Alps, and elsewhere, many geologists have less hesitancy in ascribing the sculpture of the whole basin of Lake Michigan to glacial erosion. The other factors enumerated in the quotation given above may be easily eliminated as not of significant application or of adequate amount to explain the excavation of this basin below the present lake level.

The cross-section of Lake Michigan bears the characteristics of a basin excavated by ice and not of a valley eroded by a river. It has a broad, flat bottom and abrupt walls, descending to a depth of 500 to 800 feet. Figure 82 gives a cross-section of the lake basin southeast of Sturgeon Bay. The longitudinal profile of the lake bottom, with enclosed basins and intervening swells, is also suggestive of glacial scooping, such as produces rock basins in all glaciated valleys.

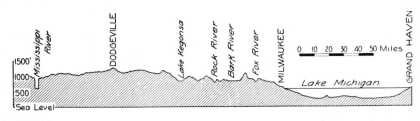

Fig. 81. Cross-section of the basin of Lake Michigan, broadened and deepened by glacial erosion, in contrast with the gorge of the Mississippi,—a stream-eroded valley in the Driftless Area.

The maximum possible amount of preglacial stream erosion is inadequate to explain the present lake basin. The weak, Devonian shales underlying Lake Michigan must have formed a lowland in preglacial time. The lowland was doubtless occupied by a master stream (Fig. 109). This stream probably flowed southward. The buried channel from the Lake Huron basin to the Lake Michigan

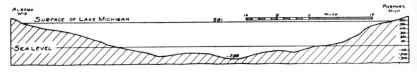

Fig. 82. Cross-section of the basin of Lake Michigan, showing the flat floor and steeper walls produced by glacial sculpture.

basin by way of Saginaw Bay slopes southwestward, as if the stream in this channel had been a tributary of the Lake Michigan river.

The known rock ledges at the southern end of Lake Michigan in Illinois and Indiana suggest that the Lake Michigan river flowed at a level no more than one to two hundred feet below the present surface of Lake Michigan or 581 feet, if indeed it was below 581 feet at all. This conclusion is corroborated by an examination of records of deep wells adjacent to Lake Michigan in Michigan and Wisconsin, as well as those in Illinois and Indiana. They show no deep buried valleys. Bedrock in a small part of southeastern Wisconsin and in much of the lower peninsula of Michigan lies slightly below lake level; but it is not certain that this is due to preglacial stream erosion rather than glacial sculpture.

The rock floor of the Mississippi River between Prairie du Chien, Wisconsin, and Dubuque, Iowa, is between 405 and 492 feet above sea level, or only about 175 feet below the present surface of Lake Michigan. As the preglacial river of the Lake Michigan basin was doubtless tributary to the Mississippi, the former stream could not possibly have eroded to a depth of 323 feet below sea level— 904 feet below the present surface of the lake. The amount of deepening of the Mississippi by glacial streams is not known. If glacial deepening of Lake Michigan be denied, then an improbable amount of subsequent warping must be assumed. There is no evidence even suggesting such an amount of warping. The relationship of the bottom of Lake Michigan to the Mississippi, therefore, suggests that in northeastern Wisconsin the preglacial Lake Michigan river flowed at or within 100 feet above or below the present lake level.

It is concluded that the preglacial stream course in the Lake Michigan basin was near present lake level at the southern boundary of Wisconsin. The bottoms of the preglacial valleys in what are now Lake Michigan and Green Bay were, accordingly, higher than 581 feet above sea level. As the bottom of the lake is (a) at a level of only 5 to 80 feet above sea level east of Door Penninsula, (b) 323 feet below sea level in the deepest portion, southeast of Sturgeon Bay, and (c) just about at sea level east of Racine in southern Wisconsin, the amount of glacial deepening, vertically, was from 500 to nearly 900 feet.

Basin of Green Bay. There are good reasons for supposing that, before the Glacial Period, the site of Green Bay was occupied by a river rather than a lake. The preglacial valley was probably at a slightly higher level than the present surface of Green Bay and Lake Michigan. If so, the glacier eroded many cubic miles of rock from the portion of Green Bay west of the Niagara upland in Door Peninsula, for it lowered the bottom of Green Bay at least 120 to 144 feet west of Washington Island and 60 to 90 feet near Sturgeon

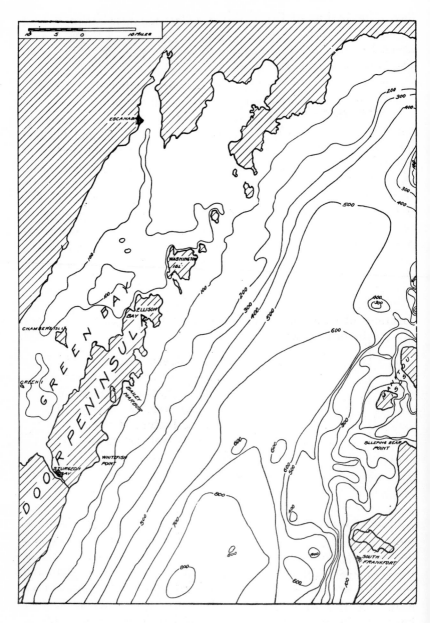

Fig. 83 The submerged hanging valley of Green Bay.

Bay. The bottom of Green Bay descends northward, as does the lowland from Lake Winnebago to the bay. The descent toward Green Bay may be partly due to structural conditions, but it may also have been caused by weakness of the Green Bay lobe toward the south, for this glacial lobe terminated near the southern boundary of the state.

A Submerged Hanging Valley. An excellent way of demonstrating that the continental glacier carved out the basins of Green Bay and Lake Michigan is to consider the junction of the floor of Green Bay with the bottom of the main lake (Fig. 83). The depth of the bay here is 100 to 144 feet and the depth in the straits north of Washington Island is 156 feet. To the east the water deepens rapidly to 576 feet (Fig. 82). This assumes that any deposits at the bottoms of Green Bay and Lake Michigan are of about the same thickness.

Junctions of main and side streams are normally even, or accordant, in regions where there have never been glaciers. This is true of the junction of the rock floors of the Wisconsin and Mississippi rivers at Prairie du Chien in the Driftless Area. The junctions of main and side valleys in glaciated regions are almost always discordant, and the side valley hangs above the main valley. This is spoken of as a hanging valley. Such discordance is produced because the larger glacier in the main valley erodes its bed more deeply than the smaller ice tongue in the side valley. Both valleys are sculptured, but the main valley is eroded the most.

Thus Green Bay was lowered by glacial erosion from above 581 feet to the present bottom, 100 to 144 feet lower, while the adjacent part of Lake Michigan was lowered over 500 feet. Since this discordant valley junction lies below lake level it is spoken of as a submerged hanging valley. Submerged hanging valleys showing discordant junctions with main fiords are well known in Alaska, as at Port Fidalgo and Orca Bay, Prince William Sound. The discordance of these two tributary fiords, with relation to a submerged channel one quarter as wide as Lake Michigan, is 600 to 700 feet and 900 to 1000 feet respectively. This is twice to three times as much as the discordance at the lip of the submerged hanging valley of Green Bay. There is no evidence of faulting to account for this discordance in northeastern Wisconsin and the known warping is absolutely inadequate in amount, but differential glacial erosion explains the facts perfectly.

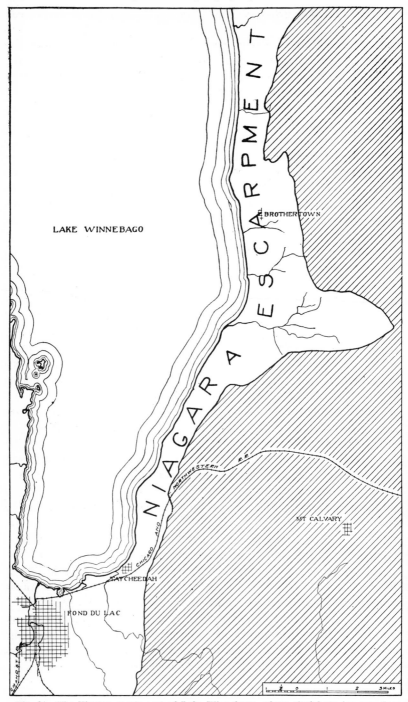

Fig. 84. The Niagara cuesta east of Lake Winnebago, where glacial erosion may have planed back the escarpment 5 or 10 miles in converting a preglacial feature like that shown in Figure 85 into the present form. (From Fond du Lac Quadrangle, U. S. Geol. Survey.)

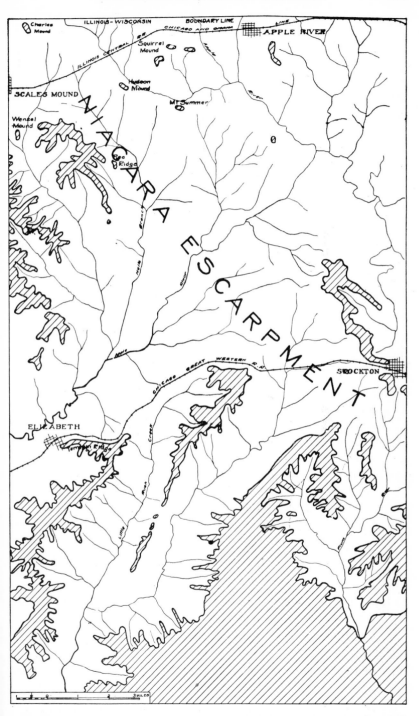

Fig. 85. The Niagara cuesta in the Driftless Area of northern Illinois, where the irregular escarpment and numerous outliers suggest the type of topography which probably existed in eastern Wisconsin before the Glacial Period. (From Elizabeth Quadrangle, U. S. Geol. Survey.)

The study of the glacial deposits of Wisconsin has long been on an adequate basis, but the recognition of the vast amount of glacial erosion has been cautiously delayed. No doubt there were marked preglacial valleys to guide the lobes of the continental ice sheet; but the basins of Green Bay and Lake Michigan, and the submerged hanging valley north of Washington Island are topographic evidences of notable glacial sculpture during the several glacial advances. This produced basins much deeper and wider than the preglacial valleys. The vast volumes of glacial drift in this and adjoining states, aside from all that was carried away by streams, furnish concrete proof of the same thing.

Absence of Richmond Shale from Lake Winnebago-Rock River Lowland. The Richmond shale in eastern Wisconsin also furnishes evidence of notable glacial sculpture. In the driftless portion of Iowa and in southwestern Wisconsin, it projects in front of the Niagara escarpment for 3 to 6 miles. In eastern Wisconsin, where it has similar thickness and is no weaker, it lies only on the slope at the base of the Niagara escarpment. The shale is 200 feet thick. This implies not only the planing back of the Richmond shale border for several miles laterally, but also vertical glacial abrasion of one or two hundred feet, even before the ice cut into the underlying Galena-Black River limestone.

Amount of Quartzite Derived from Certain Small Ledges. The quartzite blocks in the glacial drift east and south of Madison are chiefly derived from ledges near Waterloo and Portland in Jefferson County (Fig. 89). The amount of quartzite in the drift has been roughly calculated. This was compared with the area of the known ledges. As the ledges are small, and any undiscovered ledges are doubtless still smaller, it is fairly safe to accept the conclusion that 50 to 75 feet of rock were removed from these ledges by the ice sheet. The planing away of 75 feet vertically in such resistant rock as quartzite might be regarded as remarkable, were it not for the fact that the ice moved over these ledges for an exceedingly long time. Furthermore, the quartzite is well-jointed, so that the glacier was able to pluck out great blocks even before abrading them notably.

Simple Outlines of Limestone Escarpments. A consideration of the forms of the escarpments gives another line of evidence as to the great amount of glacial sculpture in eastern Wisconsin. Emphasis has been placed in the previous chapter on the unusual and generally abnormal simplicity of the Magnesian escarpment (p. 216), the Black River escarpment (p. 219), and the Niagara escarpment (p. 228). This was done in order to point out now that this simplicity is due almost wholly to glacial erosion.

It seems reasonable to assume that the glacial agencies have produced the difference between the two contrasted portions of these escarpments, one glaciated and the other driftless, for we have rock textures and structures of identical character and an equal time for identical agencies to work before the glacial invasion. Of the glacial agencies the rasping and plucking quality is the one to which we appeal. Even if the outline of the rock ledges beneath the drift is not known with the strictest accuracy, the maximum amount of possible error could not account for the difference in the glaciated and the driftless topographic forms.

Lateral Erosion of the Niagara Escarpment. In the case of the Niagara escarpment, to take a specific case, it is possible that glacial erosion wore back portions of the ledge as much as five or ten miles, in converting an irregular cliff (Fig. 85) into a simple one (Fig. 84). The Magnesian and Black River escarpments were also worn back a long distance by glacial erosion. A cliff like that at the western edge of the Niagara cuesta on Lake Winnebago, shown in Figure 76, is a perfectly normal result of glacial erosion and plucking, especially as the ice moved parallel to the escarpment. It is not a normal result of preglacial weathering and stream work.

It is, of course, recognized that the greatest amounts of lateral sculpture by the ice took place only in the most favorable locations. Where the ice moved parallel to the escarpment it planed back the ends of the salients and removed all the outliers. This was the case between the northern end of Lake Winnebago and Oakfield, Fond du Lac County. To the south, near Oconomowoc, there are a few limestone outliers. Here the ice did not move parallel to the escarpment but ascended it obliquely. The direction of striae, the axes of drumlins, and the position of the moraines tell us this, but it is expressed even more clearly in the lessened glacial erosion of this part of the escarpment.

Comparison with the Magnesian Escarpment. Again in the Magnesian escarpment west of Ripon (Fig. 86) we have the deep valley of Green Lake indenting the line of limestone cliffs. A broad salient and several small limestone hills project north of Green Lake, and there is a small outlier called Mt. Tom. This is because the ice moved across the escarpment instead of parallel to it.

Niagara Escarpment Inside and Outside the Driftless Area. The whole Niagara escarpment of Wisconsin, Illinois, and Iowa (Fig. 87) shows the effectiveness of glacial sculpture in relation to parallelism of movement and length of time. The escarpment is simple and without outliers from A to B. Here the direction of ice movement was parallel. It is slightly irregular from B to C. Here the direction was oblique. From C to D and from E to F the escarpment is deeply buried in glacial drift. It may be a little more irregular than is shown. In any event the ice moved at right angles to it, and may have accentuated the salients somewhat.

From F to G the escarpment is tremendously irregular. There are great numbers of outliers, even at distances of 60 miles. This is in the Driftless Area. There is little reason for doubting that before the Glacial Period the escarpment from A to B may have been as irregular and may have had as many outliers as are now found in the Driftless Area between F and G. From G to H the escarpment lies in the region of older drift. It is not quite as irregular as in the Driftless Area, but it is far from simple. There are four reasons for its irregularity of outline. (1) It was overridden by ice moving nearly at right angles. (2) It was close to the edge of the Driftless Area where the ice was thin and weak. (3) It was probably not glaciated more than once. (4) The glaciation took place so long ago that the escarpment may have had its irregularity renewed somewhat by the tributaries of the Mississippi.

Thus we see that the extent of irregularity or simplicity of an escarpment varies with local conditions. The simplest portion of the Niagara escarpment is east of Lake Winnebago—(a) where the ice moved parallel to the ledge, (b) where the glacier was thick and powerful, (c) where it eroded during several glacial epochs, (d) where there are no streams to cut it up subsequently, and (e) where there has not been time enough since the Glacial Period for appreciable modification. For such a favorably-situated escarpment as this, a retreat of 5 or 10 miles by glacial sculpture is subject to little doubt.

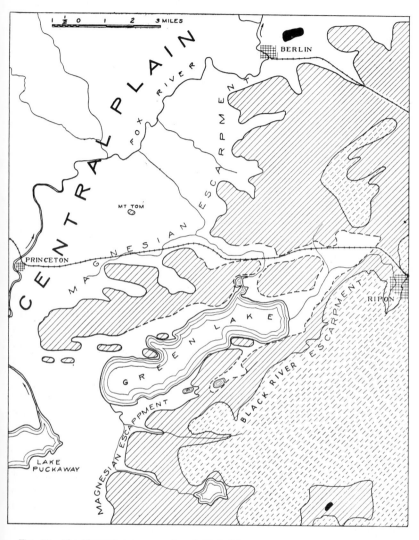

Fig. 86. The Magnesian escarpment and Black River escarpment near Ripon and Green Lake. Oblique lines, Lower Magnesian limestone; oblique dashes, Black River limestone; white, Cambrian sandstone; black, pre-Cambrian igneous rocks. The ice moved southwestward, eroding the deep valley of Green Lake, accentuating the salients in the escarpments, removing nearly all the outliers, and producing topographic forms intermediate between those shown in Figure 84 and Figure 85. Before the Glacial Period this escarpment doubtless had numerous outlines similar to those in Figure 18 and Figure 127.

Rejection of Other Possible Explanations of Simple Escarpments.
It might be argued that the very fact of existing cuestas with escarpments suggests that the glacial erosion was slight, since forms
of this type characterized the region before it was glaciated. As
a matter of fact the alternate, weak and resistant, sedimentary

Fig. 87. The Niagara escarpment inside and outside the Driftless Area. Outliers of
Niagara cuesta in the Driftless Area are abundant as at Blue Mounds, Platte Mounds,
and Sinsinawa Mound, but there are practically none in the glaciated area, where the
escarpment is abnormally simple.

formations would retain the escarpments at the borders of its
cuestas no matter how great or how little glacial erosion there
might be. Wave work, as on the Niagara escarpment in Green
Bay, might produce a steep, simple cliff. It would be a low cliff
and there would be a wave-cut bench of rock at its base.   No

wave work could affect all three of the cuestas of eastern Wisconsin. A longitudinal river in the Green Bay-Lake Winnebago-Rock River lowland would not produce a straight, simple escarpment, for the tributaries of the master stream would carve the escarpment even more deeply than now and would thus increase its irregularity. As simple an outline as that of the Niagara escarpment might have been produced by faulting, but this can be definitely proved not to have taken place in the three escarpments of eastern Wisconsin, which are identical in topographic characteristics and must have had a common origin.

Simple Escarpments Due to Glaciation. The crux of the situation is found in the comparison of the Magnesian escarpment in (a) the region of rapid movement and strong ice action, (b) the area of thin, weak ice near the limit of glaciation, and (c) the Driftless Area. In Marinette, Green Lake, and the intervening counties, the escarpment is abnormally simple; in Columbia County there are a few projecting spurs and outlying limestone-capped hills (see p. 216); in Sauk, Vernon and Monroe counties, the escarpment is most irregular. The same thing is true of the Black River and Niagara escarpments. This marked difference within the general type of cuestas points clearly to great glacial modification of parts of the escarpments of eastern Wisconsin.

Absence of Residual Soil on Cuesta Surfaces. The foregoing discussion shows the author's reasons for believing that glacial erosion caused the lateral retreat of the fronts of escarpments for distances of several miles. The amount of reduction of the surfaces of the cuestas may also be discussed in a quantitative manner.

In the Driftless Area the average thickness of the residual soil is seven feet. Upon ridges and uplands the thickness is 8 to $13\frac{1}{2}$ feet. This does not include the partly disintegrated rock below the residual clay.

Eastern Wisconsin has no residual soil or weathered rock to speak of. The places where weathered material occurs in glaciated territory are rare and easily explained. The slightly-weathered rock in small areas west of Madison show that the thin, weak ice near the border of the Driftless Area did not remove quite all the residual material. It certainly removed nearly all. We do not know how much weathered material existed above the little

patches that were left. There is weathered bedrock in the older
drift area east of Monroe in Green and Rock counties. This is
near the border of glaciation where the ice was thin and weak, and
where glaciation may not have operated more than once.

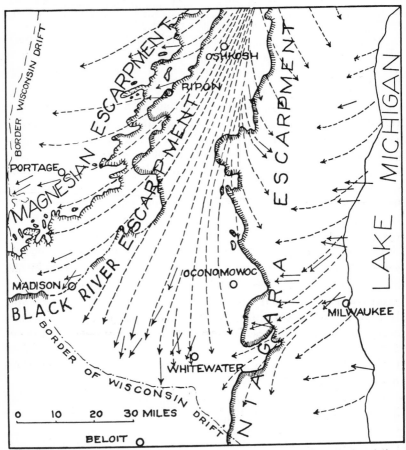

Fig. 88. Variations in the outlines of the escarpments of eastern Wisconsin in relation
to the direction of glacial movement.

Not only was all the residual soil removed from the limestones
in eastern Wisconsin, but the gradation zone of slightly-weathered
limestone was nearly everywhere planed away, leaving the firm,
unaltered rock beneath. The ice often planed down to the same
resistant bed, as at Duck Creek near Green Bay. The most con-
servative estimate gives a minimum amount of vertical glacial
erosion of 8 to 14 feet.

Scarcity of Caves and Sink Holes. That the ice abraded much more deeply than the amount just mentioned is indicated by the tiny number of caves and sink holes in glaciated eastern Wisconsin. Caves are abundant in the Galena-Black River and Lower Magnesian limestone of the driftless portions of Iowa and Minnesota, but there seem to be no caves in the Niagara there. The residual soil of the Niagara, however, is heavier than that on the Galena-Black River. Consequently we may safely assume that in preglacial time there were many caves in the Galena-Black River and the Lower Magnesian limestone of eastern Wisconsin, and that the Niagara limestone there had thick residual soil. In the Driftless Area, caves and disintegration seem to be abundant down to the limit of ground water. This is 10 to 100 feet in some places, and 100 to 300 feet in others. Sink holes in the driftless portion of the state are from 5 to 20 feet deep. Caves penetrate to a depth of 50 to 75 feet in driftless southwestern Wisconsin (p. 94). It seems logical to conclude from the relationships of residual soil and of caves that one or two hundred feet of weathered and cavernous rock have been eroded by the ice in eastern Wisconsin.

The apparent exceptions help to prove the rule regarding absence of caves in glaciated territory. There is an arch in the Silurian limestone of Mackinac Island, Mich., which has been interpreted as part of an unroofed cavern, although it may conceivably be due to wave work. The cave near Ephraim seems to have been made by wave work after the ice had retreated from the Door Peninsula. Another cave, southwest of Egg Harbor, may be of similar origin. There is a cave near Maribel, northwest of Manitowoc. The small caves or enlarged joint planes near Wilson in northwestern Wisconsin (p. 127) were near the border of the Driftless Area where the ice was thin and weak. Moreover, they are situated in the belt of older drift where the ice may not have come more than once. Small caves and solution channels along joints are found in glaciated territory in northern and central Illinois, but they have similar relationships. That there should be one cavern—the Richardson cave—in the area of Wisconsin drift near Verona, a quarter of a mile from the terminal moraine, and that the few small caves in the older drift near Wilson are only about 25 miles from the Driftless Area supports the theory of glacial removal of nearly all caves in eastern Wisconsin. It shows a gradation from the region of abundant caves in the Driftless

Area to the region of no caves in the belt of active ice movement
and repeated advances by the thick glacial lobes in eastern Wisconsin.

Absence of Marked Ridges and Valleys upon Cuesta Surfaces.
The Magnesian and Black River cuestas near Madison have notably
hilly surfaces (p. 221). This hilly region is near the western edge
of the glaciated area, where the ice was thinner and glacial erosion
was weaker than to the northeast. The continuation of these
cuesta uplands in northeastern Wisconsin is much less hilly (p.
224). The upland of the Niagara cuesta is a rather smooth surface.

It seems fair to raise the question as to whether the whole of
each of these cuesta surfaces was not much more hilly in preglacial time. A similar cuesta in Alabama is so cut up by streams as
to form a rather hilly region. This is far south of the belt of glaciation. The Magnesian cuesta in western Wisconsin is cut up by
streams until it is a hilly region of ridges and valley, with no flat-
topped interstream areas. This is in the Driftless Area. The
Allegheny Plateau in glaciated New York is decidedly less hilly
than in driftless Pennsylvania and West Virginia. Glacial erosion
wears down hills until the surfaces of glaciated cuestas are less
irregular than the corresponding surfaces of cuestas which have
never been covered by ice. For these reasons the cuesta surfaces
on the Niagara, Galena-Black River, and Lower Magnesian limestones of eastern Wisconsin seem to be abnormally simple in topography, just as their escarpments seem abnormally simple in outline.

On all these accounts the author is convinced that glacial erosion
has planed down the Lower Magnesian, Galena-Black River, and
Niagara limestones in eastern Wisconsin to depths at least one or
two hundred feet below the preglacial surface. It may be even
more.

Topographic Contrast Between Glaciated and Driftless Areas.
The Mississippi River and its tributaries have produced a distinctive topography in the Driftless Area. There is no reason for
doubting that there was topography of the same sort in eastern
Wisconsin before the Glacial Period, and that the change was
brought about by the continental glacier. This is known by the
absence of the residual soil, which must have mantled the whole

region before the first glacial invasion. It is made more emphatic by the great amount of undecayed, local limestone which we find in the glacial drift of eastern Wisconsin. It is shown especially by the character of the existing topography, which is entirely different from the preglacial topography. The evidence from the topography in eastern Wisconsin is very specific, as we fortunately have samples of the preglacial landscape preserved in the Driftless Area, where the underlying rock is similar to that of the drift-covered area farther east. The gradation in topography near and north of Madison and west of Janesville and Beloit may be partly due to structural causes. The chief cause, however, is the intermediate character of the glacial erosion. It was less than in the belt of rapidly-moving ice and long-continued, repeated glaciation. It was more than in the Driftless Area. Thus the gradation of topography helps to prove the rule.

SURFACE FEATURES DUE TO GLACIAL DEPOSITION

The Glacial Drift. The deposits left by the ice sheet are unassorted till, or bowlder clay, and stratified gravel, sand, and clay. They contain not only fragments of the local limestone, shale, and sandstone, but also igneous and metamorphic rocks imported into the region by the ice sheet. The source of this material is partly in the Lake Superior Highland in northern Wisconsin and partly in Canada. A small amount comes from the exhumed monadnocks like the quartzites and igneous rocks of the Fox River valley and the Waterloo quartzite of Jefferson County.

Bowlder Trains. Extending southwestward from Waterloo (Fig. 89) are abundant bowlders of quartzite (p. 244) scattered by the glacier in the lee of the ledges. This is known as a bowlder train. It is recognizable because the quartzite is a unique rock in this region of limestone and sandstone. The Waterloo bowlder train is more than 60 miles long. It is fan-shaped, increasing in width from a narrow band to 20 miles near Sun Prairie and Lake Mills, and 50 miles near Whitewater and Madison. The bowlders near the ledges are large and numerous, while those near the borders are small and infrequent. Smaller bowlder trains are found in the lee of the knobs of igneous rock in the valley of Fox River in the Central Plain and in the lee of the Powers Bluff monadnock of the Northern Highland in Wood County (Fig. 163).

Drift Copper. The drift in eastern Wisconsin contains fragments of native copper from the north. Masses up to 487 pounds in weight have been found in southeastern Wisconsin. Elsewhere (p. 436) there are glacial bowlders of copper up to 800 and even 3000 pounds in weight.

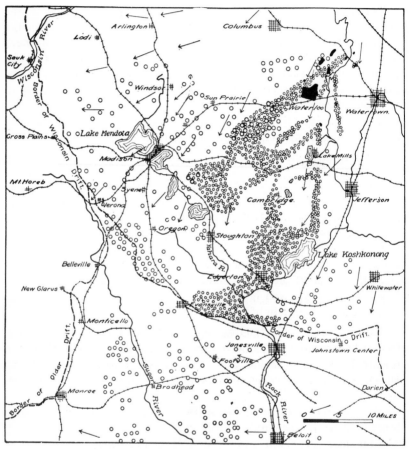

Fig. 89. The Waterloo bowlder train. Quartzite ledges shown in black, bowlder train by circles, and directions of glacial movement by arrows. (After Buell.)

Diamonds in the Glacial Deposits. A few diamonds are also found in the glacial drift. Such diamonds have been found near Eagle in Waukesha County, southwest of Oregon in Dane County, near Saukville in Ozaukee County, Burlington in Racine County, and Kohlsville in Washington County. The largest of these weighed 15 $\frac{12}{32}$ carats. Their source is as yet unknown, but it is supposed to be somewhere in Canada (Fig. 90). These glacially-transported gems have long been known. Indeed as long ago as 1670 the Jesuit fathers related a story of diamonds on some of the islands at the entrance to Green Bay.

Proportion of Local Material. Crystalline rocks form only about 13 per cent of the pebbly portion of the drift in southeastern Wisconsin, where the most careful mechanical analyses have been made, the remaining 87 per cent of the material being local limestone and sandstone. The crystalline rocks come largely from Canada and from the Lake Superior region.

Older Drift. The belt of older drift in eastern Wisconsin is confined to parts of Rock and Walworth counties. The glacial topography is less rugged than in the rest of eastern Wisconsin because of stream erosion. The surface drift is so weathered that the limestone pebbles are leached away to masses that can be picked apart with the fingers. In this respect the older drift is quite different from that of the latest or Wisconsin stage where the limestone is entirely unaltered and only a few of the basic igneous rocks are sufficiently weathered to fall apart.

Ground Moraine of the Latest Glaciation. The ground moraine which covers nearly all of eastern Wisconsin has the variable, slightly rolling topography of drift-mantled plains.

The ground moraine is made up largely of till, but may contain small areas of stratified sand and gravel. The till is deposited in the broad sheet by the melting ice, so that it is apt to be unstratified. The ground moraine covers a much larger area than the terminal moraines, which are in long narrow belts. The way in which the similar ground moraine of Iowa was deposited has been happily stated by McGee:

"The whole mass (of ice), indeed, must have lain in majestic inactivity until devoured by the hungry sun and thirsty wind. The bowlder-dotted surface * * * is its epitaph."

The thickness of the ground moraine in southeastern Wisconsin varies from a few feet on the hilltops to over 400 feet in the bottoms of the preglacial valleys. The surfaces mantled by the ground moraine have local relief of 50 to 200 feet, except where the topographic forms like terminal moraines and drumlins rise above the ground moraine.

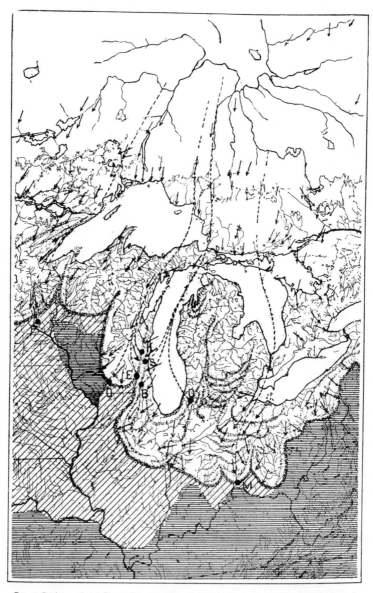

GLACIAL MAP OF THE GREAT LAKES REGION.

Driftless Areas   Older Drift   Newer Drift
Moraines   Glacial Striae   Track of Diamonds
Diamond Localities   E, Eagle   O, Oregon
K, Kohlsville O, Dowagiac M, Milton P Plum Crk. B Burlington

Fig. 90. Places where diamonds have been found in the glacial drift. (Hobbs.)

Most of the ground moraine is covered by a rather fertile clay soil, but parts of it are stony. Large areas in the ground moraine are too swampy for agriculture, or are covered by the waters of lakes. The ground moraine and terminal moraines of eastern Wisconsin appear on the soil maps as the Miami and Coloma loams.

Drumlins. A special relief feature within the ground moraine area is the class of oval hills, or drumlins, alluded to at the beginning of this chapter. There are only three portions of the United States where these specialized forms are abundant (1) eastern Wisconsin, (2) northwestern New York, and (3) eastern Massachusetts. Figure 91 shows the distribution of some of the Wisconsin drumlins. It is seen that they are confined mostly to the limestone belt and lie within 5 to 35 miles of the outermost terminal moraine. This limitation to the border region suggests that there may be a minimum possible thickness of ice which was capable of carving or building drumlins. The ice in this belt is thought to have been at least 450 to 1450 feet thick. The 1400 drumlins previously mentioned do not include anywhere near all of the oval hills of this sort in eastern Wisconsin.

Some of the drumlins rise as much as 140 feet above the adjacent plain. A few are as low as 5 feet. Their average width is about a quarter mile, and their length varies from a few rods to two miles. The material is chiefly unstratified glacial drift, but a few drumlins contain a little water-deposited material (Fig. 93). Numerous well sections show that they do not have rock cores. Their trend is north-northeast in the Sun Prairie region (Fig. 91), but in different parts of eastern Wisconsin they trend north-south, northeast-southwest, and northwest-southeast. They always have their longer axes parallel to the direction of ice flow revealed by the striae on the rock ledges. East of Fond du Lac the drumlins are of short, oval type like those near Boston, Massachusetts, while near Waterloo the drumlins are long, attenuated forms, similar to those west of Syracuse, New York. The hills at the State University, once described as "the Madison type of drumlins," are not drumlins at all, but parts of recessional moraine. There are drumlins at Madison, however, for example the hill near Lake Mendota traversed by Langdon and Gilman Streets.

There are also drumlins in the area of older drift southeast of Janesville. They have a northeast-southwest trend, in contrast

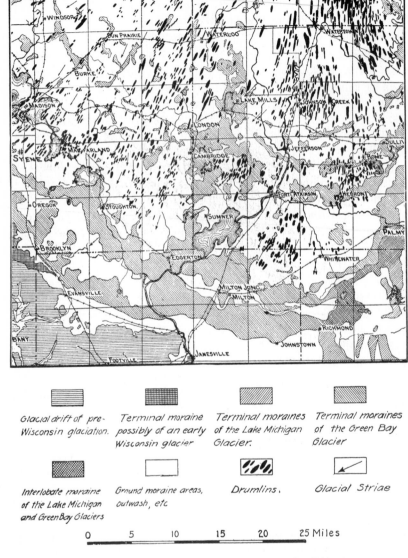

Glacial drift of pre-Wisconsin glaciation.

Terminal moraine possibly of an early Wisconsin glacier

Terminal moraines of the Lake Michigan Glacier.

Terminal moraines of the Green Bay Glacier

Interlobate moraine of the Lake Michigan and Green Bay Glaciers

Ground moraine areas, outwash, etc

Drumlins.

Glacial Striae

0   5   10   15   20   25 Miles

Fig. 91. The drumlins in southeastern Wisconsin. (Alden.)

with the general north-south trend of the younger drumlins a few miles to the north.

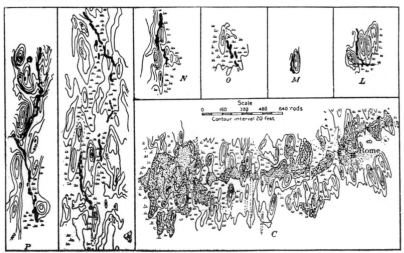

Fig. 92. Topographic maps of drumlins in southeastern Wisconsin, showing their slopes crossed by eskers (black) and a terminal moraine (dots). Contour interval 20 feet. Area C shows a moraine crossing drumlins near Jefferson. Area P shows an esker near Waterloo (Alden.)

Recessional Moraines. The distribution of the moraines built at the edge of the retreating ice sheet in southeastern Wisconsin is shown in Figures 91 and 98. Those near the Driftless Area are found at intervals which suggest a halt of the ice every 5 miles. The terminal moraine at the border of the Driftless Area varies

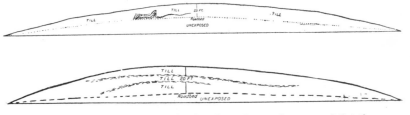

Fig. 93. Cross-sections showing that drumlins do not have rock cores, and that they occasionally contain small lenses of stratified drift, the prevailing material being unstratified till or bowlder clay. (Alden.)

in width from three-fourths of a mile to three or four miles. It stands 40 to 140 feet above the adjacent area and the local relief of its ridges and knobs above its valleys and kettles is 20 to 40 feet. It forms a striking contrast to some of the recessional moraines, for example at Madison, where the narrow, single ridges between Lakes Monona and Wingra and at the head of University Bay on Lake Mendota are less than an eighth of a mile wide. Some of the recessional moraines are discontinuous, but the great majority of them form strong upland belts of rough, wooded country.

East of Janesville near the Niagara escarpment there is thought to have been a minor tongue of the Green Bay glacier called the Delavan lobe. On account of the difficulty of distinguishing pitted outwash from gravelly terminal or interlobate moraine, there are

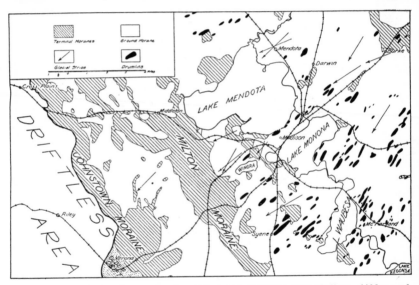

Fig. 94. Terminal and recessional moraines and drumlins near Madison. (Alden and Thwaites.)

students of the glacial geology of Wisconsin who do not fully accept the independent identity of the Delavan Lobe. Its author still adheres to it, however, and feels that the preglacial Troy valley in Walworth County caused a separate ice movement between the Green Bay and Lake Michigan lobes. The terminal moraines of the Delavan and Green Bay lobes form crescentic belts.

The Kettle Moraine. Between the Green Bay and Lake Michigan lobes was formed an interlobate deposit of unusual height and irregularity, as is shown in Figure 107. This is a part of the Kettle Moraine of Wisconsin, so called because of the deep hollows

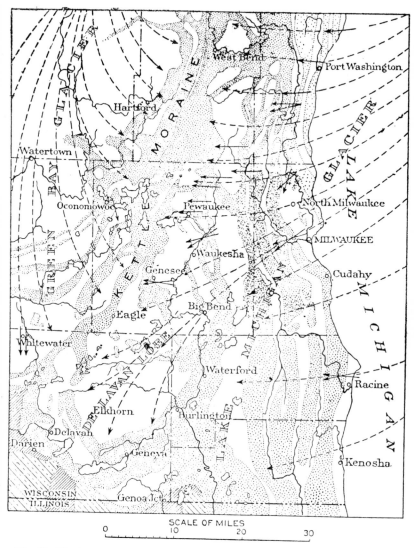

Fig. 95. The kettle moraine, an interlobate accumulation between the Green Bay lobe and the Lake Michigan lobe of the continental glacier. (Alden.)

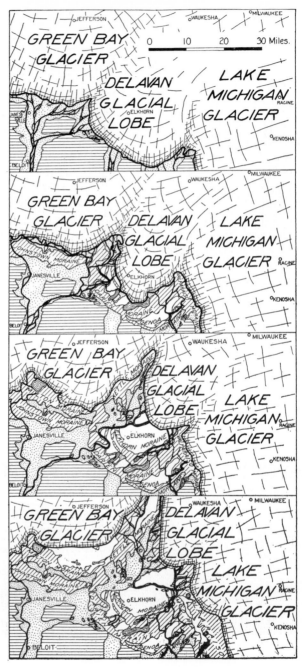

Fig. 96. Four maps to show the drainage associated with the successive stages in the retreat of the ice from southeastern Wisconsin. These streams laid down outwash deposits of glacial sand and gravel. (Based upon maps by Alden.) More recent work suggests that there may have been less definite lobation in the area between the Lake Michigan Glacier and the Green Bay Glacier.

A. GLACIAL STREAMS DEPOSITING OUTWASH GRAVEL NEAR
BORDER OF MALASPINA GLACIER, ALASKA.

B. PLAIN OF OUTWASH GRAVEL IN KEWAUNEE COUNTY, WISCONSIN.

A. TERMINAL MORAINE NEAR GILLETT.

B. STEEP RIDGES OF THE KETTLE MORAINE NEAR EAGLE.

or kettles. It rises 200 feet above the region southeast of White-water and is especially well seen near Eagle, Waukesha County. The kettles or pits are due to the melting of buried ice blocks, or to the building of irregular morainic ridges which enclose undrained depressions.

Outwash Deposits. Deposits made by streams which issued from the edge of the melting ice are found in many parts of eastern Wisconsin. They are typically developed near Janesville and Be-

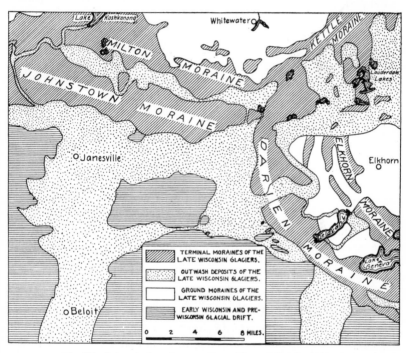

Fig. 97. Outwash deposits near Janesville and Beloit. (Alden.)

loit (Fig. 97). Outwash plains consist of low, coalescing, alluvial fans which head up against a moraine. This moraine is the outermost one which marks the edge of the ice at the time of its latest expansion, but outwash may be built at the border of any recessional moraine. Near Janesville the outwash plain slopes southward at the rate of nearly 10 feet to the mile. It has a smooth surface with slight irregularities. These are caused by projecting masses of ground moraine, by kettles, by abandoned glacial stream channels, and by the deeper ravines of postglacial streams, such as the Rock River. The latter now flows in a valley 40 to 60 feet below the level of the outwash plain. The thickness of the outwash at Janesville is 450 feet. The coarse rounded material makes stony fields at certain localities, while the finer deposits produce sandy soil in other places. A thin coating of loess and soil covers most of the outwash plain. Outwash deposits appear on the soil maps as the Plainfield, Fox, and Waukesha series.

Eskers. Allied to the outwash plains in origin are the eskers, built by glacial streams flowing beneath the ice. These are narrow, winding ridges of stratified gravel. They are not numerous. Eskers as much as six miles long are known in eastern Wisconsin. Conspicuous ones are to be seen near Waterloo in Jefferson County (Fig. 92), west of Cottage Grove in Dane County, in the southeastern part of Dodge County, the eastern part of Columbia County, and the southeastern part of Marinette County. Parts of the Rodman series of soils are upon eskers.

Gravel Seam at Ripon. A limestone quarry on the rock hill west of Ripon reveals a thin horizontal seam of glacial gravels in the solid rock. It lies beneath ten feet or more of Black River limestone, and has been seen continuously throughout an area two or three hundred feet square. The gravel seam is only an inch or two in thickness. It consists of rounded pebbles of granite, greenstone, quartzite, and limestone. Because of the lack of consolidation of this deposit of erratics it does not seem at all likely that it represents a glacial deposit of Ordovician rather than Quaternary age. The surface upon which it rests discloses no glacial striae, but is slightly weathered.

One way of explaining this deposit would be to assume that the continental ice sheet of the latest Glacial Period had slid a great slab of rock over the top of the hill, leaving it there in a horizontal position, resting upon the glacial gravels. One objection to this is that the vertical joint planes in the Black River limestone above and below the gravel seam appear to be continuous. These joint planes are usually thought to be of preglacial origin. Moreover there also appear to be glacial gravels within the St. Peter sandstone at another level on this same hill.

Another hypothesis for the explanation of the gravel seam is that the pressure of glacial ice lifted the rock layers slightly, so that glacial waters were able to force the erratic gravels into one stratum of the limestone and into certain fissures in the sandstone. It may be objected that the gravel seam is nearly everywhere of fairly uniform thickness and that there are many other layers into which the waters might equally well have forced glacial gravels. It may be felt that this second hypothesis is favored by the presence of a fold in the limestone and sandstone layers at the eastern edge of the quarry. The axis of the fold makes a considerable angle with the glacial striae at the top of the hill. Thus the fold has the right trend for a feature made by glacial pressure, though it is hard to believe that brittle limestone may be folded under such conditions. Glacial folding of solid rock has been suggested in a case near Burlington, Racine County. Near Ripon there are other minor folds which seem to be preglacial. Elsewhere in the quarry at Ripon the St. Peter sandstone fails to appear in its normal place between the Galena-Black River limestone and the Lower Magnesian limestone. This, however, is below the level of the gravel seam, and may not have anything to do with the origin of these uncemented glacial gravels in the midst of solid rock layers. The origin of the gravel seam is not yet known.

Lake Deposits. At the borders of the retreating ice sheet were temporary lakes, many of them held in between the higher drift deposits and the retreating ice. There were large bodies of water of this type in the basins of Lake Michigan and Green Bay, (Chapter XII), and small, ephemeral lakes in various parts of eastern Wisconsin. In these lakes were accumulated deposits which now form swampy tracts in some parts of the state, though there are many swamps not due to the filling of lakes.

Of the lake deposits special mention will be made of two which cover large areas. In the Niagara upland south of Milwaukee and extending nearly parallel to the shore of Lake Michigan, are low, weak, recessional moraines which have been thought to be partly or wholly accumulated under water (Fig. 98).

North of Milwaukee is a belt of red clay which seems to be a modified lake deposit. The upper portion is unstratified. The deposit broadens northward, and in Calumet and Manitowoc counties extends westward across the Niagara upland and down into the Green Bay-Lake Winnebago lowland, to a point just south of

Fond du Lac. It forms a gently-sloping plain west of Green Bay and Lake Winnebago. Its western border also extends over the Magnesian cuesta into the present Fox-Wolf valley. The lake clay runs from a few feet to over a hundred feet in thickness. It is clear that the redness is not due to weathering after the clay was deposited. It may be derived from (a) the ferruginous, red clay of the Lake Superior basin (p. 436), (b) the iron formations

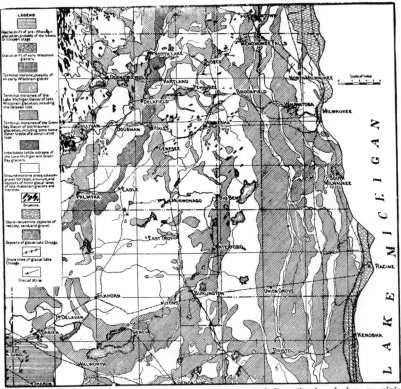

Fig. 98.   Glacial deposits in southeastern Wisconsin, including the broad, low morainic ridges near Lake Michigan and the red clay north of Milwaukee. (Alden.)

of the Menomonie and Gwinn districts in Michigan, or even (c) a sublacustrine iron range beneath the waters of eastern Lake Superior. Beneath this red clay is blue till of earlier glacial deposition.

Extensive lake deposits are doubtless hidden beneath the waters of Green Bay and Lake Michigan. There are also narrow strips of lake deposits of the higher stages of the Great Lakes, as on the coast south of Milwaukee and west of Marinette and Green Bay. The soils maps use the names Superior clay loam, Superior fine sandy loam, and Poygan fine sandy loam for these lake deposits of eastern Wisconsin. In addition to the lake clays there are ancient and modern beach deposits, described later (p. 295) in connection with the coasts of Wisconsin.

The Latest Readvance of the Ice. There is evidence that the glacier readvanced in eastern Wisconsin, after retreating far

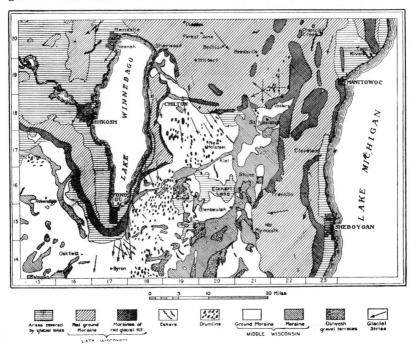

Fig. 99. Map to show the red clay moraines related to the latest readvance of the ice in eastern Wisconsin. (Alden.)

enough for the accumulation of some of the red clay just described. Part of the red clay is ridged up into terminal moraines which cross the older moraines at a low angle, as in Manitowoc County near Kiel and in Fond du Lac County east of Lake Winnebago. The southern limit of this advance seems to have been near Cudahy, south of Milwaukee, in the Lake Michigan basin, and about five miles south of the city of Fond du Lac, in the Green Bay-Lake Winnebago lowland. The crescentic terminal moraine south of Fond du Lac is made up of this reworked lake clay. In the moraines the material is red clay filled with small pebbles, so that we have here a red till. The red till and red clay make a stiff, clayey soil much disliked by farmers. If ploughed just before a rain, it makes great lumps so hard they have to be broken with a sledge hammer or axe. This soil is similar to the red clay soils of the Lake Superior lowland (p. 436). Because these red clay moraines are at an angle with older moraines, which disappear beneath them, and because the drift must have been formed in a marginal lake in front of the advancing or retreating glacier, it is clear that there was a readvance of a considerable amount during the general retreat of the glacier from eastern Wisconsin.

A Forest Bed. At Two Creeks, between Kewaunee and Manitowoc, the wave-cut cliffs of the lake shore reveal an ancient forest bed,buried beneath red till and resting on stratified red clay. It consists of logs, branches, and upright stumps. This forest bed proves that there was a period long enough for forest growth between the retreat of the ice and accumulation of the red clay and the readvance during which the red till was deposited. Similar vegetable accumulations are found in wells in the Fox River valley.

BIBLIOGRAPHY

Alden, W. C. Delavan Lobe of the Lake Michigan Glacier, Prof. Paper 34, U. S. Geol. Survey, 1904, 99 pp; Drumlins of Southeastern Wisconsin, Ibid., Bull. 273, 1905, 46 pp; Radiation of Glacial Flow as a Factor in Drumlin Formation, Bull. Geol. Soc. Amer., Vol. 22, 1911, pp. 733-734; Geological History of Green Lake County, Wisconsin, Green Lake County Training School Quarterly, Vol. 3, 1912, pp. 2-14; Criteria for Discrimination of Age of Glacial Drift Sheets, Journ. Geol., Vol. 17, 1909, pp. 694-709; Quaternary Geology of Southeastern Wisconsin, Prof. Paper 106, U. S. Geol. Survey, 1918, 356 pp.

Andrews, E. On some Remarkable Relations and Characters of the Western Bowlder Drift, Amer. Journ. Sci., 2nd Series, Vol. 48, 1869, pp. 172-179.

Bliss, J. S. Notes on Wisconsin Drift, Amer. Journ. Sci., 2nd Series, Vol. 41, 1866, p. 255.

Buell, I. M. Bowlder Trains from the Outcrops of the Waterloo Quartzite Area, Trans. Wis. Acad. Sci., Vol. 10, 1895, pp. 485-509.

Bruncken, E. Physiographical Field Notes in the Town of Wauwatosa, Wisconsin, Bull. 1, Wis. Nat. Hist. Soc., 1900, pp. 95-99.

Chamberlin, T. C.   On the Extent and Significance of the Wisconsin Kettle Moraine, Trans. Wis. Acad. Sci., Vol. 4, 1879, pp. 201-234; Quaternary Formations of Eastern Wisconsin, Geology of Wisconsin, Vol. 2, 1878, pp. 199-246; Ibid., Vol. 1, 1883, pp. 261-298; The Terminal Moraine of the Second Glacial Epoch, Third Ann. Rept., U. S. Geol. Survey, 1883, pp. 315-326.

Hobbs, W. H.   On a Recent Diamond Find in Wisconsin and on the Probable Source of this and other Wisconsin Diamonds, Amer. Geol., Vol. 14, 1894, pp. 31-37; The Diamond Field of the Great Lakes, Journ. Geol., Vol. 7, 1899, pp. 375-388; Emigrant Diamonds in America, Smithsonian Annual Rept. for 1901, pp. 359-366.

Irving, R. D.   The Quaternary Deposits of Central Wisconsin, Geology of Wisconsin, Vol. 2, 1877, pp. 608-636.

Kirch, A. B.   Geography of Dane County, Wisconsin, Unpublished thesis, University of Wisconsin, 1911.

Lapham, I. A.   On the Existence of Certain Lacustrine Deposits in the Vicinity of the Great Lakes Usually Confounded with the Drift, Amer. Journ. Sci., 2nd Series, Vol. 3, 1847, pp. 90-94.

Lawson, P. V.   Preliminary Notice of the Forest Beds of the Lower Fox River, Wisconsin, Bull. 2, Wis. Nat. Hist. Soc., 1902, pp. 170-173.

Leverett, Frank.   Preglacial Valleys of the Mississippi, Journ. Geol., Vol. 3, 1895, pp. 744-745, 758-759; Profiles Across the Bed of Lake Michigan, Monograph 38, U. S. Geol. Survey, 1899, Pl. 5 facing p. 12; Outline of History of the Great Lakes, 12th Report Mich. Acad. Sci., 1910, pp. 28-30.

Russell, I. C.   The influence of Caverns on Topography, Science, new series, Vol. 21, 1905, pp. 30-32; A Geological Reconnoissance along the North Shore of Lakes Huron and Michigan, Annual Rept. Mich. Geol. Survey, 1905, pp. 39-150; Surface Geology of Portions of Menominee, Dickinson, and Iron Counties, Michigan, Ibid., 1907, pp. 7-82.

Salisbury, R. D.   Notes on the Dispersion of Drift Copper, Trans. Wis. Acad. Sci., Vol. 6, 1886, pp. 42-50.

Smith, H. E.   Escarpments, Unpublished thesis, University of Wisconsin, 1913.

Spencer, J. W.   Origin of the Basins of the Great Lakes of America, Amer. Geol., Vol. 7, 1891, pp. 86-97; Review of the History of the Great Lakes, Ibid., Vol. 14, 1894, pp. 289-301.

Thwaites, F. T.   A Glacial Gravel Seam in Limestone at Ripon, Wisconsin, Journ. Geology, Vol. 29, 1921, pp. 57-65; The Origin and Significance of Pitted Outwash, Journ. Geol., Vol. 34, 1926, pp. 308-319; Outline of Glacial Geology, 1922, 1925, 1927, 181 pp; The Development of the Theory of Multiple Glaciation in North America, The Central District, Trans. Wis. Acad. Sci., Vol. 23, 1928, pp. 44-124.

Upham, W.   The Madison Type of Drumlins, Amer. Geol., Vol. 14, 1894, pp. 69-83.

Weidman, S., and Wood, P. O.   Reconnaissance Soil Survey of Marinette County, Bull. 24, Wis. Geol. and Nat. Hist. Survey, 1911, 44 pp.

Whittlesey, Charles. On the "Superficial Deposits" of the Northwestern Part of the United States, Proc. Amer. Assoc. Adv. Sci., Vol. 5, 1851, pp. 54-59; On the Drift Cavities or "Potash Kettles" of Wisconsin, Ibid., Vol. 13, 1860, pp. 297-301; On the Ice Movements of the Glacial Era in the Valley of the St. Lawrence, (including directions of striae in eastern Wisconsin), Ibid., Vol. 15, 1867, pp. 43-54; On the Fresh-Water Glacial Drift of the Northwestern States, Smithsonian Contributions to Knowledge, Vol. 15, 1867, 32 pp.

Winchell, N. H. The Glacial Features of Green Bay of Lake Michigan, With Some Observations on a Probable Former Outlet of Lake Superior, Amer. Journ. Sci., 3rd Series, Vol. 2, 1871, pp. 15-19.

MAPS.

U. S. Geol. Survey. See quadrangles listed on p. 233.

U. S. Lake Survey. Charts of Lake Michigan, Green Bay, and Lake Winnebago, listed on p. 236.

Wis. Geol. and Nat. Hist. Survey. See special lake maps in Bulletin 27; also published separately (Fig. 199); maps in soils bulletins referred to in this chapter.

CHAPTER XI

THE DRAINAGE OF EASTERN WISCONSIN

A Youthful System of Drainage

The poet Longfellow described the four lakes of Madison as:

"Four limpid lakes,—four Naiades
Or sylvan deities are these,
  In flowing robes of azure dressed;
Four lovely handmaids, that uphold
Their shining mirrors, rimmed with gold,
  To the fair city in the West."

It is difficult to realize that these lakes are ephemeral features of the landscape, that they were not here some scores of thousand years ago, and that in a few thousand years more they will no longer exist.

Originally there were more than four lakes near Madison. Lake Wingra is approaching extinction, although it was once twice as large. Hook Lake, east of Oregon, is now entirely swamp. The same is true of the lake of Nine Springs valley, which formerly extended east of the State Fish Hatchery for 4 miles. Nearly half of the original Lake Kegonsa is a marsh. Lake Mendota doubtless once stretched 4 or 5 miles to the northeast, with Maple Bluff, Darwin, and Mendota on a large island. Indeed the whole drainage system of eastern Wisconsin, with its crooked streams and buried valleys, is a very modern and short-lived affair, as the earth views time. The venerable streams of the state are those in the Driftless Area. The rivers, lakes, and swamps of eastern Wisconsin have all been established since the Glacial Period. In the course of time the lakes will be filled, the swamps drained, the rapids and waterfalls extinguished, and the buried valleys cleared out. Then the drainage system will return to a condition similar to that in the southeastern part of the Western Upland. Indeed the streams in the region west of Beloit and Janesville have already gone back to essentially the preglacial condition through long-continued weathering and stream erosion.

GENERAL HYDROGRAPHY

The drainage of the Eastern Ridges and Lowlands is chiefly dependent upon the larger features of rock topography, partly upon the minor topographic features due to glaciation. The chief drainage basins of eastern Wisconsin are (a) the Rock River system in the south, (b) the Fox River system in the north, and (c) minor independent streams that flow into Green Bay or Lake Michigan. The waters of the Rock flow into the Mississippi and Gulf of Mexico, while the Fox and the independent streams send their waters to the St. Lawrence and Atlantic Ocean.

THE ROCK RIVER SYSTEM

Driftless and Glacial Drainage. The chief streams of the Rock River system are the Yahara, the Crawfish, and the main Rock River on the east, and the Pecatonica and Sugar rivers in the Western Upland. Thus the drainage system is about equally divided between glaciated and driftless territory. As may be seen on the map (Fig. 100) these two parts of the system are strikingly different. Since the eastern tributaries were once like those on the west in all essential respects the conditions in the whole drainage system are summarized here.

The Pecatonica River. The Pecatonica River and its tributaries form a symmetrical, branching system (Fig. 100). No lakes interrupt its course. Its tributaries are close-set, and thoroughly drain the country through which they flow. There are no undrained interstream areas and no swamps, except in the floodplain of the main stream. The pattern of master stream and affluents simulates the form of a tree and its branches so well that the name—dendritic—is particularly applicable.

The Pecatonica rises on the southern slope of the Military Ridge west of Dodgeville at an elevation of about 1200 feet. It is a graded stream, with slope steepest at the headwaters where the conditions are those of late youth. It has no rapids and furnishes little water power. In the first ten miles it descends about 30 feet to the mile. In the next thirty miles its grade flattens to about 3 feet to the mile.

The Sugar River. The Sugar River is similar in some respects to the Pecatonica, though the drainage pattern of the Sugar is less systematically dendritic, because it drains a region in large part glaciated and because it is in a lower and less hilly area.

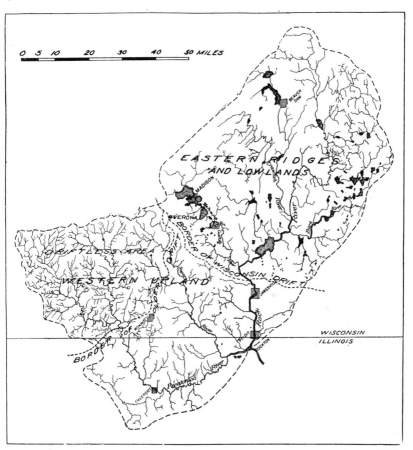

Fig. 100. The Rock River system, with (a) the dendritic drainage pattern of the Pecatonica River and its tributaries in the Driftless Area, (b) the rather systematic drainage development of the Sugar River and its tributaries in the area of older drift, and (c) the extremely youthful drainage of the Yahara and Rock rivers in the area of Wisconsin drift where lands are abundant.

Moreover, its valley was occupied by vast floods of water from the melting continental glacier. These streams carried and deposited great quantities of sand and gravel. The deposits form a floodplain, abnormally wide for so small a stream. The grade of

Sugar River is steep at the headwaters in the Driftless Area. Near Brodhead it descends at the rate of only $2\frac{1}{2}$ feet to the mile.

The eastern tributaries of Sugar River, as well as the southwestern tributaries, all rise in the uplands covered with older drift. As already stated (p. 272), this drift is now worn so thin that the drainage has been able to return to the preglacial valleys. There are minor exceptions, as in the rock gorge at Albany, near the drift dams at Dayton, and elsewhere (p. 129).

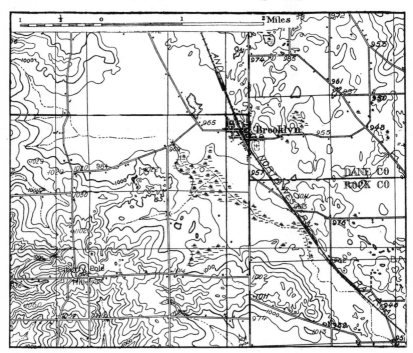

Fig. 101. Three types of topography and drainage between the Yahara and Sugar rivers. East of the railway, a poorly-drained area of Wisconsin drift in the terminal moraine of the Green Bay lobe; west of Brooklyn, a better-drained area of outwash; near Liberty Pole Hill, a well-drained area of older drift which now closely simulates the Driftless Area though doubtless once like the area east of the railway. Contour interval 20 feet. (From Evansville Quadrangle, U. S. Geol. Survey.)

The Yahara River. The Yahara or Catfish River furnishes a complete contrast to the Pecatonica. They are as different from each other as are the Mississippi and the St. Lawrence systems. The Yahara is interrupted by lakes. It has few tributaries. There are broad, undrained, interstream areas, including numerous large swamps. Its drainage pattern is not systematic but aimless. Its grade is not perfected, as in the Pecatonica, but is typical of a youthful stream. This grade is characterized by steep pitches and flat reaches.

The Yahara rises northeast of Lake Mendota at an elevation of nearly 1000 feet. The stream flows through extensive swamps, where the channel almost entirely disappears at places, only to reappear again below. From Lake Mendota to Lake Kegonsa the Yahara descends 6 feet in 8.6 miles, but only a few miles of this distance is in a channel, being for a greater distance lost in the waters of Lakes Mendota, Monona, Waubesa, and Kegonsa. Below

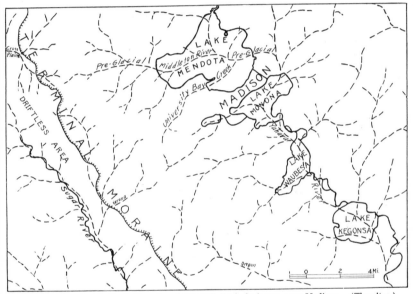

Fig. 102. Preglacial (dashed) and present (full line) drainage near Madison. (Thwaites.)

Lake Kegonsa and the city of Stoughton the grade is decidedly steeper and the valley has been eroded more deeply in the glacial deposits. From Lake Kegonsa to its confluence with the main Rock River north of Janesville the Yahara has a grade of nearly 4 feet to the mile. The lake basins which interrupt this stream are 20 to 70 feet deep, so that until the river has destroyed these lakes by cutting down their outlets or by filling their basins the stream grade cannot be perfected.

The preglacial course of the upper Yahara departs from the present course at many points. As shown in Figure 103 the departures are of small amounts. Nevertheless the modification of this stream by the continental glacier was revolutionary, for a graded, preglacial stream course has been converted by glacial erosion and deposition into a series of boat-shaped depressions which contain the present lakes.

It has already been explained that long before the Glacial Period the Wisconsin River may have flowed through the Yahara valley (Fig. 62 and p. 190).

Around the shores of the Yahara lakes, ridges of bowlders and till, called ramparts, are pushed up by the winter ice in favorable localities. The shorelines include rock cliffs, drift bluffs, beaches, barriers and spits.

In the region southwest of Madison there are abandoned stream channels which were occupied by glacial waters but now have tiny streams, far too small to have cut the channels. One of these may be seen about 3 miles east and another $3\frac{1}{2}$ miles southwest of Oregon.

Lake Wingra near Madison has been filled until it is nearing extinction, being now less than 14 feet deep. It occupies less than half the area of its former basin. This body of water was known to the Indians as Wingra, which is translated Dead Lake. Vegetation is contributing to the filling of the lake, but the chief cause for its extinction is the accumulation of marl, made up partly of the shells of small animals. This deposit is 26 to 30 feet thick in places.

The accumulation of marl, or bog limestone, in many lakes and swamps of eastern Wisconsin constitutes a valuable resource, not yet utilized. This was well appreciated by Lapham, who wrote in 1851:

"These beds constitute a great bank, not likely to be broken or to suspend payment, from which to draw future supplies of the food of plants, whenever our present soils shall exhibit signs of exhaustion."

The Main Rock River. The headwaters of the Rock River are in Fond du Lac County a few miles southwest of Lake Winnebago. Its grade is less than $1\frac{4}{5}$ feet to the mile. The course of the river

Rock River	Length in Wisconsin 154 miles	Elevation of headwaters 1000 feet	Elevation at Beloit 731 feet

lies entirely in glaciated territory and mostly within the great lowland of Black River limestone which extends north and south parallel to the Niagara escarpment. The northern part of this same lowland is occupied by Lake Winnebago, the Fox River, and

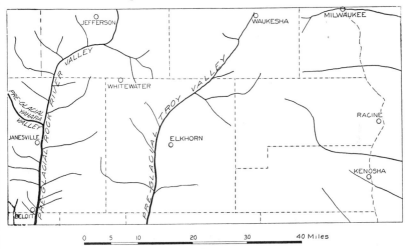

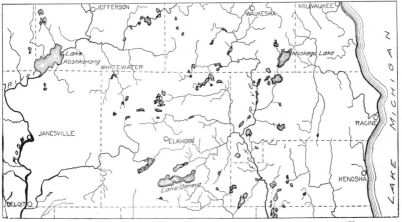

Fig. 103. Preglacial and present drainage in southeastern Wisconsin. (Upper map based partly upon data by Alden.)

Green Bay. The course of the Rock is interrupted by only three basins—Lake Koshkonong, the mill pond at Hustisford, and the Horicon Marsh. The Crawfish and Bark River tributaries of the Rock contain lakes, however, such as Beaver Lake and the Oconomowoc Lakes. Rock Lake and Lake Ripley lie on the western tributaries, and there are vast areas of swamp and woodland along the main stream, so that the run-off of the Rock River basin is somewhat regulated. Nevertheless the run-off during March is 5 to 7 times as great as that of the summer.

The Rock River near Janesville and Beloit was displaced from its preglacial course by the continental ice sheet. Its preglacial position has been worked out from well records and is known to have been east of Beloit, where it now lies to the west.

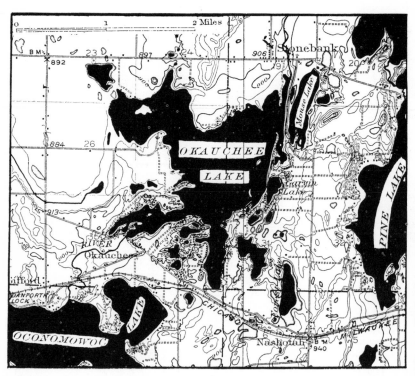

Fig. 104. Part of the Oconomowoc group of lakes in a belt of terminal moraine and glacial outwash. (From Oconomowoc Quadrangle, U. S. Geol. Survey.)

The Oconomowoc Lakes. The group of lakes at the well known summer resort near Oconomowoc, lies in a belt of terminal moraines and outwash plains. The outwash plains contain many pits or kettles, formed by the melting of buried ice blocks, and some of the lake basins are of the same origin.

Horicon Marsh and Other Swamps.   The Horicon Marsh is a filled lake (Fig. 105), occupying a shallow basin formed by glaciation.   This lake covered 51 square miles, being more than three times as large as Lake Mendota.   A dam was built here about 1842, forming a new lake, which was subsequently drained when

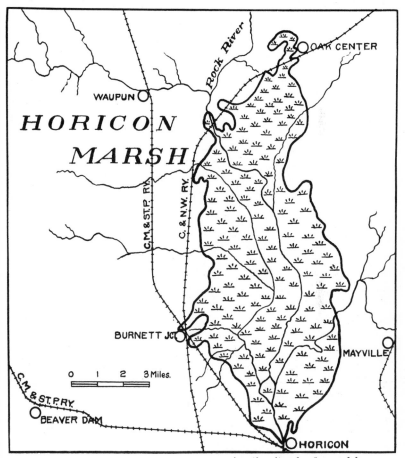

Fig. 105.   Horicon Marsh, a swamp occupying the site of a former lake.

the dam was removed in response to a court decision.   The marsh was partially drained about 1914 by a main ditch running north from the city of Horicon with several lateral ditches extending east and west from the main ditch.   Complete drainage was prevented by the backwater from a dam located in the village of

Hustisford several miles downstream from the marsh. The State is now working on a project designed to restore the waters on the marsh to the levels existing before the drainage. Borings show that the peat in the swamp is at least 5 or 6 feet thick, and is underlain by clay and gravel, so that the basin seems to have been originally much deeper and to have been filled by stream and lake deposits and then by vegetable accumulations.

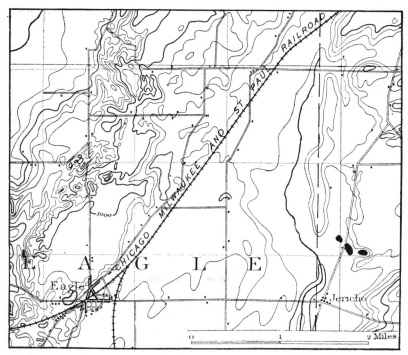

Fig. 106. Topographic map of the divide on the kettle moraine east of Rock River. Contour interval 20 feet. (From Eagle Quadrangle, U. S. Geol. Survey.)

The Scuppernong Marsh, 20 square miles, and the swamp south of Dousman, 15 square miles, are in the Bark River valley.

Other swamps and marshes abound in eastern Wisconsin, especialy in the Rock-Yahara and Fox-Winnebago systems. Peat, muck, and other swamp soils cover over 25 per cent of Waukesha and Fond du Lac counties.

The Pishtaka River. The Pishtaka, or Fox River of southeastern Wisconsin and Illinois, rises in the region just east of the Oconomowoc Lakes, having Pewaukee Lake as one of its chief reservoirs. It is separated from the lakes of the Bark River branch of the Rock only by the kettle moraine. Its grade is about $1\frac{3}{4}$ feet to the mile. The preglacial Troy valley (Fig. 103) may represent the former course of the Pishtaka; the present course probably had no preglacial counterpart.

Lakes Geneva, Delavan, and Beulah, and the Lauderdale group of lakes are among the bodies of water tributary to the Pishtaka River. Lake Geneva earns its reputation as a summer resort through the beauty of its surroundings. The lake lies in a belt of strong terminal moraine. The preglacial valley at Lake Geneva is buried beneath 220 feet of glacial drift. The hills around the lake rise 100 to 200 feet above the water surface, and the lake is 142 feet deep.

Muskego Lake and Wind Lake are broad, shallow bodies of water, now rapidly being extinguished by the immense swamps which surround them.

Abandoned Channels. Near Burlington, Elkhorn, and Lake Geneva are a number of broad, swampy valleys, far too large for the streams which now occupy them. These channels were cut by glacial waters when the ice sheet was nearby. One conspicuous abandoned outlet of the glacial floods extends from Troy to Delavan. It is partly drained today by Turtle Creek. Another series of abandoned outlets is near Burlington and Genoa Junction, including Nippersink Creek. A third, west of Vienna, Walworth County, contains Sugar Creek (Fig. 96).

The Desplaines River. The Desplaines River rises in Wisconsin and flows southward into Illinois. Its course in this state seems to be determined by the recessional moraines of the Lake Michigan glacier. In Illinois it is within 11 miles of Lake Michigan at the city of Chicago, but flows southward into the Illinois and Mississippi rivers.

THE FOX RIVER SYSTEM

Wolf River and the Upper Fox. This drainage system consists of the Fox and Wolf River headwaters, Lake Winnebago, and the lower Fox. In the headwaters, which lie in the Central Plain, are Lakes Shawano, Poygan, Buffalo, Green and Puckaway. Close to Lake Winnebago is Big Lake Butte des Morts.

Above Lake Winnebago the grade of the Fox is very gentle. The government has built 9 locks and dredged 110 miles of canal between Oshkosh and Portage. This is now largely unused because

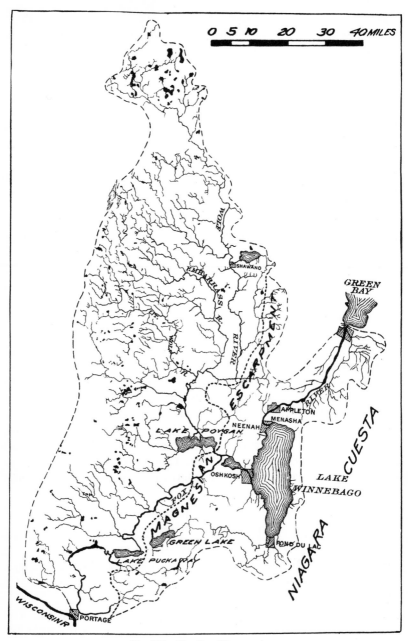

Fig. 107. The Fox River system.

| Wolf River........ | Grade above Shawano, 10 feet to the mile | Grade from Shawano to Fox River, less than ½ foot to the mile |
| Fox River.......... | Grade above Lake Winnebago, less than ½ foot to the mile | Grade below Lake Winnebago, over 5 feet to the mile |

of railway competition. The federal government has spent more than 3½ million dollars on the improvement of the Fox River, above and below Lake Winnebago. It still expends fifty or sixty thousand dollars a year on repairs, operation and improvements.

Lake Winnebago. This is the largest inland lake in Wisconsin. Lake Winnebago is 28 miles long, 10½ miles wide, and covers 215 square miles. Its maximum depth is only 21 feet. The natural dam which holds in the lake on the north is made up chiefly of glacial drift. The western shores of the lake are low. The high cliffs of the Niagara escarpment on the east are not due to wave work but to preglacial and glacial erosion of resistant limestone underlain by weak shale.

Near the south end of Lake Winnebago are extensive marshes where attempts have been made to develop peat as a commercial fuel.

The Lower Fox River. Below Lake Winnebago the old valley of the Fox is buried and the stream is in a postglacial course. In crossing the walls of its old valley, the river descends very steeply over successive rapids, furnishing a series of the best water powers in Wisconsin. Their value has been estimated at over five thousand dollars for every ten hours. The United States government has built 18 locks and 35 miles of canal between Lake Winnebago and Green Bay, and the State government spent $400,000 here between 1849 and 1852.

There are 8 rapids in the lower Fox. Four of them are responsible for the location of four large manufacturing cities,—Neenah-Menasha, Appleton, Kaukauna, and Depere. The other four have determined the situation of four smaller manufacturing centers,— Kimberly, Little Chute, Combined Locks, and Little Kaukauna or Little Rapids. The presence of these rapids in the lower course of the Fox River furnishes a decided contrast with Driftless Area streams like the Pecatonica. The lower Fox River has not only an average grade of 5 feet to the mile, in contrast with the flat grade of the upper Fox (p. 281), but the descent near Lake Winnebago is even steeper. From the city of Appleton to the foot of the rapids at Kaukauna, the river descends 134 feet in 9 miles, or nearly 15 feet to the mile. This is because it is crossing the rock ledges from its course northwest of Lake Winnebago into its buried valley, or the drift-filled valley of some preglacial stream. It has cut 50 or 60 feet into the drift, so that it flows through a gorge, with banks of red till.

The Fox River in History. Depere is at the first rapids encountered by the French explorers as they left Green Bay, and hence the first place where a canoe had to be portaged. Accordingly Père Allouez established his mission of St. Francis Xavier at this point in 1671, only 51 years after the Pilgrims landed at Plymouth

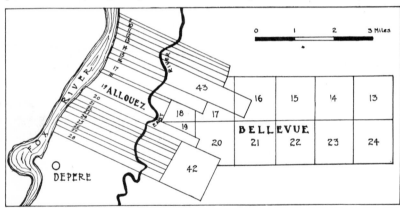

Fig. 108. The French land system, with long narrow farms fronting on the Fox River, and the American land system with square farms. (See Appendix F.)

Rock, in Massachusetts. Fathers Claude Allouez and Louis André built the mission house of St. Francis Xavier at this place during the winter of 1671-72. It was maintained as a center of missionary effort for many years. The rapids came to be known as *"Rapides des Péres"* or the Rapids of the Fathers. The name of the old village and modern city was subsequently shortened to Depere. The old name—La Baye—included both Green Bay and Depere.

The mouth of the Fox River is the harbor of the city of Green Bay. This is the oldest permanent settlement in Wisconsin. A French fort was built there by 1683, and a settlement of retired traders and soldiers dates from the early eighteenth century. When Fort La Baye was abandoned by the French in 1760, the

English sent a garrison, which in 1761 repaired the old French post and renamed it Fort Edward Augustus, for one of the royal princes. This English post was evacuated in 1763 during Pontiac's Conspiracy, and there was no fort at this place until the American troops in 1816 built Fort Howard.

The long, narrow farms, running back from Fox River, recall the French régime, since most of Wisconsin has the square farms of the township-and-range system. Only at such places as the upper Mississippi near Prairie du Chien, the lower Mississippi near New Orleans, and the lower St. Lawrence River in eastern Canada

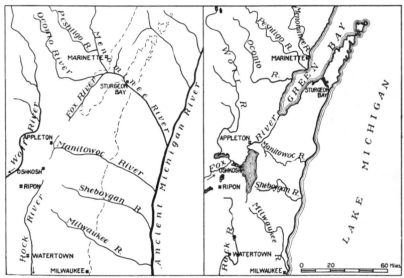

Fig. 109. Sketch map showing hypothetical preglacial and present actual drainage in eastern Wisconsin.

are these long, narrow farms preserved. They serve to recall the days when rivers, not railways, controlled travel as well as methods of living, the days of the far-flung colonial realm of New France.

Preglacial Valleys of the Fox-Winnebago Region. It seems possible that before the Glacial Period the waters of Wolf River flowed southward through what is now the upper Fox River and joined the Wisconsin near Merrimac, or perhaps west of Portage (Fig. 62).

The preglacial Fox River, lacking the Wolf River tributary, was thus a stream of much less volume than at present. It probably flowed in the middle of the lowland of Black River limestone west of the Niagara escarpment, perhaps heading north of Lake Winnebago and flowing northward to Green Bay (Fig. 109). The preglacial Manitowoc River may have drained part of the Winnebago lowland eastward through the low ground near Brillion. There is a buried stream channel, near Kaukauna. It was probably interglacial rather than preglacial, and its relations to the preglacial Manitowoc River or the preglacial Menominee River are not yet worked out. There was no lake Winnebago (Fig. 109, left-hand figure) in preglacial time, but the position of Rock River was similar to the present, except for minor differences as shown in Figure 103. The buried channel revealed by borings near Sturgeon Bay was probably interglacial rather than preglacial, since the bed of the preglacial Menominee River in the present Door County is likely to have been at least as high as the existing canal surface, if not still higher.

The most important economic result of the glacial invasion in the Fox River valley was (a) the addition of the Wolf River system to the upper Fox, giving it more than three times its former volume, (b) the formation of Lake Winnebago, and (c) the change from the preglacial course to the present course of the lower Fox with its eight rapids. These three things have made the lower Fox River more valuable for water power by adding (a) volume, (b) steadiness of flow, and (c) places of steep descent over rock. This has resulted in the development of the most important industrial district in Wisconsin.

Streams Flowing into Lake Michigan

Streams Entering Green Bay. The Menominee, Peshtigo, Oconto, and other streams flowing into Green Bay lie mostly in the Northern Highland and will be described later (Chapter XVI). All the waters entering Lake Michigan 'from Wisconsin flow down the eastern slopes of the Niagara cuesta.

Milwaukee River and Its Neighbors. The Milwaukee River, whose general course is described on page 231, rises southeast of Lake Winnebago at an elevation of about 1018 feet. Its head waters include the Cedar Lakes of Washington County. The river has a length of about 75 miles, and empties into Lake Michigan at the city of Milwaukee. Its grade averages 5 feet to the mile. The mouth is converted by the waters of the lake into an estuary,

across the mouth of which is a sand bar. This makes the best and
most important harbor in eastern Wisconsin, a harbor from which
the tonnage shipped annually exceeds that at such ports as Boston, Philadelphia, and New Orleans. The form of Milwaukee Harbor in 1836 is shown in Figure 110. The lower valley of the Mil-

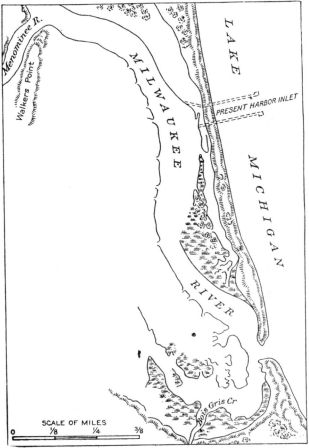

Fig. 110.   The site of Milwaukee in 1836, showing a sand bar across the mouth of the
Milwaukee River. (U. S. Army Engineers.)

waukee, and its tributary, the Menominee River, have been partly
converted into made-land.

The Kewaunee, Manitowoc, and Sheboygan rivers are similar
to the Milwaukee in most respects. The Manitowoc (p. 286) is the
only stream completely crossing the Niagara cuesta. Its channel

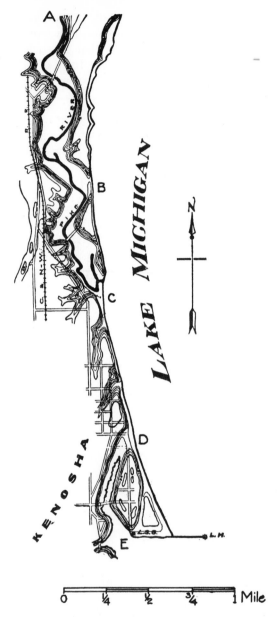

Fig. 111. Pike River, which formerly flowed from A to E. The waves cut into the side of the valley from C to D, leaving an abandoned valley from D to E. They have cut into it again at B, but the river is not yet diverted at that point. (Goldthwait.)

may have been the outlet of one of the early glacial lakes in Green Bay and the lower Fox River.

The Niagara escarpment in southern Wisconsin is so low that a canal from the Rock River to Milwaukee was planned about 1836. It was to go via the Oconomowoc Lakes to the Rock River near Jefferson. The cost was estimated at nearly a million dollars, and some construction was actually undertaken before the project was finally abandoned. The Federal government donated 200,000 acres of land and the territory of Wisconsin sold some of these lands to help finance the canal.

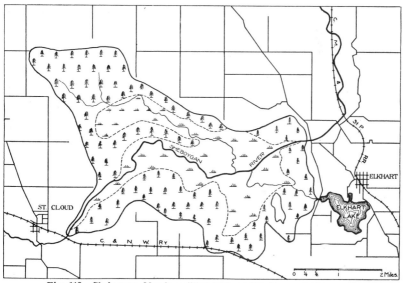

Fig. 112. Sheboygan Marsh, a filled lake. (After Sinz and Peterson.)

A Stream Diverted by Wave Work. Pike River near Racine and Kenosha flows parallel to the shore of Lake Michigan. The waves have cut into the side of the valley in two places (Fig. 111). Thus it now empties into Lake Michigan north of Kenosha, the portion of the valley near that city being without a stream. The headwaters of Pike River flow in a general north-south direction between the recessional moraines of the Lake Michigan glacier (Fig. 98).

Swamps of the Niagara Cuesta. The Sheboygan River has among its reservoirs, Cedar or Crystal Lake, Elkhart Lake, and a large swamp near Plymouth and Glen Beulah, known as Sheboygan Marsh. The latter was formerly a lake, for it has beach ridges, wave-cut cliffs, and ice ramparts. The swamp (Fig. 112) covers 15⅘ square miles. It was originally occupied by a body of water a little larger than Lake Mendota at Madison. Borings show

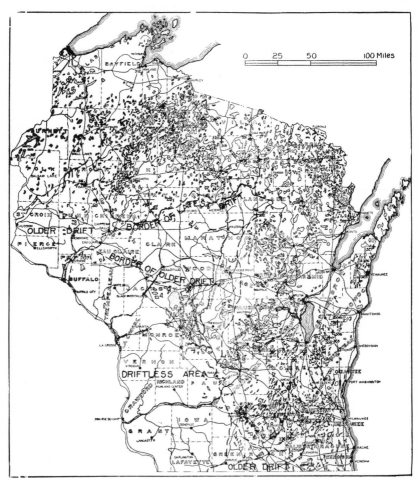

Fig. 113. The swamps of Wisconsin—dotted areas. The Driftless Area and regions of older drift are largely without swamps. (After Huels and Gillis.)

that it was at least 45 feet deep. It has 9 feet of peat at the surface, underlain by marl and clay. Elkhart Lake is a part of the original Sheboygan Lake. There was also a shallow lake in the middle of the present marsh before 1868. In that year $50,000 was expended in an attempt to drain the marsh, half of this sum being provided by the State.

The Manitowoc Swamp, east of Chilton, covers 15⅔ square miles, representing another extinct lake.

Swamps occupy a smaller proportion of the Niagara cuesta than of the Rock River-Lake Winnebago lowland to the west (p. 280). Peat, muck, and related marshy soils cover nearly 30,000 acres in Kewaunee County, or 13 per cent of the area.

PEAT RESOURCES OF EASTERN WISCONSIN

Many of the swamps alluded to in this chapter can be drained and cultivated. They are also potential sources of peat for fuel, either directly or in the form of producer gas. Other possible uses of peat include the production of chemical by-products, alcohol, fertilizers, paper, woven fabrics, artificial wood, mattresses, packing material, and dyestuffs.

In 50 peat deposits of the state which have been examined and tested, it is estimated that there were 121,000 acres, containing 151 million tons of peat, worth 155 million dollars. There may be as much as 2 or 3 billion tons of peat in Wisconsin. A large proportion of it is in the Eastern Ridges and Lowlands and the Northern Highland. A striking feature of the map showing distribution of swamp lands (Fig. 113) is their limitation to the area of latest glacial drift. Accordingly, we may say that the introduction of the drainage conditions that result in the making of swamps has furnished Wisconsin with a natural resource of great value. The peat resource is not yet utilized, the briquetting plants at Fond du Lac and Whitewater being no longer in operation.

PRAIRIES OF EASTERN WISCONSIN

There were two sorts of areas which were treeless when eastern Wisconsin was first settled. These were (a) the swamps (Fig. 113) and (b) the prairies (Fig. 114). The swamps are, of course, lowlands, but many of the prairies were on the drier uplands. The treeless areas did not cover a very large proportion of eastern Wisconsin. One of the larger prairies was north of Madison (Fig. 114) near Waunakee, Poynette, and Sun Prairie. Another was southwest of Ripon, and a third was near Janesville. The cause of the prairies has not been determined (see p. 139). The name of the principal Indian tribe encountered by the French explorers in the seventeenth century was the Mascoutin, or people of the prairies. The names of certain modern villages, such as Sun Prairie, recall the original treeless condition.

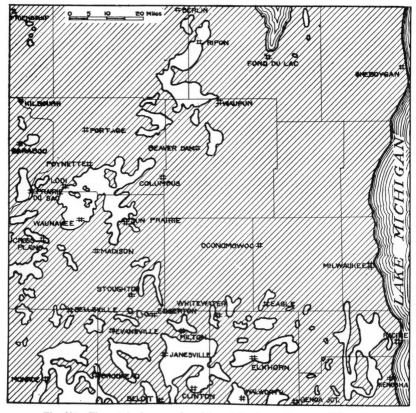

Fig. 114.   The original areas of prairie—white—in southeastern Wisconsin.

BIBLIOGAPHY

Alden, W. C.   Quaternary Geology of Southeastern Wisconsin, Prof. Paper 106, U. S. Geol. Survey, 1918, pp. 1-48.

Buckley, E. R.   Ice Ramparts, Trans. Wis. Acad. Sci., Vol. 13, 1901, pp. 141-157.

Cram, T. J.   Internal Improvements in the Territory of Wisconsin, Senate Doc. 140, 26th Congress, 1st Session, Washington, 1840, 26 pp., and large map; Survey of Neenah and Wisconsin Rivers, Senate Doc. 318, 26th Congress, 1st Session, Washington, 1840, 29 pp., and 16 maps.

Fenneman, N. M.   The Lakes of Southeastern Wisconsin, Bull. 8, Wis. Geol. and Nat. Hist. Survey, 1910, 188 pp.

Huels, F. W.   The Peat Resources of Wisconsin, Bull. 45, Wis. Geol. and Nat. Hist. Survey, 1915, 266 pp.

Juday, C.   The Inland Lakes of Wisconsin—Hydrography and Morphometry of the Lakes, Bull. 27, Wis. Geol. and Nat. Hist. Survey, 1914, 137 pp.

Lapham, I. A.   (On erosion of coast along Lake Michigan), Geology of Wisconsin, Vol. 2, 1877, pp. 231-232; Documentary History of the Milwaukee and Rock River Canal, Milwaukee, 1840, 151 pp.

Marsh, C. D.   The Plankton of Lake Winnebago and Green Lake, Bull. 12, Wis. Geol. and Nat. Hist. Survey, 1903, 89 pp.

Schwartz, G. F.   The Dimished Flow of the Rock River in Wisconsin and Illinois, and Its Relation to the Surrounding Forests, Bull. 44, Bureau of Forestry, U. S. Department of Agriculture, 1903, 27 pp.

Sinz, E. F., and Peterson, H. W.   Plans for the Draining of the Sheboygan Marsh, Unpublished thesis, University of Wisconsin, 1905, with map.

Smith, L. S.   The Water Powers of Wisconsin, Bull. 20, Wis. Geol. and Nat. Hist. Survey, 1908, 354 pp.

Thwaites, F. T.   Preglacial streams and contour map of bedrock topography near Madison, in Bull. 8, Ibid., 1910.

Thwaites, R. G.   Down Historic Waterways, Chicago, 1890, 1902, pp. 17-234.

Turner, F. J.   The Character and Influence of the Indian Trade in Wisconsin, Johns Hopkins University Studies, Vol. 9, 1891, 75 pp., also in Proc. Wis. Hist. Soc., Vol. 36, 1889, pp. 52-98.

Warren, G. K.   Report on the Transportation Route Along the Wisconsin and Fox Rivers in the State of Wisconsin, Senate Ex. Document 28, 44th Congress, 1st Session, Washington, 1876, 114 pp., including the first map of the lake in the Fox-Winnebago valley.

Whitbeck, R. H.   The Geography of the Fox-Winnebago Valley, Bull. 42, Wis. Geol. and Nat. Hist. Survey, 1915, 105 pp.

MAPS

See Chapters IX and X.

CHAPTER XII

THE WISCONSIN COAST OF LAKE MICHIGAN

THE PREDECESSORS OF LAKE MICHIGAN

The eastern border of what is now Wisconsin has had a varied and interesting history. There was a time when no lake existed, and when the present lake basin was occupied by a great river, flowing southward to the Gulf of Mexico. This was before the Glacial Period. It may have been over a million years ago.

There was a time when the region was completely buried by the continental glacier. Doubtless this condition was repeated several times. Each time the Lake Michigan glacier was retreating there was a lake at its borders. Each time the ice moved forward, after the very first advance, there was likewise a marginal lake. The waves of these lakes cut cliffs and built beaches, but every set of these shorelines, except the latest, was subsequently overridden and destroyed.

Finally the predecessors of Lake Michigan came into existence. The several stages of Lake Michigan differed decidedly from the present. The lake stood at a higher level, so that the sites of our modern cities of Racine, Kenosha, and several other lake ports, were deeply submerged (Fig. 117). The lake drained successively (a) southward into the Mississippi and Gulf of Mexico, (b) eastward into the Mohawk and Hudson rivers and New York harbor, and (c) into the Ottawa River and Gulf of St. Lawrence north of the present St. Lawrence outlet (Figs. 115, 116). The time when the later of these glacial lakes existed is not less than 20,000 to 35,000 years ago.

The Lake Michigan of today, with its present outlet and level, should not be thought of as the original lake, or one that has existed very long. The preglacial river and the iceberg-dotted lakes of the Ice Age were features of much greater duration.

THE GLACIAL GREAT LAKES

Ancient Shorelines. Our knowledge of the Glacial Great Lakes comes from the study of the abandoned shorelines which represent many different stages of these ancestors of Lake Michigan. They are at various levels and tell us of times when the waters were not only deeper, but spread over a strip near the coast of Wisconsin which is now dry land.

Causes for Changes of Level. These glacial lakes were held in by ice dams at the north, they outflowed through other outlets than the present St. Lawrence, and they fluctuated in level. The changes of level of the lake surfaces were due to two chief factors. The first of these was retreat of the continental ice sheet, resulting in changes in the positions of the ice dams from time to time and in the uncovering of outlets that were blocked by ice at earlier stages. The second factor was tilting of the land in the Great Lakes region, resulting in inclination of some of the shorelines and in the raising of certain outlets and their abandonment. A minor cause for fluctuation of the lake levels was the cutting down of the outlets. Sometimes the ice readvanced, causing the lake to rise and form a slightly higher shoreline than the one just before. Usually the lake levels kept falling from one stage to the next.

A Short History of the Great Lakes. Three of the chief stages in the history of Glacial Lake Michigan are known as Lake Chicago, Lake Algonquin, and the Nipissing Great Lakes. Glacial Lake Chicago was confined to the southern part of the basin of Lake Michigan and drained into the Mississippi by an outlet at Chicago (Fig. 115 and upper map in Fig. 116). Lake Algonquin included most of Lakes Superior and Huron as well as Lake Michigan and drained into New York harbor by the Mohawk and Hudson rivers (middle map, Fig. 116). Part of the time its waters seem to have used the Chicago outlet. Its eastward drainage reached the Mohawk outlet partly by way of Detroit and Lake Erie and partly by a now-abandoned outlet from Kirkfield, Ontario, into Glacial Lake Iroquois, which was in the basin of Lake Ontario. The Nipissing Great Lakes occupied the basis of the three upper Great Lakes (lower map, Fig. 116), with a minor outlet past Detroit into Lake Erie and a chief outlet, now abandoned, from North Bay, Ontario, down the Ottawa River to the lower St. Lawrence.

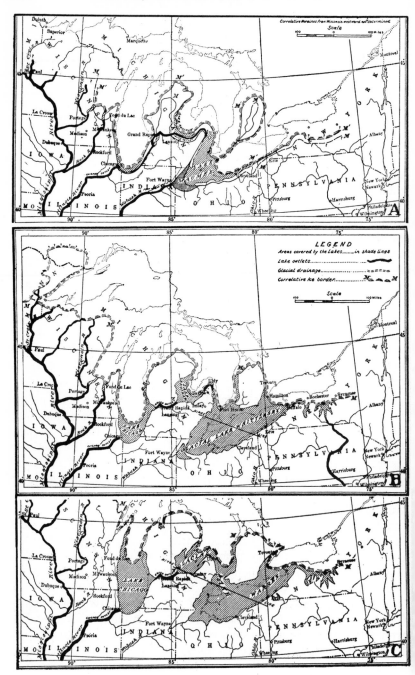

Fig. 115. Three stages in the early history of the Glacial Great Lakes. (Taylor and Leverett.)

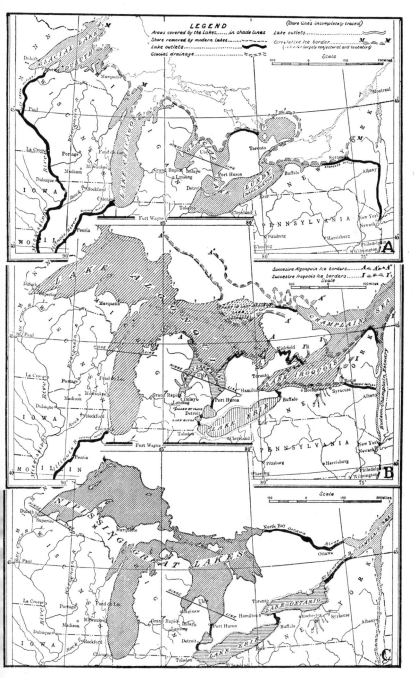

Fig. 116. Three stages in the later history of the Glacial Great Lakes. (Taylor and Leverett.)

Abandoned Shorelines of Glacial Lake Chicago. The Lake Chicago shorelines are rather well preserved in Racine and Kenosha counties and in parts of Ozaukee and Sheboygan counties. Else-

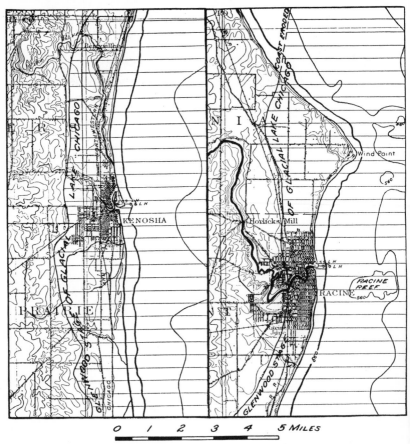

Fig. 117. Map showing the old shoreline of Glacial Lake Chicago at the Glenwood stage when the sites of the cities of Racine and Kenosha were submerged. Area of lake shown by horizontal ruling.

where they seem (a) never to have been formed because the ice sheet was still present, as to the north of Sheboygan, or (b) to have been destroyed by a readvance of the Lake Michigan lobe of the ice sheet, or (c) to have been cut away by the waves of later lakes, as is certainly the case in Milwaukee and southern Ozaukee counties. The remnants of these ancient shorelines are gravel and sand beaches or wave-cut cliffs and terraces.   Those associated with Glacial Lake Chicago are usually at one of three levels, respectively about 55, 38 and 23 feet above the present lake.   The beaches and cliffs of these three stages are called Glenwood (highest), Calumet (middle) and Toleston (lowest).   South of Milwaukee the shorelines of Lake Chicago are horizontal, but north of this point they are tilted slightly.   The readvance of the glacier already alluded to (p. 267) may possibly have taken place between the Glenwood and Calumet stages of Glacial Lake Chicago, at the time when it received the drainage of Glacial Lake Whittlesey (middle map, Fig. 115).

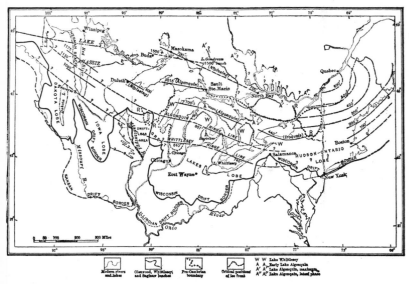

Fig. 118.   Diagram to show the position of the several hinge lines, where the beaches cease to be horizontal, in the basins of the Great Lakes.   See also Figures 115 and 117.   (Taylor and Leverett.)

Abandoned Shorelines of Glacial Lake Algonquin.   The existing beaches and cliffs in Wisconsin which represent Lake Algonquin are limited to the coast of Lake Michigan north of Two Rivers and to the shores of Green Bay.   To the south of Two Rivers they seem either to have been cut away during the Nipissing and Lake Michigan stages, or else to coincide with the Nipissing shorelines.   Near Two Rivers, the highest Algonquin shoreline is 26 feet above present lake level.   If it is still preserved to the south it must be essentially horizontal. It is 29 feet above Lake Michigan at Cormier, and 39 and 40 feet at Oconto and Sturgeon Bay. Along both coasts of the Door Peninsula and on the western shore of Green Bay the Algonquin beach is found at higher and higher levels as it is traced northward (Fig. 120). Thus it is about 40 feet above present lake level at Sturgeon Bay, and nearly 100 feet at Washington Island, the extreme northeastern corner of the state. The rate of inclination of these original-horizontal shorelines of Lake Algonquin increases toward the north, being 8 inches to the mile from Two Rivers to Sturgeon Bay and about 18 inches to the mile between Sturgeon Bay and Washington Island. Some of the shorelines split into two or more strands as they rise to the north (Fig. 120), due to the increase in rate of tilting toward the north.

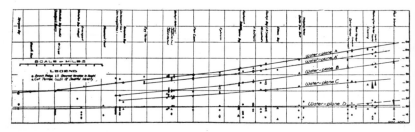

Fig. 119. The tilted water-planes in northeastern Wisconsin. Plane A passes through the highest beaches of Glacial Lake Algonquin; A', the lowest level of the same lake; B, the Battlefield beaches; C, the Fort Brady beaches; D, the beaches of the Nipissing Great Lakes. Washington Island is near the right of the diagram, Ephraim, at the middle, and Sturgeon Bay at the left. The circles and triangles indicate different types of beach ridges and wave-cut terraces. (After Goldthwait.)

The shorelines of two of the later stages of Lake Algonquin are called the Battlefield and Fort Brady beaches.

Among the soils near Marinette and Peshtigo is the Dunkirk fine sand, which probably represents an abandoned delta of Lake Algonquin.

Abandoned Shorelines of Nipissing Great Lakes. The beaches and cliffs of the Nipissing stage are similar to the earlier strand lines. The abandoned shorelines of the Nipissing stage are about 14 feet above the present level of Lake Michigan throughout most of its basin. They are nearly horizontal at the south, rising to about 18 feet at Two Rivers and 22 feet on Washington Island. They rise still higher to the north in the Lake Superior and Lake Huron basins.

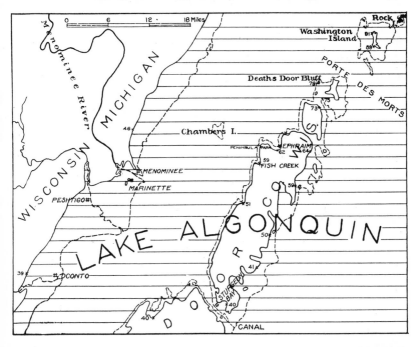

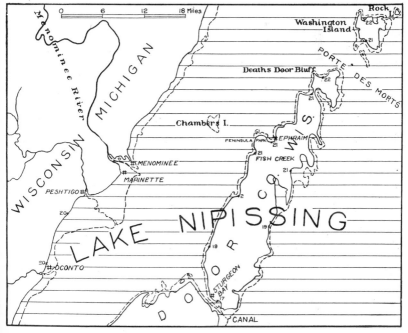

Fig. 120. Two stages of the Glacial Great Lakes in northeastern Wisconsin. Coast of Lake Michigan shown by dashed lines. Figures show heights of abandoned beaches in feet above present lake level.

Glacial Lake Jean Nicolet. During the earlier stages of Lake Chicago there was a small independent glacial lake in the Fox River valley. First there were two small lakes, which later united to form Glacial Lake Jean Nicolet. The name Glacial Lake Oshkosh has also been proposed for this body of water. It was held in by the Green Bay lobe on the north and by the hills south, east, and west of Lake Winnebago. Its outlet was first, down the Rock River from near Fond du Lac and, later, down the Fox River valley to the Wisconsin at Portage. In its latest stage it existed too short a time to leave continuous shorelines, though there is a beach about 53 feet above Lake Winnebago. The state soil survey has mapped this beach of Glacial Lake Jean Nicolet as the Superior gravelly loam. An abandoned beach 7 or 8 feet above Lake Winnebago is mapped as the Clyde fine sand. There may also be an 84 foot beach. The lake clays which accumulated in the basin of this lake before the readvance which built the terminal moraine at Fond du Lac are among the sources of the red till already described (p. 266) in the Fox River valley.

Present Shoreline of Lake Michigan and Green Bay

Regularity of the Coast. Most of the coast of Lake Michigan, at its present level, is regular in outline, the chief exception being Door Peninsula and the contiguous islands. The length of the Wisconsin coast of Lake Michigan is 381 miles. From the end of Door Peninsula to the Illinois boundary there are a few capes,—all broad—but no bays or harbors except the mouths of small rivers. The coast is still undergoing modification by wave work. This tends to cut back headlands and fill up bays, making the coast still more regular in outline.

The Coast of Green Bay. The west coast of Green Bay is low, and the adjacent country ascends gently to the westward. The deltas of Menominee, Oconto, and Peshtigo rivers form broad, low points, and the mouths of these rivers are the only harbors. Cliffs are rare and beaches predominate. The east coast of Green Bay and the shores of Washington Island are bold and rugged. This is because of the Niagara escarpment, for the waves have had little time to cut cliffs in the rock. Some of the cliffs are cut in the glacial drift and red lake clay, as at Red Banks, northeast of the city of Green Bay (Fig. 121), where there is a hundred foot bluff. It seems probable that it was here that Jean Nicolet, the first white

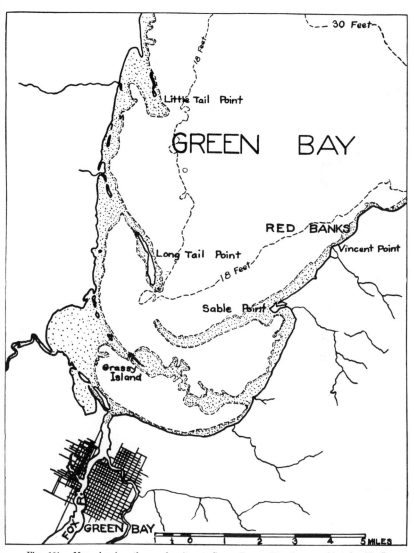

Fig. 121. Map showing the sand spits at Green Bay. (Based upon chart by U. S. Lake Survey.)

man to visit Wisconsin, landed and made his treaty with the Indians in 1634.

Sand Spits at Green Bay. There are three sand spits at the head of Green Bay (Fig. 121). They are not completed yet, like those at Superior (Fig. 190) and Ashland (Fig. 195), but are partly below lake level. The southernmost spit curves southeastward for $2\frac{1}{2}$ miles including Grassy Island, a mile and a half off shore from the mouth of Fox River. The shoals east of the steamboat channel mark its continuation. The second spit is 9 miles long. It extends nearly across Green Bay, including a spit on the west coast of the bay, then Long Tail Island, and finally a submerged sand bar which terminates in Sable Point on the east coast. This submerged bar rises within 3 to 5 feet of the surface. The third spit is only started as yet, extending southward from the west coast of Green Bay at Little Tail Point.

The Coast of Lake Michigan. The west coast of Lake Michigan has cliffs rather than beaches throughout much of the distance. Nearly all of them are sloping drift bluffs, rather than steep rock cliffs.

Coastal Lakes. The waves and currents have built bars of gravel and sand across the mouths of several deep indentations, converting bays into lakes. These coastal lakes in Door County include Kangaroo Lake south of Bailey Harbor, Clark Lake north of Whitefish Bay, and Europe Lake near the end of Door Peninsula.

Sturgeon Bay. The Sturgeon Bay gap across the Door Peninsula seems to represent the preglacial course of the Menominee River (Fig. 109), deepened somewhat by glacial sculpture, and submerged beneath the waters of Lake Michigan. It furnishes an ideal harbor and a short cut into Green Bay, now that its eastern end is connected with Lake Michigan by the government canal $1\frac{1}{3}$ miles long.

Recession of Wave-cut Cliffs. The bluffs along Lake Michigan are still being cut back. This rapid recession has destroyed some of the older shorelines. It is well shown at Pike River (p. 289) and at various places between Milwaukee and Racine, where the wave-cut cliffs retreated one to six feet a year along the whole coast lines of two counties between 1836 and 1874. At exposed points the bluffs recede even faster, for example at Racine, where the recession averaged nearly 10 feet a year for 24 years following

1840, and at Manitowoc in 1905, where the rate was 40 or 50 feet a year. This cutting helps keep the bluffs steep. The beaches at the cliff base are narrow, for the lake currents carry the eroded material away.

Reefs in Lake Michigan. There are conspicuous reefs near the shore of Lake Michigan, as at Racine, Sheboygan, Manitowoc, Whitefish Point, and Fisherman Shoal east of Washington Island. There are also minor shoals in Green Bay. The Racine Reef (Figs. 117, 122) may be described as typical of the group. It is half a mile broad by a mile and a quarter long. On the shallower part

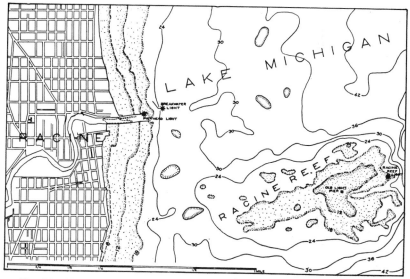

Fig. 122. Map of the reef at Racine. Depths in feet. (Based upon chart by U. S. Lake Survey.)

the water is $6\frac{1}{2}$ to 11 feet deep. It is surrounded by depths of 25 to 30 feet. The inner edge lies more than a mile offshore and the outer extremity is $2\frac{3}{8}$ miles from the mainland. The water is so shallow that it has been possible to build a lighthouse near the eastern border of the reef. There is an abandoned light pier near the center.

Whether these shoals are rock reefs, original glacial deposits, or deposits planed down below lake level by the work of waves and currents has not been determined. It does not seem certain that the original coast in the glacial deposits was as far offshore as the outer edge of the Racine Reef and other shoals. In that case there should be a continuous, wave-planed platform along the whole shore. It was once estimated that the coast from Manitowoc to Chicago had been cut back an average of 2.72 miles. The average rate of cutting was estimated at 5.28 feet a year. This gives 2720 years as the time of erosion of this coast by waves. The soundings on the detailed, recent charts do not reveal a well-developed wave-planed bench, so that this estimate is not generally accepted as a measure of the number of years since the Glacial Period.

Lake Michigan Harbors. As already stated, the harbors along Lake Michigan owe their existence to the estuary-like mouths of rivers. The artificially-deepened mouths of the rivers at Milwaukee (p. 287), Racine, Kenosha, Sheboygan, Kewaunee, Manitowoc, Port Washington, Two Rivers, and Algoma furnish lake ports which could not otherwise exist, on account of the regular character of the shoreline.

Winter Ice in Lake Michigan. The time during which the harbors of lake ports in Wisconsin are closed by winter ice varies with the local climate and local situation. Thus the harbor of Green Bay is frozen about 130 days in the year, usually from the latter part of December to the middle of April. Marinette, Sturgeon Bay, and Porte des Morts, between Door Peninsula and Washington Island, are usually closed only 107 days. Milwaukee Harbor, 130 miles farther south, is rarely ice-bound more than 14 days in the year, sometimes from Feb. 24 to March 10. Ice breakers usually keep this port and Lake Michigan open in winter, so that the extensive carferry commerce from Milwaukee, Manitowoc, and other lake ports in Wisconsin is little interrupted.

FLUCTUATIONS IN PRESENT LAKE LEVEL

Description by André. That there are periodic fluctuations in the level of Lake Michigan, Green Bay, and Lake Superior has been long a matter of observation. It was noticed at Green Bay by Father André in 1671 and Father Marquette in 1673. Father André said:

"Hitherto I have not shared the opinion of those that believe that Lake Huron is subject to an ebb and flow, in common with the Sea; because I had observed no fixed movement of the sort during the time of my sojourn on the shores of that Lake. But, after passing the so-called 'wild-oats river',[a] I began to suspect that there might really be a tide in the bay des Puans.[b] We had left our canoe in the water, in very calm weather; and the next morning were greatly surprised to find it high and dry. I was more astonished than the rest, because I bore in mind that for a long time the Lake had been perfectly calm. Thereupon, I determined to study this tide, and at the outset I reflected that the contrary, but very moderate, wind did not prevent the flow or ebb, as the case might be. I also became aware that, in the river emptying into the bay at its head, the tide rises and falls twice in a little more than 24 hours,—rising usually a foot; while the highest tide I have seen made the river rise three feet, but it was aided by a violent Northeast wind. Unless the Southwest wind is very strong, it does not check the river's course; so that ordinarily the middle flows constantly downward to the Lake, although at each end the water rises with the fixed periods of the tide. As there are but two winds prevailing on that river and on the Lake, one might easily ascribe to them these tides, were it not that the latter follow the Moon's course, a fact which cannot be doubted; for I have ascertained beyond a question that at full Moon the tides are at their highest, then they fall, and they continue to diminish as the Moon wanes. It is not surprising that this flow and ebb is more appreciable at the head of the bay than in Lake Huron, or in that of the Ilinois[c]; for were the tide to rise even but an inch in these Lakes, it would necessarily be very noticeable in the bay, which is about 15 or 20 leagues long by five or six, or more, wide at its mouth, and narrows constantly. Consequently the water, being contracted within a small space at the head of the bay, must of necessity rise much higher there than in the Lake, where it is less confined."

In 1672, Father André continued as follows:

"*The Small quantity of Paper that I have left reminds me of The promise that I made to Your Reverence last year, at The end of one of my Letters, to tell You what might seem to me* (I must not forget to tell what seems to me) *to be worthy of note in connection with The ebb and flow of our river.*[d] It is quite certain that it has its tides like those of the seas,—or, more properly speaking, of the

[a] Menominee River. It is referred to more often as the wild-rice river, or the folle avoine.
[b] Green Bay.
[c] Lake Michigan.
[d] Fox River at Depere or Green Bay.

rivers that fall into them.   The unusual severity of The winter
this year caused me to make an observation that hitherto could not
be made.   During The month of March I remarked that The highest winter tide is lower than The lowest of all The tides of the
other seasons, when neither The bay nor The river is frozen.   It
was necessary to advance a considerable distance on The river to
find water under The ice, which was a foot and a half thick; and
The surface of The ice was not higher than The low tides of summer, or The average of both The highest and lowest tides.

"I also observed that The volume of water increased in our river
during that month, in proportion as The ice in The bay of saint
Xavier[a] diminished and broke up.   This cannot be attributed to
The greater abundance of water flowing from above, because The
tide extended only as far as the foot of the rapid,—which is easily
seen at present, but not in the summer, when one does not *observe*
(perceive) that there is a rapid, because The lowest tide is generally higher than it.   These two observations have troubled me,
for I formerly believed that The winds were not the Cause of The
tide.   Were I permitted to Philosophize, I would argue against
those who attribute The formal Cause thereof to rarefaction, special or general.   For if The water rarefies and then Condenses, all
that great mass of water of the Lake of the Ilinois[b] rises in its
vast basin when it rarefies, and falls when it condenses.   And, as
water always rises as much as it falls, it would follow that, however thick The ice of The bay and of The river might be, they
would offer no more resistance than a pipe,—which, however thick
it may be, never prevents The water from rising as much as it has
fallen, for it does not press against it.   And, although it may be
said that The ice presses on The water, still it cannot be said that
it prevents The water from rising; for, while pressing on The
water, it floats on it; and The Ice should be higher than The highest tides of summer and of autumn, or of spring,—or, at least, than
The mean tides, which is not the case.   *Opposite the folle avoine,[c]*
The ice was three feet thick,—That is, where The bay Begins.   But
twelve leagues from there, as one approaches the bottom of it and
our river, The ice was about a foot and a half thick.   Your Rever-
ence knows better than I The Length and Width of The bay, so I
shall not speak to You of them.   If The cause of the ebb and flow

ª Fox River at Depere or Green Bay.
ᵇ Lake Michigan.
ᶜ Menominee River.

be attributed to the winds, there will not be much difficulty in explaining how it happens that The lowest tides at the periods when there is no ice are higher than The highest tides of winter. For it will be said that The water, impelled by a violent motion, loses its force in proportion as it strikes against ice beneath Which it Flows, and consequently less water runs into The bay. I conclude by informing Your Reverence that the ice in The bay has commenced to break up toward the bottom, and not on the side of The entrance toward The Open water of Lake ilinois,[a] where The ice was three feet thick."

Description by Marquette. "The Bay[b] is about thirty leagues in depth and eight in width at its Mouth; it narrows gradually to the bottom, where it is easy to observe a tide which has its regular ebb and flow, almost Like That of the Sea. This is not the place to inquire whether these are real tides; whether they are Due to the wind, or to some other cause; whether there are winds, The precursors of the Moon and attached to her suite, which consequently agitate the lake and give it an apparent ebb and flow whenever the Moon ascends above the horizon. What I can Positively state is, that, when the water is very Calm, it is easy to observe it rising and falling according to the Course of the moon; although I do not deny that This movement may be Caused by very Remote Winds, which, pressing on the middle of the lake, cause the edges to Rise and fall in the manner which is visible to our eyes."

The Seiches. The fluctuation described by André and Marquette is not a variation from month to month or from season to season with the rainfall and the volume of water supplied by rivers, or a variation in relation to the prevailing winds, but a tide-like rise and fall once or twice during the same day. It was noticed in the Great Lakes as early as 1636. The amount of fluctuation at Green Bay was carefully measured nearly a century ago and was found to be from a few inches to about a foot. The phenomenon is not a tide, however, like that on the sea coast and due to the moon, but a response to varying pressure of the atmosphere. It is called seiches.

The Tides. There are also actual lunar tides on the Great Lakes. Where measured at the south end of Lake Michigan and the west end of Lake Superior their range is about 3 inches.

[a] Lake Michigan.
[b] Green Bay.

Seasonal Fluctuations. Another type of fluctuation in lake level is the variation of a foot to eighteen inches from low water in January to high water in July or August. This is explained by melting ice and snow, spring rains, fluctuation in river volume and in evaporation.

Twelve Year Fluctuations. There are also groups of years when the lake is rising and others when it is falling. At Milwaukee and other points on Michigan the water surface has varied from 1 to 4¾ feet (Fig. 123) in relation to sea level.

TABLE SHOWING FLUCTUATIONS IN THE LEVEL OF LAKE MICHIGAN

High	1800	-------	1814	------	1826	------	1838	------	---a--	------
Low	-----	1811	-------	1820	------	1835	------	1841a	------	---a--
High	1861	-------	1871	------	1876	------	1886	------	1908	------
Low	-----	1868	-------	1872	------	1879	------	1896	------	1915

ª No record from 1843 to 1855.

These fluctuations in water level touch man's activities in building houses and piers, especially in the days of early settlement.

LEVEL OF LAKE MICHIGAN.

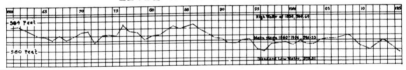

RAINFALL AT MILWAUKEE

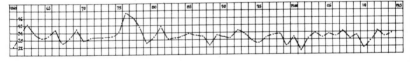

SUNSPOTS.

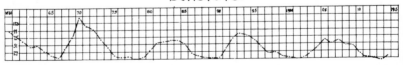

Fig. 123. The fluctuations in the level of Lake Michigan from 1860 to 1915. The variations in rainfall at Milwaukee during the same period. The periodic recurrence of abundant sun spots. It is not certain whether the variations in rainfall are related to the sun spots or not.

Solomon Juneau, the first settler at Milwaukee, remarked in 1838 that the water had never been as high as in that year, when the old Indian race-course was 6 feet under water, and that he had never seen it as low as in 1820.

It should not be thought that lake level is persistently falling or rising. It rises and falls with the fluctuations of rainfall for periods of 10 to 14 years.

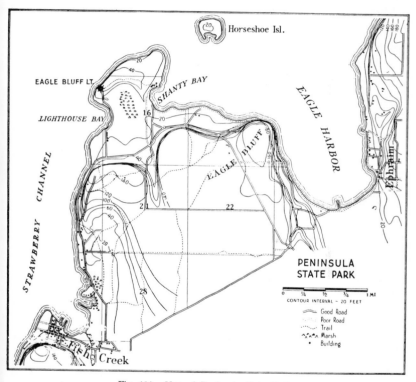

Fig. 124. Map of Peninsula State Park.

PENINSULA STATE PARK

General Description. The State Park in Door County is an especially good place to observe shorelines,—present and abandoned. Peninsula Park covers about 5⅓ square miles. It is situated on Door Peninsula east of Chambers Island, between the villages of Ephraim and Fish Creek. The park includes the whole peninsula west of Eagle Harbor, as well as Horseshoe Island.

The region includes the higher, western part of the Niagara cuesta. The escarpment rises abruptly above the waters of Green Bay with a bluff 140 feet high on the west coast. It is not an unbroken line of cliffs. There are headlands and reentrants. Eagle Bluff, on the east coast of the peninsula, has a height of 180 feet. The Niagara limestone underlies the peninsula. Corals are sometimes to be found in the higher ledges.

Shorelines. The present shorelines are (a) rock cliffs on the headlands and (b) beaches within the bays. The latter are well developed in the bay at Fish Creek, inside the park, and at the head of Eagle Harbor which lies to the east, near Ephraim.

The abandoned strand lines, likewise, include cliffs and beaches. At Fish Creek there are 8 beaches above the present one. They are made up of sand and rounded gravel or cobblestones or shingle.

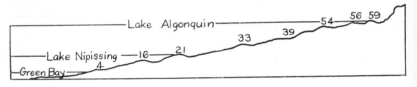

Fig. 125. Abandoned beaches at Fish Creek, Peninsula Park. (After Goldthwait.)

The highest beach is nearly 60 feet above Green Bay (Fig. 125). It was made by the waves of Glacial Lake Algonquin (p. 299). Below it are 4 other beaches, associated with the falling levels of Lake Algonquin. The main beach of Glacial Lake Nipissing is about 21 feet above present lake level, and there are 2 lower beaches.

The peninsula west of Shanty Bay at Eagle Point lighthouse was probably an island in Lake Nipissing (Fig. 124).

The tilting of the Algonquin beaches is shown between Fish Creek and Ephraim, for the highest strand line is 3 feet higher near the latter village. The upper beach of Lake Nipissing is at the same level in the two places.

The wave-cut cliffs of the headlands in Lake Algonquin and Lake Nipissing are to be seen (a) as independent precipices back of Fish Creek, (b) merged with the present cliffs on the peninsula of the state park between Fish Creek and Ephraim. At one point the waves of the glacial lake eroded a cave in the cliff (Pl. XIX, B). It is about 30 feet above the waters of Eagle Harbor.

A. WAVE-CUT BLUFF OF THE TOLESTON STAGE OF GLACIAL LAKE CHICAGO
NORTH OF MILWAUKEE.

B. CAVE ERODED IN THE NIAGARA LIMESTONE AT PENINSULA STATE PARK.

It was made during the later stages of Glacial Lake Algonquin.

A. THE LEVEL PLAIN NORTHEAST OF CAMP DOUGLAS WITH MESAS AND
BUTTES AS OUTLIERS OF AN ADJACENT ESCARPMENT.
(Photograph by D. W. Johnson.)

B. A NEARER VIEW OF ONE OF THE CASTELLATED ARID-LAND AND DRIFT-
LESS AREA FORMS SHOWN IN THE UPPER PICTURE.

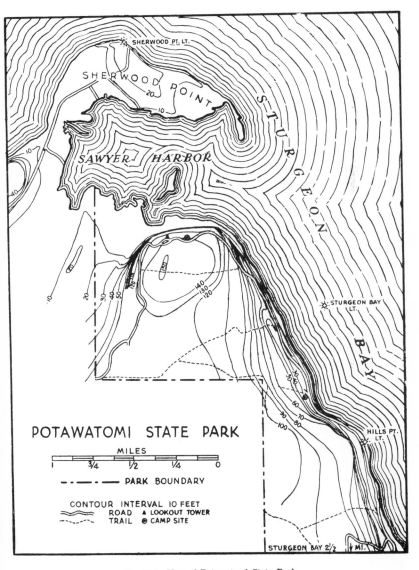

Fig. 126. Map of Potawatomi State Park.

Scenic Features. The charming scenery of the Peninsula Park, with its evergreen and hardwood forests, its open dales and woodland roads, its abandoned beaches and cliffs, its bathing beaches along the indented shoreline and islands of Green Bay, and the delightful summer climate, all make this a region that vies in attraction with the coast of Maine and the Adirondack and Canadian lakes.

POTAWATOMI STATE PARK

This park, consisting of 1100 acres, is located on the west side of Sturgeon Bay near the junction of Green Bay and Sturgeon Bay. For many years this land was held by the War Department because of the Niagara limestone quarry located upon it. The summit of Government Bluff (Fig. 126), 150 feet above the waters of Sturgeon Bay, commands a magnificent view of Sturgeon Bay, the city, and many miles of Green Bay shoreline. Within the park are abandoned shorelines of Glacial Lake Algonquin and the Nipissing Great Lakes.

BIBLIOGRAPHY

Alden, W. C.   Quaternary Geology of Southeastern Wisconsin, Prof. Paper 106, U. S. Geol. Survey, 1918, pp. 326-339.

André, Louis.   The Tide in the Bay des Puans (Green Bay), Jesuit Relations, 1671-72, Thwaites edition, Vol. 56, Cleveland, 1899, pp. 137-139; Remarkable Facts Concerning the River (that Discharges into the Bay des Puans at the Bottom of the Cove), Jesuit Relations, 1672-73, Ibid., Vol. 57, 1899, pp. 301-303.

Andrews, E.   The North American Lakes, Considered as Chronometers of Post-Glacial Time, Trans. Chicago Acad. Sci., Vol. 2, 1870, 23 pp.

Ball, J. R.   The Intercision of Pike River Near Kenosha, Wisconsin, Trans. Illinois State Acad. Sci., Vol. 13, 1921, pp. 323-326.

Case, E. C.   A Peculiar Form of Shore Ice, Journ. Geol. Vol. 14, 1906, pp. 134-137.

Dearborn, H. A. S.   On the Variations of Level in the Great North American Lakes, with Documents, Amer. Journ. Sci., Vol. 16, 1829, pp. 78-94.

Desor, E.   On the Superficial Deposits (including present and ancient beaches), Geology of the Lake Superior Land District, Part 2, Washington, 1851, pp. 232-273.

Flint, A. R.   Water Levels (in Lakes Michigan and Superior), Rept. Chief Eng., U. S. Army, 1882, Appendix 4 to Appendix SS, House Ex. Doc. 1, Part 2, 47th Congress, 2nd Session, Washington, 1882, pp. 2818-2819 and Plates 3 and 4; see also annual reports of U. S. Lake Survey.

Gilbert, G. K. Changes of Level of the Great Lakes, The Forum, Vol. 5, 1888, pp. 417-428.

Goldthwait, J. W. The Abandoned Shore Lines of Eastern Wisconsin, Bull. 17, Wis. Geol. and Nat. Hist. Survey, 1907, 134 pp; Correlation of the Raised Beaches on the West Side of Lake Michigan, Journ. Geol., Vol. 14, 1906, pp. 411-424; Ibid., Vol. 16, 1908, pp. 459-476; Ibid., Bull. Geol. Soc. Amer., Vol. 21, 1910, pp. 227-248.

Hattery, O. C. Survey of Northern and Northwestern Lakes, Bull. 24, U. S. Lake Survey, War Dept., Corps of Engineers; Detroit, 1915, 455 pp. This is an annual publication, with corrections and additions in monthly supplements.

Henry, A. J. Variations in Lake Levels and Atmospheric Precipitation, Nat. Geog. Mag., Vol. 10, 1899, pp. 403-406.

Hobbs, W. H. The Late Glacial and Post Glacial Uplift of the Michigan Basin, Publication 5, Mich. Geol. and Biol. Survey, 1911, pp. 45-46.

Lane, A. C. Level of Lake Huron, Geol. Survey of Michigan, Vol. 7, 1900, pp. 35-39 and Plate 5.

Marquette, Jacques. (On fluctuation in lake level), Jesuit Relations, 1673-77, Thwaites edition, Vol. 59, Cleveland, 1900, p. 99; Ibid., Hennepin's translation in "A New Discovery of a Vast Country in America," Part 1, London, 1698, p. 321.

Martin, Lawrence. Marginal Lakes (of the Lake Superior and Lake Michigan basins), Monograph 52, U. S. Geol. Survey, 1911, pp. 441-452.

Taylor, F. B. A Reconnaissance of the Abandoned Shore Lines of Green Bay, Amer. Geol., Vol. 13, 1894, pp. 316-327; Glacial and Postglacial Lakes of the Great Lakes Region, Smithsonian Report for 1912, No. 2201, pp. 291-327; History of the Great Lakes, Monograph 53, U. S. Geol. Survey, 1915, pp. 321, 326-328, 431, 459, etc., Pls. XIV, XVI, XVIII, XIX, XXI, XXIV.

Upham, Warren. Glacial Lake Nicolet and the Portage between the Fox and Wisconsin Rivers, Amer. Geol., Vol. 32, 1903, pp. 105-115; Glacial Lake Jean Nicolet, Ibid., pp. 330-331.

Weidman, S. The Glacial Lake of the Fox River Valley and Green Bay and its Outlets, Science, new series, Vol. 33, 1911, p. 467.

Whittlesey, Charles. On the Observed Fluctuations of the Surfaces of the Lakes, Geology of the Lake Superior Land District, Part 2, 1851, pp. 319-331.

Whiting, H. Remarks on the Supposed Tides, and Periodic Rise and Fall of the North American Lakes, Amer. Journ. Sci., Vol. 20, 1831, pp. 205-219.

MAPS

See Chapter IX, p. 233.

Fig. 127. Outliers left behind in the recession of the irregular escarpment at the western border of the Central Plain in the Driftless Area. (From Dells Quadrangle, U. S. Geol. Survey.)

CHAPTER XIII

THE CENTRAL PLAIN

THE CAMP DOUGLAS COUNTRY

A Frontier of the West. If a traveler, on his way from eastern United States to the Pacific coast, be fortunate enough to cross central Wisconsin by daylight he will pass through the village of Camp Douglas or the village of Merrillan. At Camp Douglas and Merrillan, and for many miles nearby, he may see landscape features totally unlike those anywhere else in the United States east of the Mississippi River. The hills of the region near Camp Douglas are buttes and mesas. They have the straight lines, steep cliffs, and sharp angles of an arid country rather than the soft curves of a humid region. This is the very frontier of the true West.

Camp Douglas is on the main line of the Chicago, Milwaukee, St. Paul and Pacific Railway and at its junction with the Omaha Line, a part of the Chicago and Northwestern system. It is just south of the trunk line of the Chicago and Northwestern, which passes through Wyeville and Merrillan, where it crosses the Green Bay and Western Railway. Thus a journey across Wisconsin, by any one of the three railways which carry the heaviest passenger traffic, takes one into what has been termed the Camp Douglas Country. It extends from Kilbourn through Mauston and Camp Douglas to Tomah, and from Adams or Camp Douglas through Wyeville and Black River Falls to Merrillan and Humbird.

Castellated Hills, an Escarpment, and a Level Plain. The features to be seen near Camp Douglas and Merrillan are (a) isolated, rocky hills which resemble ruined castles, (b) grotesque towers and crags of sandstone along a line of bold, irregular bluffs, and (c) an unusually-flat plain, which stretches away beyond the northern and eastern horizons. The bluffs and steep slopes on the west and south form the escarpment at the border of the Western Upland. The level country is the Central Plain of Wisconsin. Not all of the Central Plain is exactly like the Camp Douglas Country. This, however, is a representative part, and one of the most beautiful and striking.

Origin of Topographic Forms. The irregular bluffs are part of an escarpment capped by resistant rock. Sometimes the escarpment is steep and densely-wooded, sometimes it slopes more gently and supports fields and farms. The isolated castles and crags are outliers of the escarpment, left behind in its recession to the south and west under the attack of weather, wind, and streams. The flat plain has been made by the wearing down of weak and nearly-horizontal sedimentary rocks, and by deposition of unconsolidated materials upon the surface.

Arid-Land and Driftless Area Forms. The work of the wind and of the weather have played an exceedingly important part in the production of the landscape near Camp Douglas and Merrillan, because this is part of the Driftless Area. Absence of glacial erosion and of direct glacial deposition make it possible for the rather fragile rock forms produced by weathering and wind work to persist and to dominate the landscape. Hence the west-bound traveler, who has been in glaciated territory all the way from New York or Chicago to the vicinity of Baraboo or Wisconsin Dells or Adams, Wisconsin, has seen nothing of this sort before. Moreover he is getting into the edge of the belt of lessened rainfall, where the work of the wind and weather assume larger proportions. Continuing westward he will leave the Driftless Area and again pass into glaciated territory. Accordingly the arid-land and Driftless Area forms near Camp Douglas and Merrillan are not repeated short of the Great Plains in the Dakotas and Montana. Even these are partly in glaciated territory.

Coniferae and Cacti. The evergreen trees, clinging in precarious positions on the rocky buttes and mesas of the Camp Douglas Country, and the tamaracks, on the swampy, level plain, are among the first forerunners of the northern forest. They furnish a notable contrast to the open prairies near the Great Lakes, and to the deciduous trees of the East and South. With these conifers are many scrub oaks. The evergreen trees show definitely, however, that the region néar Camp Douglas is not arid. The sandy soil makes the precipitation less effective, because much of the rainfall sinks into the ground at once. Wind work is dominant, not because we are in the arid lands but because we are in a sandy part of the Driftless Area. The smaller plants on the hills at Camp Douglas, and at other places in the Driftless Area, nevertheless, include several types of dwarf cacti, such as the prickly pear. Thus the evergreens, suggesting the North, and the spiny plants, suggesting the West, mark this as a frontier region for the traveler from some parts of the East or South.

Contrast with Other Parts of the Central Plain. Large parts of the Central Plain are decidedly unlike the Camp Douglas Country. Where there is less alluvial filling, the Central Plain is more hilly. Where there has been glaciation, the buttes and mesas are few or wanting. The swamp which stretches northeastward from Camp Douglas is two-fifths as large as the state of Rhode Island (Fig. 141), but only a small proportion of the Central Plain is swampy. The soil near Camp Douglas and Merrillan is sandy so that farms are poor and settlement therefore sparse, but parts of the Central Plain are densely populated. Instead of corn or wheat, the crops are apt to be potatoes, buckwheat, oats, rye, barley, and cranberries.

The roughness of the bluffs and buttes may appeal to some travelers as grotesque rather than beautiful. The flatness of the plain, with its Pine Barrens and its swamps, may give an impression of dreariness and monotony, similar to that of the Great Plains west of the Mississippi. Other portions of the Central Plain of Wisconsin, however, have the soft swells of glacial topography and the homelike aspect that goes with waving fields of grain, prosperous farm houses, and thriving villages. In the eastern part of the Central Plain the Indians lived long before white men came to Wisconsin. Of this region the Jesuit priest Allouez wrote in 1670:

"These people are settled in a very attractive place, where beautiful Plains and Fields meet the eye as far as one can see."

Shortly afterward Father Marquette said:

"I took pleasure in observing the situation of this village. It is beautiful and very pleasing; for, from an Eminence upon which it is placed, one beholds on every side prairies, extending farther than the eye can see, interspersed with groves or with lofty trees. The soil is very fertile and yields much Indian corn."

General Description of the Central Plain

Area and Altitude. The Central Plain of Wisconsin is a crescentic belt, covering about 13,000 square miles. All of it is floored by the weak Cambrian sandstone, except in the northwest, where the removal of the sandstone has exposed the underlying Keweenawan lavas for a small area. The surface elevation ranges from

1242 feet at Cumberland, Barron County, in the western part of
the crescent, to 785 feet at Portage, Columbia County, in the central part of the plain, and 685 feet at Ellis Junction, Marinette
County, near the eastern end of the lowland.   The general slope
from place to place is very gradual indeed; for example in the 65
miles from the northern to the southern edge of the plain the grade
is only 4⅜ feet to the mile.

The local relief varies considerably, but except for a few isolated
hills it is nowhere great.   It rarely exceeds a few score or a hundred feet, even where one of the larger rivers, like the Wisconsin, Black, and Chippewa, has incised its valley in the plain.   It
is not all a continuous plain, but in many places is a region of low
hills, as northwest of Wisconsin Dells, north of Tomah, south of
Pray, and southeast of Black River Falls.   Parts of the plain are
due to (a) smooth river deposits, (b) lake-bottom accumulations,
(c) vegetation in swamps, or (d) glacial drift.   In fact there is
very little of the present surface due directly to the erosion of the
flat-lying Cambrian sandstone.

An excellent place to see the variety of features of the Central
Plain is from Saddle Mound south of Pray—Tremont—in Jackson
County.   This hill rises more than 400 feet above the surrounding
lowland, its crest being over 1400 feet above sea level.   To the
southeast lies a vast, monotonously-even plain of lake deposits.   To
the southwest is an equally level plain of non-glacial alluvium,
stretching away till it terminates at the blue wall of the Western
Upland.   In the immediate foreground are low, rounded hills of
sandstone.   Such resistant layers as cap Saddle Mound have been
removed from the adjacent hills by weathering and erosion, so that
these hills have been reduced to the stage of old age in their erosion cycle.   The nearer landscape has a marked western aspect.
One might perfectly well be in Wyoming rather than Wisconsin.
The scrub timber grows in bunchy groups.   There is no grass, and
the white sand appears between the shrubs.   The hill slopes dis-
close vertical cliffs, angular profiles, flat-topped ridges, teepee
buttes, much as on parts of the Great Plains.   Side by side are vast,
swampy plains, which need ditches to drain away the water, and
sandy plains and low slopes which need water.

The group of hills east of Millston, Jackson County, preserves a
large sample of the topography that characterized the Central
Plain before the Glacial Period.   They are high enough to rise

THE GLACIATED PART OF THE CENTRAL PLAIN, SOUTHEAST OF HANCOCK, WAUSHARA COUNTY.

THE SANDSTONE CRAGS OF THE NEILLSVILLE NUNATAK.

above the alluvial and lacustrine plain. They have moderate slopes, occasionally with steep cliffs. At one point 6 miles east of Millston there is an extensive deposit of rounded, preglacial gravels. Here all the pebbles are local sandstone and chert. Little, if any, of the Central Plain had level topography before it was smothered in the glacial outwash, lake clay, and alluvium. It was everywhere a region of slight relief, however.

Boundaries. In drawing the boundaries of this geographical province (Figs. 3, 7, 9) the author has followed the contact of the Cambrian sandstone with the pre-Cambrian crystalline rocks on the north and with the Lower Magnesian limestone on the southeast, south, and southwest, excepting in one or two areas. In west central Wisconsin the Lower Magnesian limestone has been so recently removed from the Cambrian sandstone that a considerable sandstone area has the hilly topography of the Western Upland rather than the smooth surface of the Central Plain. It has, therefore, been included in the former province. Another way of stating the matter is to say that the northeastern boundary of the Western Upland is at some places the Ordovician, Lower Magnesian limestone, at other places the Cambrian, Dresbach formation. Actually the Dresbach sandstone escarpment terminates the Central Plain northwest of the Baraboo Range. Because. of uncertainty as to whether the escarpment is everywhere determined by the Dresbach formation, or is sometimes due to a cuesta-maker in the Mt. Simon formation, it has seemed best to refer to the whole escarpment as the Magnesian escarpment (p. 54). The author would have named it the Sandstone escarpment had he not desired to indicate its identity as the continuation of the Magnesian escarpment of eastern Wisconsin. Often the escarpment contains a double-step—one on the Dresbach sandstone, the other on the Lower Magnesian limestone. This is true, for example, southwest of Mauston, Juneau County, where the whole escarpment is 430 feet high.

One student of Wisconsin physiography considers that the time is already ripe for definitely recognizing a portion of the Western Upland southwest of the Central Plain as the Franconia cuesta. The upland of this cuesta, he feels, can be traced from the Baraboo Range northwestward along the face of the higher Magnesian cuesta (see "Franconia Upland", Fig. 12) to the St. Croix River. The width of this Franconia cuesta ranges from a mile or so to more than 25 miles. It is as distinct as the uplands underlain by the Lower Magnesian limestone, and is separated from the main Western Upland by an escarpment which exposes the Jordan, Madison, Trempealeau, and Mazomanie formations. Outliers of this escarpment, where small, have conical shapes distinct from the smooth rolling uplands on the thin-bedded or shaly Franconia sandstone. At the outer edge of the Franconia cuesta is the Dresbach escarpment, the most striking line of cliffs and steep slopes in the Driftless Area. It contains the castellated hills and cliffs described at the beginning of this chapter, as well as scores of similar scenic features. Some of the outliers, especially at Camp Douglas where the sandstone is well case-hardened, stand above the Central Plain in precipitous cliffs and have flat mesa tops; others, where the rock is softer and weaker, have sloping borders and conical forms.

The same investigator emphasizes another feature at the border of the Central Plain, the Eau Claire bench. The Dresbach sandstone, the great cliff-maker of the Cambrian group at the northeastern border of the Western Upland, is nowhere much over 100 feet thick. At the bases of many of the cliffs, west of Wisconsin Dells, for example, there is a broad bench underlain by the shaly Eau Claire sandstone. Interrupted portions of this bench are also found in the Tomah region. The Eau Claire bench is said to be better marked farther north near Eau Claire than at Camp Douglas and Tomah. This bench is not a cuesta and is by no means as widely distributed, areally, as the Franconia cuesta. It affords illustrations of the protection afforded to soft and weak sandstone by a shale or other impervious rock formation, which acts as a roof does to a house. In regions like Adams County, where there is no shale in the Eau Claire formation, the Dresbach cliffs extend down into or possibly through the Eau Claire without visible topographic expression.

Neither the Franconia cuesta nor the Eau Claire bench, distinct and important as they are locally, seem to the author of this book to have sufficient areal extent to justify separate description and recognition in a work designed for state-wide reading. The Western Upland looks down upon the Central Plain with a double stepped or triple stepped escarpment.

The boundary of the Central Plain is by no means well defined at all points. It should be noted that, owing to the former lack of topographic maps, the model of Wisconsin (Plate I) does not show all of the hills of this district. The Wisconsin, Mississippi, and several other rivers have cut through the Lower Magnesian limestone to the Cambrian sandstone, but of course these river strips have not been included in the Central Plain.

Northeast of St. Croix Falls, in Burnett and Polk counties, a low part of the Keweenawan lavas would have been included in the Northern Highland had it not seemed desirable to attach the isolated western area of Cambrian sandstone to the rest of the Central Plain. This is an area of trap ridges, totally unlike the sandstone hills of the Central Plain. The ridges are obscured by glacial drift.

Topography and Origin of the Plain

The Plain Is an Inner Lowland. All the characteristics of the sandstone plain (Fig. 10) are normal to an inner lowland of a belted plain. The name inner lowland is used in connection with slightly-dissected coastal plains. Where uplift takes place in a coastal plain, made up of alternate layers of weak and resistant rock which dip gently toward the ocean, it will be carved by streams and the weather into such a form as is shown in Figure 128. An asymetrical ridge underlain by resistant rock, if of small dimensions and steep dip, is referred to as a monoclinal ridge. These large monoclinal ridges are called cuestas. Cuestas have been discussed in Chapters III and IX, in connection with the Western

Fig. 128. Diagram to show the recession of an escarpment from B to C and its former position at A.

Upland and the Eastern Ridges and Lowlands. The belt of weak rock between the inner cuesta and the partly exhumed oldland or backland (Fig. 12) is the inner lowland.

The belted plain of Wisconsin is much like a belted coastal plain, for example that in Alabama and adjacent states of the South, except in the greater distance from the ocean, the height above sea level, and, to some extent, the resulting degree of stream dissection.

Maintenance of Size While Changing Position.   The northern edge of the Cambrian sandstone is retreating southward, under the attack of streams and the weather.   We have already noted the presence of sandstone outliers on the surface of the peneplain to the north (pp. 37, 38) and of inliers of crystalline rock in the deeper valleys of the Central Plain (Fig. 11).   As the stripped area of the Northern Highland peneplain increases in size, the width of the inner lowland would be decreased, were it not for the fact that the edge of the upland of Lower Magnesian limestone to the south is also being worn back.   This increases the area of the inner lowland on that side.   That this is going on is well shown in the ragged, retreating escarpment (Figs. 18, 127) in Sauk, Juneau, Monroe, and adjacent counties.   Because of this shifting of the inner lowland down the dip of the rocks, the position, the shape, and the elevation of the inner lowland above the sea is constantly changing.   Its size has remained, and will doubtless continue, fairly constant.

Difference Between Inner Lowlands and Peneplains.   A part of the inner lowland of Wisconsin near Camp Douglas, Juneau County, has sometimes been referred to as a peneplain or a baselevelled plain.   It is exceedingly level and it is a plain, but to refer to it as a peneplain involves unnecessary complications, (see page 55).   It is a plain, and it is a normal thing for an interior lowland of moderate age in weak rocks to be a plain.   Moreover there will be an equally perfect plain in the inner lowland of a lower level after it has shifted southward far beyond the site of Camp Douglas; but neither that plain nor the present one is a peneplain.

As a matter of fact, that particular locality owes its extreme levelness to valley-fill of glacial outwash, associated lake deposits, and some preglacial alluvium.   This is true superficially of much of Adams, Monroe, Juneau, and Jackson counties.   The lowland itself is due to the weak sandstone, rather than to the surface veneer of sand and gravel which makes it level.

Variations of Width and Elevation.   There is a notable difference of width and elevation in various portions of the Central Plain.

In eastern Wisconsin the Cambrian sandstone dips rather steeply and is thinner than in the central part of the state.   Accordingly the inner lowland is narrow.   Its width in Marinette County is only a little over five miles, but in Clark, Jackson, and Monroe counties, where the sandstone dips less steeply and is thicker, the width of the inner lowland is 50 miles. The width of the lowland seems to be chiefly determined by the erosion surface as related to the dip of the sandstone.

The elevation of the Central Plain, as already stated, varies considerably. In Marinette and Oconto counties heights of 801 feet above sea level at Gillett and 685 feet at Ellis Junction constitute the lowest part of the sandstone plain. The lowness is not wholly due to greater activity by streams and glaciers in removing the sandstone here than at Plainfield, Waushara County, where the level of the plain is 1108 feet, or Camp Douglas, Juneau County, where it is 935 feet, or Cumberland, Barron County, where it is 1,242 feet. The inclination of the surface of the peneplain, and of the Cambrian sandstone which rests upon it, is chiefly responsible for these differences, especially for the lowness of the narrow plain in Oconto and Marinette counties. As a result of greater warping of the buried peneplain there and the steeper dip of the sandstone, the inner lowland stands at a different level from the continuation of the same plain, which is 400 to 600 feet higher in central and northwestern Wisconsin.

On divides near the edge of the pre-Cambrian peneplain the Central Plain is high, while in river valleys near the edge of the limestone upland on the other side it is low. These differences are due to variations of rate of weathering and stream erosion, but the differences of elevation cited in the last paragraph are due largely to structural control preceding the erosion.

The Buttes and Mesas. Rising above the sandstone plain in places are numerous, usually flat-topped ridges and hills, often bearing the name mound. Since the lowland is not a peneplain, the name monadnock is not appropriate in describing these buttes and mesas. They are very abundant in the region north and west of Wisconsin Dells, in the vicinity of Camp Douglas, and in many other portions of the Central Plain. Most of them lie within the Driftless Area. The name mound is, however, inappropriate for forms whose outlines are rarely rounded. These so-called mounds have simple rock structures, flat tops, and cliffed sides. They are exactly like those hills of the Great Plains and arid southwestern part of the United States which are called mesas—Spanish for table—if they are large, and butte, if small. A few of these so-called mounds in Wisconsin are conical, but the overwhelming majority of them are flat-topped. Bald Bluff, also called Glover Bluff, Liberty Bluff, and Lime Bluff, is situated east of Friendship in Marquette County. It differs from most of its neighbors in the Central Plain in possessing complex structures rather than simple horizontal or gently dipping sedimentary rocks. Whether ice push or preglacial faulting is responsible for these complexities, has not been fully determined. The craggy sides of the mounds often look, from a distance, like ruined castles and towers, as at Roche à Cris in Adams County north of Friendship. Roche à Cris stands about 300 feet above the adjacent plain, its crest being about 1185 feet above sea level. It is a long, narrow, flat-topped ridge bordered by sheer precipices. Thus it is probably the steepest hill in Wisconsin. It is also one of the most conspicuous and beautiful. Friendship Mound, its southern neighbor, rises 85 feet higher, but is a much less striking topographic feature. Looking north, east, south, or west from Roche à Cris, one sees scores of sandstone crags and towers, such as Pilot Knob, Mosquito Mound, Bald Bluff, Long Mound, Bear Bluff, Rattlesnake Mound, and Dorro Couche. Their white battlements punctuate the monotonously-even, green plain which stretches eastward to the terminal moraine of the Green Bay lobe and westward to the escarpment of the Western Upland. One may well imagine that he is upon an island in Glacial Lake Wisconsin (p. 337) and that the waters of this vast inland sea still cover the expanse of tree-clad plains and swampy clearings at his feet.

A dozen miles to the west are Petenwell Peak and Necedah Mound. The former is a craggy ridge of sandstone, its summit well-nigh inaccessible. The latter is a rounded knob of quartzite, easily ascended. Necedah Mound is a partly-exhumed monadnock of the pre-Cambrian peneplain (p. 38). It is a mound rather than a bluff or butte. Petenwell Peak is a more modern hill.

Three stages in the erosion cycle are well represented by Friendship Mound, Roche à Cris, and Petenwell Peak. The first will eventually be reduced to the stages represented by the other two, but not until Petenwell Peak has been completely destroyed by weathering and erosion. The Elephants Back near Wisconsin Dells is much like Friendship Mound. The crags west of New Lisbon and near Camp Douglas (Pl. XX) represent the same stage as Petenwell Peak.

Allied to these forms· are the still smaller, isolated crags and pinnacles like Stand Rock at the Dells of the Wisconsin, and many others (p. 351). Unfortunately, not all of this interesting area has been mapped topographically.

The crags and castellated hills at Camp Douglas (Fig. 18) are isolated remnants, left by the retreat of the escarpment to the southwest. The height of these buttes and mesas above the adjacent sandstone plain is 100 to 300 feet. The walls are everywhere steep and portions of them are precipices. The smaller buttes and mesas have flat tops of fair size.

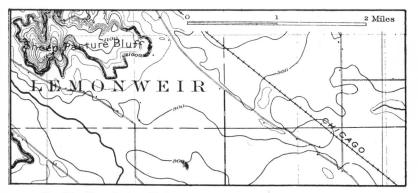

Fig. 129. Sheep Pasture Bluff, a mesa in the Central Plain. Contour interval 20 feet. (From the Dells Quadrangle, U. S. Geol. Survey.)

Sheep Pasture Bluff, in Juneau County, southeast of Mauston, is a typical mesa. Its dimensions are one-half to one mile by one and five-eighths miles. It rises, as the topographic map (Fig. 129) shows, to an elevation of 300 feet above the surrounding plain. The lower portion of its wall is a 100 foot cliff, probably in the Cambrian, Dresbach sandstone. Other mesas and buttes are Bruce Mound at Merrillan, the mesa at Humbird, and the castellated hills to the east of Black River Falls, south of Wyeville, west of New Lisbon, south of Mauston, and near Camp Douglas.

The sandstone outliers extend a long distance east and northeast of the Magnesian escarpment. The buttes east of Roche à Cris include Mosquito Mound at Bancroft, south of Stevens Point, Bald Bluff east of Friendship, and many others. Bald Bluff is about 25 miles from the Magnesian escarpment to the east and that to the west. It may equally well have been left behind by the recession of either one. Mosquito Mound, however, is 50 miles from the escarpment to the southwest and only 35 miles from the escarpment to the southeast. It is doubtless related to the Magnesian cuesta of Winnebago or Green Lake County (Fig. 130).

Cause of Mesas and Teepee-Buttes. The capping material which forms the flat tops of these buttes and mesas is one of several resistant sandstone layers, which is better cemented than the average. In the Cambrian sandstone the Dresbach formation is often a cliff-maker because heavy-bedded and soft. That certain layers in such a soft and relatively-weak rock as the Cambrian sandstone should stand up in places in precipitous cliffs, irregular crags, and towers is due partly to the porosity of the rock, partly to its lack of limy and shaly beds, and its thick-bedded character. Weathering and wind work are going on rapidly around the borders of many of these tabular hills. The sandstone breaks down along vertical joints, the rock falls to pieces, and is blown and washed away. The rimming cliffs are retreating, so that large mesas are being made smaller, and small mesas are being converted into buttes.

When the resistant cap is removed by weathering and the prevalent wind work, the buttes soon wear down to conical hills. These are well seen in the hills south of Pray (p. 320) as well as in Monroe County east of Sparta, in a valley of the Western Upland, where there are many teepee-buttes, rising 100 to 150 feet above the adjacent plain. They are apparently produced in weak sandstone after the wearing away of the protective cap of more resistant sandstone or limestone. The next stage is the complete destruction of the hill, and this is soon reached after the removal of the capping layer.

THE CHANGES DUE TO GLACIATION

Area Glaciated. The ice of the continental glacier advanced over the Central Plain of Wisconsin from both the northeast and the northwest; but an intermediate portion of the inner lowland in Wood, Portage, Adams, Juneau, Monroe, and Jackson counties was not overridden by the glacier. In this portion of the Driftless Area, except for the low areas mantled by stream and lake deposits, are found the topographic forms which must have characterized the whole geographical province before the Glacial Period. These enable us to tell something regarding the measure of glacial modification which has taken place elsewhere by erosion and deposition.

Glacial Erosion. Except in the sandstone counties west of Green Bay, the depth to which the continental glacier eroded the surface of the Central Plain was relatively less than in many other parts of Wisconsin. This is because the glaciated part of the Central Plain was mostly outside the regions of rapid glacial movement and near the edges of the Green Bay, Chippewa, Superior, and Minnesota lobes, where their erosive power was comparatively weak. If this inner lowland had been in a path of rapid ice movement the weak sandstone would doubtless have been eroded to a great depth, as was the sandstone of the Lake Superior basin (p. 434).

Though slight compared with the profound glacial sculpture of the Lake Michigan basin, the ice erosion in the Central Plain was by no means negligible. We know, for example, that the continental glacier must have removed hundreds of outlying mesas and buttes, and thousands of smaller pinnacles similar to Stand Rock, for these forms are now virtually limited to the Driftless Area. Before the Glacial Period they must have existed in vast numbers throughout the whole Central Plain. Likewise, we may be certain that along the Magnesian escarpment of eastern Wisconsin thousands of projecting spurs and isolated crags of sandstone capped by limestone were removed by glacial erosion. These ephemeral forms are plentiful in the Driftless Area (p. 91), as in the borders of the Western Upland near Camp Douglas. A markedly smaller number of outliers is found where the edge of the glacier was thin and weak, as in southwestern Columbia County near Okee and Lodi and at the border of the older drift near Neillsville. They are almost wanting in the regions of more active ice movement, for example near Knapp, Dunn County, west of Ripon, Green Lake County, or near Shawano, Shawano County.

Where the weak sandstone lay in the direct path of a rapid current of the glacier, the ice eroded deeply and scoured out lake basins. Green Lake, for example, is partly in the sandstone plain and partly in the limestone cuesta. Its valley was eroded to a depth of 300 to 450 feet by glacial erosion. The lake is 237 feet in depth, being the deepest inland lake in Wisconsin. How much of this erosion was accomplished by streams in preglacial time is not known, but it is certainly a small proportion of the whole.

The effect of glaciation upon the Central Plain was suggested in 1891 by G. K. Gilbert, who said:

"At Kilbourn City, Wis., the line of travel leaves glaciated territory and enters the Driftless Area of the Upper Mississippi Basin. Thence to La Crosse the topography and the constitution of the surface material stand in sharp contrast to the corresponding features of the region farther east. Rock exposures, which have been rare eastward, and altogether wanting over considerable areas, are here of almost constant occurrence wherever the surface has any considerable relief. Frequently, too, butte-like hills or fantastically carved, castellated towers of sandstone give some indications of the extent of the subaerial erosion the region has suffered. From the presence of these bold eminences within the Driftless Area, rising 200 or 300 feet above the more or less completely base-leveled plain on which they rest, and from their absence in the area covered by ice, instructive inferences may be drawn as to the work effected by the ice in the country over which it passed."

This is well proven by the numerous castellated hills and crags in the 50 miles of Driftless Area between Mosquito Mound, Portage County, and the Magnesian escarpment at Camp Douglas (Fig. 130), in contrast with the almost total absence of such outliers in the 35 miles of glaciated territory between Mosquito Mound and the Magnesian escarpment of Winnebago County.

A view from the summit of Mt. Morris, Waushara County, furnishes a striking contrast to the prospect from Roche à Cris. The latter is in the Driftless Area, and, as already stated (p. 326) overlooks a monotonously-even plain with scores of castellated mounds. Mt. Morris is in the midst of glaciated territory. Its flat top is bestrewn with granite bowlders, carried there by the ice. Its moderately-sloping sides reveal ledges of sandstone, but there are none of the crags and pinnacles that must have characterized the hill before the Glacial Period. Mt. Morris rises 250 feet or so above the adjacent plain. The plain is not smooth, however, but slightly irregular, with gently-undulating ground moraine, rougher terminal moraines, and outwash plains containing deep kettles and many valleys eroded since the outwash was deposited. The most striking contrast with the view from Roche à Cris is that almost no rock hills are to be seen from Mt. Morris. There is an exceedingly small number of limestone hills, as at the group near Bald Bluff in the edge of the terminal moraine. There are a very few exhumed monadnocks of pre-Cambrian rock, as near Spring Lake and Red Granite. There is Mt. Tom near Princeton, close to the Magnesian escarpment (Fig. 86). Where the driftless, Magnesian escarpment of Juneau County has scores and scores, and perhaps hundreds, of outliers, the glaciated continuation of the same escarpment in Green Lake County has outliers by ones and twos. Mt. Morris lies midway in the 35 miles from the Magnesian escarpment on the east to the terminal moraine on

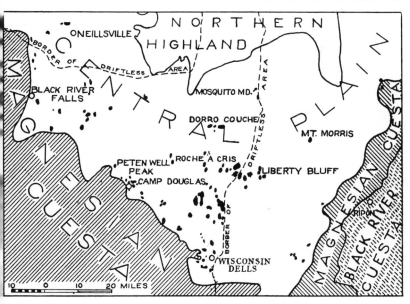

Fig. 130. Map showing—in black—some of the numerous outlying mounds in the Driftless Area in contrast with the small number in the glaciated region of central Wisconsin where they have been removed by glacial erosion. The former city of Kilbourn is now called Wisconsin Dells. Liberty Bluff is also called Bald Bluff.

the west. In this distance it is the only outlier. From the terminal moraine westward to the driftless portion of the Magnesian escarpment the outliers are exceedingly abundant, proving that it is glacial erosion which has removed the outliers in the eastern portion of the Central Plain.

Bald Bluff and the adjacent hills repeat the same story. They overlook a vast expanse of country devoid of rock hills. The rare exceptions, like Bald Bluff, were left because they lay near the thin, weak edge of the ice. If these outliers reveal any rock ledges whatever, as on the hill northeast of Bald Bluff, the ledges are confined to the southern or western slopes, where glacial plucking has produced them. Crags and pinnacles are entirely lacking.

Glacial Deposits. The drift deposits of the Central Plain are similar to those in the Western Upland (p. 123) and the Eastern Ridges and Lowlands (p. 253), except that the glacial deposits of the plain are decidedly more sandy. Such of the features of preglacial topography as escaped destruction by ice erosion, are often entirely extinguished beneath the ground moraine and terminal moraines, though in some places the preëxisting topography has not been greatly altered. The linear development of the sandstone bluffs often suggests that they represent divides between the preglacial stream valleys.

In contrast with the numerous glacial lakes in eastern Wisconsin, there are relatively few lakes in the Central Plain. This is chiefly because the depressions produced by glacial erosion and by irregularities in the drift are sandy-bottomed, and, therefore, do not generally hold water. It is also partly due to the lack of moraines. Where moraines are more abundant, lakes do occur, however, as in Burnett, Polk, Barron, and adjacent counties, in the northwestern part of the province, and parts of Waushara County in the east.

Older Drift. The older drift now preserved in the Central Plain was deposited chiefly by the ice of the Chippewa, Superior, and Minnesota lobes. There is no older drift along the borders of the Green Bay lobe. In many places the older drift is thin and weathered. Much of its surface bears a mature, erosion topography, rather than an irregular, glacial topography. It is much like the older drift of the Western Upland (p. 126) and the Northern Highland (p. 404).

It seems certain that the continental glacier covered little, if any, more territory in the Central Plain than is shown in Figure 28. At Necedah, Juneau County, there are linear markings on the quartzite of the mound. These resemble glacial striae but really are grooves called slickensides. They extend down into the solid rock ledges, and are not of glacial origin. Near Wisconsin Dells there are unquestionable glacial striae between the outermost terminal moraine and the Wisconsin River. It is not yet certain, however, that the granites which bear the striae are parts of a ledge rather than large, iceberg-rafted erratics. The widespread deposits of outwash, lake clay, and peat in the Driftless Area may possibly conceal patches of older drift in parts of the Central Plain which we now consider never to have been overridden by ice. Lack of glacial removal of the castellated bluffs, mounds, buttes, and pinnacles prove, however, that the Juneau, Adams, Monroe, and Jackson County portions of the Central Plain were never glaciated.

Ground Moraine of the Latest Glaciation. The till sheet of the Wisconsin stage of glaciation in the eastern portion of the Central Plain is thick, concealing nearly all the rock ledges. Its surface features are similar to those in the adjacent Fox-Winnebago valley (p. 255). Parts of Outagamie, Waupaca, Waushara, and Winnebago counties contain deposits of red clay. These are similar to those already described (p. 265) as having been laid down in a glacial lake and subsequently ridged up into ground moraine or terminal moraine by the readvance of the Green Bay lobe. This clay is mapped by the soil survey as the Superior series. The sandy till of ground moraine and terminal moraines appears on the soils maps as the Coloma sand and loam. The limestone pebbles are often entirely leached away in the surface layers, although limestone may be abundant at greater depths. This makes it clear that the presence of chert, quartz, and other insoluble silicious materials and the absence of limestone is not of itself an absolute proof that a deposit is older drift, rather than of Wisconsin age.

The Humbird and Neillsville Nunataks. A few castellated hills rise through the morainic deposits as nunataks. These are hills which were once completely surrounded by the ice, though never overridden. They are limited to the border of the glaciated region where the ice was thin.

The castellated bluff or mesa of iron-stained sandstone at Humbird, Jackson County, is one of the most accessible of these nunataks. The Chicago and Northwestern Railway passes close by its base. The ice of the continental glacier wrapped completely around the foot of this nunatak. The ice sheet terminated less than a mile to the southwest. Hence it was so thin at this point that it never rose to any considerable height on the slopes of the mesa. This is demonstrated also by the castellated and craggy character of the nunatak.

The Neillsville nunatak (Pl. XXII) in Clark County is of the same character as the one at Humbird. Both are exactly like the mesas and buttes of the Driftless Area, except that they are surrounded by glacial debris. At Neillsville a long, narrow ridge is made up of a series of castellated, sandstone crags and towers. On its lower slopes are glacial bowlders, but none are found on top.

The question has been raised as to whether the Neillsville ridge might not have been glaciated and subsequently restored to its preglacial form by weathering. It seems certain that there has been far too little time for this. A period long enough for such profound weathering would also be more than ample time for the destruction of all the glacial drift in the surrounding region, or its burial beneath a thick covering of sandstone talus. As a matter of fact the sandstone detritus nearby is very slight in amount and the erratics are exceedingly well preserved, even for older drift. On the northern slope of the mound is a marked, terminal moraine. No hesitancy is felt in asserting that the Neillsville ridge is a true nunatak.

This nunatak furnishes a rough measure of the slope of the continental glacier in central Wisconsin. This castellated ridge is about 10 miles from the outermost stand of the ice. Its crest is between 1400 and 1500 feet above sea level. No erratics are found within 200 feet of the top. The elevation of the terminal moraine to the southwest is between 900 and 1000 feet. If we assume that the ice rose steeply to a height of 50 or 100 feet at the border of the Driftless Area and then sloped gradually northeastward to the Neillsville nunatak, its surface gradient was not more than 15 or 20 feet to the mile.

Terminal Moraines. The terminal and recessional moraines in the Central Plain are of the normal sort. The kettle moraine at the eastern border of the Driftless Area is an exceptionally broad, irregular accumulation, well seen (a) north of Baraboo and Wisconsin Dells (Fig. 132), (b) east of Hancock, and (c) near Amherst Junction east of Stevens Point. South of Waupaca the recessional moraines of the Green Bay lobe form conspicuous ridges. John Muir described the kettles in one of the recessional moraines, northeast of Portage, as "formed by the melting of large detached blocks of ice that had been buried in moraine material."

Near the St. Croix Dalles, Polk County, there are two sets of terminal moraines. The outer, or eastern, moraines are made up of reddish drift from the Superior lobe of the Patrician ice sheet. The western moraines consist of younger, gray drift from the

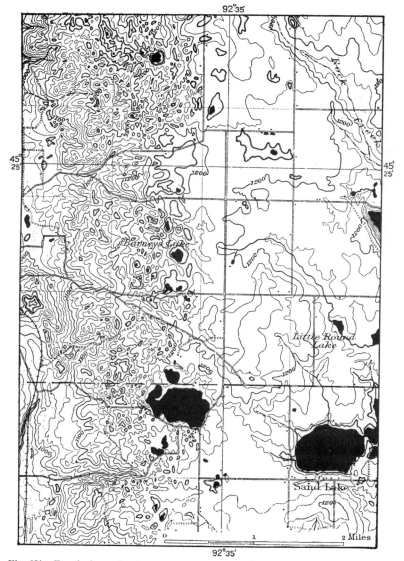

Fig. 131. Terminal moraine topography, on the left, and pitted outwash on the right. Contour interval 20 feet. (From St. Croix Dalles Quadrangle, U. S. Geol. Survey.)

Minnesota lobe of the Keewatin ice sheet. Remarkable as it may seem, the glacier was moving toward the northeast (see p. 454).

Outwash Deposits. The deposits made by streams from the melting ice sheet cover large areas in the Central Plain (p. 131). The streams laid down sand and gravel, not only in the actually glaciated region, but in the otherwise driftless area as well. Some of these glacial outwash deposits have been described as interglacial alluvium, but this conclusion appears doubtful. These grav-

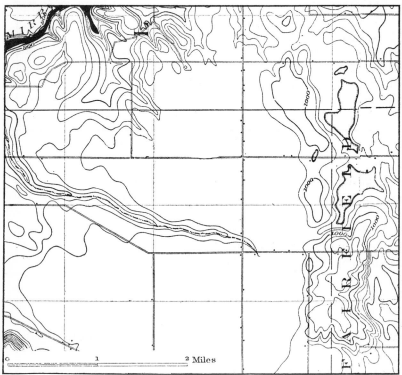

Fig. 132. Outwash plain built by glacial streams from the terminal moraine between Baraboo and Wisconsin Dells. Contour interval 20 feet. (From Briggsville and Dells Quadrangles, U. S. Geol. Survey.)

el, sand and clay deposits attain thicknesses of 100 to 200 feet or more, as at Necedah. Many of them have smooth, gently-sloping surfaces, as in the region between Wisconsin Rapids and Wisconsin Dells and to the northwest of Baraboo. Opposite certain gaps in the terminal moraines are broad alluvial fans of outwash gravels. These indicate some of the later stream channels occupied by the glacial rivers which deposited the outwash.

Near some of the rivers the outwash has been gullied by ravines and cut into great terraces, as in the valley of the Wisconsin River south of Wisconsin Rapids and east of Necedah, along the Chippewa River in Eau Claire and Dunn counties southwest of Eau Claire, in parts of Waushara County, and in Marinette County along the Menominee River. Near the St. Croix Dalles there are pitted outwash plains adjacent to the red moraines (Fig. 131), and terraced valley train gravels along the St. Croix River.

Lake Deposits. The deposits of former glacial lakes cover parts of the Central Plain, in the driftless as well as the glaciated area. These lakes were apparently short-lived, for they produced few well-defined shorelines. Deltas were built at many stream mouths. The existence of these bodies of water is proved by the finding of lake-bottom sediments, as near Grantsburg, Burnett County, and Menomonie, Dunn County, where the clays are used in making brick. The Grantsburg lake is discussed later (p. 456).

In the eastern part of the Central Plain there are extensive deposits of lake clay—Superior series and Poygan series on the soils maps.

Deposits on the Floor of Glacial Lake Wisconsin. The largest area of lake deposits is on the bed of Glacial Lake Wisconsin. This former body of water, already described in general, (Fig. 133 and p. 130), left lake deposits over an area of more than 1,825 square miles. In the southern part of its basin are reddish silts and sandy clays—Superior series of soils. These lake deposits are covered in many places by glacial outwash, by dune sand, by peat and muck, or by alluvium. In the northern and northwestern part of the basin, the lake deposit is white quartz sand. This is well seen in the embankments bordering the extensive drainage ditches of Wood, Juneau, and Jackson counties. The presence of sand rather than clay as the dominant lake deposit is due to the nearness to the lake shore and to the large area of crystalline rocks on the north. The glacial streams flowing into Glacial Lake Wisconsin from the north deposited their coarsest detritus near shore. The sand was carried offshore in suspension and deposited evenly over the bottom of the glacial lake. The clay was deposited still farther to the south. Similarly the glacial torrents flowing into this lake from the east deposited their coarse materials in the outwash plains east of the lake, as in the great gravel pit southeast of the village of Grand Marsh, Adams County. The finer detritus was carried offshore and spread over the lake bottom as a mantle of limy clay.

Ice-Rafted Erratics. The deposits of this glacial lake include isolated, erratic bowlders of granite, greenstone, and other crystal-

Fig. 133. Sketch map of Glacial Lake Wisconsin. The name of Kilbourn has been changed to Wisconsin Dells, and Grand Rapids to Wisconsin Rapids.

line rocks, for example in Juneau County west of Wisconsin Dells, and in Sauk County west of Baraboo (pp. 123, 130). Our best evidence of the height of the lake surface comes from these ice-rafted erratics. Near Baraboo and Wisconsin Dells they do not occur above a level of about 960 feet. At favorable localities at or below that elevation they are rather abundant. This is true, for example, on the quartzite and sandstone slopes of the Baraboo Range near North Freedom, northwest of Baraboo, south of Delton, and west of Wisconsin Dells.

Beach Deposits. To the northwest, rounded sandstone and chert bowlders are found in a beach deposit on Mile Bluff south of Mauston. No erratic bowlders have been found here. The deposit is less than 980 feet above sea level. Erratics have been found to the southwest near Reedsburg, where there was a bay of the glacial lake (Fig. 133).

To the northeast, erratics occur on the slopes of a mound 5 miles south of Friendship, Adams County. They are at an elevation of about 980 feet above sea level.

There are erratics in the gravel deposit at the northern end of Necedah Mound, and perhaps also on Petenwell Peak. The highest are more than 960 feet above sea level.

There are numerous erratics on the quartzite mound 4 miles northwest of Babcock, Wood County. These are at an elevation of nearly 1000 feet. Some of the best beach deposits of Glacial Lake Wisconsin, are on islands, as in the cases just cited.

Lake Deposits, in Relation to Tilting. The levels at which erratics and beach gravels are found, show that the shorelines of Glacial Lake Wisconsin may not be horizontal at present, but rise toward the north. They appear to increase in altitude from 960 feet near Baraboo and Wisconsin Dells to almost 1000 feet, 55 miles to the north, near Babcock and City Point. If the elevations cited were maxima and the tilting uniform, the rate of inclination of the deformed lake surface would be about 8 inches to the mile. It may be only half that amount. Before the verity and rate of tilting can be stated positively, however, it will be necessary to find many more occurrences of erratics on the shores and islands of Glacial Lake Wisconsin and to determine their elevations not barometrically but by levelling. It seems possible that the shorelines in the southern half of the lake basin are horizontal. In this case the rate of tilting in the half of the basin north of the hinge line may be as much as one and one-third feet per mile.

Erratics and beach gravels are lacking on many of the shores and islands, doubtless being covered by talus and dune sand, as well as by the lake clay and sand. Foreign bowlders are not usually to be found on the surface of the lake clay and lake sand, except near the southwestern part of the basin. This is interpreted as meaning that icebergs did not float out into the lake at its later stages because the earlier ice cliffs north and northeast of the lake were masked by outwash gravels. Near Wisconsin Dells and Baraboo, however, icebergs seem to have been discharged up to the very end of the lake's existence.

Lake Deposits, in Relation to Outlets. The deposits of Glacial Lake Wisconsin also extend out into the part of the Central Plain west of the Black River divide. The outlet of this body of water was westward down the East Fork of Black River from Scranton, Wood County, to Hatfield, Jackson County, and thence southward down the Black River. That this was the outlet was determined by visiting the headwaters and lower courses of three eastern tributaries of Black River—Robinson Creek, Morrison Creek, and the East Fork of Black River. The first two were eliminated as possible outlets because (a) their headwaters were at too high a level, (b) their valleys were no wider than could be made by the present streams, and (c) these valleys contain no erratic material.

The headwaters of the East Fork of Black River, on the other hand, are lower than the headwaters of Morrison and Robinson creeks. The divide region south of Scranton and City Point— 1000 feet—retains no sign of a glacial outflow channel. It is a level swamp with thick deposits of peat. Westward near Pray, however, and still farther west near Hatfield, this stream has an extremely broad valley, now occupied by a very small stream. The valley contains abandoned channels. More important, it contains erratic material in the form of granite and greenstone bowlders, as well as extensive sand deposits. The erratics might be thought to have come from the older drift to the north, but they are not weathered. Thus it appears likely that they came from Glacial Lake Wisconsin to the east. This conclusion appears necessary because there are no erratic bowlders whatever in the valleys of Robinson and Morrison creeks, which are the only other streams heading on low divides. At the southeast the valleys near Baraboo and Wisconsin Dells were blocked by glacial ice; but, when the ice melted back from the latter, the waters of Glacial Lake Wisconsin rapidly cut through the terminal moraine until the lake was drained out to the southeast and ceased to exist.

Deposits Made Since the Lake Was Drained. The swamps shown in Figures 113 and 141 were formed as a result of the accumulation of the more impervious deposits of this glacial lake. After it was drained, outwash gravels and other stream deposits were spread over some areas within the lake basin.

Wind Deposits. The wind-blown deposits of glacial time are discussed in Chapter VI. They are found in both drift-covered and driftless areas in part of the Central Plain. In Juneau County the finer loess predominates. In Adams County the dune sand covers wide areas.

Improvement of Soils. On the whole, the soil of the glaciated portion of the Central Plain was improved by the importation of the drift with limy rock flour, limestone bowlders, and various crystalline rocks from the region outside. The sandy soil of the driftless portion of the plain and the areas covered with outwash or dune sand are far less productive than the part of the plain with limy, sandy, and stony glacial till, or with loess and lake clay.

BIBLIOGRAPHY

Alden, W. C. Quaternary Geology of Southeastern Wisconsin, Prof. Paper 106, U. S. Geol. Survey, 1918, 356 pp; on the outlet of Glacial Lake Wisconsin, p. 223.

Allouez, Claude. (On the eastern part of the Central Plain) Jesuit Relations, Vol. 54, 1669-71, Thwaites edition, Cleveland, 1899, p. 231.

Berkey, Charles. Geology of the St. Croix Dalles, Amer. Geol., Vol. 20, 1897, pp. 345-383; Ibid., Vol. 21, 1898, pp. 139-155, 270-294; A Guide to the St. Croix Dalles for Excursionists and Students, Minneapolis, 1898, 40 pp; Laminated Interglacial Clays of Grantsburg, Wis., Journ. Geol., Vol. 13, 1905, pp. 35-44.

Chamberlin, R. T. Glacial Features of the St. Croix Dalles Region, Journ. Geol., Vol. 13, 1905, pp. 238-256; Older Drifts in the St. Croix Region, Ibid., Vol. 18, 1910, pp. 542-548.

Chamberlin, T. C. Geology of Eastern Wisconsin, Geology of Wisconsin, Vol. 2, 1877, pp. 98-405; Historical Geology—Cambrian Age, Ibid., Vol. 1, 1883, pp. 119-137; (on Glacial Lake Wisconsin) Ibid., pp. 284-285; The Terminal Moraine of the Second Glacial Epoch, 3rd Annual Report, U. S. Geol. Survey, 1883, pp. 315-322, 381-388.

Chamberlin, T. C., and Salisbury, R. D. The Driftless Area of the Upper Mississippi Valley, 6th Annual Report, U. S. Geol. Survey, 1885, pp. 20-322.

Clark, V. B. Geography of the Potsdam Sandstone Area in Wisconsin, Unpublished thesis, University of Wisconsin, 1910.

Coapman, Lillian. Geography of Columbia County, Wisconsin, Unpublished thesis, University of Wisconsin, 1913.

Davis, W. M. Physical Geography, Boston, 1898, pp. 136-137, 197-198, and Figs. 85 and 123.

Edwards, E. C. The Petrography of a Portion of Chippewa and Eau Claire Counties of Wisconsin, Unpublished thesis, Universtiy of Wisconsin, 1920.

Ekern, G. L., and Thwaites, F. T. The Glover Bluff Structure, a Disturbed Area in the Paleozoics of Wisconsin, Trans. Wis. Acad. Sci., Vol. 25, 1930, pp. 89-97.

Ellis, R. W. Glacial and Post-Glacial Phenomena of the Portage Region, with Special Reference to their Bearing on the Southwestward Drainage of Lake Winnebago, Unpublished thesis, University of Wisconsin, 1910.

Fenneman, N. M. (On the Camp Douglas Country), Annals Assoc. Amer. Geographers, Vol. 4, 1914, p. 106.

Fischer, C. W. Geography of the Middle Wisconsin River Basin, Unpublished thesis, University of Wisconsin, 1922.

Gilbert, G. K. (On Glacial erosion in the Central Plain), Geological Excursion to the Rocky Mountains, Compte Rendu de la 5me. Session, Washington, 1891, Congrès Géologique International, Washington, 1891, p. 290.

Hinn, Helen B. The Geography of the Wisconsin River Valley, Unpublished thesis, University of Wisconsin, 1920.

Hobbs, W. H., and Leith, C. K. The Pre-Cambrian Volcanic and Intrusive Rocks of the Fox River Valley, Wisconsin, Bull. 158, University of Wisconsin, 1907, pp. 247-277.

Irving, R. D. Geology of Central Wisconsin, Geology of Wisconsin, Vol. 2, 1877, pp. 413-636; On the Nature of the Induration in the St. Peters and Potsdam Sandstones, and in Certain Archaean Quartzites in Wisconsin, Amer. Journ. Sci., 3rd Series, Vol. 25, 1883, pp. 401-411.

Leverett, Frank. Moraines and Shorelines of the Lake Superior Basin, Prof. Paper 154A, U. S. Geol. Survey, 1929, Plate I.

Marquette, Jacques. (On the eastern part of the Central Plain), Jesuit Relations, Vol. 59, 1673-77, Thwaites edition, Cleveland, 1900, p. 103.

Martin, Lawrence. The Lowland Plains of the Lake Superior Region, Monograph 52, U. S. Geol. Survey, 1911, pp. 108-110; The Pleistocene of the Lake Superior Region, Ibid., pp. 427-459; The Physical Geography of Wisconsin, Journ. Geog., Vol. 12, 1914, pp. 229-231.

Muir, John. (On kettles and glacial lakes), The Story of my Boyhood and Youth, Boston, 1913, pp. 98, 117-118.

Ruggles, D. Geological and Miscellaneous Notice of the Region around Fort Winnebago, Michigan Territory, Amer. Journ. Sci., Vol. 30, 1836, pp. 1-8.

Salisbury, R. D., and Atwood, W. W. (On the Central Plain near Baraboo and Camp Douglas), Bull. 5, Wis. Geol. and Nat. Hist. Survey, 1900, pp. 6-13, 71-72, 129-130.

Shumard, B. F. (On the Central Plain near the Wisconsin River), Owen's Report of a Geological Survey of Wisconsin, Iowa, and Minnesota, Philadelphia, 1852, pp. 517-520.

Smith, Guy-Harold. The Influence of Rock Structure and Rock Character upon the Topography in the Driftless Area, Unpublished thesis, University of Wisconsin, 1921.

Strong, Moses. Geology of the Mississippi Region, Geology of Wisconsin, Vol. 4, 1882, pp. 7-98.

Thomas, K. Glacial Gold in Wisconsin, Eng. and Mining Journ., Vol. 74, 1902, p. 248.

Thwaites, F. T. On Franconia Cuesta and Eau Claire Bench (unpublished data).

Thwaites, F. T., Twenhofel, W. H., and **Martin, Lawrence.** The Sparta-Tomah Folio, U. S. Geol. Survey (in preparation).

Trainer, D. W., Jr. Molding Sands of Wisconsin, Bull. 69, Wis. Geol. and Nat. Hist. Survey, 1928, 103 pp.

Upham, Warren. Age of the St. Croix Dalles, Amer. Geol., Vol. 35, 1905, pp. 347-355.

Weidman, S. A contribution to the Geology of the Pre-Cambrian Igneous Rocks of the Fox River Valley, Wisconsin, Bull. 3, Wis. Geol. and Nat. Hist. Survey, 1898, 63 pp; Preliminary Report on the Soils and Agricultural Conditions of North Central Wisconsin, Bull. 11, Ibid., 1903, 64 pp; Reconnoissance Soil Survey of Part of Northwestern Wisconsin, Bull. 23, Ibid., 1911, 100 pp; Reconnoissance Soil Survey of Marinette County, Ibid., 1911, 44 pp; Geology of North Central Wisconsin, Bull. 16, Ibid., 1907, pp. 396-681.

Whitson, A. R., and **others.** Soils Reports: Wisconsin Geol. and Nat. Hist. Survey, Bulls. 11, 23, 24, 29, 38, 47, 49, 52A, 52B, 52C, 54B, 54C, 54D, 55, 60B, 60C, 61D.

Wooster, L. C. Geology of the Lower St. Croix District, Geology of Wisconsin, Vol. 4, 1882, pp. 101-159.

MAPS

U. S. Geological Survey. The Dells, Briggsville, Portage, Denzer, Baraboo, Poynette, St. Croix Dalles, Strum, Black River Falls, Tomah, Kendall, Mauston, Neshkoro, and Ripon Quadrangles (Fig. 197); river maps (Fig. 199).

Wis. Geol. and Nat. Hist. Survey. Lake maps (Fig. 199); soils, maps, see references in this chapter and Fig. 205.

URING THE CIVIL WAR the Red River campaign in *Louisiana in 1864 was saved from disaster by a Wisconsin soldier, a native of Kilbourn or Wisconsin Dells, who made use of the knowledge of rivers which he had gained in the lumbering industry on the rivers of Wisconsin.*

"The special honors of the Red River expedition were　*　* won by Lieutenant-Colonel Joseph Bailey of the Fourth Wisconsin. While the fleet was above the rapids at Alexandria, the stage of water fell, making it impossible for the vessels to descend, a perilous situation which encouraged the enemy to swarm upon the banks and seriously to threaten the little navy with destruction. Bailey was serving　*　*　* as chief engineer, and proposed the construction of a huge dam, by which the water in the river should be raised to a sufficient height; then, the obstruction being suddenly broken in the centre, the entrapped vessels might escape upon the outrushing flood. The scheme was familiar enough to Wisconsin lumbermen, who in this manner still artificially 'lift' stranded rafts of logs; but his army colleagues laughed at Bailey, although he was given three thousand men for the purpose, and told to amuse himself with this visionary experiment. His first requisition was for the 'lumber boys' of the Twenty-third and Twenty-ninth Wisconsin, who appreciated what was needed in this backwoods engineering scheme, and soon trained their fellows to the task. Bailey's sappers worked unwearyingly through the first eight days of May, and on the mórning of the twelfth the great gun-boats plunged through the boiling chute, thus triumphantly escaping the clutches of the discomfited Confederates, who had thought the expedition an easy prey. Admiral Porter frankly acknowledged that the fleet owed its safety entirely to the Wisconsin engineer's 'indomitable perseverance and skill.' "　　R. G. Thwaites.

CHAPTER XIV

THE DRAINAGE OF THE CENTRAL PLAIN

THE DELLS OF THE WISCONSIN

A Youthful, Postglacial Gorge. Near Wisconsin Dells (Kilbourn) the Wisconsin River flows through a narrow, steep-sided gorge known as the Dells. This is one of the most attractive, small, scenic features of the state, for the banks of the river are clothed in picturesque, coniferous forest; the course of the river is cut deeply in solid sandstone, diversified by unusually well-developed cross-bedding and conspicuous joint planes. Weathering and stream erosion have carved the river banks into bold cliffs, sharp chasms, and striking, isolated rock pillars. One may observe, as the geologist, Norwood, did in 1847 "singular and beautiful effects. Architraves, sculptured cornices, moulded capitals, scrolls, and fluted columns are seen on every hand; presenting, altogether, a mixture of the grand, the beautiful, and the fantastic."

The Dells of the Wisconsin represent an exceptional phase of river development in the Central Plain. The wider and less beautiful river above and below the Dells is described later (p. 353). This section of the Wisconsin is a youthful, postglacial gorge.

General Description. The gorge of the Wisconsin is seven and one-fifth miles in length, the portion below the city of Wisconsin Dells being one-third of the total length and known as the Lower Dells. The gorge above the city is usually spoken of merely as The Dells. In the Dells the river is 52 to 1000 feet wide, in contrast to 1500 or 2000 feet in the valley of the Wisconsin above and below the gorge. With this constriction the river becomes both deeper and swifter. The grade of the stream in the Dells is not very steep, however, the water descending only 10 feet in a little over seven miles. The gorge walls rise 80 to 100 feet above the river.

The Dells probably did not exist till the end of the last glacial epoch. There is no evidence to show that it is either an interglacial gorge or a rejuvenated preglacial gorge.

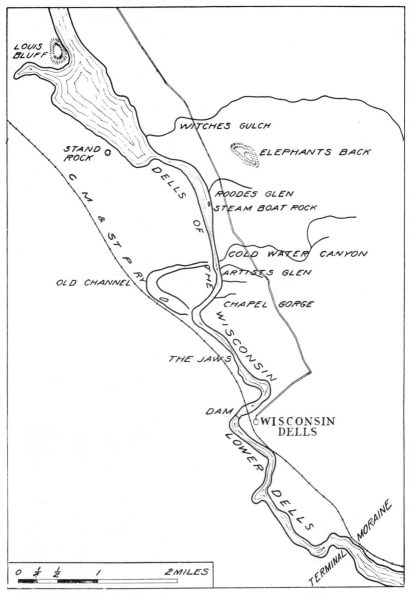

Fig. 134. Sketch map showing the Dells of the Wisconsin River and the location of the tributary gorges.

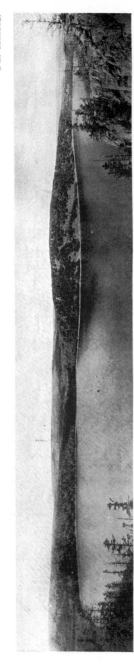

A. DEVILS LAKE STATE PARK IN THE BARABOO RANGE.

Mature pre-Cambrian topography at left of center and youthful post-Cambrian topography at right. The lake is a postglacial feature.

B. THE DELLS OF THE WISCONSIN RIVER. A POSTGLACIAL GORGE CUT IN CAMBRIAN SANDSTONE.

A. SUGAR BOWL.

B. WHIRLPOOL CHAMBER.

C. COLDWATER CANYON.

D. NAVY YARD.

AT THE DELLS OF THE WISCONSIN.

On the western side of the Dells are two abandoned stream channels (Figs. 134, 135). Their origin is to be explained in connection with the early history of Artists Glen and the loop which extends west opposite the hotel at Coldwater Canyon (p. 349). On the eastern side are four, prominent, tributary gorges and several smaller ones. These four are Witches Gulch, Roodes Glen, Coldwater Canyon, and Artists Glen.

Witches Gulch and Coldwater Canyon. The northernmost of the tributaries to the Wisconsin at the Dells is called Witches Gulch. It is a narrow, steep-sided gorge of the same sort as Watkins Glen in New York. Coldwater Canyon is like Witches Gulch in all respects. Three things are mainly responsible for the pleasing variations in the forms of these gorges—(a) relative weakness or resistance of rock layers, (b) the joint planes, (c) the work of running water, acting upon these two.

Variations in Width of Gorges. The most striking illustration of the first of these is the variation in width of the gorges. Witches Gulch is about 100 feet wide near the mouth, and 40 or 50 feet wide near the head. At points between, it is constricted to less than 2 feet. There are places where it is the shape of an hourglass, being 10 to 20 feet wide at the top and bottom and only 2 or 3 feet in the middle. All this seems to be due to the presence of alternate weak and resistant sandstone layers. The narrow, upper part of this gorge is 20 or 30 feet higher than the mouth, so that the stream profile cuts across a series of unequally-resistant beds, with resulting variations in the form of the gorge.

Influence of Joint Planes. The influence of the joint planes is shown in the trend of the gorges. They are not straight, but extend east and west with a series of right-angled turns. It is clear in many places that these angular turns are due to the control of the stream course by joint planes, of which there are two systems at right angles.

Pot Holes and Cascades. The work of running water in cutting the gorges is especially manifested by the cascades or waterfalls and by the pot holes. The latter are circular or oval in form. Some of them are in the present stream course and their origin is shown by the eddies which swirl around in them, often at the bases of little waterfalls. Within the pot holes are masses of sand and rounded pebbles. These are slowly swirled around by the eddying water, more, of course, in the high water of the spring months than during the summer when the stream is nearly dry. They erode the bottom and sides of the pot hole. By such erosion all pot holes have been made. Among the pot holes in Coldwater Canyon, the one called the Devils Jug is especially large and perfect. There are also remnants of pot holes at various levels on the sides of the gorges, proving beyond the possibility of doubt that Witches Gulch, Coldwater Canyon, and the other tributaries of the Wisconsin have been produced by the erosion of running water and are not gaping cracks due to faulting, or to any other violent cause.

The waterfalls are tiny affairs, but at their bases stream erosion is very effective. The crests of the falls are also worn down by the attrition of sediment-laden running water.

Joint planes aid in the recession of these cascades. The work of running water is effectual not only in pot holes and cascades but also in the reaches between falls. In such sections of the gorges, the stream undercuts the rock and widens its channel.

It is chiefly by these three types of stream work that Witches Gulch and the other gorges have been cut out in the solid rock. They are being widened above by frost action and other weathering but these are, thus far, relatively ineffective.

Effect of the Dam at Wisconsin Dells. The latest episode in the history of Witches Gulch, Coldwater Canyon, and Roodes Glen has been a submergence of the mouths of the gorges, due to the building of the dam at Wisconsin Dells. This has raised the level of the water several feet in Witches Gulch and the adjacent gorges. The level was increased about 16 feet in the main Dells near the dam. By this submergence the erosion of the lower portions of all these tributary gorges has been decreased or stopped entirely.

The author has visited the Dells repeatedly, both before and since the building of the dam. It is his opinion that the scenery has not been impaired in any respect by the raising of water level in the main stream or tributary gorges. He misses certain familiar features. He regrets the leaving of dead trees in the mouths of the gorges and in the new lake near Stand Rock. These should be removed by the power company that built the dam. For every detail lost, as the lower cascades of Witches Gulch or the bases of rocks at the Navy Yard, something new and equally interesting and beautiful has been created. These include the deepening of the water that makes possible the launch trip into the mouth of Roodes Glen, and many other features.

Stream Diversions. Artists Glen seems to be subject to less active, present deepening than the other gorges. Its mouth is above the level of Wisconsin River, even since the dam was built. Because of this discordance of stream grade Artists Glen is one sort of hanging valley. This brings up two problems.

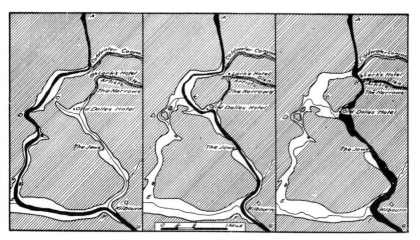

Fig. 135. Three stages in the development of drainage at the Dells of the Wisconsin. The name Kilbourn has been changed to Wisconsin Dells.

The first has to do with the question as to whether the presence of tributary gorges almost exclusively on the eastern side of the Dells does not mean that Witches Gulch, Roodes Glen, Coldwater Canyon, Artists Glen, Chapel Gorge, Glen Eyrie, and the smaller gorges were excavated chiefly by glacial waters. There is relatively little water in some of them today, but when the edge of the continental glacier stood at the terminal moraine only 4 miles to the east, there was ample water for cutting these gorges. Of course we know that Glacial Lake Wisconsin (p. 337) must have submerged the site of the Dells for a time; but both before and after this lake existed there may still have been good-sized glacier-fed streams at work here.

The other question raised by the discordant or hanging junction of Artists Glen with the main Wisconsin has to do with the origin of the abandoned channel west of the Larks Hotel or Dells Inn. It is clear that the Wisconsin used to turn westward just below the mouth of Coldwater Canyon (at B, middle map, Fig. 135), looping back to the present channel about three-quarters of a mile down stream at Allen's Landing where the old hotel stood. The stream in Artists Glen is then supposed to have flowed southward through what is now the narrowest portion of the Dells just above the Navy Yard. Subsequently the Artists Glen stream cut into its north bank and the Wisconsin cut into its own south bank so that the narrow strip of rock between the two streams was eventually cut through, presumably in a period of high water in the spring.

As soon as the main Wisconsin River was diverted into the narrow channel of Artists Glen it quickly cut down, for it gained velocity because of the constriction and steeper grade. It cut down more rapidly than the earlier stream in Artists Glen could erode. Perhaps also the loss of the glacial waters diminished the latter stream's erosive power. For these two reasons the mouth of the Artists Glen stream was left hanging above the main channel of the Dells.

From this it is clear that the old channel west of Coldwater Canyon is not the incised meander of a river in old age, as is sometimes suggested, but an exceedingly youthful form due to stream capture.

Moreover, this is the second time that the Wisconsin has been thus diverted. In the left-hand map of Figure 135 it appears that the river formerly had the course AEG. West of Allen's Landing at the old hotel a tributary headed at C. It cut back at the headwaters till it tapped the Wisconsin, diverting it to the course CF. Before the Artists Glen diversion, while it was still flowing through the channel from C to Allen's Landing, one of its small tributaries cut back into the abandoned channel DE, reversing the direction of flow for a short distance.

Features of the Main Gorge in the Dells. The main gorge is like its tributaries in most respects. Thus it shows the influence of rock texture, as where High Rock and Romance Cliff at The Jaws form a constriction in a resistant portion of the sandstone. The result of jointing is illustrated in the section called the Navy Yard, at the Giant's Hand, and the several boat caves. It appears on a small scale at Chimney Rock in connection with weathering, and on a large scale at Steamboat Rock above Coldwater Canyon and at the Inkstand and Sugarbowl in the Lower Dells in association with river erosion as well as weathering.

THE NARROWS, DELLS OF THE WISCONSIN.
(Copyright by Underwood & Underwood, N. Y.)

STAND ROCK, WEST OF THE DELLS OF THE WISCONSIN IN THE DRIFTLESS AREA.

(Copyright by Underwood & Underwood, N. Y.)

The constriction at the Narrows, or Black Hawk's Leap, does not appear to be due to resistant rock there, but rather to the recency of the stream diversion (Fig. 135). Here the channel is only 52 feet wide. It is said to be 40 to 80 feet deep, and the river is often spoken of as running on edge. Because of this narrowness it was possible to build the toll bridge which spanned the Narrows from 1848 to 1866. It was the first bridge across the Wisconsin River in this part of the state.

Gorges Near the Dells. The picturesque gorges near Delton and Wisconsin Dells are also postglacial stream channels, similar to the eastern tributaries of the Wisconsin in the Dells. Among these are the Mirror Lake Gorge, the Congress Hall Gorge, and Taylors Glen. They are in the region which was not reached by the continental glacier, but are surrounded by deposits of Glacial Lake Wisconsin and of glacial streams. It seems probable that they were not cut by glacial waters and have been eroded chiefly in postglacial time. They are directly related to glaciation, however; Dell Creek, for example, was diverted from a quite different preglacial course (p. 194) by the building of the terminal moraine. Mirror Lake is artificial, being held up by a dam.

Stand Rock and Vicinity. On the western side of the Dells are Louis Bluff, Stand Rock, and several other interesting, natural features (Fig. 29). These are in the Driftless Area and seem to be chiefly due to preglacial weathering and wind work. Here the gorge of the Wisconsin broadens out into a wide valley, containing isolated rock hills such as Louis Bluff and the sandstone mounds to the north. These mounds and the Elephants Back mound east of the Wisconsin are outliers, left behind in the recession of the Magnesian escarpment (Fig. 127).

Stand Rock is an isolated column of sandstone, situated close to a sandstone cliff just north of the Dells. The column of Stand Rock is about 45 feet high and 6 or 8 feet in diameter. It is capped by a layer of resistant sandstone, some 20 feet in diameter. As the capping layer is at just the same height as the adjacent cliff and within 6 feet of it, and as the uppermost sandstone layer projects in a cornice-like ledge, it is clear that the projecting ledge and the isolated slab-capped column of Stand Rock are due to the same cause. This is the action of frost, of alternate contraction and expansion with heat and cold, the wedging action of roots, and the other agencies which we usually speak of as weathering.

Fig. 136. The Hornets Nest, a rock pillar produced by weathering and wind work at the Dells of the Wisconsin. (Drawing by Chicago, Milwaukee, St. Paul, and Pacific Railway.)

Gravity causes the weathered rock on the face of the cliff to fall. The swirling action of the wind transports grains of sand away. Thus the resistant, upper layer of sandstone is undermined and left projecting. Visor Ledge is of this origin. Such projecting ledges occasionally fall and lie at the base of the cliff. As the cliff slowly recedes, however, some parts are removed faster than others. Accordingly the cliff has an irregular outline, with salients projecting and reentrants indenting the valley wall. Some of the larger indentations are gullies made by small wet-weather streams. Stand Rock is the isolated end of such a salient, and there are other columns in various stages of formation. The Hornets Nest (Fig. 136) and the feature called Luncheon Hall are archways where the capping rock layer is still connected with the main cliff. The rock known as The Anvil, situated at the end of the salient between the Hornets Nest and Stand Rock, is a feature of the same sort, only the capping rock layer has fallen from the column. Thus we have the Hornets Nest, Stand Rock, and The Anvil—three features illustrating progressive stages in the destruction of a receding Driftless Area cliff.

Just west of Stand Rock is Squaws Bed-Chamber, a narrow cave which extends back into the cliff for 50 feet or more. It has a tiny branch cavern on the left side. Both cave and branch have been made by weathering along well-developed joint planes in the weak sandstone.

All of these features in the neighborhood of Stand Rock are of the sort that can exist only in the Driftless Area. They are relatively fragile and would certainly have been eroded away or buried in glacial deposits if the continental ice sheet had advanced west of the present course of the Wisconsin River.

The Wisconsin River Outside the Dells

The Middle Section of the Wisconsin River. The lower course of the Wisconsin has already been described (p. 185). The upper course will be discussed in Chapter XVI. The river leaves the Northern Highland at Wisconsin Rapids and flows for a short distance in a valley whose sides are Cambrian sandstone and whose floor is pre-Cambrian crystalline rock. The stream then begins its course across the Central Plain and here the valley is relatively shallow. Sandstone ledges are rarely present in the stream bed except in a few rapids and postglacial gorges, such as the Dells of the Wisconsin. There are narrows near Dekorra, but no ledges in the stream course. It is clear from Figure 58 that this middle portion of the Wisconsin has had a different history from those to the north (Fig. 164) and to the southwest. From Wisconsin Rapids to Portage and Prairie du Sac the river has few tributaries, the dentritic pattern is replaced by an aimless pattern, and there are large, undrained, interstream areas. In 181 miles the river descends $138\frac{1}{2}$ feet or at the rate of only $1\frac{7}{10}$ feet to the mile.

The cause of this flattening of the river grade, which is only half as steep as in the Northern Highland, is in a measure related to the normal grading of a stream. The middle and lower courses are always less steep than the tumultuous headwaters. The change of grade is also partly due to the complications of glaciation. Before the Glacial Period the Wisconsin had a more direct southward course from Stevens Point to Portage. Well borings reveal a buried preglacial Wisconsin valley extending north and south in the region between Portage and Wisconsin Dells (Fig. 62). Part of the flattening of the stream grade is due to lengthening of this middle portion of the Wisconsin by its being diverted westward in the great bend (Fig. 58) between Stevens Point and Portage, where crowded west by the moraine.

The absence of tributaries and of a well-defined valley may also be due to the short time during which the Wisconsin has occupied this course. The undrained interstream areas are also related to the residual soil of the Cambrian sandstone and the sandy outwash, dune sand, and lake deposits west of the terminal moraine. The levelness of this region is also a factor. The lack of eastern tributaries of any length is explained by the nearness of the divide on the terminal moraine.

The present course of the Wisconsin across the Central Plain, therefore, is in large part a postglacial course. The river occupies the place in the Driftless Area to which it was diverted by the outwash deposits of the streams from the melting ice. Roche à Cris Creek and the other eastern tributaries between Stevens Point and Portage are shrunken descendants of these glacial rivers.

The grade of the Wisconsin descends less steeply than that of the plain of outwash gravels through which it flows. Consequently the river is in a shallow trench near Wisconsin Rapids but flows almost on the surface of the plain north of Wisconsin Dells. At Stevens Point and Wisconsin Rapids there are several terraces. At

Nekoosa the present floodplain is separated from the glacial outwash plain by a steep bluff 55 feet high. At Petenwell Peak, near Necedah, there are two low terraces. The increase in number of terraces toward the north may be related to the tilting already described (pp. 163, 295, 300). It is not yet known where the hinge line crosses the Wisconsin River.

Lemonweir and Yellow Rivers. Some of the western tributaries of the Wisconsin, like Yellow River, are also diminished representatives of greater glacial streams. Others, like the Lemonweir, have never had glacial complications in their history, except in the lake or outwash deposits over which they flow.

The Relationships of the Wisconsin and Fox Rivers. At the city of Portage occurs an unusual relationship of two rivers. The Fox River, which flows northeastward to Green Bay and the St. Lawrence drainage, is only about a mile and a half from the Wisconsin River (Fig. 137). Both streams are on a flat, swampy plain, where the Indians and the early explorers, either portaged their canoes or floated them across at high water. A little ditch which would float canoes across the portage was dug in early days, perhaps as early as 1766. The present canal was started in 1849 by the United States government. The Fox-Wisconsin waterway was soon completed, so that a steamboat was able to go from the Mississippi to the Great Lakes in 1856 through the Fox-Wisconsin canal at Portage.

Floods sometimes inundate the plain at Portage, so that a very little of the water of the Wisconsin flows into the Fox. It is feared that the whole Wisconsin may sometime be diverted into the channel of the Fox. As the latter is 3 feet lower and descends northward with a steeper grade than the Wisconsin, there is basis for believing that if the Wisconsin were given this steeper grade it would quickly incise a deep channel in the unconsolidated swamp accumulations and glacial deposits, not only below but above Portage as well. Such a stream diversion would then be permanent, unless the state or some private companies went to the great expense of turning the Wisconsin back into its channel. If the diversion took place it would give the paper mills and other power plants of the Fox too much water, resulting in serious damage to the factories in such places as Oshkosh, Neenah-Menasha, and Appleton on Lake Winnebago and the Lower Fox. The towns of the lower Wisconsin, on the other hand, would be deprived of nearly

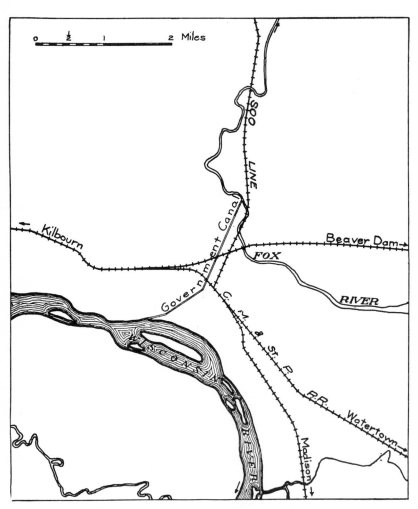

Fig. 137. The portage between the Wisconsin and Fox Rivers.

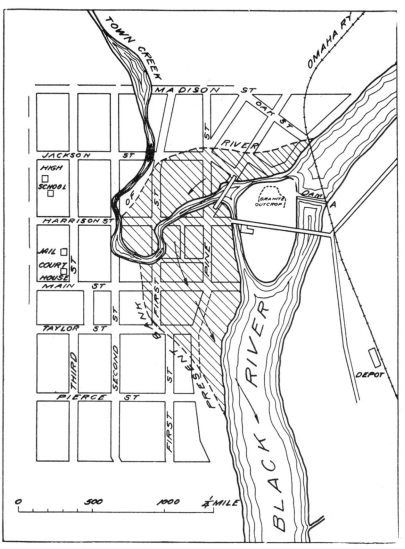

Fig. 138. The city of Black River Falls and the area devastated by the flood in 1911.
(After Pence.)

all their river water. This would result in such hardships as the abandonment of the new $5,000,000 hydro-electric power plant at Prairie du Sac, and possibly to future problems regarding disposal of sewage at Sauk City, Spring Green, Boscobel, and other cities.

FOX RIVER SYSTEM

Preglacial Wolf River. The reversal in direction of drainage in the upper valley of Fox River has already been indicated (Fig. 62 and p. 285). Where the Fox now flows northward, it seems probable that the Wolf River flowed southward. This reversal is due to the accident of glacial deposition and to the establishment of a southward flow by one of the outlets of Glacial Lake Jean Nicolet (p. 302).

Present Wolf River. Lake Shawano, in the upper course of the Wolf River, and Lake Poygan, near its junction with the Fox in Winnebago County, are among the largest lakes in the Central Plain. The attractive, small bodies of water include the Waupaca Chain O'Lakes. All these are entirely postglacial, though related in some cases to preglacial stream valleys. The Rat River Marsh near Lake Poygan covers 310 square miles.

Near Waupaca a railway grade has been built across a peat bog. The weight of the gravel fill has caused the peat to rise each side of the railway till now the track seems to be in a low cut.

Lakes in the Course of the Upper Fox. The present gentle grade of the Upper Fox, with its slope of less than 6 inches to the mile, is interrupted by two stretches where the river widens and the current slackens. These are Lakes Buffalo and Puckaway (Fig. 107). The former is a shallow, crescent-shaped lake about 12 miles in length and three-fourths of a mile or less in width. Lake Puckaway is about 7 miles in length and a mile and a half in width. It is separated from Green Lake (p. 329) by a terminal moraine. The rock floor of the preglacial valley at Lake Puckaway is buried to a depth of over 330 feet. Along the lower Fox are extensive swamps in a well-developed floodplain. The swamps and lakes regulate the flow of the lower Fox River, so that its floods are diminished. The volume of water available in summer at the mills below Lake Winnebago is greatly increased by the lakes and swamps in the Upper Fox and Wolf rivers.

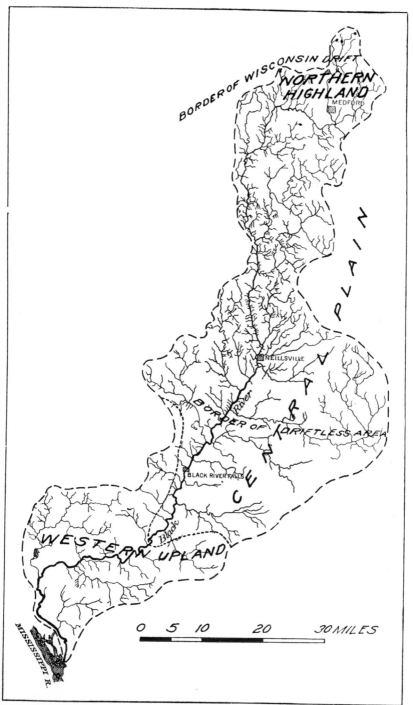

Fig. 139. The Black River, showing the similarity of drainage patterns in the area of older drift and in the Driftless Area.

BLACK RIVER

The Black River in the Central Plain is almost entirely in the region of older drift (Fig. 139). Consequently it has few lakes and swamps and is subject to severe floods. At Black River Falls, near the border between the Central Plain and the Western Upland, the Black River flows through a shallow, steep-sided trench, at the bottom of which the city is located. During a severe flood in October, 1911, the river left its channel and inundated the business portion of the city, destroying houses and streets (Fig. 138), eroding one bank of the river to a notable extent (Pl. XXVII), and doing about $2,000,000 worth of damage. Forty-two acres of land in the business district were washed away.

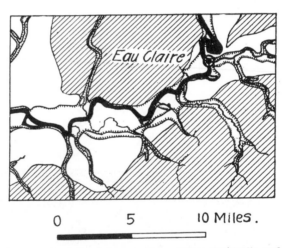

Fig. 140. Terraces along the Chippewa River. (After Chamberlin and Salisbury.)

CHIPPEWA RIVER

The Chippewa River within this geographical province is chiefly notable for the gentle grade of the river, the moderate slopes of its valley, and the series of terraces carved in the outwash gravels (Fig. 140). The paper mills and furniture factories at the city of Eau Claire are made possible by the water power developed at rapids there. This is one of the few extensive water powers of the Central Plain that is situated away from the Fall Line at its northern border (p. 361).

A. THE CITY OF BLACK RIVER FALLS BEFORE THE FLOOD IN 1911.

B. VIEW FROM THE SAME POINT AS THE UPPER PHOTOGRAPH SHOWING
THE DEVASTATION WROUGHT BY THE FLOOD.

Pl. XXVIII.

THE DEVILS CHAIR, A PINNACLE IN THE GORGE OF THE ST. CROIX RIVER
AT INTERSTATE PARK.

ST. CROIX RIVER

Most of the St. Croix valley in the Central Plain is rather broad, but at one point the river flows through a deep, steep-sided, postglacial gorge, known as the Dalles of the St. Croix. This is near the city of St. Croix Falls, Polk County. The Dalles are cut in the well-jointed lavas of the Keweenawan, exposed by the removal of the weaker Cambrian sandstone. The features here are summarized in the description of the Interstate Park (pp. 364-366).

OTHER STREAMS IN THE CENTRAL PLAIN

The drainage of the Central Plain by smaller streams is similar to that along the rivers just described. Some of the drainage is simple, some of it is complicated by lakes which interrupt the r'ver course, as at Noquebay Lake, Marinette County, and the lake region of Polk, Barron, and Burnett counties.

The crescentic sandstone lowland of the Central Plain is not drained by any single master stream and its tributaries, nor is the drainage within it longitudinal. It is crossed by a series of transverse streams, of which the Wisconsin River is the largest and most important. The others include fhe Black and Chippewa and part of the St. Croix on the west, and the Wolf-Fox, Oconto, Peshtigo, Menominee, and smaller streams on the east. This transverse drainage is characteristic of the inner lowland of a belted plain. If the sandstone lowland were really a peneplain (p. 324) it would probably be drained to a greater extent by longitudinal streams as a result of readjustment.

FALL LINE CITIES OF THE CENTRAL PLAIN

There is a line of cities in central Wisconsin whose situation was determined by water power at one type of locality. The larger cities in this group are St. Croix Falls, Chippewa Falls, Eau Claire, Wisconsin Rapids, Stevens Point, and Waupaca. The locality where water power occurs is at or near the northern edge of the Central Plain. Water power is present here because the streams are leaving the resistant rocks of the Northern Highland and hence have rapids or low cascades. Accordingly these were places where there were great sawmills in the heyday of Wisconsin lumbering. These are places where pulp mills, paper mills, and furniture factories are situated today. A line drawn between these cities is a Fall Line. In most respects it is identical with the Fall Line that separates the Piedmont Plateau from the Atlantic Coastal Plain, passing Trenton, Philadelphia, Baltimore, Washington, Richmond, Petersburg, Columbia, Augusta, Macon,

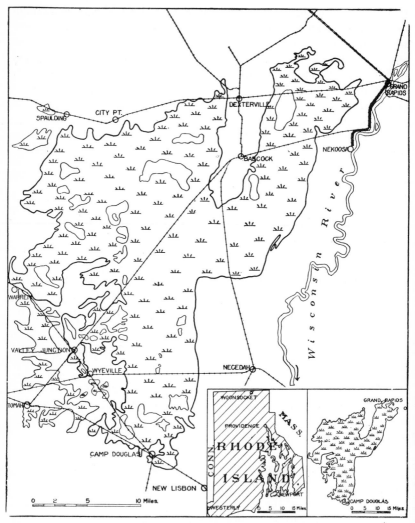

Fig. 141. The Great Swamp of central Wisconsin. Grand Rapids is now called Wisconsin Rapids.

and Montgomery. Our Fall Line marked the head of steamboat navigation in Wisconsin the day when the St. Croix, Chippewa, Wisconsin, and Wolf rivers were counted navigable streams.

SWAMPS OF THE CENTRAL PLAIN

Causes of Swamps. There are vast areas of swamps in the Central Plain. As the area is underlain by the Cambrian sandstone there should be a few swamps, for porous sandstone usually allows water to percolate freely. The glaciation of the eastern and northwestern portions of this geographical province, however, has resulted in the bringing of quantities of less permeable soil, the till of the ground moraine being sometimes clayey, sometimes sandy, but always less porous than the original residual soil of a never-glaciated sandstone lowland. West of the terminal moraine the clayey deposits of Glacial Lake Wisconsin form less porous soils than those originally there. Outwash deposits are often made up of porous sand and gravel, but they are sometimes so deposited as to help make the undrained or poorly-drained swamp areas. Lake bottom sand underlies vast areas of swamp in Wood, Juneau, and Jackson counties, but it is underlain in turn by finer and less porous sand and by impervious clay.

The Great Swamp of Central Wisconsin. Although the marshes, bogs, and other swamps of Wisconsin are chiefly in the area of latest glaciation (Fig. 113), the largest swamp in the state happens to be in the Driftless Area. This is the Great Swamp of central Wisconsin (Fig. 141). It lies in the area between Wisconsin Rapids, Camp Douglas, and Black River Falls, covering about 300,000 acres, exclusive of strips of swamps along three rivers to the southeast. It is more than twice as large as Milwaukee County, and 38 per cent as large as the whole state of Rhode Island (Fig. 141). There are still other good sized swamps nearby. The scarcity of roads in parts of Jackson and Juneau counties is a good index of the sparsity of population that goes with this swampy condition.

TABLE SHOWING PERCENTAGE OF WET LANDS IN 3 COUNTIES
OF THE CENTRAL PLAIN

County	Acres of peat, muck, etc.	Percentage of county
Juneau	176,320	35
Waushara	68,480	16
Columbia	59,712	12

Swamp Utilization. The whole state contains 3970 square miles of peat lands and 3655 square miles of other wet soils which are largely over-flow lands along streams. A considerable portion of these wet lands of the state are already in woods or forests which are adapted to that condition. It has been urged that this use be extended. A large amount of this swampy land lies in the Central Plain, perhaps ¾ of a million acres. The amount in 3 counties is shown in the table, page 363.

Such swamps are a source of peat. They may be ditched or drained and made into grazing or farm land. Some of them are so situated as to make excellent cranberry bogs. An unusual industry, found in parts of this swampy area, is the gathering of sphagnum moss, used by florists. Considerable quantities are shipped each year from City Point. Peat has not yet come into use in Wisconsin for fuel. Experiments along this line were once conducted at Tomah, as well as in eastern Wisconsin (p. 291).

INTERSTATE PARK

The states of Wisconsin and Minnesota have set aside an Interstate Park at the Dalles of the St. Croix, near St. Croix Falls, Wis., and Taylor Falls, Minn. It covers 730 acres, of which 580 are in Wisconsin and 150 in Minnesota.

The rock ledges here are ancient lava flows, of which seven may be identified, rising like giant steps above the river. The lava, or trap, is well-jointed, so that there are vertical precipices (Pl. XII, B) and isolated crags (Pl. XXVIII) along the St. Croix River.

The river is in a postglacial gorge from a short distance north of the Interstate Park nearly to Osceola. Before the Glacial Period the St. Croix River probably lay to the west and followed a different course all the way to the Mississippi (p. 202). In cutting the gorge it has swirled about and produced exceptionally fine pot holes, roughly circular bowls, cut in the trap by rolling stones, kept in motion by the swirling waters of eddies. These range in size from shallow holes a few inches in diameter to gigantic wells 5 to 25 feet in diameter and as deep as 80 feet. The walls are worn smooth, but are somewhat uneven, due to unequal hardness of the rock. The existence of these pot-holes at all elevations from the river level up to 100 feet above the river is evidence that this gorge was cut by running water.

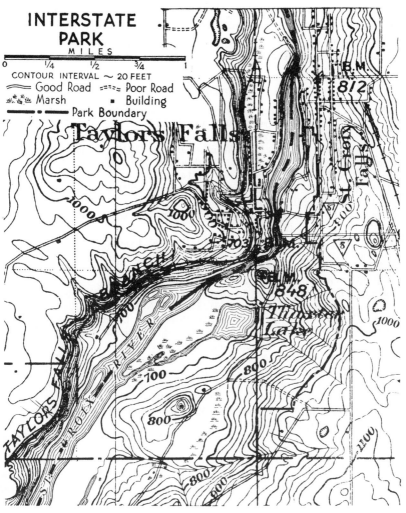

Fig. 142. Map of Interstate Park.

On the rock ledges glacial striae may be found. The glacial deposits from (a) the Labrador and (b) the Keewatin ice sheets (p. 334) may be distinguished because the former are red, the latter gray. The outwash gravels laid down by glacial streams were subsequently partly removed. This leaves a system of terraces, of which five may be distinguished. Some of them are determined in position by rock ledges. The main street of the city of St. Croix Falls is on the next to the highest of these terraces.

BIBLIOGRAPHY

Alden, W. C.   Quaternary Geology of Southeastern Wisconsin, Prof. Paper 106, U. S. Geol. Survey, 1918, pp. 29-48.

Cox, H. J.   Frost and Temperature Conditions in the Cranberry Marshes of Wisconsin, Bull. T., U. S. Weather Bureau, 1910, 121 pp.

Fenneman, N. M.   The Lakes of Southeastern Wisconsin, Bull. 8, Wis. Geol. and Nat. Hist. Survey, 1910, pp. 148-174.

Huels, F. W.   The Peat Resources of Wisconsin, Bull. 45, Wis. Geol. and Nat. Hist. Survey, 1915, 263 pp.

Irving, R. D.   Geology of Central Wisconsin, Geology of Wisconsin, Vol. 2, 1877, pp. 413-424, 570.

Jones, E. R.   The Larsen Marsh, 3rd Biennial Report of the Conservation Commission of the State of Wisconsin, Madison, 1912, pp. 68-70.

Juday, C.   The Inland Lakes of Wisconsin, Bull. 27, Wis. Geol. and Nat. Hist. Survey, 1914, pp. 92-96, 100-112.

Lapham, I. A.   An Early Journey through Sauk County (describing the Dells and Devils Lake in 1849), Baraboo News, Jan. 4, 1912.

Norwood, J. G.   (On the Dells of the Wisconsin), Owen's Geological Reconnoissance of the Chippewa Land District, Senate Ex. Doc. 57, 30th Congress, 1st Session, Washington, 1848, p. 108.

Pence, W. D.   Failure of the Dells and Hatfield Dams and the Devastation of Black River Falls, Engineering News, Vol. 66, 1911, pp. 482-489.

Salisbury, R. D. and Atwood, W. W.   (On the Dells of the Wisconsin), Bull. 5, Wis. Geol. and Nat. Hist. Survey, 1900, pp. 69-71.

Smith, L. S.   The Water Powers of Wisconsin, Bull. 20, Wis. Geol. and Nat. Hist. Survey, 1908, 341 pp.

Van Hise, C. R.   The Origin of the Dells of the Wisconsin, Trans. Wis. Acad. Sci., Vol. 10, 1895, pp. 556-560.

Warren, G. K.   Report on the Transportation Route along the Wisconsin and Fox Rivers in the State of Wisconsin, Senate Ex. Doc. 28, 44th Congress, 1st Session, Washington, 1876, 114 pp., accompanying atlas, Plates 8 and 9.

Whitson, A. R., and Jones, E. R.   Drainage Conditions of Wisconsin, Bull. 146, Wis. Agricultural Experiment Station, 1907, 47 pp.

MAPS

See Chapter XIII, p. 343.

CHAPTER XV

THE NORTHERN, OR LAKE SUPERIOR, HIGHLAND

THE LOST MOUNTAINS OF WISCONSIN

Far back in the geological past, perhaps 6 or 7 hundred million years ago, Wisconsin was part of a mountainous region which covered all this state and much territory outside. It had peaks and ridges similar to those in the Alps, except that the latter has many peaks which are sharpened by the enlargement of glacial cirques. We know of this former mountainous condition from a study of the rocks and the topography of today (Fig. 143). A tree may be cut down, but the size of the stump, the number of annual rings, the kind of bark, the spread of the roots, the slash from its branches, and the nature of adjacent trees, tell a definite story as to the kind and size of tree that once grew there.

So remnants of rock folds reveal the fact that there were once lofty ridges and deep valleys in northern Wisconsin. The types of folds tell us that the ridges were parts of a mountain range more like the Alps or Rockies than the Appalachians. The granites show that erosion has cut down to igneous rocks such as are formed only by deep-seated cooling of molten intrusives, often beneath the arch of a lofty mountain range. The gneisses and schists suggest the former presence of tremendous pressure and some heat. The trap rocks indicate that lava flows emerged at the surface in the later stages of the mountain history. The fossils in the overlying sedimentary rocks show that these mountains are among the oldest in the world.

These lofty mountains were attacked by weather, wind, and streams, by solution underground, by plants and animals at the surface, as mountains are being attacked today. They were gradually worn down, till nothing remained but a low, undulating plain with occasional hills. This we call a peneplain. The destruction of the mountains took a long time, of course; but time enough was available. The rivers carried sand and mud and dissolved mineral

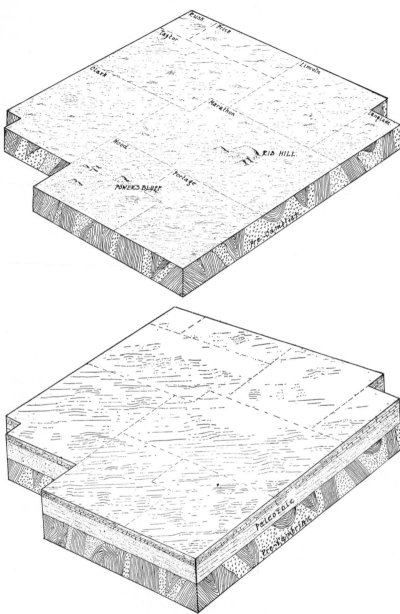

Fig. 143. Block diagrams showing part of the ancient mountain region of northern Wisconsin worn down to a peneplain, with the Rib Hill and Powers Bluff monadnocks rising above it. After it sank beneath the sea and was uplifted again the peneplain lay buried beneath sedimentary rocks of Paleozoic age. (After Weidman.)

matter from the mountains into the sea. There it was deposited as sandstone, shale, and limestone. The mountains were uplifted again and worn down again, repeating this history several times.

Eventually Wisconsin and the adjacent region sank beneath the ocean, probably remaining submerged for long ages. While it was sinking, the hills that rose above the surface of the peneplain may have been little islands in the sea for a short time. Waves may have beaten against their shores, making beaches. There seems to be no good reason for doubting that the whole of Wisconsin was finally submerged. Subsequently it was uplifted and submerged several times.

Two hundred million years or so ago this part of the United States was uplifted for the last time, and has since remained dry land. It was then a low plateau or plain similar to the coastal plain of Alabama or Texas. The peneplain on the site of the ancient lofty mountains of Wisconsin was completely hidden beneath the limestones and sandstones.

The work of weather and streams recommenced, and continued till the state was fashioned into something similar to its present form. In the northern part of Wisconsin the worn down mountains have again been revealed. This may not have taken place until after the Seneca and Windrow stream gravels were laid down (page 53), perhaps in the Cretaceous period, fifty or sixty million years ago. Throughout all parts of Wisconsin except the Northern Highland, the Baraboo Range, the Barron Hills, and such places, the worn-down pre-Cambrian mountains lie deep beneath the present surface. We know the visible portion of these worn-down, buried, and exhumed mountains as the peneplain of the Lake Superior Highland, or, within the state, as the Northern Highland of Wisconsin.

TOPOGRAPHY OF THE NORTHERN HIGHLAND

Position and Altitude. The Lake Superior Highland belongs to a great upland area that stretches northward in Canada to Labrador and Hudson Bay. The part of it south of Lake Superior is the Northern Highland of Wisconsin and Michigan. In this state it covers about 15,000 square miles. Its altitudes are somewhat as follows, all elevations being given in feet above sea level.

This table shows that the area has a strong southward slope, and, as the highland is shield-shaped and is gently arched, it also

TABLE SHOWING ELEVATION OF DIFFERENT PARTS OF THE
NORTHERN HIGHLAND

Northwestern border 1100-1200 feet	Northern border 1500-1700 feet	Northeastern border 700-800 feet
Southwestern border 900-1000 feet	Southern border 1000 feet	Southeastern border 850-950 feet

has east and west components of slope. The slant of a medial line from the northern to the southern border is at the rate of less than six feet to the mile.

Typical Portion. A portion of this highland in Marathon County, near the village of Marathon, may be described as typical of the whole. It is shown in Figure 144. It lies in the Driftless Area and, therefore, represents a portion of the highland not at all modified by glacial erosion and deposition, but shaped entirely by weathering and stream erosion.

As the topographic map shows, this is a moderately hilly region, the tops of the hills reaching a general elevation of 1,300 to 1,400 feet. The deepest valleys are cut down to 1,100 or 1,200 feet, so that the local relief is only about 200 feet. The hills are so moderate in slope that practically all roads are laid out in the rectangular system of the township and section lines. The valleys branch in dendritic or tree-like fashion. There are no lakes and practically no swamps. The area of this map is underlain by granite, with subordinate amounts of syenite, schist, gneiss, and quartzite. An examination of the whole Marathon Quadrangle, from which Figure 144 was made, shows that the surface of this part of the upland slants southward at the rate of $5\frac{7}{10}$ feet to the mile. A view from a hilltop in this highland area shows an even skyline in every direction.

Great Area of the Peneplain. In the peneplain of northern Wisconsin there are two distinctive kinds of topography, related directly to the underlying rocks. These are (a) the upland plains and (b) several types of ridges. The plain covers a great expanse in northern Wisconsin. It is rather smooth upland where homogeneous rocks, like granite, reach the surface. In some places other homogeneous rocks, besides granite, have been worn down to this smooth condition.

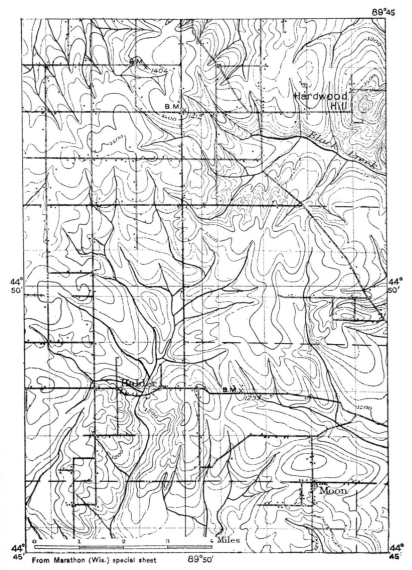

Fig. 144. A typical portion of the pre-Cambrian peneplain in the Driftless Area. Contour interval 20 feet. (U. S. Geol. Survey.)

The great extent of the smooth upland plain is very well indicated by the trend of railways in the Northern Highland. In most mining regions the transportation problem is complicated by the expense of railway construction. This is true of nearly all mining, except that of coal and some lead and zinc. Much of our iron, copper, gold, silver, and other precious metals are found in youthful, rugged mountains. The iron and copper of the Lake Superior region also occurs in mountains, but they are old mountains, now worn down to a peneplain. Accordingly the railways leading from the mines to the lake-ports and to the markets are constructed without great expense.

All railways in the Northern Highland, whether leading to mines or merely crossing the upland, are predominantly straight. Thus the new Soo Line between Chicago and Duluth-Superior crosses the Northern Highland from northwest to southeast with an exceedinly straight course, made possible by the low relief of the peneplain. It is 172 miles in a straight line from Marshfield, at the southern border of the peneplain, to the city of Superior. The distance by rail, 180 miles, is only eight miles longer than a bee line. Another division of the Soo Line—Minneapolis, St. Paul, and

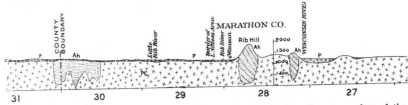

Fig. 145. Cross-section of part of the Northern Highland where the even surface of the peneplain truncates folded, metamorphic rocks (Ah) and reveals igneous masses of pre-Cambrian rock (PC) which were originally deeply buried. Area overlain by glacial drift (P). Rib Hill monadnock underlain by resistant quartzite.

Sault Ste. Marie Railway—crosses the peneplain from east to west. Its length in the Northern Highland is 205 miles. This is only ten miles longer than a perfectly straight line. The same thing is true of practically all the railways,—witness, among others (a) the Northwestern Line between Hurley and Rhinelander, or the Eagle River line of the same road between Eland Junction and the Michigan boundary, (b) the Soo Line—old Wisconsin Central—between Abbotsford and Penokee Gap, (c) the Chicago, Milwaukee, St. Paul and Pacific between Wisconsin Rapids, Woodruff Junction, and the state line, and (d) the Omaha lines of the Northwestern System

THE EVEN SKYLINE OF THE PENEPLAIN IN THE NORTHERN HIGHLAND NORTHWEST OF WAUSAU.

A. MODEL OF THE LAKE SUPERIOR REGION.
(Copyright, 1909, by the University of Wisconsin.)

B. PENEPLAIN OF THE NORTHERN HIGHLAND, SHOWING A TERMINAL MORAINE
EAST OF STEVENS POINT.

between Chippewa Falls and Superior, or between Ashland and Spooner. These railways are crooked only for short distances. These are where there are (a) high morainic hills or numerous bodies of water,—as near Lac Court Oreilles on the Soo Line—or (b) deep river valleys,—as on the Wisconsin Valley Division of the St. Paul between Merrill, Wausau, and Knowlton.

The smoothness of the upland plain of the Northern Highland is also evidenced by the route that was planned for the old Fort Wilkins and Fort Howard wagon road. It was straight for 33 miles from Fort Howard, near Green Bay, to Shawano. Then it turned and ran across the Northern Highland through Crandon to State Line, Vilas County, a distance of 100 miles, straighter than the flight of Arctic water fowl.

Ridges and Hills of the Insequent Pattern. Where the peneplain is cut up by streams the hills are of an insequent or inconsequent pattern. This is because the underlying rocks are homogeneous in their resistance to erosion. The hills of granite and other igneous rocks, and the hills underlain by certain kinds of metamorphic rocks present this insequent pattern. This is also true of the ridges in parts of the Western Upland, where the sedimentary rocks are nearly horizontal. The topographic features are without marked trend. Their forms are not a consequence of any marked difference in resistance of the rock, hence the name inconsequent or insequent. The ridges of one neighborhood all rise to about the same level. They branch irregularly with the branching of the streams. This condition is well shown in Figure 144, where the drainage pattern is of the tree-like, or dendritic, pattern.

Fig. 146.   Cross-section of tilted lava flows of trap rock.   (After Strong.)

Trap Ridges. In other parts of the peneplain the hills are chiefly related to underlying rocks which are not homogeneous over large areas, but along narrow belts. For example, in northwestern Wisconsin, the Keweenawan lava flows have been acted upon by weathering and erosion so that they form parallel ridges (Fig. 146), usually with a steep slope on one side and a gentle slope on the other. Where the rocks dip in only one direction the structure is referred to as a monocline. In northwestern Wisconsin the lava flows and intervening sedimentary layers are usually inclined at angles of less than 30°. These low, flat monoclines are often made up of alternating weak and resistant layers, the resistant layers forming the crests and back slopes of the ridges, the weak layers forming the bases of the escarpments and the intervening lowlands, as in the Douglas, St. Croix, and Minong copper ranges of Douglas, Bayfield, Washburn, Burnett, Polk, and adjacent counties. Neighboring ridges usually rise to about the same level.

Figure 181 shows the angular, trellis-pattern of drainage in a region of inclined, alternate weak and resistant trap layers. The longitudinal sections of the streams are on the weak layers and here the valleys are broad. The transverse sections are where the streams pass through narrow water gaps in trap or conglomerate. The longitudinal branches in a system of trellis drainage are spoken of as subsequent streams. They are so called because they develop after the consequent streams, which are determined by original slopes on various rock structures.

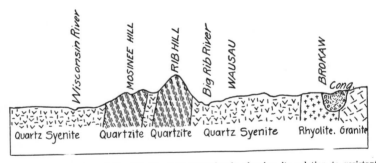

Fig. 147.   Cross-section of the Rib Mountain Monadnock, showing its relation to resistant quartzite and weaker igneous rocks.   (Weidman.)

Monadnock Ridges Rising Above the Peneplain. The ridges in the Northern Highland are of two different kinds. One has crests rising to about the same level and to about the level of the gently-sloping peneplain, as in the trap ridges and ridges of the insequent pattern. The other has ridge crests rising distinctly above the peneplain level. These are known as monadnocks. The monadnocks stand up above the level of the Lake Superior Highland because the rocks are much more resistant than those in the surrounding areas. The Penokee Range is a good example of a monadnock ridge of this character. The Flambeau Ridge and other quartzite hills of Barron and Chippewa counties also illustrate this condition well. Other monadnocks scattered throughout the area include Silver or McCaslin Mountain, Thunder Mountain, Marinette County, and Powers Bluff, Wood County. There are few, if any, monadnocks among the trap ridges.

RIB MOUNTAIN STATE PARK

Located on the summit of Rib Mountain, about 4½ miles southwest of Wausau, is Rib Mountain State Park (Fig. 148). Rib Mountain, whose highest point is 1940 feet above sea level, is one

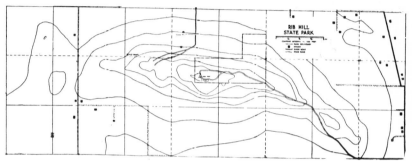

Fig. 148. Map of Rib Mountain State Park. Contour interval—100 feet.

of the most prominent monadnocks in Wisconsin. It is a quartzite ridge, rising 640 feet above the general peneplain level on the adjacent granites and truncated folds of metamorphic rocks. Its top is about 800 feet above the Wisconsin River near Wausau. It is about three miles long and over a mile wide. Its slopes, covered with angular quartzite blocks, rise at the rate of 1000 to 1200 feet to the mile. A short distance away are two other quartzite monadnocks—Mosinee Hills and Hardwood Hill. The park, which includes an area of 160 acres surrounding the highest known elevation in the state, can be reached by highway.

The Penokee Range. The Penokee-Gogebic Iron Range of Wisconsin and Michigan is about 80 miles long and half a mile to a mile wide. It is a long, narrow, monadnock, similar to Rib Mountain, though of much greater magnitude.

Irving described it in 1880 as follows:—

"The Penokee Range is, in general, much higher than the ridge to the northward of it, and, from the few high points where the thick forest does not prevent, Lake Superior can readily be seen;

while from the lake, at a distance of from ten to fifteen miles from the shore, the crest of the range shows as a blue line against the sky, coming to an abrupt end towards the west, where it drops

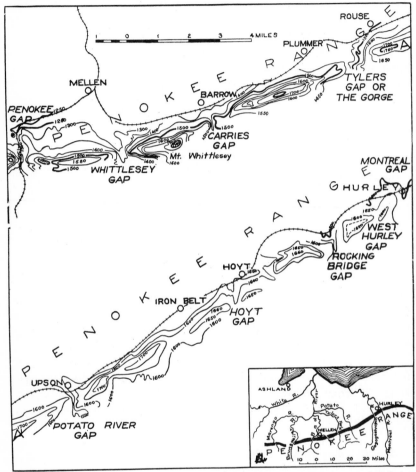

Fig. 149. Contour map of the Penokee Range and its water gaps. Contour interval 50 feet. (Compiled from maps by Irving.)

down suddenly some 200 to 300 feet. From a point on the Wisconsin Central road about two miles south of Penokee Gap * * * a heavy windfall * * * permits one to see the crest of the Penokee Range to the northward trending a number of miles east and west and alternately swelling into high peaks and sinking into gaps. * * * From the top of a mass of rock * * * one sees the range for some eight or ten miles east from Bad river rising abruptly 200 to 500 feet into a narrow serrated crest, whose highest points in sight are nearly 1,200 feet above Lake Superior."

The Penokee Range is a monoclinal ridge, with a steep dip to the north. Its crest is formed in some places by the harder portions of the Huronian iron formation, in other places by resistant

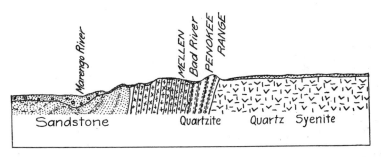

Fig. 150. Cross-section showing the relation of the Penokee monadnock to resistant quartzite and other rocks and its height above the peneplain.

quartzite, quartz slate, slaty schist, gneiss, or other metamorphic rocks. Between the trap ridges and the Penokee Range is a great valley, through which the Soo railway runs between Hurley and Mellen. This is a subsequent lowland, worn down on the site of relatively-weak, slate layers (Fig. 151). The crest of the range rises 100 to 300 feet above the peneplain to the south and 100 to 600 feet above the broad valley to the north. In some places the range is broad and gently rounded; in others, it is narrow, steep-sided and serrated. The highest point on the crest of the range is Mt. Whittlesey, 1,866 feet high and the third highest point in Wisconsin. It terminates in a 200 foot cliff of slate.

The Penokee Range is a more conspicuous elevation than such monadnocks as Rib Mountain, Flambeau Ridge, or Baraboo Range. It is longer than the Barron Hills, but not so wide. It appears like a mountain from both sides, whereas the Niagara escarpment and the Military Ridge are steep on only one side. The Penokee Range is similar to the Blue Ridge in the Appalachians, but it is not so high. The western quarter of the range is interrupted and discontinuous.

The Penokee Water Gaps. A portion of the Penokee Range in Wisconsin is broken by many gaps.

In each of these gaps, except three, streams flow northward across the Penokee Range,—the Bad River in Penokee Gap, Tylers Fork in The Gorge, the Potato River in its gap, the Gogogashugun River in Rocking Bridge Gap, and the Montreal River in Montreal Gap.

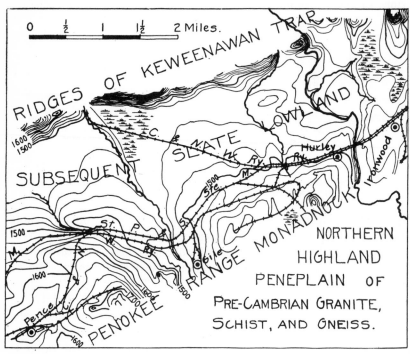

Fig. 151. The broad lowland north of the Penokee Range. Contour interval 20 feet.
(After Irving and Van Hise.)

The streams are all of moderate size, often with waterfalls and rapids in the narrow gorges where they cross the resistant rocks of the Penokee Range.

Penokee Gap (Fig. 152) shows very clearly that all these gaps are made by stream erosion. There has been faulting at Penokee Gap, but the faulting has nothing essential to do with the topographic features. To make a gap by faulting would necessitate two parallel breaks in the rocks, but there is only one. It would be

TABLE SHOWING ELEVATIONS OF THE EASTERN GAPS IN THE
PENOKEE RANGE

Name	Distance from last-named gap, in miles	Elevation of gap, in feet
Penokee Gap		1280
First wind gap	1	1480
Whittlesey Gap	2¼	1320
Devils Creek Gap	3¼	1460
Second wind gap	1½	1700
Tylers Gap or The Gorge	3¼	1460
Potato River Gap	4¼	1480
Hoyt Gap	5	1600
Third wind gap	1½	1620
Rocking Bridge or Gogogashugun Gap	2¼	1460
West Hurley Gap	1¾	1580
Montreal Gap	1	1460

essential, also, that there be vertical displacements along the fault, whereas the throw of the fault at Penokee Gap is horizontal or slightly inclined. Finally, the stream gorge should follow the course of the fault, but the river in this case leaves the fault both at the northern and southern ends of the gap. The faults are of ancient date. The streams are, relatively, newcomers.

Moreover the gorges at other gaps in the Penokee Range are identical in general topography with the one first described, though faulting is usually absent. There is a fault at the gap of Potato River, but nine other gaps have no faults. Thus we see that, although faulting may have provided weak rock along the gaps of the Bad River and the Potato River, and thus facilitated erosion there, the gaps are in no sense due to faulting.

Past History of the Water Gaps. The gorges which cross the Penokee Range are preglacial. They are unusually numerous. In the 30 miles between the Montreal and Bad rivers there are 9 water gaps. In the Blue Ridge of Pennsylvania it is unusual to find half this number of water gaps in an equal distance, and there are many stretches of more than 30 miles with no gaps whatever. For so many streams to cut across the resistant rocks of this monadnock is inconsistent with the usual history of peneplain drainage. It is perfectly consistent, however, with the history of a monadnock on a peneplain which has been buried and exhumed.

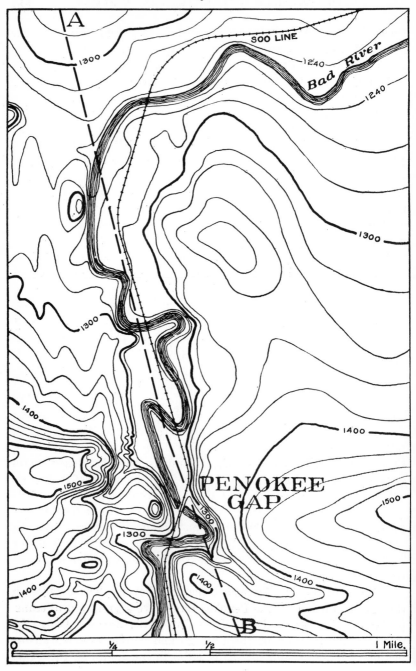

Fig. 152. The Bad River in Penokee Gap, showing a fault line at A—B. Contour interval 20 feet. (After Irving and Van Hise.)

Under such conditions, many parallel streams, flowing upon the Paleozoic sediments, might be superimposed upon underlying resistant rocks of this pre-Cambrian monadnock. This, of course, adds another to the many lines of proof that the Northern Highland was completely submerged during the early Paleozoic.

The water gaps to the west of Bad River are wider than those to the east, or we may say that the Penokee Range itself is made up of disconnected ridges. It seems probable that pre-Cambrian stream erosion rather than preglacial stream erosion or glacial sculpture, has produced the lower ridges and broader gaps in this region of flatter dip and less-continuous quartzite and iron formation. The Marengo or Maringouin River has a gorge with a depth of 250 feet in places, but Brunsweiler River crosses the Penokee Range by the broadest of the gaps.

The Future of the Penokee Water Gaps. As time goes on, streams always adjust themselves to conditions of grade and rock structure. The larger or more favored streams capture and divert the weaker ones. A case of this sort has been cited at Buffalo River in the Driftless Area, (Fig. 65 and p. 201). Accordingly we may expect that the Bad River, or Tylers Fork, or the Montreal River, may divert the neighboring streams from their water gaps, as the Potomac River has done along the Blue Ridge in Maryland and Virginia. Each of the three is a fairly large river, and, more important, each has cut its gap down to a low level (see table on p. 379). In time, the Bad River should be able to capture and divert all the streams of the Penokee Range, for its water gap is 180 feet lower than any of the others, except Whittlesey Gap. After such stream diversions, the abandoned water gaps will be spoken of as wind gaps. As the successful streams cut their water gaps lower and as the adjacent peneplain and subsequent valley are worn lower, these wind gaps will be left at their present levels because they have no streams. This process was already in progress in pre-Cambrian or preglacial time, as is shown by the presence of three wind gaps (p. 378) high up above the surrounding country.

Relation of Gaps to Transportation and Towns. These passes in the barrier formed by the Penokee Range are a great aid in transportation. Just as the Devils Lake Gap in the Baraboo Range determined the position of the Chicago and Northwestern Railway in this monadnock and for many miles to the south, so the Penokee Gap of Bad River controlled the route of the Soo Line for miles to the south. It is not clear why the old Wisconsin Central —now the Soo Line—was built through Penokee Gap rather than the Whittlesey Gap. The latter is directly opposite the city of Mellen where the railway turns northward to Ashland. Probably it was because the floor of the Whittlesey Gap is at least 40 feet higher and has an exceedingly steep descent toward Mellen.

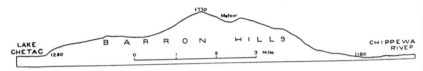

Fig. 153. Cross-section of the Barron Hills, a monadnock on the pre-Cambrian pene-plain. (Based upon leveling under the direction of E. F. Bean.)

Farther east, at Montreal River the Hurley Gap is traversed by the Chicago and Northwestern Railway, and a branch of the same railway goes through the gap at Hoyt.

A large number of the iron mining towns are located opposite gaps in the Penokee Range. As the smooth upland of the peneplain to the south has its forests cleared, its swamps drained, and its soil cultivated, these villages and cities will have a decided advantage because of the highways which will be built through the water gaps.

The Barron Hills. The monadnock group of Barron, Rusk and Sawyer counties is known as the Barron Hills. It is more than 25 miles long and 10 miles or less in width, trending northeast and southwest. It is a rolling upland, 300 to 600 feet above the adjacent peneplain. The highest point, near Meteor, Sawyer County, is 1,770 feet above sea level. This is one of the six highest points in Wisconsin.

The quartzite of the Barron Hills has been thought to be somewhat younger than the Baraboo, Rib Mountain, Flambeau, Waterloo, and several other pre-Cambrian quartzites, but this does not interfere with the interpretation of the Barron Hills as a monadnock on the pre-Cambrian peneplain.

Flambeau Ridge and Other Monadnocks. A narrow quartzite monadnock near the boundary of Chippewa and Rusk counties is called Flambeau Ridge. It is about 3½ miles long, and rises three to four hundred feet above the surrounding country. Near the eastern end is a wind gap, partly filled with glacial drift. This abandoned valley is probably the preglacial gorge of the Chippewa River. The gap, like those in the Penokee Range (p. 381) and Baraboo Range (p. 57), suggests that the Flambeau Ridge was completely buried beneath the Cambrian sandstone, so that the river was superimposed upon this buried obstacle.

Powers Bluff, Wood County, is a quartzite monadnock, rising about 300 feet above the adjacent peneplain. Thunder Mountain and McCaslin Mountain in Forest, Marinette, and Oconto counties, are similar monadnocks, with 400 to 500 feet of local relief.

The Peneplain Cycle. The monadnocks rising above the general level of the peneplain are the most conspicuous eminences in the old mountain mass of which the existing Northern Highland of Wisconsin is a remnant. The upland plain, however, represents the larger part of the former mountains, now worn down to a peneplain.

This peneplain was, of course, very long in process of formation. The period of time occupied in the wearing down of a part of the earth's crust, like the ancient mountains of northern Wisconsin, and making it nearly a plain or peneplain is alluded to as a cycle. The first erosion cycle in Wisconsin which is represented by a topographic surface of today was completed when the Northern Highland of the state was a peneplain (Fig. 143). Cutting, filling, warping, and possibly faulting, may have been going on at the same time during the latter part of this erosion cycle.

That there were earlier cycles of erosion is shown by the great unconformities of the pre-Cambrian. Several of them represent base-levelled surfaces, or peneplains, which now extend deep beneath the surface.

Warping of the Peneplain. Since the end of the peneplain cycle two important things have taken place. First, the peneplain has been lowered beneath the sea and partly, and probably wholly, covered with marine sediments, including the Cambrian sandstone (Fig. 143). Secondly, it has been uplifted again. In connection with this uplift it has been warped, so that the surface from which the sandstone has been removed is no longer approximately level. When the peneplain cycle was completed the surface had inequalities of only a few score and perhaps one or two hundred feet.

Now, however, the peneplain surface has such differences of altitudes as are shown in the table on page 370, or in the following section, which shows the east-west arching that attended the uplift of the peneplain. The places listed are 35 to 50 miles apart, and all the elevations are given in feet above sea level.

TABLE SHOWING EAST-WEST ARCHING OF THE PENEPLAIN

Locality	Webster	Couderay	Fifield	Woodruff Junction	Gagen	Armstrong Creek	Girard Junction	Wausaukee
Elevation	978	1260	1454	1615	1645	1427	1041	746

In connection with this warping the peneplain was also given a southward inclination, as is indicated in the following table. The places listed are 10 to 30 miles apart in a general north-south direction, the elevations given being, as before, in feet above sea level.

TABLE SHOWING SOUTHWARD INCLINATION OF THE PENEPLAIN

Locality	Elevation
Land o' Lakes	1708
Star Lake	1679
Arbor Vitae	1627
Woodruff Junction	1615
Hazelhurst	1592
Goodnow	1513
Irma	1507
Near Merrill	1500
Near Wausau	1400
Jnuction City	1142
Wisconsin Rapids	1000

Dissection of the Peneplain. As a result of the uplift and warping of the peneplain and the removal of the sandstone (Fig. 159) from what we now call the Northern Highland of Wisconsin, the peneplain was further modified by having stream valleys cut into it. In this respect it may be spoken of as a slightly dissected peneplain. But only part of the peneplain is dissected, or cut up by stream valleys, for the highland of northern Wisconsin does not constitute the whole of the ancient peneplain, as will be explained later. All of the highland of Northern Wisconsin, however, is more or less dissected.

The valleys which have been cut in its surface are typically represented by the valley of the Wisconsin River at Wausau. There the depth of the valley bottom below the general peneplain level is 250 to 300 feet. The valley is one to two miles wide. If we note carefully all areas which rise to or above 1,300 or 1,400 feet, taking that as the elevation of the peneplain at Wausau, it is ap-

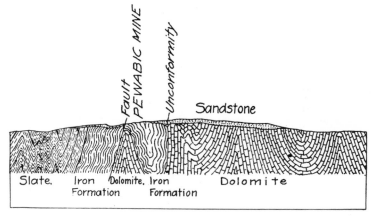

Fig. 154. Horizontal deposits of Cambrian sandstone lying upon the truncated pre-Cambrian layers near Iron Mountain, Mich. (Van Hise and Bayley.)

parent that, close to the river, very little of the original peneplain surface is left. As we go back from the river the interstream areas are broader because the peneplain remnants are very much larger. The same thing is true as we go up the Wisconsin River toward its headwaters, where the peneplain is much less cut away by stream erosion. At and near Wausau the peneplain has not only been cut into by the Wisconsin River, but by small tributaries. The same thing is true in the area near Marathon, shown in Figure 144. The valleys of the Rib River, Black Creek, the stream near Halder, and their tributaries, have been incised in the peneplain, which is represented only by the hilltop areas at elevations of 1,200 to 1,400 feet.

That this stream dissection was not accomplished before the peneplain was lowered beneath the sea is proved by the absence of the Cambrian sandstone from most of the stream valleys. Close to the borders of Florence and Marinette counties, for example, in the Menominee mining district of Michigan (Fig. 154), the sandstone lies upon the peneplain surface at elevations

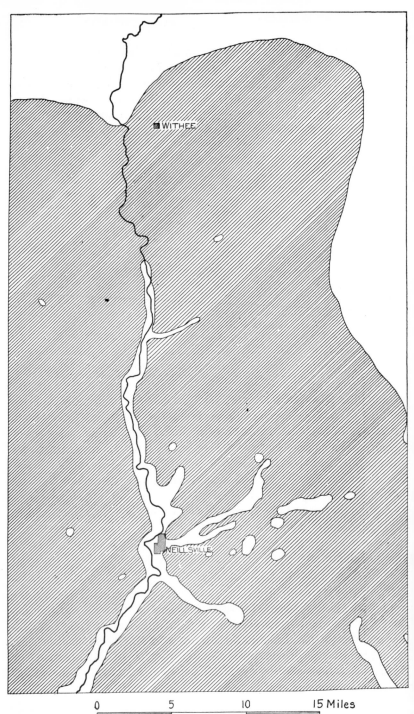

Fig. 155. A valley strip or inlier of pre-Cambrian—shown in white—where the buried peneplain has been revealed by stream erosion in the area of Cambrian sandstone—oblique ruling. (Based upon map by Weidman.)

of 940 to 1,300 feet. This is east of the city of Iron Mountain, where the Menominee River has subsequently been incised to depths of at least one or two hundred feet. Even here, however, it is clear that the peneplain was by no means a perfectly level surface originally, for the positions of patches of sandstone on the slopes of hills indicate the bluffy character of the peneplain surface.

A Fossil Land Surface. After a cycle of erosion has advanced to the peneplain stage it may be interrupted in one of two ways. (1) The region may be uplifted, in which case the peneplain will be dissected and eventually destroyed.   (2) The region may be submerged, resulting in the burial and preservation of the peneplain. The fossil land surface of the buried and preserved peneplain of Wisconsin was again uplifted, however, and it now consists of two quite different parts.   One is the Northern Highland, from which the sediments have been stripped away, so that the streams have incised shallow valleys.   The other part of the Wisconsin peneplain is still buried, and this may now be described.

It includes all the remainder of the state.   Only small portions of it are visible.   These visible portions of the peneplain are (a) inliers or narrow strips along stream valleys, and. (b) exhumed monadnocks.

Inliers. The narrow valley strips are like the one near Neillsville and Black River Falls, (Fig. 155).   This appears at Hemlock, Clark County, about 18 miles north of Neillsville.   It is separated from the main peneplain by only about 9 miles of sandstone covering.   It extends southward from Neillsville down the Black River valley 22 miles to Black River Falls, Jackson County.   Its width is only a mile or two, in contrast to a length of over 40 miles.   Its form is that of a tree and its branches, because it has been exposed by the removal of the sandstone from the beds of the main river and its tributaries.   The side valley extensions are mostly confined to the northern tributaries, where the sandstone is thinnest.   This inlier or peneplain strip extends southward from an elevation of less than 1,200 feet at Hemlock to about 997 at Neillsville, and less than 800 feet at Black River Falls.   Its slope is more than 8 feet to the mile.

Exhumed Monadnocks. The exhumed monadnocks (Fig. 11), also fossil-like in their preservation, are typified by the Baraboo Range.   This is discussed fully in Chapter III (pp. 55-59).   It may

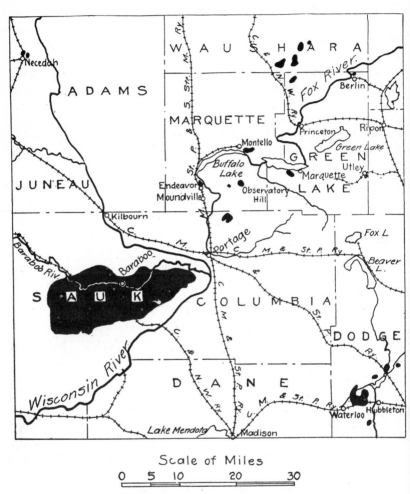

Fig. 156. Sketch map of the Baraboo Range and other exhumed monadnocks of the pre-Cambrian peneplain. (Weidman.) Later geological work shows still other monadnocks.

be briefly described here as an east-west ridge, standing 400 to 800 feet higher than the adjacent plain of sandstone to the east, and known by well borings in the vicinity to rise 900 to 1,300 feet above the surface of the buried peneplain. Its superior height has resulted in its being revealed by weathering and erosion, while the adjacent portions of the peneplain are still deeply buried. Like organic fossils, often injured in process of exhuming, the Baraboo Range has been slightly modified since it began again to see the light of day as a fossil monadnock.

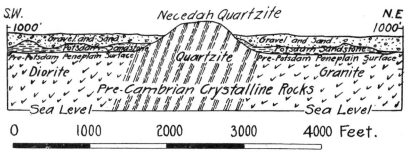

Fig. 157. The partly-exhumed monadnock at Necedah, rising above the buried peneplain. (After Weidman.)

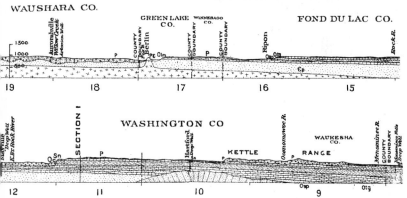

Fig. 158. Cross-sections showing the partly-exhumed monadnock at Berlin and the fossil monadnock near Hartford which is not yet uncovered.

Similar, smaller, lower, partly-exhumed monadnocks are (a) the Necedah Mound in Juneau County (Fig. 157), (b) a large number of knobs in and near the Fox River valley (Fig. 156), including the granite hill at Montello, (c) the low hills of quartzite in Jefferson County, northeast of Waterloo, and (d) the group of hills east of

Black River Falls, and (e) the Hamilton Mounds in the northeast part of Adams county. As such partly-exhumed monadnocks are like the Baraboo Range in history, it has been proposed that they be called baraboos.

Wells in the glacial drift reveal the presence of additional monadnocks in other parts of Wisconsin. Still other monadnocks are known to project up into the overlying sedimentary rocks, but are not yet revealed by erosion, as near Hartford, Washington County (Fig. 158), and near Mt. Calvary east of Lake Winnebago.

Indeed it seems probable that the large buried monadnocks, for which the names Fond du Lac Range and Waterloo Range have been suggested (Fig. 11), may be no less extensive than the Baraboo Range.

The Hidden Peneplain. In addition to the small visible portions of the peneplain there is a broad area still completely buried. This is known from well borings, which encounter the peneplain at various levels in different parts of the state (Figs. 11, 159). They suggest (1) that it is a surface of moderate relief, except where buried or partly exhumed monadnocks rise above it, (2) that it is warped, and (3) that it inclines southward, just as it does in the exposed area of the Northern Highland.

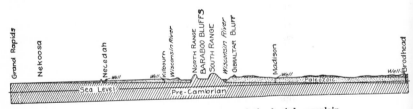

Fig. 159. The southward inclination of the buried peneplain.

The slight relief is shown in wells at Madison, which encounter the buried peneplain surface at elevations of from 21 to 139 feet above sea level. Two of these wells show a maximum difference of 91 feet in the elevation of the surface of the pre-Cambrian in a horizontal distance of 900 feet, indicating that the peneplain had slight relief and was by no means a dead-level plain. This is also shown at Necedah (Fig. 157), where four wells encounter the peneplain surface about 192, 202, 203 and 229 feet respectively below the present surface of Cambrian sandstone. This shows a pre-Cambrian relief of only 37 feet at points 1 to 2 miles apart. A fifth well reaches the pre-Cambrian 310 feet below the present surface, so that the maximum relief is 118 feet.

The southward inclination of the buried peneplain (Fig. 159) is revealed in a generally north-south section from the edge of the Northern Highland at Wisconsin Rapids to the southern boundary of the state. The heights given below are in feet above sea level.

This shows a little steeper southward tilt, for the buried peneplain, than on the exposed peneplain to the north. The grade, about 9 or 10 feet to the mile, is steeper than that of the present Wisconsin River, which itself now has a distinctly steeper grade than is normal for a river in old age on a peneplain. The southeastward and southwestward slopes of the peneplain are steeper than that to the south.

This is too steep a descent to be the original slope of a stream-eroded peneplain. It seems to show that the peneplain has been warped. Its surface now lies 814 feet below sea level at Platte-

TABLE SHOWING DEPTHS AT WHICH THE BURIED PENEPLAIN
SURFACE IS NOW FOUND

Locality	Depth below surface, in feet	Elevation above sea level, in feet
Nekoosa	0	925
Necedah	310	595
Wisconsin Dells	450	478
Madison	730	120

TABLE SHOWING SLOPES OF BURIED PENEPLAIN TO THE
SOUTHEAST AND SOUTHWEST

Locality	Descent in feet	Grade, in feet per mile	Locality	Descent in feet	Grade, in feet per mile
Shawano to Green Bay	945	21	Green Bay to Casco Jc	682	38
City Pt. to Richland Center	900	13	Richland Center to Platteville	885	22

ville, in southwestern Wisconsin, more than 600 feet below sea level at Two Rivers, and upwards of 2000 feet below sea level at Racine and Kenosha on the coast of Lake Michigan. This is one of the reasons for concluding that the pre-Cambrian mountains, and the peneplain made by their wearing down, covered all of Wisconsin as well as parts of adjacent states. The warping suggests that there was no Isle Wisconsin (pp. 394, 395) or projecting peninsula in northern Wisconsin, but that the Paleozoic sediments were laid down upon an essentially level surface. Instead of warping, however, there may have been extensive faulting (Fig. 11). In southeastern Wisconsin the buried pre-Cambrian surface is thought to be more than 2000 feet below sea level. If this is true, the surface upon which the Paleozoic sediments were laid down may have had much more substantial relief than we had previously considered.

Are There Several Peneplains on the pre-Cambrian? The writer has assumed that there is only one peneplain in the area of pre-Cambrian rocks in Wisconsin. It is not certain, however, that this is the case. Figure 11 shows that the fossil landscape beneath the Paleozoic rocks preserves a topographic surface which is steeply inclined in northeastern Wisconsin and more gently inclined in the parts of the state to the south and west. Figure 159 also suggests that the buried peneplain is more steeply inclined in places than is its exhumed continuation. The question, therefore, presents itself whether there may be several peneplains on the pre-Cambrian. If there are two or more, they intersect each other at small angles. If there is only one, the differences in angle of slope of the present surfaces is to be explained by differential warping during uplift. Various complications related to amounts of dissection of surfaces, to points where rates of inclination change, and to other factors make it difficult to determine at present whether there is a single pre-Cambrian or early Paleozoic peneplain in Wisconsin or more than one.

Buried Soils of Peneplain Surface. At Madison, where it lies 818 feet below the present surface, the buried peneplain is mantled by weathered rock and perhaps residual soil. This kaolinized rock is also found, with a thickness of 10 to 12 feet, beneath the edge of the mantle of sandstone where the buried peneplain emerges at Wisconsin Rapids, Stevens Point, Black River Falls, and elsewhere.

Borders of the Northern Highland. In drawing the boundaries of the Northern Highland province it has seemed best to follow the edge of the Cambrian sandstone, where it rests upon the pre-Cambrian rocks. This border is an irregular one, being high here and low there, advancing in a salient here and receding in a reentrant there, as the retreating edge of the thin sandstone mantle determines. The variations in height are only a few hundred feet. The distances between points of salients and heads of embayments run up to 20 or 25 miles. Most of the exposed pre-Cambrian peneplain in the adjacent sandstone plain lies along the river valleys, as in the long narrow tongue which extends southward from Wisconsin Rapids with a width of only three or four miles. Since most of the embayments are along stream valleys, the larger number of the salients rest on divides.

The southern boundary of the pre-Cambrian at surface (Fig. 11) incorporates the most recent data. The most striking feature of this border is the forty mile embayment extending southerly from Stevens Point. This is not followed by a single modern river valley. Before the Glacial Period, however, the Wisconsin followed this more direct course from Stevens Point to Portage. The embayment as mapped is based upon well records since this valley is filled with glacial drift. It should be noted that the boundary of the pre-Cambrian northeast of this embayment is now known to be located farther southeast than is shown in Fig. 10. At Gillett the pre-Cambrian is under 412 feet of drift. At Black Creek granite underlies 512 feet of drift.

With the exception of the faulted border in Douglas, Bayfield, and Ashland counties (see Chapter XVII) the borders of the Northern Highland are all in process of modification by the removal of the Cambrian sandstone. The sandstone border does not usually form a retreating escarpment. It is so weak that its edge does not form any significant topographic feature. In places, however, the border is a low sandstone escarpment facing the peneplain, as northeast of Chippewa Falls, west of Wisconsin Rapids, and near the boundary between Barron and Rusk counties.

Sandstone Outliers. Resting upon the surface of the peneplain, at various distances from the border of the Cambrian sandstone, are detached sandstone masses, or outliers, usually bearing the name, mound (Fig. 160). Some of these outliers are a mile or less, others 15 miles or more from the border. They are usually of small size. Their height varies from 15 or 20 to 200 feet. They often stand up as prominent, flat-topped hills, partly because of the slightly greater resistance of the sandstone. They are conspicuous chiefly because of the monotonously level surface of the peneplain upon which they rest.

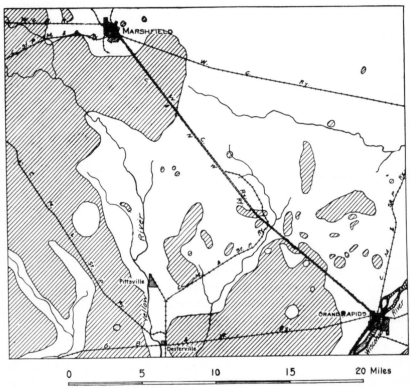

Fig. 160. Sandstone outliers—ruled areas—resting upon the surface of the pre-Cambrian peneplain. (After Weidman.) Grand Rapids is now called Wisconsin Rapids.

Their presence is proof of the former extension of the Cambrian sandstone over the surface of the peneplain. They suggest that the peneplain was originally completely covered by this sandstone and some of the other sedimentary rocks of the Paleozoic. This suggestion is supported by (a) the known occurrence of sandstone within 100 feet of the maximum height of the peneplain, (b) the absence of shoreline features at the borders of the sandstone, (c) the absence of adequate erosion features in the high central portion of the peneplain, such as would have been made if the central area had remained unburied, and (d) the natural relationship of the existing embayed border of the main mass to the outliers of sandstone on the surface of the peneplain.   In the mind of the writer there is no doubt whatever that the whole Northern Highland of Wisconsin was formerly buried beneath the sedimentary rocks of at least the lower part of the Paleozoic.   The present, exposed peneplain has been exhumed by a process of erosion, whose continuation will eventually add very materially to the area of the geographical province under discussion.

Moreover the Paleozoic rock formations have such relations to one another that it seems clear that they do not bevel out, but continue parallel to each other.

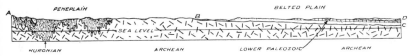

Fig. 161.   The peneplain of the Northern Highland (A-B) and its buried extension (B-C).

A reason for the lack of adjustment between many of the streams and the weaker structures in the peneplain, entirely aside from the episode of glaciation, appears to be that these are not the rivers which carved the peneplain from the ancient mountain mass.   They are fairly modern streams, superimposed upon the peneplain from original consequent courses on the formerly overlying sedimentary rocks of the Paleozoic.

Age of the Peneplain.   It has already been shown that the peneplain in the Northern Highland province is an ancient feature.   It was produced in an earlier cycle, then submerged and buried, then uplifted and partly exhumed in the present cycle, during which it has been slightly dissected by stream erosion.   Obviously the peneplain is younger than the pre-Cambrian, since its surface truncates structures of Archean, Huronian, and Keweenawan age. It is also patent that it is older than the Upper Cambrian sandstone, since this formation rests upon the peneplain surface.   This places the age of the peneplain in late Keweenawan or early Cambrian time, but it will serve all purposes if we speak of it as the pre-Cambrian peneplain.

GLACIATION OF THE NORTHERN HIGHLAND

Glacial Lobes and the Areas They Affected. The lobes of the continental ice sheet which affected the Northern Highland of Wisconsin (Fig. 28) were (a) the Green Bay lobe of the Lake Michigan Glacier, (b) the Chippewa lobe of the Lake Superior Glacier, possibly with an intermediate Keweenaw lobe, (c) the western extension of the Lake Superior Glacier, and (d) lobes from the Keewatin center. These lobes represented borders of the continental ice sheet. They came from the Labrador, Patrician, and Keewatin glacial centers which are shown in Figure 27. The easternmost lobes from the Keewatin center are generally called the Minnesota lobe and the Rainy Lake lobe. Keewatin ice may even have affected northern Wisconsin in places where it did not actually enter the state.

The ice of the continental glacier came into northern United States many times, withdrawing from it only to advance again. The complications of the Pleistocene Glacial Period are very great. They need not be entered into in this book, however, because field work on the glacial geology of Wisconsin has not yet advanced to such a stage that we can say with certainty what happened in Wisconsin during the three or four early major stages of the Glacial Period. The older drift, alluded to as a unit in connection with discussion of other parts of Wisconsin, will be treated in northern Wisconsin also as if it were one entity. It should be understood, however, that glacial deposits of Nebraskan, Kansan, Iowan, or Illinoian age might conceivably be represented in parts of northern Wisconsin. It seems desirable to state here, moreover, that the ice which covered northern Wisconsin during the several periods of deposition of older drift, as well as during the periods of deposition of the drift of the Wisconsin stage of glaciation, came at different times from different' centers of glaciation, and that the latest observations in northern Wisconsin are pointing towards the conclusion that there was a pendulum-like swing in the whole group of periods of glacial advance.

The significance of the reference to a pendulum-like swing may be grasped by looking at Figure 27 while reading the following statements. The ice which laid down the Nebraskan drift and the Kansan drift, and probably that which laid down the Iowan drift, is thought to have come from the Keewatin center. The

Illinoian drift probably came from the Labrador center. It is possible, of course, that some of the older drift came from the Patrician center as well as from the Keewatin center. In any event, it was the glacial centers west of Hudson Bay that are likely to have sent the first glacial lobes into northern Wisconsin. Following the deposition of the older drift, and after tremendously long interglacial lapses of time, came the latest advances of the continental ice sheet. These are spoken of, collectively, as the Wisconsin stage of glaciation, both in the State of Wisconsin and elsewhere in northern United States. It appears likely that the earlier Wisconsin deposits in northern Wisconsin came from the Labrador center, that the intermediate group came from the Patrician center, and that the latest ones came from the Keewatin center. This attribution of glacial deposits to the Labrador rather than the Patrician, or the Patrician rather than the Keewatin, or the Keewatin rather than the Labrador center, is based primarily upon the directions of glacial striae and upon the lithological composition of the glacial drift.

When the glacial geology of the Lake Superior region is known in sufficient detail, we may be able to say with certainty what we merely suspect now. (1) Continental glaciation is likely to have first reached Wisconsin from the Keewatin center or the Patrician center west of Hudson Bay. (2) The ice from the Labrador center came later. (3) At the closing stages the ice came from the Patrician center, and finally from the Keewatin center. This applies to the whole Middle West, as well as to northern Wisconsin. When this is worked out in its entirety, we shall have demonstrated the pendulum-like succession alluded to above, with the swing of refrigeration, (a) beginning at the west, (b) continuing in the east, and (c) returning again to the west. The reason for this cannot yet be stated. It appears likely that general continentality of climate, superimposed upon glacial climate, will enter largely into the explanation. That is merely another way of saying that, when glacial conditions became inevitable, they arrived sooner in the region west of Hudson Bay than in the region east of Hudson Bay, since the latter has a climate partaking more of the equability associated with nearness to the ocean. For the same reason, there is likely to have been a longer duration of glacial climate in the region west of Hudson Bay than in that to the east. That may explain why the latest of the glacial advances into Wisconsin which we have thus far experienced, came from the Keewatin ice sheet.

The Green Bay, Chippewa, and Superior lobes advanced till they covered the whole Northern Highland with the exception of a very small area in the valley of the Wisconsin River near Wausau, Stevens Point, and Wisconsin Rapids.

It is thought that there were several partial or complete withdrawals of the ice sheet from the highland. At the last oscillation, known as the stage of Wisconsin glaciation, an even larger area of the Northern Highland was left unglaciated, comprising many hundred square miles in parts of Marathon, Wood, Clark, Taylor, Lincoln, and Langlade counties. This area may have been left partly unglaciated in several glacial oscillations earlier than the last or Wisconsin stage. In our discussion it will be spoken of simply as the region of older glacial drift. The remainder and predominating portion of the Northern Highland was completely covered by the continental glacier at all stages of the Glacial Period.

In relation to the effects of these various episodes in the glacial history of the Northern Highland it is possible to divide the province into three contrasting areas. The first, and smallest, is part of the Driftless Area, the second is a region of older drift, and the third, and largest, is a region of latest, or Wisconsin drift.

Driftless Area. Within the Driftless Area (see also Chapter IV) the Northern Highland is a normal peneplain, with residual soil produced by the weathering and decay of the underlying rocks. These residual soils are, of course, not the original weathered materials of the peneplain surface. They are new residual soils, largely or wholly produced after the sandstone was removed from the exhumed surface of the previously buried peneplain.

The streams drain the area thoroughly, so that there are no swamps or lakes. In general they are well graded in relation to the peneplain surface, so that waterfalls and rapids are not commonly found in the stream courses. Most of the rivers and creeks branch and subdivide with the tree-like, or dendritic, pattern characteristic of long-continued stream adjustment (Fig. 144). The only exception to this is found where streams from the glaciated area have brought in deposits and covered the otherwise driftless surface. The rock is almost everywhere near the surface.

Fig. 162. Block diagrams showing part of the Northern Highland of Wisconsin after the Paleozoic covering (Fig. 143) had been removed, and again after glacial deposits had buried the preglacial topography, except in the Driftless Area, and had given the streams their present courses. (After Weidman.)

Area of Latest Glaciation. The portion of the Northern Highland in the area of Wisconsin glaciation forms a striking contrast with the Driftless Area. There is no residual soil. Instead, there is a transported, glacial soil (Fig. 162). Rapids and waterfalls are abundant in the streams. There are large undrained interstream areas. Lakes and swamps are found everywhere. The drainage pattern is most irregular, resembling nothing systematic, as is perfectly normal for so youthful a drainage system.

It is interesting to note that the second highest point in Wisconsin, so far as now known, is not a conspicuous rocky peak but a morainic hill. It is about 1891 feet high and is situated east of Ogema, Price County. The fifth among the high points in the state is a broad morainic hill west of Crandon, Forest County. This has an elevation of about 1850 feet. The railway from Crandon to Pelican makes a long detour to avoid the morainic country of which this hill forms a part. It is higher than Blue Mound or the Baraboo Range. It is possible that the morainic hills of northern Wisconsin may contain a point even higher than Rib Mountain.

Materials in the Drift. Mechanical analyses of the drift have been made in about 2,500 square miles of the Northern Highland. They show that the igneous rocks—granites, porphyries, gabbros, and fine greenstones—constitute 65 to 70 per cent of the rock materials in the drift. The remainder is schist, quartzite, sandstone, iron formation, other metamorphic and sedimentary rocks, and quartz sand. This forms a striking contrast with the drift of southeastern Wisconsin. In the region south of Madison and Milwaukee the igneous rocks form a small percentage of the whole (p. 255). Crystalline rocks, including both igneous and metamorphic rocks, make up only 13 per cent of the drift. The remaining 87 per cent consists chiefly of local limestone and sandstone.

This contrast is, of course, due to the difference in the bed rock of northern and of southeastern Wisconsin. In northern Wisconsin the pre-Cambrian rocks are chiefly granites, gabbros, traps, gneisses, and schists. Where quartzite ledges outcrop there is abundant quartzite in the drift, the amount sometimes being as much as 20 to 70 per cent of the whole. Where iron formation occurs the drift nearby reveals it, even if the ledges are entirely buried. The amount is sometimes 20 or 30 per cent of the drift. By this means, indeed, iron mines may possibly be located by drilling in lands completely mantled by the drift.

In the northeastern part of the peneplain the glacier imported some limestone, but the drift of the Northern Highland is usually without limy constituents.   One advantage is that spring water and well water throughout the Northern Highland is prevailingly soft.

Terminal Moraines. The form of the peneplain surface has been slightly modified by the glacial deposits. Three principal sorts of topographic forms are found, (a) the terminal or recessional moraines, (b) the ground moraine, and (c) the outwash deposits.

In various parts of the Northern Highland the thickness of the glacial material in terminal moraines varies from 75 to 100 feet. It has a probable maximum of 350 feet, in the Wisconsin Valley moraine north of Merrill, Lincoln County, and perhaps as much as 500 or 600 feet west of Ashland, Bayfield County.

The material is, variably, unassorted till or stratified sand and gravel. The till, or bowlder clay, is made up of fine clay, sand, and subangular, striated bowlders of various sorts, including foreign bowlders from the region around Lake Superior. It is unassorted because deposited directly by the melting ice. The stratified sand and gravel is material carried by streams from the melting glacier and is, therefore, assorted. The glacial scratches have been removed from bowlders in the outwash. Many of them have been rolled along by the streams until they are rounded instead of subangular. The surface form of the terminal and recessional moraines is sometimes a smooth, broad-topped ridge, sometimes a hilly mass of knobs and kettles, the latter often containing lakes and small swamps.

There are interlobate, kettle moraines in at least two parts of the Northern Highland. One is well developed in Oconto County at the junction of the Green Bay and Keweenaw-Chippewa lobe. The other is in Bayfield County, between the Chippewa and Superior lobes. It is related to the massive terminal moraine which extends westward from the point where the Chicago and Northwestern Railway crosses the Michigan state line in Vilas County to Patzu in Douglas County, Wis., and Belden in Pine County, Minn.

The terminal and recessional moraines are well represented by the area shown in Figures 162 and 163. Much of the terminal moraine at the border of the Wisconsin drift is kettle moraine, with deep pits, high knobs, and strong ridges. The terminal mo-

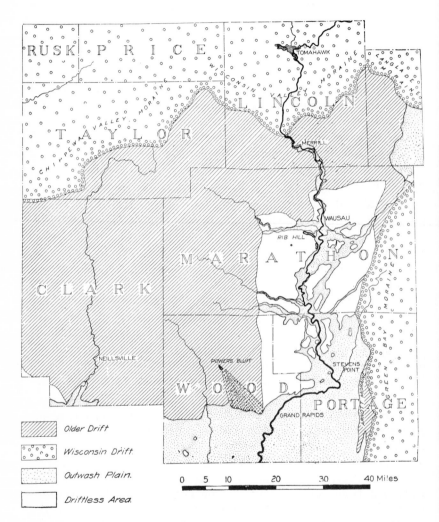

Fig. 163. Map showing the distribution of the deposits of glacial drift in the Northern Highland near Wausau. The dashed lines showing the bowlder train at Powers Bluff indicate that the ice which deposited the older drift here moved southeastward. The outwash plain south of Wisconsin Rapids (Grand Rapids) is made by coalescing valley trains from the north. (After Weidman.)

raines of Florence and Marinette counties are narrow, and represent the same readvance which deposited the red drift of eastern Wisconsin (p. 267) and northwestern Wisconsin (p. 396), including the massive terminal moraine referred to above.

Where the ice rode up over the Barron Hills monadnock the moraine has an irregular border (Fig. 34), similar to that on the Baraboo Range east of Devils Lake.

Ground Moraine. The ground moraine covers a wide area, in contrast with the terminal moraine, which is found in narrow strips. Its thickness is from a few inches to 100 feet or more, and the material throughout much of the area is unassorted till. The topography of the area of ground moraine varies. In places there is only a thin veneer of ground moraine with abundant bare rock ledges. Elsewhere there may be thick ground moraine which completely buries the rocky surface of the peneplain. There is apt to be a rolling surface, sometimes with broad swells and shallow sags, the latter often containing enormous swamps (p. 416). A group of drumlins in Vilas County has a trend of 30° to 40° west of south.

Outwash Deposits. The outwash deposits cover vast areas in the Northern Highland. The thickness of the sand and gravel deposited by streams from the melting ice often exceeds 30 or 40 feet. In one case in the Namakagon valley in Washburn County it is over 160 feet. Some of these outwash deposits cross the Driftless Area. The stratified gravels form outwash plains of great extent, as (a) in Vilas and Oneida counties, (b) near Eagle River, (c) in Douglas, Bayfield, and Washburn counties, where there are extensive pine barrens with dunes, and (d) in Wood and Portage counties, near Wisconsin Rapids. In many localities the glacial gravels are in valley trains in the bottoms of the stream courses which trench the peneplain, as in Lincoln and Marathon counties, between Merrill and Wausau.

As one looks down upon the barren plains adjacent to Malaspina Glacier, Alaska, one sees scores of braided glacial streams, branching and reuniting in a silvery network (Pl. XVII, A.). On approaching these streams one sees that they are heavily laden with sediment, which gives them a coffee color, in contrast to which the muddy Missouri River is pale and anemic. When wading across one of the smaller glacial streams—the larger torrents are absolutely impassable except with a boat or on horseback—one finds that good-sized cobblestones and bowlders are being rolled along the bottom. These stones, and the gravel and coarser sand, fill up the stream channel so that the glacial rivers are constantly shifting in position. Hence a broad area, adjacent to the streams from the melting ice, is being built up with an accumulation of glacial outwash or outwash gravel.

The outwash plains in Wisconsin are usually smooth. There are slight irregularities due to channels of the glacial streams which built up the outwash. Other irregularities are the dry or lake-filled pits and kettles, produced by the melting of buried ice blocks during the retreat of the glacier. Kettle lakes are abundant in places. In parts of northern Wisconsin the pitted outwash is so irregular as to simulate terminal moraine.

The valley trains are also apt to have irregularities in the form of terraces, cut by the glacial streams as their load was decreasing, and by postglacial erosion of present rivers. Such terraces are well developed (a) in the Wisconsin valley near Wausau, Stevens Point, and Wisconsin Rapids, (b) in the valley of the Brule and Pine rivers near Florence, and (c) along the Namakagon River between Cable and Trego. The upper terrace levels often have kettles due to slumping of buried ice masses, as in the valley of the Menominee River in Florence and Marinette counties.

Eskers. Another type of glacial stream deposits are eskers, formed in tunnels beneath the ice. They are sinuous ridges of rounded gravel, and may be found, among other places, in northern Florence County.

Area of Older Drift. In the region of older drift the conditions are intermediate between those in the Driftless Area and in the area of latest glaciation. The thickness of the older drift varies from 5 to 170 feet. There is no thick residual soil, as in the Driftless Area, but the transported soil is somewhat weathered, some of the minerals being changed to clay 10 or 15 feet below the surface. To cite one instance: buried soil 2 feet in thickness is reported from one well which goes through 58 feet of the younger drift before encountering the soil layer on the surface of the older drift.

The lakes have practically all been filled or drained, but some of the stream systems still show glacial characteristics. In general the surface is one of erosional rather than depositional characteristics.

Within the area of older drift is the Powers Bluff bowlder train (Fig. 163), which has a length of over 4 miles.

Application of Results of Glacial Occupation. The Northern Highland, on the whole, has been profoundly affected by the glacial occupation. The soil is changed and, in general, is stonier and more sandy. This results in vast areas that are better suited to forest than to crops (pp. 416-418), especially as large areas are swampy. The lakes are a source of steady water supply for the rivers that flow from this highest part of the state, as well as an asset in the lumbering industry and an attraction to fishermen and summer visitors. The rapids and waterfalls will furnish invaluable water power in addition to that already utilized (p. 420). The iron deposits of the region, though less eroded during the glacial occupation than the soft ores of the Mesabi region of Minnesota, are more difficult to find than in a region of residual soil, being often deeply buried beneath the transported glacial soil. It has been asserted that the glaciated part of the Northern Highland has not benefited by glaciation, as in southern Wisconsin (p. 139). Judging by the difference between agriculture in New England and in the never-glaciated Piedmont Plateau of Pennsylvania and Virginia, this may be the case.

BIBLIOGRAPHY

Aldrich, H. R. Geology of the Gogebic Iron Range of Wisconsin, Wis. Geol. and Nat. Hist. Survey, Bull. 71, 1929, 279 pp.

Bean, E. F. Analyses of Glacial Drift in 87 townships, or over 3000 square miles, in Washburn, Sawyer, Barron, Rusk, Chippewa, Bayfield, Ashland, Price, and Oneida counties,—Published in Hotchkiss, Bean, and Wheelwright's Mineral Lands in Part of Northwestern Wisconsin, Bull. 44, Wis. Geol. and Nat. Hist. Survey, 1915, pp. 62-74; Methods of Mapping Glacial Geology in Northern Wisconsin, Annals Assoc. Amer. Geographers, Vol. 5, 1915, p. 144.

Berkey, C. P. Geology of the St. Croix Dalles, Am. Geologist, Vol. 20, 1897, pp. 345-383; Vol. 21, pp. 139-155, 270-294, 1898; Laminated Interglacial Clays of Grantsburg, Wisconsin, Journ. Geology, Vol. 13, 1905, pp. 35-44.

Brooks, T. B. Geology of the Menominee Iron Region, Geology of Wisconsin, Vol. 3, 1880, pp. 431-663.

Brown, E. D. The Geography of the Chippewa Valley, Unpublished thesis, University of Wisconsin, 1922.

Chamberlin, T. C. Geology of Eastern Wisconsin, Geology of Wisconsin, Vol. 2, 1877, pp. 248-256; Historical Geology—Pre-Laurentian History,

Laurentian Age, Huronian Age, Keweenawan Period, Geology of Wisconsin, Vol. 1, 1883, pp. 45-118.

Clark, A. C., and Chamberlin, T. C.   Superficial Geology of the Upper Wisconsin Valley, Geology of Wisconsin, Vol. 4, 1882, pp. 717-723.

DuReitz, T. A.   The Deformation of the Pre-Cambrian Peneplain of North America, Geol. Fören., Stockholm Förhandl., Bd. 47, 1925, pp. 250-257.

Edwards, Ira.   The Sea Caves of Devils Islands, Milwaukee Pub. Museum, Yearbook, 1924, Vol. 4, 1926, pp. 94-98.

Ellsworth, E. W., and Wilgus, W. L.   The Varved Clay Deposit at Waupaca, Wisconsin, Trans. Wisconsin Acad. Sci., Vol. 25, 1930, pp. 99-111.

Grant, U. S.   The Copper-Bearing Rocks of Douglas County, Wisconsin, Bull. 6, Wis. Geol. and Nat. Hist. Survey, 1901, 83 pp.

Hinn, Helen B.   The Geography of the Wisconsin River Valley, Unpublished thesis, University of Wisconsin, 1920.

Hotchkiss, W. O. Bean, E. F., and Aldrich, H. R.   Mineral Lands of Part of Northern Wisconsin, Bull. 46, Wisconsin Geol. and Nat. Hist. Survey, 1929, 212 pp.

Hotchkiss, W. O., Bean, E. F., and Wheelwright, O. W.   Mineral Lands in Part of Northwestern Wisconsin, Bull. 44, Wis. Geol. and Nat. Hist. Survey, 1915, 367 pp.

Irving, R. D.   On Some points in the Geology of Northern Wisconsin, Trans. Wis. Acad. Sci., Vol. 2, 1874, pp. 107-119; On the Age of the Copper-Bearing Rocks of Lake Superior and on the Westward Continuation of the Lake Superior Synclinal, Amer. Journ. Sci., 3rd Series, Vol. 8, 1874, pp. 46-56; Kaolin in Wisconsin, Trans. Wis. Acad. Sci., Vol. 3, 1876, pp. 3-30; Geology of Central Wisconsin, Geology of Wisconsin, Vol. 2, 1877, pp. 461-524; Note on the Age of the Crystalline Rocks of Wisconsin, Amer. Journ. Sci., 3rd Series, Vol. 13, 1877, pp. 307-309; Note on the Stratigraphy of the Huronian Series of Northern Wisconsin, Ibid., Vol. 17, 1879, pp. 393-398; Geological Structure of Northern Wisconsin, Geology of Wisconsin, Vol. 3, 1880, pp. 1-25; Geology of the Eastern Lake Superior District, Ibid., pp. 53-238; The Copper-Bearing Rocks of Lake Superior, Monograph 5, U. S. Geol. Survey, 1883, 446 pp; Ibid., 3rd Annual Rept., U. S. Geol. Survey, 1883, pp. 89-188; Archean Formations of the Northwestern States, Ibid., 5th Annual Rept., 1885, pp. 175-242; Classification of Early Cambrian and Pre-Cambrian Formations, Ibid., 7th Annual Rept., 1888, pp. 365-454.

Irving, R. D., and Van Hise, C. R.   Crystalline Rocks of the Wisconsin Valley, Geology of Wisconsin, Vol. 4, 1882, pp. 627-714; The Penokee Iron-Bearing Series of Michigan and Wisconsin, Monograph 19, U. S. Geol. Survey, 1892, 474 pp; Ibid., 10th Annual Rept., Part 1, 1890, pp. 341-507.

Keyes, Charles.   Lake Superior Highlands; their Origin and Age, Journ. Geol., Vol. 23, 1915, pp. 569-574.

King, F. H.   Geology of the Upper Flambeau Valley, Geology of Wisconsin, Vol. 4, 1882, pp. 585-621; Physical Features and Climatic Conditions of Northern Wisconsin,—in W. A. Henry's Northern Wisconsin—A Hand-Book for the Homeseeker, Madison, 1896, pp. 24-40.

Lapham, Increase A. The Penokee Iron Range, Trans. Wis. State Agricultural Society, Vol. 5, 1859, pp. 391-400.

Leverett, Frank. Moraines and Shore Lines of the Lake Superior Region, Prof. Paper 154A, U. S. Geol. Survey, 1929, pp. 16-17, 29-31, 39, and Plate 1; Surface Geology of the Northern Peninsula of Michigan, Publication 7, Mich. Geol. and Biol. Survey, 1911, 86 pp.

Martin, Lawrence. Physical Geography of the Lake Superior Region, Monograph 52, U. S. Geol. Survey, 1911, pp. 85-117; The Pleistocene, Ibid., pp. 427-459; The Physical Geography of Wisconsin, Journ. Geog., Vol. 12, 1914, pp. 229-230.

Murphy, R. E. The Glacial Geology of an Area in Northwestern Wisconsin, Unpublished thesis, University of Wisconsin, 1926; The Geography of the Northwestern Pine Barrens of Wisconsin, Trans. Wis. Acad. Sci., Vol. 26, 1931, pp. 69-120.

Osgood, Wayland. Geology of the Gogebic Range in the Vicinity of Potato River Gap, Unpublished thesis, University of Wisconsin, 1923.

Percival, J. G. The Primary Rocks, Annual Report of the Geological Survey of the State of Wisconsin, Madison, 1856, pp. 103-111.

Russell, I. C. The Surface Geology of Portions of Menominee, Dickinson, and Iron Counties, Michigan, Annual Rept., Mich. Geol. Survey, 1907, pp. 7-82.

Stehr, R. A. A Study of Pro-Glacial Lake Deposits in T. 48 N., R. 9 W., and the W ½ T. 48 N., R. 8 W., Wisconsin, Unpublished thesis, University of Wisconsin, 1925.

Strong, Moses. Geology of the Upper St. Croix District, Geology of Wisconsin, Vol. 3, 1880, pp. 367-428.

Strong, Moses, and others. Quartzites of Barron and Chippewa Counties, Geology of Wisconsin, Vol. 4, 1882, pp. 573-581.

Sweet, E. T. Geology of the Western Lake Superior District, Geology of Wisconsin, Vol. 3, 1880, pp. 310-362; Notes on the Geology of Northern Wisconsin, Trans. Wis. Acad. Sci., Vol. 3, 1876, pp. 40-55.

Thwaites, F. T. Glacial Geology of Part of Vilas County, Wisconsin, Trans. Wis. Acad. Sci., Vol. 24, 1929, pp. 109-125; The Origin and Significance of Pitted Outwash, Journ. Geol., Vol. 34, 1926, pp. 308-319; The Buried PreCambrian of Wisconsin, Bull. Geol. Soc. Amer., (abstract), Dec. 1930, p. 32.

Van Hise, C. R. Excursion to Lake Superior, Sketch of Pre-Cambrian Geology South of Lake Superior, with References to Illustrate Localities, Congrès Géologique International, Compte Rendu de la 5me. Session, Washington, 1891, pp. 493-512; An Historical Sketch of the Lake Superior Region to Cambrian Time, Journ. Geol. Vol. 1, 1893, pp. 113-128; A Central Wisconsin Baselevel, Science, new series, Vol. 4, 1896, pp. 57-59; Principles of Pre-Cambrian North American Geology, 16th Annual Rept., U. S. Geol. Survey, Part 1, 1896, pp. 571-874; Iron Ore Deposits of the Lake Superior Region, Ibid., 21st Annual Rept., Part 3, 1901, pp. 305-434.

Van Hise, C. R., and Bayley, W. S. Folio 62, Geologic Atlas of the United States, Menominee Special Folio, U. S. Geol. Survey, 1900.

Van Hise, C. R., and Leith, C. K. The Geology of the Lake Superior Region, Monograph 52, U. S. Geol. Survey, 1911, 626 pp.

Weidman, Samuel. The Pleistocene Succession in Wisconsin (abstract), Science, new series, Vol. 37, 1913, pp. 456-457; Bull. Geol. Soc. America, Vol. 24, 1913, pp. 697-698; The Pre-Potsdam Peneplain of the Pre-Cambrian of North Central Wisconsin, Journ. Geol., Vol. 11, 1903, pp. 289-313; The Geology of North Central Wisconsin, Bull. 16, Wis. Geol. and Nat. Hist. Survey, 1907, 681 pp; Soil Surveys of Marinette County, of North Central Wisconsin, and of part of Northwestern Wisconsin, Ibid., Bulls. 11, 23, and 24, 1903 and 1911.

Whitson, A. R., and others. Soils Reports: Wisconsin Geol. and Nat. Hist. Survey, Bulls, 11, 23, 24, 31, 32, 43, 47, 50, 52A, 52B, 52C, 54C, 54D, 55.

Whittlesey, Charles. Magnetic Iron Beds of the Penokie Range,—in Owen's Report of a Geological Survey of Wisconsin, Iowa, and Minnesota, Philadelphia, 1852, pp. 444-447; The Penokie Mineral Range, Wisconsin, Proc. Bost. Soc. Nat. Hist., Vol. 9, 1865, pp. 235-244; On the Ice Movements of the Glacial Era in the Valley of the St. Lawrence (includes direction of striae in northern Wisconsin), Proc. Amer. Assoc. Adv. Sci., Vol. 15, 1867, pp. 43-54.

Wooster, L. C. Transition from the Copper-Bearing Series to the Potsdam Amer. Journ. Sci., 3rd Series, Vol. 27, 1884, pp. 463-465.

Wright, C. E. The Huronian Series West of Penokee Gap, Geology of Wisconsin, Vol. 3, 1880, pp. 239-301; Geology of the Menominee Iron Region, Ibid., pp. 667-741.

For further references on the stratigraphy and economic geology of the Lake Superior region see Van Hise and Leith's "Pre-Cambrian Geology of North America" (Bull. 360, U. S. Geol. Survey, 1909), and "The Geology of the Lake Superior Region" (Monograph 52, Ibid., 1911).

MAPS

U. S. Geological Survey. Marathon, Wausau, Robbins and Three Lakes Quadrangles, Wisconsin; See also Iron River and Menominee Special Quadrangles, Michigan (Fig. 197); and 26 sheets of special river profile maps by L. S. Smith (Fig. 199).

University of Wisconsin. Model of the Lake Superior region (Plate XXX,A).

Wisconsin Geological Survey. Contour maps of the Penokee Range (Fig. 202); soils maps of various counties (Fig. 205); profiles on township maps in Bulletin 44 (Fig. 202).

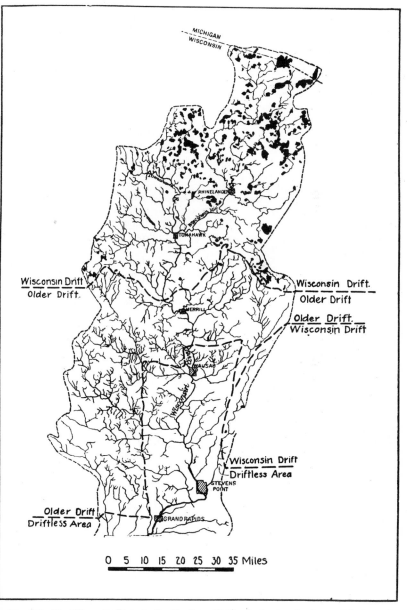

Fig. 164. The Wisconsin River in the Northern Highland, showing the types of drainage in the latest or Wisconsin drift, in the area of older drift, and in the Driftless Area.

Fig. 165. Lake and marsh district in northern Wisconsin. (Hobbs, after Fairbanks).

*N*EARLY ALL OF THESE LAKES, so far as observed, possess the characteristics peculiar to those of broad, morainic belts. They are beautiful sheets of water, clear, soft and deep, encircled by bold, fantastic rims, and dotted with tree-clad island cones of such varied beauty in the autumn season, that as one toils in unexpectedly upon them up the rapids of the narrow shaded rivers, he forgets his fatigue and revels in an exquisite garden of foliage plants. Sometimes a fringe of white cedar lies upon the water's edge; higher up a wreath of white birch, then a belt of poplar, and, capping the rounded hilltops, maple and yellow birch, throughout all of which there is a generous setting of rich green white and Norway pines." F. H. King, in 1879.

CHAPTER XVI

THE LAKES AND STREAMS OF THE NORTHERN HIGHLAND

GEOGRAPHY OF AN ANCIENT BATTLEFIELD

A score of years or more before William Penn founded the City of Brotherly Love, a war was in progress just south of Lake Superior. We know how, in one battle at least, the combatants took advantage of the geography of the region. In their strategy they were aided by the vast extent of land covered by lakes and swamps. This was very well indicated by Perrot in 1656-62.

It was a fight between the Huron Indians and the Sioux. The battle took place in northwestern Wisconsin or nearby in Minnesota, in a country that was "nothing but lakes and marshes, full of wild rice; these are separated from one another by narrow tongues of land, which extend from one lake to another not more than thirty or forty paces, and sometimes no more than five or six. These lakes and marshes form a tract more than fifty leagues square."

In this battle 3,000 Sioux drove 100 Huron Indians into a swamp. There "they could not do better than to hide among the wild rice, where the water and mud reached almost to their chins." Then, as Perrot says, the Sioux, or Scioux, "bethought them of this device: they stretched across the narrow strips of land between the lakes the nets used in capturing beavers; and to these they attached small bells * * * * . They divided their forces into numerous detachments, in order to guard all the passages, and watched by day and night, supposing that the Hurons would take the first opportunity to escape from the danger which threatened them. This scheme indeed succeeded; for the Hurons slipped out under cover of the darkness, creeping on all fours, not suspecting this sort of ambuscade; they struck their heads against the nets, which they could not escape, and thus set the bells to ringing. The Scioux, lying in ambush, made prisoners of them as soon as they stepped on the land. Thus from all that band but one man escaped."

THE NORTHWESTERN LAKE DISTRICT

The glaciated portion of the Northern Highland abounds in just such lakes and swamps as Perrot described. They lie in two groups—one in northwestern Wisconsin at the headwaters of the St. Croix and Chippewa rivers, and the other in the extreme northern part of the state at the headwaters of the Flambeau branch of the Chippewa, and the Wisconsin, Wolf, and Menominee rivers.

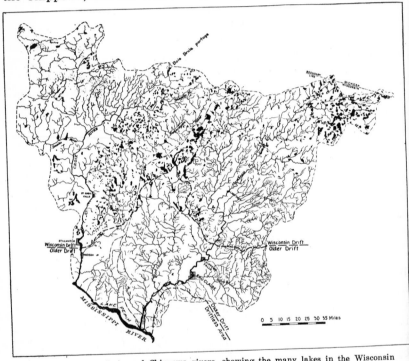

Fig. 166.   The St. Croix and Chippewa rivers, showing the many lakes in the Wisconsin drift and the different type of drainage in the older drift.

The numerous lakes of Washburn, Burnett, and adjacent counties are in the region of Wisconsin drift. Among the larger and better-known of these bodies of water are Upper St. Croix, Namakagon, Court Oreilles, Chetac, and Chetek lakes. Some of them have no outlets. Most of these lakes drain into the Chippewa and St. Croix rivers. Marshes are also abundant. The peat swamps occupy two-thirds of some townships in western Burnett County. There are (a) open marshes with low timbered knolls, (b) heavily timbered swamps, and (c) wet swamps subject to overflow.

THE UPPER ST. CROIX AND CHIPPEWA RIVERS

The headwaters of these streams lie in the northwestern lake district, but the part of the glaciated region, northwest of Chippewa Falls (Fig. 166), has relatively few lakes. Its stream pattern resembles that of the region of older drift, between Chippewa Falls and the Mississippi River. Many of the tributaries of the St. Croix and Chippewa contain rapids and waterfalls, as at Burnett Falls, where there is a vertical descent of 26 feet.

As is shown in Figure 166, there are many more lakes east of the St. Croix River in Wisconsin than there are to the west in Minnesota, though both areas are in the region of latest glaciation. This is apparently because the Northern Highland is more hilly than the Central Plain of Cambrian sandstone, and also because there were more halts of the ice near the terminal moraine which marks its outer limit in Wisconsin. The lake-covered land is, therefore, rougher country than the region west of the St. Croix.

The St. Croix River rises close to the headwaters of Bois Brule River of the Lake Superior drainage system, and the western branch of the Chippewa is close to the north-flowing Bad River and its tributaries.

HISTORIC HIGHWAYS OF NORTHWESTERN WISCONSIN

The narrowness of the portage from the St. Croix to the Bois Brule River made this an important highway of communication between Lake Superior and the Mississippi. Thus it was so familiar to the Indians that the earliest white explorers learned of it at the very beginning and utilized it frequently. This was also true of the route from the Chippewa valley to Chequamagon Bay by way of the several headwaters of Bad River. Du Luth crossed the Bois Brule portage in 1680, Carver traveled that way in 1767, Le Sueur in 1693, and Schoolcraft in 1832. It was once planned to build a canal along the St. Croix-Brule route. The Bois Brule pass, originally the outlet of one of the glacial ancestors of Lake Superior (p. 458), is now traversed by two important railways.

THE HIGHLAND LAKE DISTRICT

Number of Lakes. The Highland Lake District of northern Wisconsin (Fig. 167) lies in Vilas, Oneida, and adjacent counties. There are hundreds of lakes in this group. Their total number and

closeness of position may be inferred from the fact that, although
one of the largest of these bodies of water, Trout Lake, covers
only 6½ square miles, the 346 lakes and ponds of Vilas County occupy 140 square miles, or over 15 per cent of the area of a county

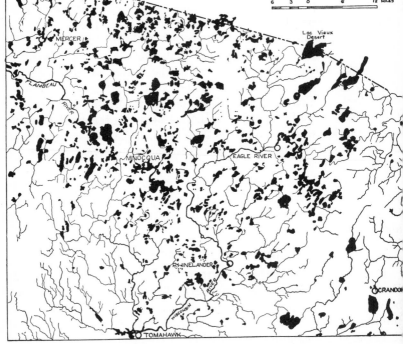

Fig. 167. The Highland Lake District at the headwaters of the Wisconsin, Chippewa,
and Menominee rivers, see Figures 164, 166, and 171.

nearly as large as the state of Rhode Island. The northern part of
Oneida County is equally lake-covered, as is the southeastern part
of Iron County. In few parts of the world are there more lakes
to the square mile. Parts of the state of Minnesota and of the
province of Ontario, northwest of Lake Superior, and part of
southern Finland east of the Gulf of Bothnia, furnish the only
parallels.

Pike Lake, Lac du Flambeau, Minocqua Lake, Tomahawk Lake,
Arbor Vitae Lake, Star Lake, Pelican Lake, and the lakes near
Eagle River are a few of the many well-known summer resorts.

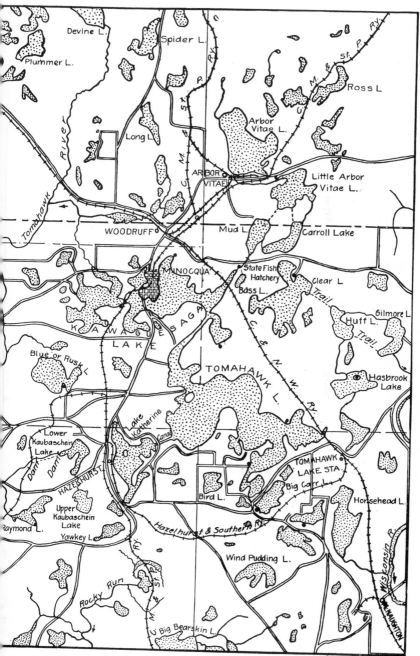

Fig. 168. Map showing lakes—dotted areas—near Minocqua. (Juday.)

Origin of Lakes. Figures 164 and 166 shows nearly all the lakes in northern Wisconsin. They are, as a rule, small lakes, closely spaced, irregular in outline, and connected by streams which have the most irregular courses. All this is typical of lakes in a glaciated region. These bodies of water are all glacial but the origins of the lake basins are diverse. Some are in shallow depressions in the ground moraine, some are held in by recessional moraines, and great numbers are in hollows in the outwash gravel plains. The smaller hollows are kettles formed at the close of the Glacial Period by the melting of buried ice blocks. Few, if any, are in glacially-excavated rock basins, for this part of the state has the rock ledges deeply buried by glacial drift.

Muskegs and Other Swamps in Northern Wisconsin. Open swamps or marshes in northern Wisconsin often go by the Indian name *muskeg*. There are also cranberry and blueberry swamps and drier marshes and swamps not called muskeg, as well as level, tree-covered tamarack swamps and hummocky, cedar swamps. Some of the marshes are filled lakes, but a larger number are merely regions of poor drainage due to glacial accumulations. Marshes cover about 425 square miles in Vilas, Oneida, and adjacent counties. This is about 21 per cent of the area. In Douglas, Bayfield, Sawyer, and adjacent counties to the west and in Marinette County to the east they occupy a smaller proportion of the Northern Highland.

The Flambeau and Manitowish marshes are among the largest of these flat expanses of grassy muskeg. Each of them covers 15 to 25 square miles. They are irregularly circular, while the muskegs to the east in Marinette County are long, narrow swamps, trending northeast-southwest between the morainic ridges.

The swamps are usually monotonously level. The surfaces are often covered with peat and decayed vegetation. Hardpan underlies the peat and it is this that causes the marshes, for water easily escapes from sand-floored depressions. There is often sand below the hardpan. The muck in certain swamps is poorly consolidated and some of the muskegs are quaking bogs, especially near the borders of lakes.

It is the presence of this 21 per cent of marsh land in the lake country of northern Wisconsin, together with the sandy soil of the outwash plains and the hilly topography of the terminal moraines that makes a large area in Vilas, Oneida, Iron, Forest, and adjacent counties far better adapted to being a forest reserve than a region of farms. Half the area is either swampy or has poor soil,

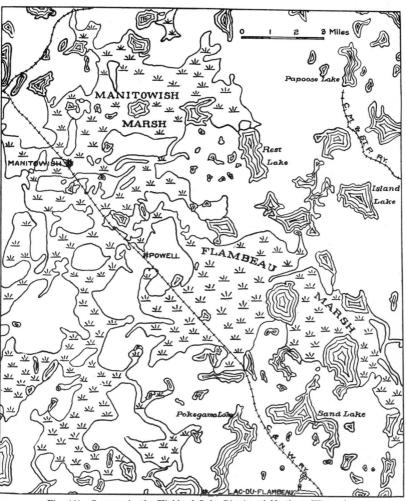

Fig. 169. Swamps in the Highland Lake District of Northern Wisconsin.

and only a little over a sixth of it has good soil. There are 780 clearings or farms in 2,004 square miles, but the cleared land covers only 17 square miles. Thus less than one per cent of the swampy, sandy, and irregular portion of the Northern Highland in Vilas, Oneida, and adjacent counties, is under cultivation, and only a sixth to a half of it is worthy of farm development. The peat in these swamps is a valuable resource (p. 291) not yet utilized. The swamps will require drainage ditches before the land could be used for crops, and after this the peat soil must be pulverized and fertilized. As these swamps are important reservoirs, which help regulate the flow of the Wisconsin and other rivers, it seems to the author far more profitable not to drain the swamps, but to keep the whole swampy area as part of the State Forest Reserve. In 1910 only about 3 per cent of Iron, Vilas, and Forest counties was in farms.

THE UPPER WISCONSIN RIVER

Effect of Glaciation. The drainage system of the upper Wisconsin (Fig. 164) may be divided into two parts, north and south of the vicinity of Merrill. The drainage of the northern area is characterized by many lakes, and crooked, systemless stream courses. The southern part has systematic dendritic drainage and no lakes. The terminal moraine of the Wisconsin valley is the boundary line between the two, and it also forms the southern half of the divide between the Wisconsin and Fox River systems.

The Wisconsin River rises in Lac Vieux Desert, on the boundary between Wisconsin and Michigan, at an elevation of 1,650 feet above sea level. Undoubtedly the lakes and muskegs of northern Wisconsin regulate the flow of the Wisconsin and prevent much greater spring floods than would otherwise occur. The Wisconsin flows through the Northern Highland to a point about 15 miles south of Wisconsin Rapids, leaving this geographical province at an elevation of a little less than 920 feet above sea level. In this distance of about 220 miles the river has a grade of about 3½ feet to the mile. In the glaciated northern half of its course the river is in a shallow valley; in the southern half, which lies partly in the Driftless Area and partly in the region of older drift, the river is in a steep-sided, flat-bottomed trench. Here it flows over a broad valley train of glacial outwash gravels, which have been terraced by late glacial, or postglacial, stream erosion.

Rapids and Portages. The steepest descent in the course of the Wisconsin is at Grandfather Falls north of Merrill, where there is no cataract, but a descent of about 90 feet in a mile and a half of rapids. This is in the glaciated area, and there are many other rapids in this postglacial section of the river. At these rapids the early explorer, traveling by canoe, had to portage his goods and his boat around the impassable or dangerous waters. At similar rapids on the Flambeau River east of Butternut, where a portage was not thought necessary, Moses Strong was drowned in 1877, while working for the State Geological Survey. Rapids of this character introduced notable difficulties in the driving of logs in the days of Wisconsin lumbering. The portages at rapids were often the sites of the earliest sawmills and towns. They play an important part in the geography and history of the region.

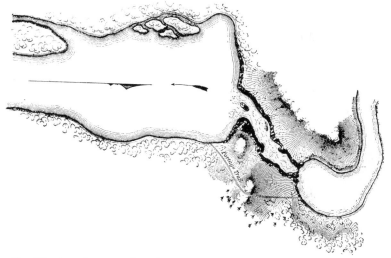

Fig. 170. A portage around rapids at Sturgeon Falls on the Menominee River. (T. J. Cram.)

Water Powers and Undeveloped Sites. The following table shows the extent to which the water power sites in Wisconsin have been developed and an estimate of the undeveloped sites, which may be developed when the need for power in the locality requires it. Many of the developed water powers are small mills used for grinding feed and flour. Approximately 96% of the theoretical horse power available 50% of the time at the 464 plants in the state is found in 161 plants of over 100 installed wheel horse power. The installed wheel horse power in the existing plants is about 38% more than the theoretical horse power available 50% of the time at the same plants.

The construction of water power plants has been less rapid in the state in recent years than formerly. This is probably due to two causes, first the more cheaply developed sites have been developed, and second the recent material reduction in the cost of full power due to more efficient generating plants has made the development of water powers less attractive.

| POWER DAMS NOW DEVELOPED | | | | | UNDEVELOPED SITES | |
River System or Drainage Basin	Drainage area Sq. Miles	Power Dams Including Tributaries	Developed Head Feet	Theoretical Horse Power 50% Time	Possible to Develop Feet	Theoretical Horse Power 50% Time
Black	2,270	13	302	3,352	----------	----------
Chippewa	9,573	37	703	79,878	248	30,570
Flambeau	1,983	8	174	22,320	308	26,830
Fox (L. Michigan)	6,400	39	491	55,332	----------	----------
Fox (Illinois)	1,000	17	226	656	----------	----------
Lake Michigan	----------	28	428	1,135	----------	----------
Lake Superior		8	447	4,872	----------	----------
Menominee	4,000	12	488	85,570	90	20,320
Milwaukee	840	22	288	1,566	----------	----------
Mississippi	----------	36	416	1,906	----------	----------
Oconto	934	5	100	5,383	57	3,550
Peshtigo	1,123	6	274	13,868	291	13,110
Rock	3,500	66	741	11,701	----------	----------
St. Croix	7,576	34	856	27,968	101	30,060
Sheboygan	412	7	71	408	----------	----------
Wisconsin	12,200	89	1,166	134,996	213	49,948
Wolf	3,650	37	449	5,567	473	23,385
Totals	----------	464	7,620	456,478	1,781	197,773

Of the above horse power at developed plants 100,730 theoretical horse power and 137,644 installed wheel horse power is located on boundary streams.

THE MENOMINEE RIVER

The larger portion of the Menominee River is in the Northern Highland. As is shown in Figure 171, there are relatively few lakes at the headwaters of the Menominee compared with the great number to the west in the Wisconsin headwaters.

The Menominee flows along the boundary between Wisconsin and Michigan in a fairly deep, steep-sided valley (Fig. 170), the greater part of which has been incised below the level of the peneplain since the covering of Cambrian sandstone was removed.

During the Glacial Period a great deal of valley train gravel was deposited by streams from the retreating ice of the Green Bay lobe. The several sets of southward deflections of the river have been ascribed to the influence of various positions of the ice edge. The level surface of the valley trains of the Menominee and its tributaries are broken by kettles (see p. 416) and, near the river, by stream-carved terraces. In places the river has cleared

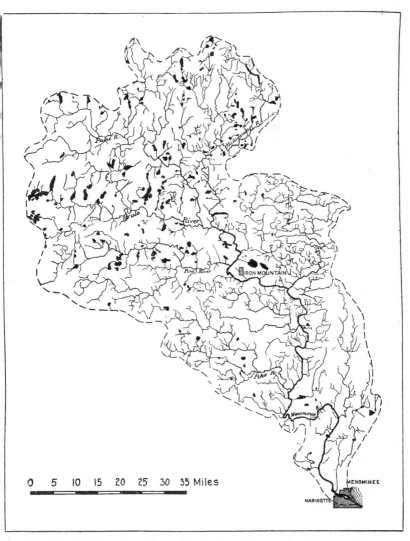

Fig. 171. The Menominee River, with abundant lakes in the headwater region.

all the glacial drift from the rock and descends over waterfalls and rapids, such as Lower Quinnesec Falls—64 feet—and Big Quinnesec Falls—54 feet.

The lumbering industry made great use of the Menominee River, as well as the other streams of northern Wisconsin, for floating logs from the forest to the sawmill. The best areas of white pine (Fig. 6) were in the Northern Highland, where these streams rise. The logs were cut in winter and were rafted down the rivers during the high water of the spring floods. Marinette, at the mouth of the Menominee River, is one of the sawmill towns.

Other streams of the Northern Highland, such as the Peshtigo and the Wolf rivers (Fig. 107), show similar relationships of lakes, rapids, and waterfalls to glaciation. High Falls on the Peshtigo have a descent of 60 feet.

WISCONSIN-MICHIGAN BOUNDARY IN RELATION TO RIVERS AND LAKES

As provided in 1836-46 the boundary between the states of Wisconsin and Michigan west of Green Bay, follows portions of the Menominee River, its tributary the Bois Brule, and the Montreal River. Between the Bois Brule and the Montreal are two straight portions of the boundary, not following any river but joining at a low angle in Lac Vieux Desert (Fig. 172). It was first thought that both the Montreal and the Brule-Menominee flowed out of Lac Vieux Desert. Then the lake was assigned to one of these streams, later to the other, and finally it was discovered that it belonged to neither, but to the Wisconsin River (Fig. 167).

Within the Brule and Menominee rivers, however, are many islands; and it has not always been agreed as to which side of a given island the main channel follows. This has sometimes resulted in discussion as to which of two towns in different states on opposite sides of the river should build or keep bridges in repair. The obvious case in point has to do with the discussions prior to the building of bridges between Marinette, Wis., and Menominee, Mich., in 1863-67 and 1894-95.

Moreover there has not been unanimity as to whether the stream which flows between the cities of Hurley, Wisconsin, and Ironwood, Michigan, is the main Montreal River (Fig. 172). Michigan maintained that the Gogogashugun fork of the Montreal, 3 miles west of Hurley, is the main stream. The establishment of this claim would deprive Wisconsin of valuable mines in the Peno-kee-Gogebic iron range.

Each of these state boundary problems, together with one other, was formally raised in 1923 when Michigan brought suit against Wisconsin in the Supreme Court of the United States. The additional item of dispute had to do with the meaning of the phrases

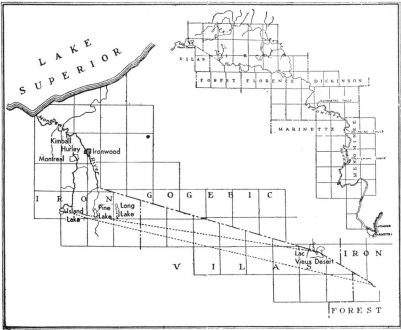

Fig. 172. Two general maps showing the portion of the Michigan-Wisconsin boundary between Green Bay and Lake Superior. Quinnesec Falls are shown in the map on the right. Wisconsin possesses all the islands in the Menominee River south of these falls. The southernmost of the two dashed lines in the map on the left indicates the extent of this part of Michigan's claim in the boundary suit of 1923-26.

"the most usual ship channel of the Green Bay of Lake Michigan," and "the main channel of Green Bay" (see Appendix B, page 482, 483). This involved the islands in and at the mouth of Green Bay. Michigan claimed Chambers Island, the Strawberry Islands, Whaleback Shoal, several reefs, and some 418 square miles of the waters of Green Bay (Fig. 173). She asserted that she was the rightful owner, also, of Washington, Rock, Detroit, Pilot, and Plum islands in the mouth of Green Bay north of the Porte des Morts or Death's Door, as well as of St. Martin's Shoal and adjacent waters in Lake Michigan.

In all Michigan claimed 1,291 square miles of land and water adjacent to a boundary upwards of 330 miles in length, or as long as from Milwaukee to Columbus, Ohio. The population involved was something like 6,900 persons. The agricultural, grazing and forest lands in the portion of the disputed area between Lac Vieux Desert, Island Lake at the head of the Gogogashugun branch of the Montreal River, the confluence of this stream with the main Montreal River, and the land boundary which was marked in 1847 (Fig. 172), were appraised by Michigan in 1923 at some sum in excess of $10,000,000. The assessed valuation of one city and 5 adjacent towns in the disputed area was $7,300,000. The islands in the Menominee River numbered at least 131, to say nothing of islands in its tributary, the Bois Brule. A single one of these islands was assessed in 1925 at $805,000. The waterpowers in the Menominee have considerable potential worth. The fishing grounds in Green Bay possess substantial value, as does the waterfront property on Lake Michigan, Green Bay, and the lakes of the disputed area in northern Wisconsin.

This suit, together with preceding attempts at boundary determination, cost Michigan, Wisconsin, and the United States upwards of $100,000.

In 1926 the Supreme Court handed down its Opinion, denying all of Michigan's claims and clarifying the situation as to the islands in the Menominee River and the Bois Brule. Those in the lower portion of the Menominee, between Marinette, Wis., and Quinnesec Falls, are to belong to Wisconsin. All those above that point are to belong to Michigan. The dividing point (Fig. 172) lies southeasterly of Iron Mountain, Dickinson County, Michigan, and not far east of the north end of the boundary between Menominee and Florence counties, Wisconsin. Michigan appears to have lost her case largely because, for 60 to 90 years, Wisconsin had had possession of or administrative control over the lands and islands concerned. Wisconsin actually acquired clear title to at least 14 or 15 islands in the Menominee River which she seems never to have taxed nor administered; she retained all the Green Bay islands and all the areas along her northern frontier which she possessed at the beginning of the suit. These lands were equal collectively to the combined area of 4 independent countries in Europe—Andorra, Liechtenstein, San Marino, and Monaco.

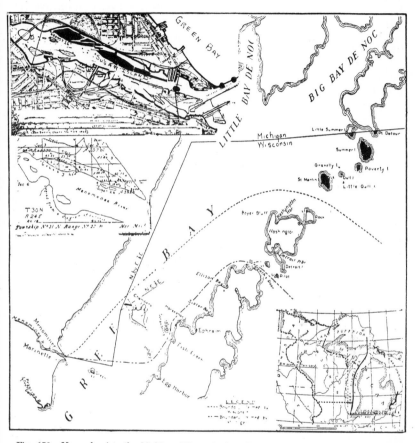

Fig. 173. Maps showing the Michigan-Wisconsin boundary in Green Bay and at the mouth of the Menominee River. On the main map the dashed line in Deaths Door represents the boundary claimed by Michigan in 1923, the dashed line north of Washington Island and Rock Island represents the boundary claimed by Wisconsin, and the full line north of Poverty Island is the state boundary awarded by the Supreme Court of the United States in 1926. North of Chambers Island the abbreviation "N.E.⅞E." stands for the words "northeast seven-eighths east", the abbreviation "N.byE.⅞E." stands for the words "north by east seven-eighths east". The latter were inserted in the boundary decree when the former were probably intended. The small map in the lower right corner shows the relationship of the southern tip of Lake Michigan to the site of Toledo at the western end of Lake Erie. In lieu of this site, Michigan was given the Upper Peninsula in 1836. The upper left map shows the relation of "Sugar Island" and "Grassy Island" to the cities of Marinette, Wis., and Menominee, Mich., and to other islands in the Menominee River. The other small map on the left is part of a General Land Office plat, made in 1848 when "Sugar Island" was not divided into lots. All the islands shown in solid black on these maps were newly acquired by Wisconsin in 1926.

Later in the year 1926, at the request of the two states, the Supreme Court issued a Decree, fixing the Wisconsin-Michigan boundary in detail. By virtue of certain unfortunately chosen words and phrases in this instrument the technical phraseology of the boundary description is such that at present, in May, 1931, Wisconsin is the *de jure* possessor of 707 square miles of water and 4,255 acres of land which she has never before possessed nor administered, which were not even in dispute in 1923-26, and which no one considers that the Supreme Court of the United States intended to take away from Michigan and award to Wisconsin.

The lands referred to include St. Martin, Gull, Little Gull, Gravelly, Poverty, and Summer islands, the south half of Little Summer Island, and the tip of the Point Detour peninsula in Delta County, Michigan. These 6½ islands seem to be devoid of population except lighthouse keepers and their families. They belong to Wisconsin because a Michigan attorney, in framing a Decree for the Supreme Court of the United States, used the phrase "north by east" where the word "northeast" was intended (Fig. 173). Here, as may be appropriately pointed out in an educational bulletin, a superfluous word, one word, a two-letter word, the word *by*, technically transferred 707 square miles of water and 4,255 acres of land from Michigan to Wisconsin.

Other lands, which Wisconsin is probably as much surprised to possess as Michigan is to lose, are in the archipelago of the so-called "Sugar Island" (Fig. 173), situated in the Menominee River near its mouth, opposite Marinette, Wisconsin, together with the so-called "Grassy Island" and 3 small islands between. In this archipelago Wisconsin now has title to "Sugar Island", and to the western half of "Grassy Island", because Michigan lawyers and surveyors were unfamiliar with the failure of the General Land Office of the United States to divide "Sugar Island" into lots in 1848 or subsequently.

Michigan and Wisconsin seem to have officially acquiesced, in 1927, in all these unintended and unexpected outcomes of the boundary suit. In that year, without being cognizant of these complications, each state passed laws appointing boundary commissions to mark the limits fixed by the Supreme Court in 1926. They went still further in 1928, in connection with arrangements for a new interstate bridge at Marinette-Menominee, and caused the Congress of the United States to consent to an agreement which includes a clear implication that "Grassy Island" lies in Wisconsin.

Michigan, for her part, is possessed of 33 square miles of the waters of Green Bay opposite Menominee,, Michigan, and Marinette, Wisconsin (Fig. 173), where fishermen from the latter city have been wonted to fish without paying license fees to Michigan.

In order to iron out all these technical complications there must be further but simple boundary proceedings in the Supreme Court of the United States in 1931 or 1932. The complications are distressing but in no sense insoluble. The great gain from the boundary suit of 1923-26 is that the islands in the Menominee and Bois Brule rivers are no longer left in ambiguity and that state neighborliness is completely restored.

BIBLIOGRAPHY

Cram, T. J. Report on the Survey between the State of Michigan and the Territory of Wisconsin, Senate Doc. 151, 26th Congress, 2nd Session, Washington, 1841, 16 pp., and 5 maps; Report on the Survey of the Boundary between Michigan and Wisconsin, Senate Doc. 170, 27th Congress, 2nd Session, Washington, 1842, 12 pp., and map.

Henry, W. A., King, F. H., and others. Northern Wisconsin—A Hand-Book for the Homeseeker, Madison, 1896, 192 pp.

Horton, A. H., and others. Surface Water Supply of the Upper Mississippi Valley, Water Supply Paper 325, U. S. Geol. Survey, 1914, pp. 125-127, 137-138.

Irving, R. D. Drainage of the Eastern Lake Superior District, Geology of Wisconsin, Vol. 3, 1880, pp. 80-88; River Systems and General Surface Slopes of Central Wisconsin, Ibid., Vol. 2, 1877, pp. 413-424.

Juday, C. Inland Lakes of Wisconsin, Bull. 27, Wis. Geol. and Nat. Hist. Survey, 1914, pp. 113-120.

King, F. H. Hydrology and Topography of the Upper Flambeau Valley, Geology of Wisconsin, Vol. 4, 1882, pp. 607-611.

Martin, Lawrence. Physical Geography of the Lake Superior Region, Monograph 52, U. S. Geol. Survey, 1911, pp. 90-91, 435, 454-456, 459; The Michigan-Wisconsin Boundary Case in the Supreme Court of the United States, 1923-1926, Annals Assoc. Am. Geog., Vol. 20, 1930, pp. 105-163.

Miller, George J. The Establishment of Michigan's Boundaries, Bull. Amer. Geog. Soc., Vol. 43, 1911, pp. 339-351.

Norwood, J. G.   Narrative of Explorations Made in 1847 between Portage Lake and the Head-Waters of Wisconsin River,—in Owen's Geological Survey of Wisconsin, Iowa, and Minnesota, Philadelphia, 1852, pp. 277-293.

Owen, D. D.   (Description of parts of the Black, Chippewa, and St. Croix Rivers), Ibid., pp. 151-165.

Perrot, Nicolas.   (On lakes and swamps in northern Wisconsin), Mémoire sur les moeurs, coustumes et relligion des Sauvages de l'Amérique Septentrionale, 1656-62, written about 1715-18, published in Paris in 1864, —translation in Vol. 16, Collections Wis. Hist. Soc., 1902, pp. 17-19.

Roth, Filibert.   On the Forestry Conditions of Northern Wisconsin, Bull. 1, Wis. Geol. and Nat. Hist. Survey, 1898, 78 pp; also published as Bull. 16, U. S. Dept. Agr., Washington, 1898.

Smith, L. S.   Water Powers of Wisconsin, Bull. 20, Wis. Geol. and Nat. Hist. Survey, 1908, 341 pp; Water Powers of Northern Wisconsin, Water Supply Paper 156, U. S. Geol. Survey, 1906, 137 pp; see also Water Supply Paper 207, U. S. Geol. Survey, 1907.

Stewart, C. B.   Storage Reservoirs at the Headwaters of the Wisconsin and their Relation to Stream Flow, Wisconsin State Board of Forestry, Madison, 1911, 60 pp.

Strong, Moses.   Drainage of the Upper St. Croix District, Geology of Wisconsin, Vol. 3, 1880, pp. 372-381.

Sweet, E. T.   Hydrographic Features of the Western Lake Superior District, Geology of Wisconsin, Vol. 3, 1880, pp. 315-323.

Weidman, S.   Physiographic Geology of North Central Wisconsin, Bull. 16, Wis. Geol. and Nat. Hist. Survey, 1907, pp. 577-578, 610-631.

Whittlesey, Charles.   Physical Aspect of the Bad River Country,—in Owen's Geological Survey of Wisconsin, Iowa, and Minnesota, Philadelphia, 1852, pp. 431-435; Description of the Country between the Wisconsin and Menominee Rivers, Ibid., pp. 452-461.

MAPS

See Chapter XV, p. 408.

CHAPTER XVII

THE LAKE SUPERIOR LOWLAND

A Lake Set Deeply in a Highland

One of the most striking features of Lake Superior, the largest body of fresh water in the world, is its steeply-rising walls. The other Great Lakes of North America have gently-sloping walls. Lake Superior is bounded by steep escarpments. It is a lake set deeply in a highland.

On the southwest is the Bayfield Peninsula and Douglas Copper Range of Wisconsin, 400 to 600 feet above the surface of the lake. On the south are the Penokee-Gogebic Iron Range, the Porcupine Mountains, and Keweenaw Point in Wisconsin and Michigan, 750 to 1400 feet above Lake Superior. On the northwest, north, and northeast is the Lake Superior Highland of Minnesota and Canada, 800 to 1000 feet above the lake. It is only on the southeast, between Sault Ste. Marie and Marquette, Michigan, that Lake Superior has a low-lying shore like the coasts of Lakes Michigan, Erie, and Ontario. This southeastern coast rises 300 feet or less above Lake Superior.

The Lowland is a Graben

The portion of Lake Superior basin lying at the western end of Lake Superior in Wisconsin and Minnesota is a graben, similar to those of Central Africa, southwestern Germany, eastern France, and the lowlands of Scotland, though different in certain respects. Such lowlands are often spoken of as rift valleys. The history of the Lake Superior basin has been long and complicated. Its salient features may be summarized as follows:

1. The Lake Superior basin originally came into existence as a trough or syncline, probably without a lake.
2. The syncline was peneplained and ceased to exist as a topographic feature.

3. A graben was made by faulting, at least at the western end of Lake Superior. The term graben, literally a grave, is a technical one, used by geologists for depressions between parallel fault lines.

4. The graben was buried beneath sedimentary rocks and, for a long time, obliterated.

5. It has been partly exhumed by stream erosion and glacial sculpture, and the lake has been formed.

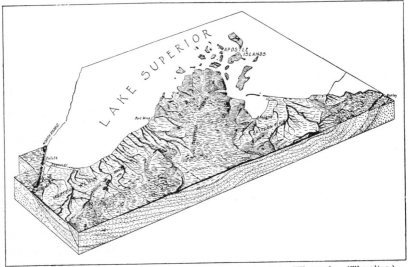

Fig. 174. Block diagram of the Lake Superior lowland in Wisconsin. (Thwaites.)

Thus the Lake Superior basin is now a lowland because of the dropping down of a block of the earth's crust in a rift, or graben, fault (Fig. 175). Subsequent sedimentation and erosion in the district are quite as important as the ancient faulting. The synclinal folding has not been a factor in relation to topography, since a long time before the Lake Superior peneplain was made.

General Description. The Lake Superior lowland in Wisconsin is part of the larger province just described,—the basin of Lake Superior. It occupies portions of Douglas, Bayfield, and Ashland counties in the northwestern corner of the state. Its area is about 1250 square miles, not including the 2400 square miles more in Wisconsin that is submerged beneath the waters of Lake Superior. Its altitude ranges from less than 1000 feet above to about 300 feet below sea level, and it rises 150 to 350 feet above and goes 600 to 900 feet below the level of Lake Superior, which stands 602 feet above sea level.

Boundaries of the Lowland. The author has drawn the southern boundary of this geographical province of Wisconsin by following the highest abandoned beach line of Lake Superior. This boundary also agrees fairly well with the topographic boundary provided by the escarpments at the edge of the Lake Superior sandstone, where it abuts upon the older igneous rocks. In topography this province is chiefly a plain.

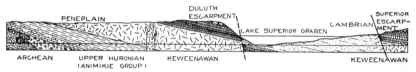

Fig. 175. Graben or rift valley at the western end of Lake Superior.

The floor of the western end of Lake Superior before it was downfaulted and otherwise modified, was doubtless at the level of the Lake Superior highland or peneplain, in Wisconsin, Michigan, Minnesota, and Ontario. The position of the two fault lines is now marked by escarpments. One of them, not in Wisconsin, extends southwestward through Duluth and Fond du Lac, Minnesota. The other, a lower scarp, extends northeast and southwest in Wisconsin from the Minnesota line towards the Apostle Islands, where it is lost beneath the glacial drift. The lowland of Chequamegon Bay near Ashland is not known to be of fault origin, for the Superior fault may merge into a fold near the Apostle Islands; but the plain at Ashland is separated from the Northern Highland province in Wisconsin by a low, sloping wall. The escarpment south of Superior, sometimes called the Douglas Copper Range, or the South Range, has about 350 feet of local relief and slopes northward at the rate of 160 to 300 feet to the mile.

Relationships of Escarpments. The escarpments which form the borders of this rift valley are probably of fault origin, but they have been much modified. The Duluth escarpment in Minnesota, which may originally have been determined by a nearly vertical fault, has been planed back by stream work and glacial erosion to a flaring wall, sloping at the rate of 450 to 1000 feet to the mile. Likewise, the Superior escarpment in Wisconsin, though originally determined by a fault line, is now modified by the work of streams and glaciers to a gently-sloping wall. The fault at the Superior escarpment is a thrust fault, a break inclined at a high angle,— with the rock layers of the present lowland sliding down under the strata of the present upland. This interpretation of the fault in no way interferes with the theory of origin of the basin of western Lake Superior as a rift valley. The escarpment southeast of Ashland, sloping even more gently, is not known to be of fault origin at all.

All these escarpments are of such form as to make it clear that they are not produced in the peneplain cycle, nor could the even upland of the peneplain have been produced at a time when the

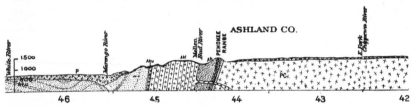

Fig. 176. Cross-section showing the escarpment which separates the Lake Superior Lowland from the Northern Highland, and the thick covering of glacial drift (P) upon the pre-Cambrian rocks (Akl, Aku, Ah, PC).

present lowland existed (Fig. 175). If so, the peneplain, which extends up to the very edges of the escarpments, would have been deeply trenched by stream erosion. As it is, the short streams which flow down the face of the escarpment are in deep gorges and have rapids and waterfalls, as in the Bois Brule River of Douglas County, the Montreal River on the boundary between Wisconsin and Michigan, and many others. These gorges and waterfalls, however, are postglacial features, That the escarpments are so little trenched by streams and that the streams extend so short a distance back into the peneplained upland would be an evidence that the escarpments and the lowland were recently formed, were it not for the recent episode of glaciation and for the still earlier burial of the escarpment beneath the Lake Superior sandstone. This suggests that the escarpment and the lowland are very old.

Age of the Lowland. The age of the Lake Superior lowland may be determined by the relationships of the fault scarps of the Lake Superior basin to the pre-Cambrian igneous and metamorphic rocks and the later sedimentary rocks of the region.

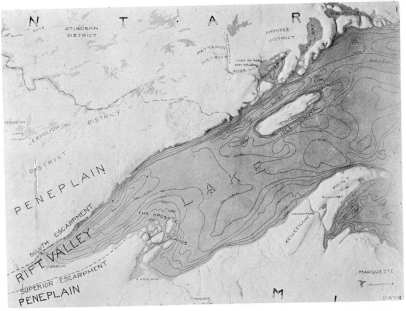

A. MODEL OF THE WESTERN END OF LAKE SUPERIOR, SHOWING THE RIFT
VALLEY IN RELATION TO THE PENEPLAIN AND ESCARPMENTS.

B. WATERFALL WHERE THE AMNICON RIVER CROSSES THE FAULT LINE
ESCARPMENT SOUTHEAST OF SUPERIOR.

Sandstone in foreground, cascade on trap rock.

FALLS OF THE MONTREAL RIVER.

The peneplain all around the basin of Lake Superior is terminated abruptly by steep escarpments. Of these escarpments the one extending northeastward from Duluth, the one extending from Superior to Bois Brule River, and the one on the southeastern side of Keweenaw Point in Michigan have been determined by geologists to be of pre-Cambrian or early Cambrian age. The faults cut rocks of Archean, Huronian, and Keweenawan age. They also cut the Lake Superior sandstone. There is some evidence that movement on the faults south of Lake Superior took place during the deposition of the Lake Superior sandstone, which is possibly non-marine and may be of late Keweenawan or early Cambrian age. The presence of marine Cambrian sediments within the Lake Superior basin east of Keweenaw Point definitely fixes its pre-Cambrian age. It is, therefore, thought that the pre-Cambrian peneplain which forms the Northern Highland of Wisconsin used to be connected with the peneplain on the northwest in Minnesota and on the north and northeast in Canada. The connecting feature was a portion of peneplain which included the site of the present Lake Superior lowland and the remainder of the basin of Lake Superior. It seems probable that the faulting and folding at the close of the Keweenawan resulted in the down-faulting and down-warping of a portion of this peneplain. Thus the existing rift valley near Superior and Ashland, and to the northeast in the basin of what is now Lake Superior, was produced before or at the time when the region was buried beneath late Keweenawan or early Cambrian deposits. Accordingly the escarpments were buried and preserved, before they had been greatly modified by stream erosion. Subsequently the uplifted region has had the larger portion of the weak sediments removed by streams and glaciers; but remnants of the Lake Superior sandstone rest against the fault scarps, as in Douglas and Bayfield counties, and below the level of Lake Superior. The resistant pre-Cambrian rocks were less eroded during the time of removal of the Lake Superior sandstone, so that the carving of stream courses in the escarpments by swiftly-flowing rivers has just commenced.

GLACIAL MODIFICATION OF THE LOWLAND

Glacial Erosion. The modification of the Lake Superior lowland, and the escarpments which bound it, by the continental glacier are of two sorts,—erosion and deposition. This work was accomplished chiefly by the Superior lobe of the Labrador ice sheet (Fig. 28). During certain stages of the Glacial Period the Lake Superior lowland was also invaded by ice from the Keewatin and the Patrician centers west of Hudson Bay.

Sculpture by the continental ice sheet has notably modified the rift valley of northwestern Wisconsin. It is becoming more and more evident from observations in Alaska, the Alps, Norway, New Zealand, and elsewhere, that glacial ice can erode deeply and can in time cut down and carve out profound basins. During the time when the Paleozoic sediments were being removed from the peneplain of the Northern Highland, streams were doubtless removing the weak sandstone from the basin of Lake Superior.

It is not clear whether Lake Superior existed before the first invasion by the continental glacier, but probably it did not. The depth to which these preglacial streams could erode near their headwaters was limited by the obstacles to downcutting between the headwaters and the sea. The preglacial master stream from the western end of the Lake Superior basin probably flowed eastward to some point between Marquette and Sault Ste. Marie, Michigan. There it joined the Lake Michigan River (Fig. 109) and flowed southward, through what is now the basin of Lake Michigan to the Illinois and Mississippi rivers. There is no good reason for believing that the preglacial Lake Superior River flowed southwestward to the Mississippi in the region north of St. Paul, Minnesota. In any event this river was probably unable to erode below some such level as the surface of Lake Superior, a little over 600 feet above the sea level. No broad preglacial channels are known to cut into the sandstone around the borders of Lake Superior. A narrow buried channel, 200 feet or more in depth, extends from the eastern end of Lake Superior to Lake Huron. It is not yet known whether this buried valley is preglacial rather than glacial or interglacial in age. Accordingly it seems reasonable to conclude that all of the scouring out of the buried rift valley below approximately the present level of Lake Superior was the work of the continental ice sheet. There is no way of determining how much of the erosion above the present lake level was glacial rather than preglacial. Glacial erosion certainly lowered the bottom of the rift valley to a depth of 600 to 900 feet. Its bottom now lies at levels between that of the surface of the ocean and 300 feet below śea level (Pl. XXXI, A). In other parts of the rift valley the deepening was less, but everywhere there was profound glacial sculpture.

The conditions here are especially favorable for the glacial cleaning out of the sandstone and shale-filled rift valley. The sediments are weak, while the igneous and metamorphic rocks of the enclosing walls are resistant. The ice of the continental glacier, therefore, moved freely through the lowland and eroded much more deeply than if there had been no weak sediments to be removed from the buried rift valley.

To what extent the escarpments were eroded is not so easy to say. To the west in Minnesota, the St. Louis River seems to still retain parts of thé preglacial course by which it diverted some

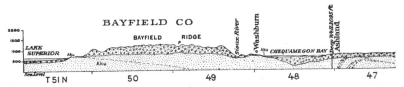

Fig. 177. Cross-section of Bayfield Peninsula, showing the great thickness of the glacial deposits. Aku, Bayfield, and Oronto sandstones.

of the headwaters of the Mississippi (Fig. 179). This suggests relative weakness of glacial erosion in the resistant pre-Cambrian rocks of the escarpment. On the other hand, there is no definite trace of a system of drainage in the Apostle Islands. They are made up of comparatively weak sandstone and might be thought to represent hilltops between preglacial valleys. Glacial erosion seems to have deepened and widened channels between islands, to have eroded hilltops—now island crests—and to have modified the region sufficiently so that the preglacial drainage pattern is completely lost. It is, therefore, apparent that the ice sculpture was much more effective in the weak sandstone rift valley than upon the resistant rocks of the bordering escarpments. The Lake Superior basin owes a notable part of its exhumation, and all, or nearly all, of its present depth below lake level to erosion by the continental ice sheet during the several glacial epochs.

Morainic Deposits. The retreating glacier deposited very little in the way of moraines in the Lake Superior lowland of northwestern Wisconsin. There may be moraines concealed beneath lake level, but the chief visible deposit of this character is the kettle moraine in Bayfield County and its extension southward into Minnesota in the Nickerson and Kerric morainic systems. There is also a narrow terminal moraine which reaches Lake Superior west of the Montreal River in Iron County.

The kettle moraine of Bayfield County is interlobate,—an accumulation made between the Superior and Chippewa lobes of the retreating glacier (Fig. 28). It is called a kettle moraine because of the many kettles and knobs which make portions of its surface exceedingly irregular. The glacial drift is of great thickness (Fig. 177), perhaps as much as 600 feet in depth. The drift on the escarpment slopes is thin. On the Apostle Islands, and below the level of the base of the escarpment the lowland is completely mantled by stratified lake and stream deposits. Waterlaid sandy morainic ridges on the clay plains beneath the level of Glacial Lake Duluth (p. 457) run up to 300 feet or more in height.

The Superior clay, as it is called on the soils maps, covers more than 1000 square miles. In the soil and subsoil of this belt 82% of the material is clay and silt. Much of it is red.

Drift Copper. Among the materials in the glacial drift of the Lake Superior Lowland are masses of native copper carried westward by the ice itself and in icebergs. One such mass found near Ontonagon, Michigan, weighs about 3000 pounds, and one from the Bayfield Peninsula near La Pointe, Wisconsin, 800 pounds. One of the Jesuit fathers observed in 1669 that there were such bowlders of copper in the Apostle Islands. He relates that the squaws often found copper fragments of 20 to 30 pounds weight in digging holes in the sand to plant their corn (p. 446), and suggests the transportation of this copper by floating ice, though he did not imply that this was glacial ice.

The Origin of the Plain of the Lake Superior Lowland. The surface of these lake deposits forms the present plain of the Lake Superior Lowland (Fig. 174). The plain south and west of Superior, and south of Ashland, slopes gradually toward the lake, its grade varying from 10 to 50 feet to the mile. The surface, however, is not everywhere that of a plain, except near Superior and Ashland, where it furnishes the level sites of these cities. Farther from the lake it has been deeply trenched by postglacial streams, forming ravines 40 to 100 or more feet deep. These have so dissected the plain as to make portions of it very hilly indeed. The clay has been more extensively dissected than the sand because the water sinks into the latter and erodes it very little. This stream cutting has necessitated vast expense for railway bridges and culverts. The

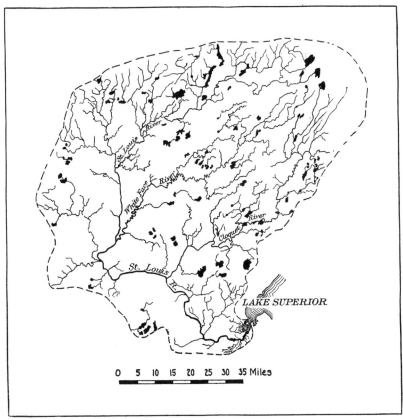

Fig. 178. The St. Louis River in Wisconsin and Minnesota.

highways which cross the streams are continually going up and down hill, and the clay soil makes sticky roads. The soil is not as well adapted for farming as for grazing and the production of hay, and the topography of the stream-sculptured portions of the plain even less so.

DRAINAGE OF THE LAKE SUPERIOR LOWLAND

The St. Louis River. The St. Louis River (Fig. 178), most of whose drainage basin is in Minnesota, flows eastward into Lake Superior after descending the Duluth escarpment near Fond du Lac, Minnesota. It forms part of the boundary between Wisconsin and Minnesota. From Fond du Lac to Superior, it is an estuary with little current because of the drowning of its valley beneath the waters of Lake Superior (p. 464).

Fig. 179. St. Louis River before and after the preglacial stream capture which, as there is little reason to doubt, resulted in the diversion of one of the Mississippi headwaters.

The valley of the St. Louis is one of the historic waterways of the Northwest. It was long used by the Indians and was traversed by Sieur Duluth in 1679, by Schoolcraft in 1820, and by many others.

The volume of the St. Louis is much greater than it was in pre-glacial time and before, as most students consider, the stream captured and diverted certain headwaters of the Mississippi, as is shown in Figure 179. There is a moderate number of glacial lakes (Fig. 178) among the Minnesota headwaters of the St. Louis, so that its run-off is partly regulated, but not to so great an extent as is desirable. The water power, resulting from the rapids and falls where the St. Louis river descends the Duluth escarpment, is now developed a mile and a half west of the Wisconsin state line. By the transmission of this electric power to the city of Superior, a distance of only 14 miles, a great deal of manufacturing is possible at this lake port.

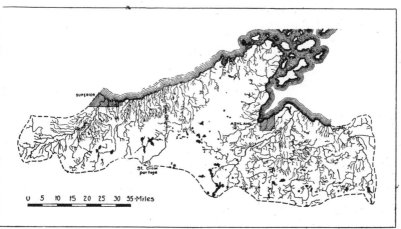

Fig. 180. The Nemadji, Bois Brule, Bad, and Montreal rivers.

The Nemadji River. As is shown in Fig. 180, the Nemadji River, which flows northeastward into Lake Superior, has a short course from the Superior escarpment to the lake. The river has two parts, (a) the torrential, steep course on the escarpment, and (b) the flatter grade over the plain of lake clays and glacial stream deposits. In this latter portion the Nemadji and its tributaries are flowing in steep-sided ravines 25 to 100 feet deep, which decrease in depth as the lake is approached. There are relatively few glacial lakes on the headwaters of the Nemadji River, whose steep slope enables it to discharge rapidly in seasons of great precipitation or of melting snow, giving it a small volume during the summer months. Accordingly, artificial reservoirs would be necessary if water power were to be developed where the Nemadji descends the Superior escarpment. This stream can never be as useful in this respect as the St. Louis, which is longer and has a greater volume. The cataract of the Black River, a tributary of the Nemadji in Pattison State Park, is the highest waterfall in Wisconsin.

Bois Brule, Bad, and Montreal Rivers. The short streams which flow northward to Lake Superior from the Northern Highland descend steeply over the escarpment and then flow with gentler grades to the lake. As is shown in Figure 180, these streams have a few glacial lakes at their headwaters. The rapids in the Bois Brule necessitated many portages in the days of canoe travel over the Brule-St. Croix pass. In the headwaters of the Bad, White, and Montreal rivers there is a trellis pattern of drainage, especially in the tributaries which rise in the east-west valley of weak slate north of the Penokee Range (Fig. 151) and between the trap ridges to the north (Fig. 181).

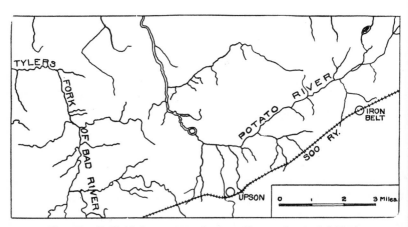

Fig. 181.   Trellis drainage among the trap ridges southeast of Ashland.

The waterfalls at or near the escarpment are among the highest in the state, including Copper Falls of Bad River (Plate XXXIII), the falls at the junction of Tylers Fork and Bad River (Plate XXXIV), both in Copper Falls State Park, and the falls of the Montreal River (Plate XXXII). The native name for Montreal River was Kawasiji-wangsepi, said to mean *white falls river*. Superior Falls Dam, with a head of 127 feet, utilizes the final drop of the river over sandstone ledges near Lake Superior. Saxon Dam, about $3\frac{1}{2}$ miles up-stream, has a head of 117 feet. The

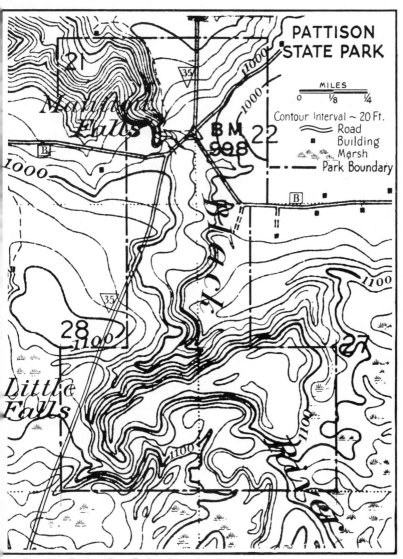

Fig. 182. Pattison State Park. (From Superior Quadrangle, U. S. Geol. Survey.)

stream-cut gorges formed in postglacial time are deep and steep-sided (Plate XXXIII) where these rivers are flowing through rock, but are broader and have gentler slopes in the clay belt near Lake Superior. None of these streams have deltas.

In connection with all these waterfalls much hydro-electric power can be and has been developed. This is of importance, both in relation to future manufacturing at such places as Ashland, Port Wing, and Superior, and for power used in mine operation.

PATTISON STATE PARK

Manitou and Little Falls are the centers of interest in this park (Fig. 182), which is 13 miles south of Superior on State Trunk Highway 35. At Manitou Falls the Black River, a tributary of the Nemadji, drops over the steep northern slope of the Douglas Range in a fall of 160 feet in a horizontal distance of not more than 150 feet. At the falls the gorge walls are of trap. Below the falls, the river flows for about a mile in a narrow canyon with sandstone walls rising from 100 to 170 feet above the river. Below the sandstone gorge the valley is somewhat wider, and is bordered by clay banks from 60 to 100 feet in height. The waterfall and gorge indicate that, after the continental ice sheet withdrew, the Black River cascaded down a slope much like the one now occupied by Highway 35. This stream soon cut a channel through the loose sand, gravel, and clay. There was a waterfall near the north end of the present sandstone canyon. This waterfall gradually worked back into the sandstone and finally into the trap, producing the gorge and falls of today (Frontispiece).

Little Falls are a little over a mile in a straight line, about a mile and a half by the river, from Manitou Falls. At Little Falls there is a perpendicular fall of 31 feet over a trap ledge.

The park includes part of the escarpment which forms the southern border of the Lake Superior Lowland, as well as several of the abandoned beaches of Glacial Lake Duluth. It lies within the Nickerson terminal moraine.

COPPER FALLS STATE PARK

Copper Falls State Park (Fig. 183) includes within an area of 520 acres two remarkable waterfalls—Copper Falls of Bad River and the falls at the junction of Tylers Fork and Bad River. Copper Falls is a picturesque drop of 29 feet into a narrow, trap-walled gorge. About one-fourth of a mile downstream Tylers Fork enters the 65 foot gorge of Bad River by a fall of 30 feet (Plate XXXIV).

COPPER FALLS OF BAD RIVER.

TYLERS FORK FALLS

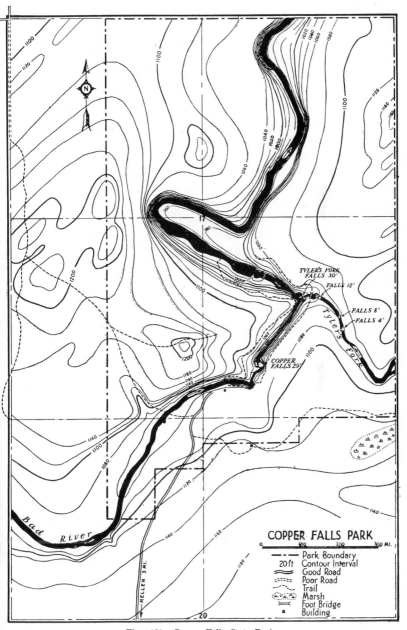

Fig. 183. Copper Falls State Park.

A short distance up-stream there are three smaller falls with intervening rapids, making a total fall of 70 feet, all in Keweenawan trap. Below the mouth of Tylers Fork the Bad River gorge is 100 feet deep with walls of sandstone and conglomerate. About one-fifth of a mile below the falls the rock ledges disappear and the river flows through a colorful gorge cut in red clay.

This clay was laid down in the bottom of Glacial Lake Ashland, one of the predecessors of Glacial Lake Duluth. The waters of the lake first named stood 1,123 feet above sea level or 521 feet above that of Lake Superior. Its outlet led westward to Pike Lake and Muskeg in Bayfield County, thence southward into Glacial Lake Brule which drained into the St. Croix River by the outlet subsequently occupied by the river from Glacial Lake Duluth (see p. 457). A terminal moraine occupies the uplands of this state park, and upon the trap ledges nearby are glacial striae. These point slightly east of south and appear to have been made by the Patrician ice sheet from west of Hudson Bay rather than by a lobe of the Labrador ice sheet from the northeast.

The park is easily accessible, being only three miles north from Mellen on a good road. Within the park good trails and safety rails enable the visitor to see places of scenic and geological interest.

BIBLIOGRAPHY

Agassiz, Louis. The Terraces and Ancient River Bars, Drift, Boulders, and Polished Surfaces of Lake Superior, Proc. Amer. Assoc. Adv. Sci., First Meeting, 1848—Philadelphia, 1849, pp. 68-70; On the Origin of the Actual Outlines of Lake Superior, Ibid., pp. 79-80; Lake Superior, Boston, 1850,—The Erratic Phenomena about Lake Superior, pp. 395-416; The Outlines of Lake Superior, Ibid., pp. 417-428.

Desor, E. Drift of the Lake Superior Land District, in Foster and Whitney's Report on the Geology and Topography of a portion of the Lake Superior Land District, Part 1, House Ex. Doc. 69, 31st Congress, 1st Session, Washington, 1850, pp. 186-218; On the Superficial Deposits of this District, Ibid., Part 2, Senate Ex. Doc. 4, Special Session, Washington, 1851, pp. 232-273.

Grant, U. S. Copper-Bearing Rocks of Douglas County, Bull. 6, Wis. Geol. and Nat. Hist. Survey, 1900, 83 pp; Junction of the Lake Superior Sandstone and Keweenawan Traps in Wisconsin, Bull. Geol. Soc. Amer., Vol. 13, 1901, pp. 6-9.

Irving, R. D. On Some Points in the Geology of Northern Wisconsin, Trans. Wis. Acad. Sci., Vol. 2, 1874, pp. 107-119; Geology of the Eastern Lake Superior District, Geology of Wisconsin, Vol. 3, 1880, pp. 53-214; The Copper-Bearing Rocks of Lake Superior, Monograph 5, U. S. Geol. Survey, 1883, 446 pp.

Irving, R. D., and Chamberlin, T. C. Observations on the Junction of the Eastern Sandstone and the Keweenaw Series on Keweenaw Point, Bull. 23, U. S. Geol. Survey, 1885, 124 pp.

Leith, C. K. Relations of the Plane of Unconformity at the Base of the Cambrian to Terrestrial Deposition in Late Pre-Cambrian Time, Compte Rendu de la XIIe Session, Canada, 1913, Congrès Géologique International, Ottawa, 1914, pp. 335-337.

Leverett, Frank. Moraines and Shore Lines of the Lake Superior Region, Prof. Paper 154 A, U. S. Geol. Survey, 1929, 72 pp.

Lyell, Charles. Lake Superior,—Principles of Geology, tenth edition, Vol. 1, 1867, pp. 421-422.

Martin, Lawrence. The Basin of Lake Superior, Monograph 52, U. S. Geol. Survey, 1911, pp. 110-117; The Pleistocene, Ibid., pp. 427-459; Journ. Geog., Vol. 12, 1914, pp. 230-231.

Owen, D. D. Formations of Lake Superior, Report of a Geological Survey of Wisconsin, Iowa, and Minnesota, Philadelphia, 1852, pp. 187-193.

Smith, L. S. The Water Powers of Wisconsin, Bull. 20, Wis. Geol. and Nat. Hist. Survey, 1908, pp. 250-260.

Sweet, E. T. Geology of the Western Lake Superior District, Geology of Wisconsin, Vol. 3, 1880, pp. 305-362.

Thwaites, F. T. Sandstones of the Wisconsin Coast of Lake Superior, Bull. 25, Wis. Geol. and Nat. Hist. Survey, 1912, 109 pp.

Van Hise, C. R., and Leith, C. K. Keweenawan Series of the Lake Superior Region, Monograph 52, U. S. Geol. Survey, 1911, pp. 366-426.

Whitson, A. R., and others. Soils Reports, Wisconsin Geol. and Nat. Hist. Survey, Bulls. 31, 32, 72A.

Whittlesey, Charles. Geological Report on that Portion of Wisconsin Bordering on the South Shore of Lake Superior, Surveyed in the Year 1849, —in Owen's Geological Survey of Wisconsin, Iowa, and Minnesota, Philadelphia, 1852, pp. 425-444; On the Fresh-Water Glacial Drift of the Northwestern States, Smithsonian Contributions to Knowledge, Vol. 15, 1867, 32 pp; Physical Geology of Lake Superior, Proc. Amer. Assoc. Adv. Sci., Vol. 24, 1876, pp. 60-72.

MAPS

U. S. Geological Survey. Superior Quadrangle, Wisconsin; Duluth Quadrangle, Minnesota, (Fig. 197).

U. S. Lake Survey. Survey of the Northern and Northwestern Lakes: Charts 9, 96, 961, 962, 964, 966; various scales from 1: 15,000 to 1: 5,000,000, (see Fig. 199).

CHAPTER XVIII

THE WISCONSIN COAST OF LAKE SUPERIOR

Copper in Relation to Former Lake Levels

"Near that place are some islands, on the shores of which are often found Rocks of Copper, and even slabs of the same material."

Thus wrote Father Dablon in 1669. The alluded to the Apostle Islands near Chequamegon Point. He continued as follows:

"Last Spring we bought from the Savages a Slab of pure Copper, two feet square, and weighing more than a hundred livres (100 pounds, troy weight). It is not thought, however, that the mines are found in the Islands, but that all these Copper pebbles probably come from Minong (Isle Royale), or from the other islands which are the sources of it, borne upon floating ice or rolled along in the depth of the water by the very impetuous winds,—particularly by the Northeast wind, which is extremely violent.

"It is true that on the Mainland (Bayfield Peninsula), at the place where the Outaouaks raise Indian corn, half a league from the Water's edge, the women have sometimes found pieces of Copper scattered here and there, of the weight of ten, twenty, or thirty livres. It is in digging up the sand to plant their corn that they make these chance discoveries."

This is the first discussion of the physical geography or mineral resources of the Wisconsin portion of Lake Superior region. The copper alluded to was carried to the Bayfield Peninsula and Apostle Islands, as already explained (p. 436), by icebergs, floating in a glacial lake. Certainly it was not rolled along the bottom of the lake. The copper of Keweenaw Point, Michigan, which occurs in ledges as well as in drift bowlders, was mentioned by Sagard in 1636. Still earlier, it was known to the aboriginal inhabitants, who mined it and fashioned it into idols, and even into implements and weapons.

SUBMERGENCE AND THE STATE BOUNDARY

It was provided in 1846 that the western boundary of Wisconsin should extend due south from the St. Louis River to the main branch of the St. Croix, starting at the first rapids in the St. Louis above the Indian village, according to Nicollet's map. When the boundary was actually marked, six years later, the rapids in question were difficult to locate. Upon the map alluded to, the Indian village is shown, but with no rapids at just that point. A Chippewa Indian assured the surveyor that there were rapids nearly opposite the Indian village only a few years before. The surveyor accepted the statement as evidence that the lake level was rising. Nevertheless he proceeded up stream to the first rapids of that day, where he located his boundary. This resulted in giving Wisconsin a strip of territory a quarter mile wide and over 40 miles long that might otherwise have been in Minnesota; but the decision was just, because the submergence of the rapids had commenced centuries before the boundary was even proposed.

The fluctuation in lake level that causes the submergence of St. Louis River is exceedingly slow. Such changes began thousands of years ago when Lake Superior was very different. It then submerged the sites of the present cities of Superior and Duluth (Fig. 189) more than 500 feet. Its outlet was down the St. Croix River to the Mississippi. It terminated at the east in a vertical ice cliff far higher and grander than the snowy Antarctic barriers of today.

In 1916 the State of Minnesota brought suit against the State of Wisconsin in the Supreme Court of the United States. This suit was dismissed upon motion of the complainant. The case was renewed and carried on, however, during the next 5 years. There had been earlier judicial glances at the problem of the waters between Superior and Duluth. These were incidental to the case of *Wisconsin versus Duluth* in 1877 and to that of *Norton versus Whiteside* in 1915. In 1911-12 a joint commission from the legislatures of Wisconsin and of Minnesota had played about the fringes of the question. The Minnesota-Wisconsin boundary suit of 1916-22 in the Supreme Court of the United States had to do with areas of water, between Superior, Wisconsin, and Duluth, Minnesota, in the Bay of Superior, Lower St. Louis Bay, Upper St. Louis Bay, and the St. Louis River from Upper St. Louis Bay to the falls. The suit was popularly thought to concern itself chiefly with an area of a little over 126 acres, if one may speak in terms of acres of a water area containing improvements due to the work of man. As a matter of fact the Supreme Court was asked to ascertain and establish the State boundary line all the way from Lake Superior to the point where it turns southward as a land boundary, a distance of about 27 statute miles. Minnesota claimed to the middle of each of the bays named above. Wisconsin did not seriously dispute this claim so far as Superior Bay and Lower St. Louis Bay were concerned. She maintained, however, that in the waters to the westward, i. e., above Grassy Point, the boundary followed a sinuous course, at some points near the Minnesota shore, at other points equally near the Wisconsin shore.

This case arose primarily because (a) United States Army engineers had laid out a dock line in 1890-99 and had subsequently dredged a deep channel through these waters, (b) Minnesota corporations had built expensive docks out toward this artificial channel, crossing the line which Wisconsin had always maintained and understood to be the State line, and (c) the city of Superior, Wisconsin, assessed and attempted to tax the outer ends of these docks. The real estate and improvements in question were placed at a valuation of $168,500 and are understood to have been worth approximately a third of a million dollars. Minnesota was taxing the property; Wisconsin was attempting to establish the right to do so with respect to the portion that lay within her own limits.

The deep channel was dredged in 1893. Wisconsin found that its border crossed and recrossed the new waterway and bisected existing docks and sites of probable future docks, so that portions of them were or would be divided between the two States. The boundary was established by Congress in 1846-48, so far as Wisconsin is concerned, and in 1857-58 so far as Minnesota is involved. To these States the federal government had said in part, using opposite orders of description:

TO WISCONSIN

Your boundary runs "through the center of Lake Superior to the mouth of the St. Louis River; thence up the main channel of said river to the first rapids in the same, above the Indian village, according to Nicollet's map. . . ."

TO MINNESOTA

Your boundary follows "the boundary line of the State of Wisconsin, until the same intersects the Saint Louis River; thence down said river to and through Lake Superior . . . . "

Nicollet's map, published in 1843, is on too small a scale to help in ascertaining exactly what the Congress may have meant in 1846-58 with respect to the parts of the boundary which caused the difficulties in 1916.

Minnesota contended that the "mouth of the St. Louis River" was southwest of the so-called Big Island. There, she thought, the St. Louis lost the width, banks, channel, and current which are characteristic of a river, and acquired the width, form, and flow-regime of a lake or a bay on the border of a lake: Wisconsin held that the mouth of the river was midway between the two great sand spits called Wisconsin Point and Minnesota Point. The issue thus raised has to do with the question, apparently never before determined judicially for students of physical geography, as to whether a drowned and estuarine portion of a river valley is part of the river or part of a lake into which the river flows. In 1920 when the Opinion in this case was handed down, Wisconsin's contention was upheld, as to this point, for the Supreme Court said:

"The defendant insists, and we think correctly, that [the mouth of the St. Louis River] is at the junction of Lake Superior and the deep channel between Minnesota and Wisconsin Points—'The Entry'".

Minnesota contended further that the "main channel" of the St. Louis River lay midway between the shores, essentially along the line of the dredged channel referred to above. Wisconsin held that the main channel was to be identified by the line of the deepest soundings in the St. Louis River above Grassy Point. Each State presented a great deal of testimony and many documents and maps in support of its contentions.

The outcome of the suit with respect to this point is perplexing, and not too encouraging in connection with the avoidance of future disagreements between the States of Minnesota and Wisconsin.

In the Opinion, dated March 8, 1920, the Supreme Court seems to leave the defendant, Wisconsin, in the position of winner of the suit. (a) In the waters east of Grassy Point the boundary is determined essentially as Wisconsin and Minnesota agreed to be equitable and convenient for shipping. (b) The court awarded to Minnesota a little less than the whole 126 acres in which both States were particularly interested. (c) From a point near the

west end of the 126 acres to the west end of the water boundary
between Wisconsin and Minnesota the State line is fixed in the
position contended for by Wisconsin,—that is, it does not follow
the dredged channel but weaves back and forth across it.   Here
is the pertinent portion of the Opinion:

"A decree", declares the court, "will be entered declaring and
adjudging as follows: That the boundary line between the two
States must be ascertained upon a consideration of the situation
existing in 1846 and accurately disclosed by the Meade Chart.
That when traced on this chart the boundary runs midway between
Rice's Point and Connor's Point and through the middle of Lower
St. Louis Bay to and with the deep channel leading into Upper
St. Louis Bay and to a point therein immediately south of the southern extremity of Grassy Point; thence westward along the most
direct course, through water not less than eight feet deep, eastward of Fisherman's Island and as indicated by the red trace
'A, B, C', on Minnesota's Exhibit No. 1, approximately one mile, to
the deep channel and immediately west of the bar therein; thence
with such channel north and west of Big Island up stream to the
falls".

The Meade chart is a map entitled *St. Louis River*, by George
G. Meade, published in two parts in 1861 on the scale of 1:16,000.
Fisherman's Island was a body of low land, now almost wholly
gone; it appears without name on the Meade chart south of the
site of the coal dock of the Zenith Furnace Company.   The words
"deep channel" appear to allude to the line of deepest soundings on
the Meade chart and not to the still deeper and uninterrupted
18-foot dredged channel of today.   It is instructive to compare the
Meade chart of 1861 with chart No. 966 of the U. S. Lake Survey,
scale 1:24,000, issued June 13, 1930.   Upon this 1930 chart the
present State boundary is shown in detail, except in Superior Bay
where it is best represented upon the Superior Quadrangle, scale
1:62,500, issued by the U. S. Geological Survey in 1917.

Subsequently the Supreme Court, Minnesota, and Wisconsin
agreed, in an interlocutory decree, dated Oct. 11, 1920, to have a
commission of three men run, locate, and designate the river boundary awarded in the Opinion, locating it by courses and distances,
tied to proper and adequate monuments on the land.   This permanent marking of the boundary, costing, incidentally, less than six-

teen thousand dollars, included the placing of many two-ton concrete monuments with bronze tablets. Upon certain structures within improved or partly improved properties crossed by the State boundary line, the commissioners placed bronze tablets, appropriately marked, and cemented into the structures.

The boundary commissioners construed the Opinion and interlocutory decree of the Supreme Court in such a way as to produce a State boundary which crosses the modern dredged channel at a number of points, running overland across Tallas Island, the subject of an earlier suit in federal courts. The boundary bisects the three docks of the McDougal-Duluth Shipbuilding Company at Riverside, Minnesota.

The facts do not seem to be popularly known either at St. Paul or at Madison, but the outcome of the work of the boundary commission, which was made official in a final decree of the Supreme Court of the United States on February 27, 1922, may be said to include the following. (a) At the outer ends of the docks west of Grassy Point where, in 1916, threatened taxation of part of the property of the Zenith Furnace Company brought on this suit, Wisconsin still possesses land attached to the Minnesota shore as well as two or more small islands; these lands are separated from the rest of Douglas County, Wisconsin, by the whole breadth of the St. Louis River, here half a mile wide. (b) A vessel proceeding westward from Duluth-Superior to the steel plants beyond Spirit Lake, and to Fond du Lac, Minnesota, must cross the State boundary half a dozen times; this may involve notable problems of future enforcement of the laws and regulations of the two States and of the federal government. (c) Wisconsin is formally and legally possessed of the submerged land upon which now stand the outer parts of the docks or piers of the McDougall-Duluth Shipbuilding Company; nearly 300 feet of Pier A, 125 feet of Pier B, and an outer corner of Pier C lie in Wisconsin; she could assess and tax these today and in the future. (d) The situation, lamentably, is analogous to that which precipitated the suit in 1916, as well as previous discussions and proceedings. Wisconsin, however, has no right to tax the existing docks of the Zenith Furnace Company. But every ambiguity about the legal position of the State boundary in the St. Louis River seems to have been removed.

In Lake Superior, however, the precise positions of the State boundaries are uncertain. The Minnesota-Michigan boundary departs from the United States-Canada frontier near Pigeon Point, Minnesota, where there are wholesale ambiguities. These, in turn, directly affect the Wisconsin-Minnesota and Wisconsin-Michigan boundaries in Lake Superior. These water boundaries extend for 130 to 135 miles easterly from the mouth of the St. Louis River and for about 90 miles northerly from the mouth of the Montreal River. It is to be hoped that each of the three States concerned may soon deem it advisable to define and adjust all its boundaries, as the Supreme Court of the United States recommends. This should be accomplished, not through expensive and vexatious judicial determination, but through amicable interstate agreements confirmed by Congress. These agreements ought to be undertaken and completed before difficulties of federal or state law enforcement raise an issue.

GLACIAL LAKES IN THE SUPERIOR BASIN

Nature of the Lakes. The ancestors of the present Lake Superior have already been alluded to in the description of glacial lakes in eastern Wisconsin (p. 295), and in the discussion of accumulations of lake clay near Superior and Ashland (p. 436). These glacial lakes were held in between the land and the margin of the Superior lobe of the continental ice sheet. The lake levels were determined by the heights of gaps in the surrounding highland through which the water of these lakes might drain. As the Superior lobe melted back toward the northeast the successive lakes enlarged in area. As the ice vacated more and more of the basin of Lake Superior they decreased in altitude, as lower outlets were exposed.

Glacial Lake Nemadji and its Neighbors. At the early stages there were small independent glacial lakes, south, west, and northwest of Lake Superior. These included Lakes Brule, Ashland, Nemadji, Ontonagon, Upham, St. Louis, Aitkin, and others. One of them, Glacial Lake Nemadji, is represented diagrammatically on Figure 184, in order to show the relationships between such predecessors of Glacial Lake Duluth, the retreating ice border, and the cols which became outlets and thus determined the levels of the lake surfaces. Glacial Lake Brule was in Douglas and Bayfield counties, Wisconsin. Glacial Lake Ashland lay just north of the Copper Falls State Park. Glacial Lake Ontonagon was in the upper peninsula of Michigan with an outlet which entered Wisconsin north of Hurley and passed through Saxon. The other glacial lakes were in Minnesota. Glacial Lakes Upham and St. Louis had at least one outlet that led into Wisconsin. Glacial Lake Nemadji drained down the Kettle River of eastern Minnesota into the St. Croix of western Wisconsin.

The Larger Glacial Lakes. We are interested chiefly, however, in the main stages of the marginal lakes at the borders of the continental glacier. These stages in the Lake Superior basin have generally been called Lake Duluth, Lake Algonquin, and the Nipissing Great Lakes. Like all such ancient and vanished bodies of water, the ancestors of Lake Superior are identified by abandoned beaches and deltas, lake-bottom deposits, and outlet channels.

The history of the deglaciation of the Lake Superior basin, even the last episodes, is more complicated than this, however. One must glance at a few of its own complications in order to understand the latest drift sheets in the northern and eastern peripheries of Wisconsin. The glacial history of the basin of Lake Superior is intimately associated with that of the basin of Lake Michigan and those of the other Great Lakes. Partly upon the basis of observable deposits, partly upon knowledge of the behavior of water bodies at the borders of existing glaciers in Alaska and elsewhere, and partly upon a philosophical basis, it is possible to tentatively state, in brief outline, the drainage history of the northern and eastern borders of Wisconsin during the time since the latest or Wisconsin glacier began to retreat. The maps (Figs. 184-188) have not been redrawn to illustrate all these stages of glacial lake history and drainage, but are here reprinted as in the earlier edition of this book. Seven stages of pronounced characteristics may be tentatively identified.

Glacial Lake Rouge and Glacial Lake Milwaukee. The first of these is the stage of the red lake in front of the retreating Wisconsin glacier (see page 267), which we may conveniently call the blue glacier. It probably came from the Patrician center. The interesting time is that when the ice border had receded far enough to uncover the divides, so that glacial waters began to accumulate between them and the ice front, but before the ice in the Lake Superior basin and the ice in the Lake Michigan basin had retreated far enough to admit of coalescence of independent lakes. The body of water in front of this blue glacier in the Lake Superior basin may come to be called Glacial Lake Rouge, that in front of the glacial lobe of the Lake Michigan basin, Glacial Lake Milwaukee. Their outlets were the same as those of Glacial Lake Duluth and Glacial Lake Chicago some thousands of years later. Glacial Lake Rouge was a red lake, Glacial Lake Milwaukee, a milky-white lake. The former acquired its color from suspended reddish clay and silt stained with iron which was derived from the several rock formations in the Lake Superior basin.

Glacial Lake Sioux. With the continued retreat of the lobes, these two glacial lakes coalesced into what may come to be called Glacial Lake Sioux. Its outlet was probably at Chicago. When the continental glacier retreated far enough so that Glacial Lake Rouge coalesced with Glacial Lake Milwaukee to form Glacial Lake Sioux, the milky waters of the latter were stained by the ferruginous clays and silts so that Glacial Lake Sioux was also a red lake. The deposits in its bottom were red clays and sandy silts, sprinkled with iceberg-rafted bowlders. It included the present Green Bay and much of Lake Michigan as well as part of Lake Superior.

Readvance of the Ice Sheet From the Northeast. A readvance of the continental glacier plowed up the lake bottom deposits of Glacial Lake Sioux, producing the red drift (a) of eastern Wisconsin north of Milwaukee and (b) of northern Wisconsin west of Ashland; in contrast with the substage of Wisconsin glaciation alluded to as the blue glacier, the ice lobes of this stage may conveniently be called the red glacier. It probably came from the Labrador center. The interval between the termination of Glacial Lake Sioux and the readvance was sufficiently long for the growth of forests (p. 268).

Readvance of another Glacial Lobe From the Northwest. During the period when this red glacier had completed its advance and was retreating to the northeast so that an early stage of Glacial Lake Duluth was formed in front of it, there came an independent glacial advance from the Keewatin center to the northwest. This may be called the gray glacier. It actually entered Wisconsin only in the parts of Burnett and Polk counties between Osceola, St. Croix Falls, and Grantsburg, but another portion of its low border lay upon the Minnesota upland about 30 miles northwest of the city of Superior. The waters from the melting of this gray glacier were mingled with those from the melting of the red glacier and caused complications in the history of the glacial lakes within the State of Wisconsin. It was during the retreat of this gray glacier to the northwest, that Glacial Lakes St. Louis and Aitkin came into existence, the former draining southeastward into the St. Louis River, the latter draining southwestward into the Mississippi.

Previous publications seem to have failed to take account of one important glacial lake which lay outside the limits of Wisconsin but drained into the St. Croix River, and another glacial lake which was chiefly within the State. These lakes were dammed by a great lobe of the gray glacier from the Keewatin center. This lobe extended across the Mississippi between St. Paul and Clearwater, Minnesota, advancing northeastward to Grantsburg, Wisconsin. At the same time, other portions of this glacier lobe appear to have stood northwest of Clearwater, with land-ending ice fronts at points west, north, and northeast of Brainerd, Minnesota. As these points lay within the basin of the Mississippi and as the melting ice must have supplied water, there should have been one of two consequences: either (a) the glacier waters flowing southward from the vicinity of Brainerd went underneath the glacial lobe from Clearwater to St. Paul or (b) there must have been a damming of the river. If this be true, the Mississippi was diverted eastward, and then southward, on the west borders of Burnett and Polk counties, Wisconsin. The latter possibility appears to be the simpler and more likely to have represented the facts. Under these conditions, then, a glacial lake lay in and east of the present Mississippi, held in by an ice dam near Clearwater, and backing up the valleys of the Mississippi, St. Francis, and Rum rivers, past St. Cloud and Princeton. It was the outlet of this short-lived glacial lake which flowed eastward and then southward along the Wisconsin boundary. If the glacial lobes north of Brainerd and northeast of St. Paul had not been contemporaneous, no such lake need have been formed, but there is no reason to suspect that they were not contemporaneous, and characteristic deposits prove that the lake existed.

The second glacial lake which necessarily existed through the presence of the gray glacier northeast of St. Paul, lay in the valley of the St. Croix River in Wisconsin and the adjacent part of Minnesota. The St. Croix flows southwestward. The gray glacier lay as a dam in its way. There must have been a glacial lake in the region north of Grantsburg. The lake bottom clays of this body of water were recognized and described many years ago. Glacial Lake Grantsburg, as it may come to be called, doubtless received the waters of the outlet of Glacial Lake Duluth at certain times, and these united waters flowed southward and southwestward through Polk and St. Croix counties into the Mississippi.

This fourth of the stages in the later history of the Glacial Great Lakes in the basin of the present Lake Superior closely involved some of the early phases of Glacial Lake Duluth, which eventually had the extent and outlet indicated on Figure 185, and concerning which other details are given several paragraphs below.

The Detached and Stagnant Ice Mass in Eastern Lake Superior. The next stage in the Lake Superior basin was that of the time when the enlarged Glacial Lake Duluth was draining southward into Glacial Lake Chicago through the low land between Au Train and Little Bay de Noc, Michigan, and a large isolated block of stagnant ice occupied the eastern part of the basin of Lake Superior (Fig. 186). That this ice block was stagnant was shown by the fact that it was detached from the main Labrador glacier. That such detachment actually took place is demonstrated by the presence of abandoned beaches of small glacial lakes northeast of Lake Superior in such places as the Michipicoten district of Algoma, Ontario. This stage of stagnation, which admittedly possesses a basis which is partly philosophical and speculative, was postulated in 1911 by the writer in a publication of the United States Geological Survey and has received support in recent years through the demonstration of the presence of similar detached ice blocks of great magnitude in such regions as the Connecticut valley of New England.

The Immediate Ancestors of Lake Superior. The last two stages of the glacial history of northern and eastern Wisconsin are described below under the headings *Glacial Lake Algonquin* and the *Nipissing Great Lakes*. They were the immediate predecessors of the modern Lake Superior. While it is thought that Glacial Lake Sioux may have had an extent and outlets similar to those of the early and southward-draining phase of Glacial Lake Algonquin, just as it is thought that Glacial Lake Rouge had an extent and outlet strictly analogous to Glacial Lake Duluth, it does not appear so probable that there was a pre-red-glacier stage similar to that of the Nipissing Great Lakes.

One must appreciate these generalizations concerning advancing, retreating, and readvancing glaciers, as well as existing, disappearing, reappearing, and fluctuating glacial lakes, in order to understand why red glacial drift overlies bluish glacial drift in eastern and northern Wisconsin. Weathering and oxidation of iron did not cause this striking contrast of color. The distribution and surface forms of the red drift would, in any event, make it impossible to explain the phenomenon upon a basis involving oxidation and other processes of weathering.

One might consider, momentarily, two other hypotheses. One is that a great belt of ruddy iron formation and associated red rocks lies submerged beneath eastern Lake Superior, occupying the area between the Slate Islands, Michipicoten Island, and the region eastward toward the Michipicoten district of Algoma, Canada. This hypothetical source of red drift was concealed, one might say, beneath bluish, grayish, and other non-red sedimentary rocks. The Illinoian glaciation did not erode down to it. The early stages of the Wisconsin glaciation likewise failed to reach it. The later Wisconsin uncovered it and used its pigments to stain the red drift of the readvance (see page 267). This would be interesting. It is not impossible. Nothing suggests that it is true. A second hypothesis would assume that glaciation from the Labrador center brought the blue drift of the Illinoian and the early Wisconsin, traversing no areas of ruddy rock, and that subsequent glaciation from the Patrician center brought the red drift into eastern Wisconsin from the central and western parts of the Lake Superior basin, where there is plenty of ruddy rock. This is by no means impossible, though the distribution of red drift east of Lake Michigan presents certain difficulties.

It is simpler, however, to reject these two hypotheses and to adhere to that which involves the red glacial lakes and the transportation of the coloring matter by glacial waters, which we know to have moved eastward from western Lake Superior to Green Bay and Lake Michigan.

We must now go back and amplify certain features of Glacial Lake Duluth and its successors in northern Wisconsin.

Glacial Lake Duluth. Glacial Lake Duluth (Fig. 185) started as a small lake, but was enlarged by the eastward retreat of the ice dam until it was over a third as large as Lake Superior. Its surface stood 450 to 700 feet higher than the present level of Lake Superior, as is proved by abandoned beaches and deltas, and by lake clays and sands on the hill slopes. Its outlet (Fig. 189) was southward through the gap near Upper St. Croix Lake. It probably did not come into existence as soon as Glacial Lake Chicago,

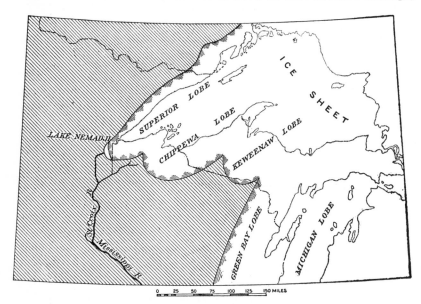

Fig. 184.   Glacial Lake Nemadji and the Kettle River outlet.

(p. 298), perhaps being contemporaneous initially with the Whittlesey-Chicago stage (middle map, Fig. 115) and continuing till the Lundy-Chicago stage (upper map, Fig. 116).

Glacial Lake Algonquin. Glacial Lake Algonquin was a very large lake (Fig. 187), occupying all of the basins of Lake Superior, Michigan, and Huron, and overlapping their borders. Of course there was slow change from the Lake Duluth (Figs. 185 and 186) to the Lake Algonquin stage, as there had previously been from the Lake Nemadji to the Lake Duluth stage. In the Superior basin the level of Glacial Lake Algonquin stood 300 to 350 feet above the present lake. Its outlets at different stages were probably:

(1) from the south end of what is now Lake Michigan, through the Chicago River to the Illinois and Mississippi, and so to the Gulf of Mexico;

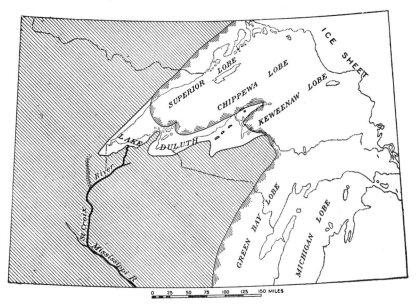

Fig. 185. Glacial Lake Duluth and the Bois Brule outlet.

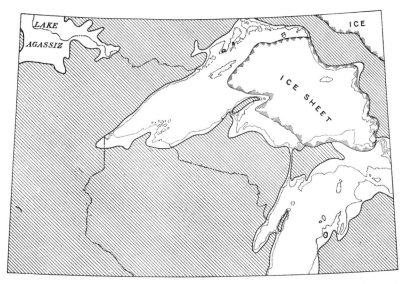

Fig. 186. Glacial Lake Duluth at an hypothetical intermediate stage with drainage southeastward into Glacial Lake Chicago and a detached ice mass in the eastern part of the Lake Superior Lowland.

(2) from the south end of what is now Lake Huron, through the St. Clair River;

(3) from the extreme east end of what is now Georgian Bay in Lake Huron, through the Trent River—Kirkfield outlet—and into Glacial Lake Iroquois, which occupied the basin of the present Lake Ontario. From here the water drained eastward by way of the Mohawk and Hudson rivers to the Atlantic Ocean at New York City.

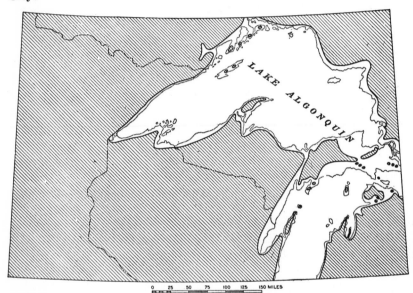

Fig. 187. Glacial Lake Algonquin in the Lake Superior Lowland. For outlet at this stage see middle map, Figure 116. The form of this glacial lake is now known to have been different in certain respects.

Nipissing Great Lakes. Lake Nipissing, in the Lake Superior basin, (Fig. 188) was similar to Lake Algonquin in shape but was shallower. It, therefore, overspread less of the borders of Lakes Superior, Michigan, and Huron than its predecessor. Its surface was a few feet higher than Lake Superior at the mouth of the Montreal River on the boundary between Wisconsin and Michigan, but here the Nipissing beach is now cut away by Lake Superior. At Bayfield it is about 4 feet above Lake Superior. Near Washburn on Chequamegon Bay the level which we think to represent the former surface of Lake Nipissing now dips below the present lake. At the city of Superior it may be 25 feet or so below the present level of Lake Superior. Its outlet was northeast of Lake Huron (Fig. 116), and its waters drained into the arm of the ocean which then occupied the St. Lawrence valley.

Each of the changes mentioned above was separated by numerous intermediate stages, here omitted. Each of them was caused by the retreat of the continental glacier toward the northeast and the consequent enlargement of the lake basins. The changes in lake levels were due chiefly to the uncovering of lower and lower outlets, previously blocked by the ice.

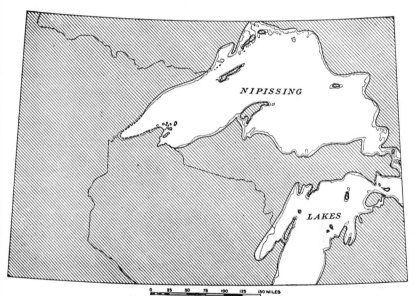

Fig. 188. The glacial lake in the Lake Superior Lowland at the stage of the Nipissing Great Lakes. For outlet at this stage see lower map, Figure 116.

There is proof that the land was being uplifted during the latter part of this glacial lake history, for some of the lake beaches are not horizontal, but rise toward the north. After Glacial Lake Nipissing had persisted for a long time, this uplift raised the outlet northeast of Lake Huron. The present outlet of Lake Ontario, near the Thousand Islands, was not raised so high, and the waters of the Great Lakes began to flow by the present St. Lawrence River to the ocean. The St. Lawrence outlet had previously been submerged, and then had been higher than the Nipissing outlet. It was not until the Nipissing outlet was abandoned and the St.

Lawrence outlet established that Lakes Superior, Michigan, and the other Great Lakes, began the present episode of their history.

The Abandoned Outlets. Only one abandoned major outlet of the Glacial Great Lakes is in Wisconsin. This is the valley occupied by the Upper St. Croix Lake near Solon Springs. It should not be confused with the long, narrow, Lower Lake St. Croix, occupying the mouth of the same river between Prescott, Wis., and Stillwater, Minn. This outlet is a steep-walled valley at the south end of a long, narrow bay, which led southward from Lake Duluth (Fig. 189). It is now followed for a short distance near Solon Springs and Gordon by the Chicago, St. Paul, Minneapolis, and Omaha Railway and is crossed nearby and then paralleled by the Minneapolis, St. Paul, and Sault Ste. Marie Railway.

This outlet lies nearly 50 feet lower than the Kettle River outlet of Lake Nemadji, but was not occupied by a stream from that lake because it was then blocked by ice.

The beaches leading into the outlet from either side are 526 to 532 feet above Lake Superior and the outlet is only 420 feet above the lake. It has been suggested that this does not indicate a depth of over 100 feet of water in the great stream which formerly flowed here, but a great amount of erosion by the stream. Streams emerging from lakes are commonly free from sediment. They are thought of as unable to erode their outlets, as for example in the case of the Niagara River between Lake Erie at Buffalo and the crest of Niagara Falls. Since the outlet of Lake Duluth is in glacial drift rather than in solid rock, it is possible, however, that there was erosion, as has taken place at the St. Clair outlet of Lake Huron. The St. Croix outlet was also occupied during certain of the later stages of Lake Duluth when lower beaches were formed.

The outlets of still lower stages of Lake Duluth, also represented by beaches, are not yet worked out. Two suggestions have been made. One is that there was drainage northward from Lake Superior to Lake Nipigon, and thence westward into Glacial Lake Agassiz. The other is that there are Lake Duluth outlets across the upper peninsula of Michigan to Glacial Lake Chicago.

The Abandoned Shorelines. Above the borders of the Wisconsin shore of Lake Superior are the abandoned shorelines from which we know the series of events just narrated. These are gravel and sand beaches, gravel deltas, and wave-cut cliffs. They are not at all unlike the present coast lines, except that they are high above the present lake. Some of them are overgrown with vegetation. The Nipissing beaches are usually broader and the wave-cut cliffs higher than those of the Duluth and Algonquin stages. Indeed it seems possible that the present stage of Lake Superior has not yet existed as long as did the Nipissing Great Lakes.

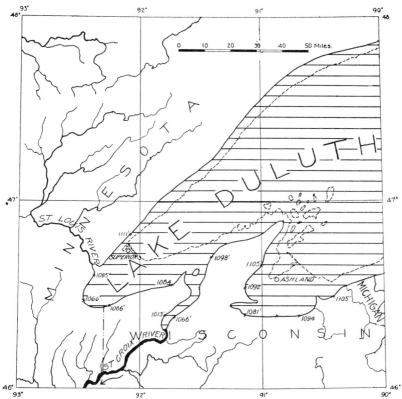

Fig. 189. Glacial Lake Duluth and the Bois Brule—St. Croix outlet. Figures give elevations of abandoned beaches in feet above sea level. (After Leverett, whose revision of these figures changes the altitudes of the Wisconsin portions of the highest beach of Glacial Lake Duluth so that they range from 1070 feet southwest of Superior to 1115 feet southeast of Ashland.)

It is clear from Figure 189 that the beaches of Glacial Lake Duluth ascend gradually northeastward, being 45 feet higher above Lake Superior near the Montreal River than at Superior. This slope is at the rate of about 2 feet to the mile, but the slope varies from place to place. The abandoned beaches of Lake Algonquin and of the Nipissing Great Lakes are also inclined, though not at the same rate.

LAKE SUPERIOR

General Features. The modern Lake Superior came into existence after the Nipissing Great Lakes and with the establishment of the present St. Lawrence outlet. The coast line of Lake Superior in Wisconsin is a little over 150 miles long. This does not include minor sinuosities, nor the shores of the Apostle Islands. These islands, more than 24 in number, seem to have acquired this name when there were thought to be only twelve islands.

The lake surface is subject to minor fluctuations, as on Lake Michigan (p. 309).

The Drowned River Valleys. The tilting of the land which had caused the inclination of the beaches and the submergence of the western part of the shoreline of the Nipissing Great Lakes may still be in progress. This was first noticed more than 75 years ago and was afterwards studied more carefully, so that we now know something about the rate at which the tilting is going on. To say that a line 100 miles long and trending approximately north-south, is being tilted at such a rate that the southern end of it will be four or five inches below the northern end after the lapse of a century seems to indicate a very slow movement. It may turn out, however, that the tilting is spasmodic, with intervals of movement interrupted by intervals of rest. However this may be, the slow tilting has sufficed to submerge the stumps of trees near Superior, where the trees, of course, grew above lake level. The tilting has submerged rapids during the lifetime of some of the Indians, as indicated at the beginning of this chapter.

The tilting has drowned the valleys of streams on the southern and southwestern side of Lake Superior (p. 447). Of these drowned valleys, or estuaries, Flag Lake at Port Wing may be mentioned, but the best illustration is the lower course of the St. Louis River and the Bay of Superior. Stream erosion had previously proceeded far enough so that the St. Louis had a meandering course on the plain of lake clays (upper map, Fig. 190). These meanders were intrenched in the clays when the lake level fell and the stream cut into these weak deposits. Subsequently they were drowned by the canting of the waters of Lake Superior into the river valley. This is well shown at Spirit Lake. The river is navigable for a long distance by the lake steamers, and smaller boats can go up it 17 miles to Fond du Lac, Minnesota. Thus a

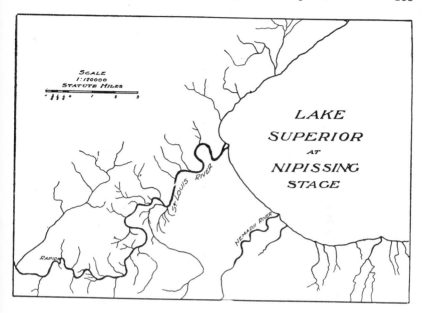

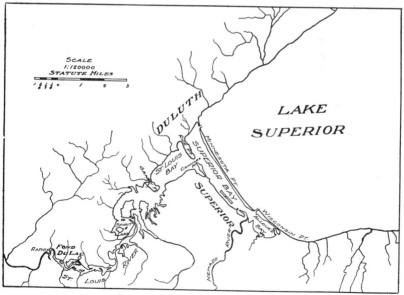

Fig. 190. The drowned valley of the St. Louis River at Superior (lower map), and the river before the submergence (upper map).

great commercial advantage, the use of the Bay of Superior as a harbor, has been made possible by this drowning of the mouths of the small rivers on the south side of Lake Superior.

Present Sea Cliffs and Benches.   The present shorelines of Lake Superior are the product of wave-cutting in some places, of deposition by waves and by currents at other localities.   Some of these shorelines are cut in solid rock, but the greater number are in the unconsolidated deposits of glacial drift.

The rock coast shows marked evidence of modification by the work of waves at the present level of Lake Superior.   These are best developed where the sandstone is made up of comparatively thin layers.   Of the rock shores, mention may be made of two types of coast: one is the low coast near the mouth of Montreal River and a few miles to the west near Clinton Point, where the rock ledges are practically at lake level and the wave-cut cliffs only a few feet high.   The opposite extreme is the high cliff of the type developed in the Apostle Islands.   Here the sandstone, which is red or brown and distinctly coarser-grained than that at Clinton Point, lies in an essentially-horizontal position.   The waves have, therefore, been able to carve fairly-high precipitous rock cliffs in the sandstone, the height being 10 to 30, and in extreme cases, 60 feet. There are many wave-formed arches and caverns (Pl. XXXV, A) of the type known as sea caves.   These are best seen on Devils Island.

Pillared Rocks in the Apostle Islands.   The Pillared Rocks were described in 1848 by Norwood, as follows:

"Between Madeline island and Bark point the red sandstone shows itself on the lake shore for nearly the whole distance.   At Point Detour, and many other places, the bluffs are perpendicular and from forty to sixty feet in height, and are overlaid by beds of sand and red marl.   Beyond the mouth of the small river opposite Oak island, the rock has been worn, by the incessant action of the waves, into most singular and interesting architectural forms.   Among these the pillar and arch predominate.   These *Pillared Rocks* extend for many miles and are interesting, not only on account of their picturesque appearance, but also as illustrating the means by which the lake is gradually enlarging its southern boundary.   Some of the arches are circular, but most of them are pointed.   In the space of two hundred yards, at one point,

A. WAVE-CUT ARCH AT SQUAW BAY, BAYFIELD COUNTY.

B. BEACH ON THE COAST OF LAKE SUPERIOR.

VALLEY CUT IN THE LAKE CLAY ON THE SOUTH SHORE OF LAKE SUPERIOR.

I counted over fifty arches, all possessing great regularity, and resting upon pillars almost as symmetrical as though they had been subjected to the chisel of the artisan. Through these arches the waters of the lake dash with every swell, and their unceasing play has hollowed out numerous caverns of great depth. Two caves were particularly noticed, each more than an acre in extent, and supported at intervals by pillars of all sizes, from twelve feet to half the number of inches in diameter, and forcibly reminding one of the descriptions of the celebrated cave of Elephanta. Regular architraves, friezes, and cornices are constantly seen, but it is only occasionally that a pillar shows a base, as they are sunk beneath the waters of the lake. Some of the arches are large enough to permit the passage of a Mackinaw boat. There is generally from twenty-five to forty feet of sandstone resting on the arches, the layers being nearly horizontal, and supporting a capping of majestic forest trees, giving to the whole scene an indescribably grand and picturesque appearance."

Destruction of Islands. The Apostle Islands have been greatly modified by wave erosion. Steamboat Island has been completely destroyed since the coming of white men to the region, while another island has been cut in two.

Bluffs and Beaches. The common wave-cut bluffs in the unconsolidated glacial till or lake clay and sand is a more gently sloping form. The slope is sometimes as much as 40°, but is generally less. Some of these bluffs are very high, as in the 200 foot cliff at the northern end of Oak Island.

Associated with these cliffs are sand and gravel beaches (Pl. XXXV, B). Such beaches skirt the shores of the larger part of the Lake Superior coast of Wisconsin. There are also barrier beaches, which have been built from headland to headland and which hold in lagoons (Figs. 190, 193).

The Sand Spits at Superior. The Bay of Superior and Duluth Harbor are formed by four sand spits (Fig. 191). The two on the east separate the port from Lake Superior, while the two on the west separate it from the estuary of the St. Louis River. Among the shore deposits now being formed in Lake Superior none are more striking than the two great bars or spits which extend across the head of Lake Superior at Duluth—Minnesota Point and Wisconsin Point. Their ends are separated by a narrow channel which formed the only entrance to the Bay of Superior until the Government dredged the canal near the Duluth shore. These bars have a total length of about 10 miles. Minnesota Point is $6\frac{1}{2}$ miles long, Wisconsin Point $3\frac{1}{4}$ miles. Their width varies from a little over an eighth of a mile to less than a hundred yards.

Fig. 191. The sand spits at the head of Lake Superior. (Gilbert.)

They have been built up above the water by a combination of two causes. The first and more important is the interference of the shallowing lake bottom with the passage of waves, causing waves to be overturned on the site of the present Minnesota and Wisconsin Points. The overturning stirs up the deposits at the bottom of the lake and causes the waves to heap up material at this locality. The continued accumulation of material along this narrow line has gradually built up a deposit that approached the surface of the water and was augmented by the deposits of the second kind. These are the materials derived from the shores of the lake, transported outward along the submerged embankment, under the influence of the shore currents. The combination of these two agencies soon carried the spits a great distance out from the lake shores, and they were eventually built up above lake level.

Drill holes put down near the Government canal at Duluth and at the new jetties between Wisconsin Point and Minnesota Point have shown that the points are built upon a base of fine lake clay. This is overlain near the shore by very coarse material, which is replaced a short distance out by fine sand. The sand goes 60 to 90 feet below present lake level. On the Minnesota side no pebbles are found on the present beach at a greater distance from shore

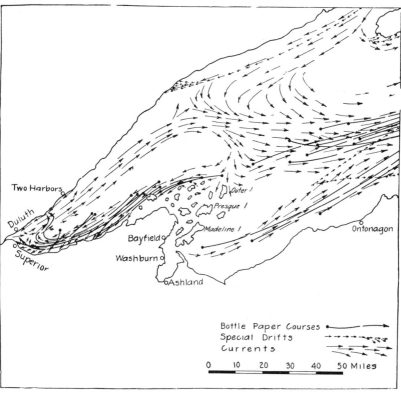

Fig. 192. Currents at the western end of Lake Superior. (Harrington.)

than three-quarters of a mile. This shows that the contribution of the coarse, along-shore drift in the middle of the point is not very great, and that the larger part of the material is washed up by the waves. It is augmented by the material drifted along the beaches. The higher parts of these points, which rise 20 or 30 feet above the lake level, consist of very fine sand, built up into sand dunes by the wind. Upon these dunes, evergreen trees have been able to grow.

About a mile back from Minnesota and Wisconsin Points, another pair of spits—Rice Point and Connors Point—has been built. They separate St. Louis Bay, where most of the ore boats are loaded, from Superior Bay. These were doubtless formed as spits at an earlier date, in a manner exactly similar to the outer spits, though they have never been connected. The outer bars could not have then existed.

Still farther up St. Louis River there are projections from the sides of the valley, like Grassy Point and others. In some respects they are similar to the spits, though of an entirely different origin. In the post-Nipissing tilting of the lake waters into the valley of St. Louis River, portions of the low spurs on the valley sides were drowned. Parts of these spurs still emerge from the water and resemble spits (Fig. 190).

Origin of the Sand in the Spits. The sand which is carried along shore to the spits is directly related to the currents in western Lake Superior. These currents make a great eddy west of the Apostle Islands. The water moves southwestward on the southern shore and northeastward on the northern shore, but, in the latter case, with a southwest-setting counter current for a short distance east of Duluth.

The presence of the sand which forms these spits calls for special explanation. It has been shown by a study of the soundings in the western part of this lake basin that the prevailing bottom material is clay and mud. The shore material also is chiefly glacial clay, for there are no sandstone ledges on the coast west of Port Wing and Orienta on the south shore. The ledges of the north shore are mostly basic Keweenawan rocks, which contain practically no quartz. The acid volcanic rocks which might supply some sand, outcrop on the coast northeast of Minnesota Point for less than two miles. They have been little modified by wave work at the present level of Lake Superior. The narrow strip of sand between these two clay belts, one above, the other below lake level, is therefore, seen to be derived from glacial sands and from the along-shore drift of sandy material. Its probable origin is 25 or 30 miles to the northeast in the sandstone ledges near Port Wing, where the abundant sand below lake level begins and continues without interruption to Minnesota Point. A small part of it may come from rivers like the Iron, Bois Brule, Poplar, and Amnicon, which cross sandstone ledges near the escarpment, though these streams are drowned at the mouths and, therefore, should supply little but silt to Lake Superior. As the rate of movement of the surface current is only 6 to 24 miles per day, and the rate of movement of the sand along shore is infinitely slower, except in time of storms, it is clear that the accumulation of the sand in these great spits has occupied a period of many thousands of years.

Spits at Port Wing. Port Wing, in Bayfield County, also has a pair of spits (Fig. 193). They partly close the mouth of the estuary made by the drowning of the mouth of Flag River. Within the bay the submerged channel may still be traced by soundings. It is remarkable that the estuary has not been completely filled, in view of the sand accumulation which makes the spits at the bay mouth.

Spits in the Apostle Islands. Farther east, near the end of the Bayfield Peninsula, there is a broad sand spit below lake level, connecting the mainland west of Point Detour with Sand Island (Fig. 194). It is over 2 miles long, a mile wide, and has water 6 to 8 feet deep where the depth is normally 40 to 60 feet. Its proximity to the mouth of Sand River, however, does not indicate that this is delta as well as shore accumulation, for this stream is like its neighbors in having no delta. The mouths of the rivers which enter Lake Superior on the south are practically all without deltas because of the tilting of the waters of the lake into the valley on this side. The soundings in Lake Superior show sand deposits which trail southwestward from the islands with the prevailing lake currents. There is a similar sand deposit extending southwestward from Bark Point, where sandstone ledges outcrop at lake level.

There are other small spits, above and below lake level, on the shores of some of the Apostle Islands. Many of them point southwestward. There are also sand bars which unite pairs of islands, as at York Island and Rock Island. There are converging spits which enclose lagoons, as on the southwest sides of Michigan and Outer Islands.

On Madeline Island, the largest of the Apostle archipelago, the head of Big Bay on the eastern side is barred off by a barrier beach. This also is the case on the southwestern side of the same island between the modern village of La Pointe and the Old Mission. There is a similar barrier in Bark Bay, Bayfield County.

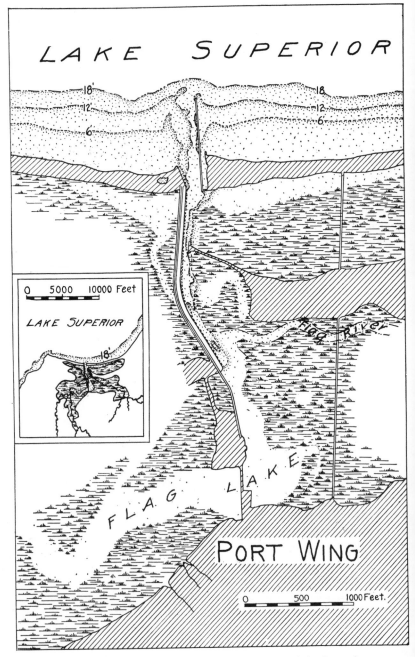

Fig. 193. The sand spits at Port Wing. (After U. S. Lake Survey.)

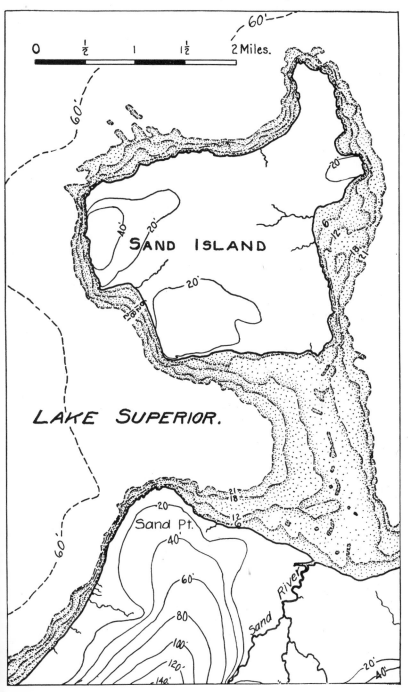

Fig. 194. The submerged sand bar or tombolo, which connects one of the Apostle Islands with the mainland. Depths and heights in feet. (After U. S. Lake Survey.)

Chequamegon Point. East of the Apostle Islands is Chequamegon Point, a long, narrow sand spit to which the harbor of Ashland owes its protection. It extends nearly across the bay (Fig. 195). Chequamegon Point was formerly a narrow spit nearly 8 miles long, reaching within $1\frac{1}{2}$ miles of Madeline Island and within $2\frac{3}{4}$ miles of the mainland of Bayfield Peninsula. About 1840, again in 1870, and again in 1891, a gap was formed during a severe storm, making the outer 4 miles of the spit into an isolated bar, called Long Island. This gap, known as the Sand Cut, has water only $1\frac{1}{2}$ to 4 feet deep. Inside Chequamegon Point is an older sand spit called Oak Point. Nearby Radisson and Groseilliers

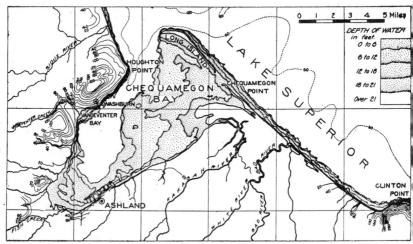

Fig. 195. Chequamegon Point and the great deposit of sand in the bay at Ashland.
(After U. S. Lake Survey.)

portaged across from Lake Superior to Chequamegon Bay in 1661. It is assumed that the break in Chequamegon Point did not then exist, for otherwise the Indians would not have taken the French *coureurs de bois* across the portage.

The order of events in the formation of Chequamegon and Oak points has been worked out as follows: Following the first stage, when there was no sand accumulation here, Oak Point bar is thought to have been formed (2nd stage, Fig. 196). Soon it was connected with the mainland by Bad River bar (3rd stage, Fig. 196). Subsequently outer Chequamegon Point was formed on the site of the present Long Island (4th stage, Fig. 196). Then the whole point was connected by the large Chequamegon Point (5th stage, Fig. 196). Finally the Sand Cut breach was made (Fig. 195). This latter episode has been interpreted as evidence that this and all the adjacent sand spits are wasting under the influence of higher wave attack with the southward tilting of the lake waters.

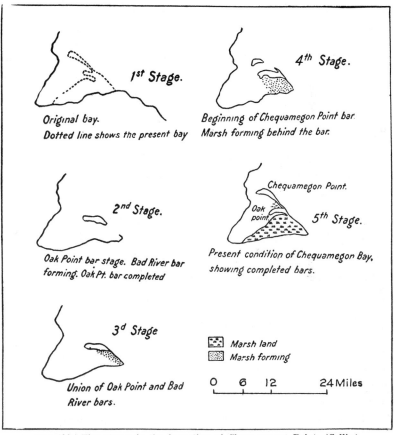

Fig. 196. Five stages in the formation of Chequamegon Point. (Collie.)

La Pointe, where Father Allouez established the mission of St. Esprit in 1665, was on the mainland west of Chequamegon Bay, between Washburn and Ashland. There Allouez found a village of 800 Indians, and there Father Marquette lived in 1669. This should not be confused with the Old Mission of La Pointe or the present village of La Pointe, both on the west shore of Madeline Island, opposite Bayfield. At none of these three places called La Pointe is there a conspicuous peninsula or point, comparable to Chequamegon Point. It was the latter that originally gave the name to the mission and the region of La Pointe du St. Esprit (p. 22).

Lake-Bottom Deposits. West of Chequamegon Point there is an extensive marsh and shallow water area which has filled the eastern two-thirds or three-fourths of Chequamegon Bay. This is quite as important a deposit as the visible sand spits. Its formation has accompanied the building of the spits, but it is nowhere above lake level, except where the deposition has been of sufficient amount to allow marsh plants to raise the deposit above lake level and make a swamp. This process is much slower than in warmer southern waters.

Between the mouth of Montreal River and the Apostle Islands the bottom of Lake Superior is shown, by a study of the soundings on lake charts, to be prevailingly sandy, though with small areas of clay in the deeper water. This sand is the source of the spits. As in Wisconsin Point at Superior, it has been built up above lake level by the interference with waves on the shallow bottom, as well as by alongshore transportation.

This great accumulation of sand in water 50 to 400 feet deep forms a notable contrast with the prevailing clay and mud bottom of Lake Superior. It is clearly related to the friable sandstone ledges of the Apostle Islands and to the currents in the lake. The drift of bottles and of wrecks (Fig. 192) shows that the surface waters flow northeastward from the Apostle Islands and from Chequamegon Point. The observations here are incomplete, but the abundant, submerged deposits of sand south of the archipelago and the lack of sand to the west suggests that there may also be strong southward currents through the Apostle Islands. There may also be a counter-current close to shore and extending westward from the Montreal River to Chequamegon Point. There are no rock ledges at lake level, except at the mouth of Montreal River and near Clinton Point, and the sandstone here is more shaly and much less friable than in the Apostle Islands. Accordingly the chief source of the sand of this spit may be in the latter locality, though much is also derived from the sandy glacial drift of the lake shore. Further observations of drifting bottles and a petrographic study of the sand of the spit may throw light upon this question.

Ice in Lake Superior. Winter ice closes Lake Superior to navigation for several months each year. Ashland Harbor is usually closed for 145 days, Dec. 1 to Apr. 25; Outer Island in the Apostle group for 157 days; and Duluth-Superior for 96 to 133 days. This is a decided contrast to Milwaukee (p. 306).

Relation of Sand Bars to Commerce. The relation of the sand spits to the protection of the estuary harbor of Duluth-Superior, of the harbor of Ashland at Chequamegon Bay, and to a less extent the harbor of Port Wing, is a potent influence in connection with the shipping of iron ore, lumber, and grain from these ports. This is especially important at Duluth-Superior, whose harbor (Fig. 191) ships a greater tonnage than Chicago, Milwaukee, Cleveland, Fort William, Port Arthur, Buffalo, or any other lake port, and, in spite of the ice embargo of winter, probably more than any American seaport except New York.

BIBLIOGRAPHY

Bayfield, H. W. Outlines of the Geology of Lake Superior, Trans. Lit. and Hist. Soc. of Quebec, Vol. 1, 1829, pp. 1-43.

Collie, G. L. The Wisconsin Shoreline of Lake Superior, Bull. Geol. Soc. Amer., Vol. 12, 1901, pp. 197-216.

Dablon, Claude. (On drift copper in and near the Apostle Islands), Jesuit Relations 1669-71, Thwaites edition, Vol. 54, pp. 161-163.

Gilbert, G. K. The Topographic Features of Lake Shores, 5th Annual Rept., U. S. Geol. Survey, 1885, pp. 93-95; Recent Earth Movements in the Great Lakes Region, Ibid., 18th Annual Rept., Part 2, 1898, pp. 595-647.

Harrington, M. W. Surface Currents of the Great Lakes, Bull. B, U. S. Weather Bureau, Dept. of Agriculture, 1895.

Hattery, O. C. Survey of Northern and Northwestern Lakes, Bull. 24, U. S. Lake Survey, War Dept., Corps of Engineers, Detroit, 1915, 455 pp. This is an annual publication, with corrections and additions in monthly supplements.

Irving, R. D.   Coastal Features of the Eastern Lake Superior District, Geology of Wisconsin, Vol. 3, 1880, pp. 70-76.

Leverett, Frank.   Outline of History of the Great Lakes, 12th Annual Rept., Mich. Acad. Sci., 1910, Superior Basin, pp. 21, 25-28; Moraines and Shore Lines of the Lake Superior Region, Prof. Paper 154A, U. S. Geol. Survey, 1929, 72 pp., and "Map Showing Areas Formerly Covered by Glacial Lakes of Northern Minnesota, Northern Wisconsin, and Northern Peninsula of Michigan", Plate 2, (in pocket); see also blue slip of credits and corrections, by George Otis Smith, attached to page 1.

Martin, Lawrence.   Marginal Lakes of the Lake Superior Region, Monograph 52, U. S. Geol. Survey, 1911, pp. 441-453; Postglacial Modifications, Ibid., pp. 455-459.

Nicollet, J. N.   Map in Senate Doc. 237, 26th Congress, 2nd Session, Washington, 1843.

Norton, Executor, versus Whiteside, 239 U. S., pp. 144-155.

Norwood, J. G.   (On the Apostle Islands and south coast of Lake Superior), —in Owen's Report of a Geological Reconnoissance of the Chippewa Land District of Wisconsin, Senate Ex. Doc. 57, 30th Congress, 1st Session, Washington, 1848, pp. 75-77.

Radisson, Pierre Esprit, sieur de, (On Chequamegon Point in 1661), The Fourth Voyage of Radisson, Collections Wis. Hist. Soc., Vol. 11, 1888, p. 72.

Schoolcraft, H. R.   Native Cooper of Lake Superior, Amer. Journ. Sci., Vol. 3, 1821, pp. 201-216.

Stuntz, G. R.   On Some Recent Geological Changes in Northeastern Wisconsin, Proc. Amer. Assoc. Adv. Sci., Salem Meeting 1869, Vol. 18, 1870, pp. 205-210.

Taylor, F. B.   A Reconnoisance of the Abandoned Shorelines of the South Coast of Lake Superior, Amer. Geol., Vol. 13, 1894, pp. 365-383; Monograph 53, U. S. Geol. Survey, 1915, pp. 321, 327-328, 431, 456, 460, 509, 510.

Upham, Warren.   The Western Superior Glacial Lake and the Later Glacial Lakes Warren and Algonquin, 22nd Annual Rept., Minn. Geol. and Nat. Hist. Survey, 1894, pp. 54-66; Origin and Age of the Laurentian Lakes and of Niagara Falls, Amer. Geol., Vol. 18, 1896, pp. 169-177.

U. S. Supreme Court.   Minnesota-Wisconsin Boundary Case of 1916-22, Transcript of Record, 2 volumes, Washington, 1918, 1074 pp; U. S. Reports, Cases Adjudged in the Supreme Court of the United States, Vol. 252, New York, 1920, pp. 273-285; Vol. 254, New York, 1921, pp. 14-16; Vol. 258, Washington, 1923, pp. 149-158.

Whittlesey, Charles.   The Apostle Islands, Geological Report on that Portion of Wisconsin Bordering on the South Shore of Lake Superior,—in Owen's Geological Survey of Wisconsin, Iowa, and Minnesota, Philadelphia, 1852, pp. 437-438; On the Cause of the Transient Fluctuations in Level in Lake Superior, Proc. Amer. Assoc. Adv. Sci., Portland Meeting, 1873, pp. 42-46. On Owen's geological map of Wisconsin, Iowa, and Minnesota, published in 1851, Whittlesey shows the border of the red clays and marls.

Williams, F. E. Recent Sedimentation in the Western Great Lakes, Unpublished thesis, University of Wisconsin, 1910.

Wisconsin versus Duluth, 96 U. S., pp. 379-388.

MAPS

See Chapter XVII, p. 445.

APPENDIX A

AREA OF WISCONSIN

The land area of Wisconsin is 55,256 square miles, the total of the county areas in Appendix C. The water area is 810 square miles, making a total of 56,066. This figure is based upon careful computations published by the United States Geological Survey, and based upon the joint work of the Federal Census Office, General Land Office, and Geological Survey. Their method of measurement is described as follows (Bulletin 302, U. S. Geol. Survey, 1906, p. 6) ; Ibid., Bulletin 817, 1930, p. 248.

"The areas of all square degrees included entirely within a State or the United States are taken from tables of such areas. Where a square degree is crossed by a boundary line, so that only part of it is included, both the part included and that excluded are measured from the best maps by planimeter, and the correctness of the measurement is tested by comparing their sum with the tabular area of the square degree."

The 810 square miles of water surfaces represent the area of rivers and inland lakes. It is believed that this figure is too small, perhaps representing only half the water surface of rivers and inland lakes.

In addition Wisconsin has 2,378 square miles of the surface of Lake Superior and 7,500 square miles of the surface of Lake Michigan, aside from unsettled complications due to the status of the Michigan-Wisconsin boundary in Green Bay in 1931.

Its total area may be summarized as follows:

	Square miles
Total area, including rivers, inland lakes and parts of the Great Lakes	65,944
Area of land surface	55,256
Area of rivers and inland lakes	at least 810
Area of Wisconsin portion of Lake Superior	2,378
Area of Wisconsin portion of Lake Michigan	7,500
Area of land and inland waters	56,066

The geographic center of the state is 9 miles southeast of Marshfield. The state ranks twenty-fourth in area in the United States.

APPENDIX B

BOUNDARIES OF WISCONSIN

The best general description of the limits of the state is "The Boundaries of Wisconsin" by R. G. Thwaites, (Collections Wis. Hist. Soc., Vol. 11, 1888, pp. 451-501). See also M. M. Strong's "History of the Territory of Wisconsin from 1836 to 1848," Madison, 1885, 637 pp; H. C. Campbell's "Wisconsin in Three Centuries, 1634-1905," 4 vols., New York, 1906; and Lawrence Martin's "Michigan-Wisconsin Boundary Case in the Supreme Court of the United States, 1923-26," Annals Assoc. Amer. Geographers, Vol. 20, 1930, pp. 105-163.

The following is quoted from the Thirteenth Census of the United States, (Vol. 3, Population, 1910, Washington, 1913, p. 1047).

"Wisconsin was named from its principal river. The significance of the word, which is of Indian origin, is not positively known, but among the meanings given are 'wild, rushing river,' 'gathering of the waters,' and 'great stone or rocks.'

"The first explorers of the region now constituting Wisconsin were the French. In 1634 Jean Nicollet, sent out by the governor of New France to promote trade with the Indians, landed where the city of Green Bay now stands, and ascended the Fox River to a point about 20 miles west of Lake Winnebago. Twenty years later two fur traders, Radisson and Groseilliers, ascended the Fox and may have descended the Wisconsin to its junction with the Mississippi. In 1669 a mission was established on the Fox River a few miles above its mouth, and about the mission grew up the town of Depere, the first permanent settlement within the present limits of Wisconsin.

"In 1763, at the close of the French and Indian war, the French possessions east of the Mississippi were ceded to England. At the close of the Revolution the territory northwest of the Ohio and east of the Mississippi was ceded by Great Britain to the United States. The former country, however, did not at once relinquish its hold, and, although its outposts in that region were evacuated in the summer of 1796, it was not until the close of the war of 1812 that it ceased to exercise some degree of control in the territory between Lake Michigan and the Mississippi.

"In 1787 the region bounded by Pennsylvania, the Ohio River, the Mississippi River, and the Great Lakes was organized as the Northwest Territory, the claims of Massachusetts, Connecticut, and Virginia, based on their early charters, having been ceded to the United States between 1781 and 1786. In 1800 the present area of Wisconsin was included in the newly organized territory of Indiana; in 1809 it was made a part of Illinois territory; and in 1818, when Illinois became a state, it was added to Michigan territory.

"Wisconsin was organized as a separate territory in 1836. At this time it included, in addition to the area of the present state, the region now constituting Minnesota, Iowa, and those portions of North and South Dakota lying east of the Missouri and White Earth Rivers. In 1838 that part of Wisconsin territory situated west of the Mississippi River and a line drawn north from its source to the Canadian boundary was organized as the territory of Iowa. In May, 1848, Wisconsin, with boundaries as at present, became a state of the Union."

The name of the state has also been spelled "Wiskonsin, Ouisconsin, Misconsing, etc." (see R. G. Thwaites' book entitled "Wisconsin, the Americanization of a French Settlement", American Commonwealth Series, Boston, 1908, p. 233).

The boundaries of Wisconsin are described as follows in Bulletin 817, U. S. Geol. Survey, 1930, pp. 194, 197-200; see also the Constitution of the State of Wisconsin, Article II, (Wisconsin Blue Book, Madison, 1929, pp. 698-699).

"As originally constituted its area comprised all that part of the former Territory of Michigan which lay west of the present limits of the State of Michigan. The limits are defined in the act for its organization [in 1836] as follows (see U. S. Statutes at Large, Vol. 5, p. 11):

'Bounded on the east, by a line drawn from the northeast corner of the State of Illinois, through the middle of Lake Michigan, to a point in the middle of said lake, and opposite the main channel of Green Bay, and through said channel and Green Bay to the mouth of the Menomonie river; thence through the middle of the main channel of said river, to that head of said river nearest to the Lake of the Desert; thence in a direct line, to the middle of said lake; thence through the middle of the main channel of the Montreal river, to its mouth; thence with a direct line across Lake Superior to where the territorial line of the United States last touches said lake northwest; thence on the north, with the said territorial line, to the White-earth river; on the west, by a line from the said boundary line following down the middle of the main channel of White-earth river to the Missouri river, and down the middle of the main channel of the Missouri river to a point due west from the northwest corner of the state of Missouri; and on the south, from said point, due east to the northwest corner of the State of Missouri; and thence with the boundaries of the States of Missouri and Illinois, as already fixed by acts of Congress.'

"In 1838 all that part of the territory lying west of the Mississippi and a line drawn due north from its source to the international boundary—that is, all that part which was originally comprised in the Louisiana Purchase and the Red River drainage basin south of the 49th parallel—was organized as the Territory of Iowa.

"When the Territory of Wisconsin was organized it was supposed that there was an almost continuous water-boundary line between Michigan and Wisconsin from Green Bay to Lake Superior. Congress in 1838 ordered the running and marking of this boundary (5 Stat. L. 244), but it was soon discovered that the line could not be run as described, for the head of the Montreal River is more than 50 miles from the Lake of the Desert (now called Lac Vieux Desert), which was supposed to be its source. It was therefore recommended that the boundary location be changed to the position later described in the Wisconsin enabling act of 1846 (9 Stat. L. 56-7) and in greater detail in the Michigan constitution of 1850, which reads as follows. 'Through Lake Superior to the mouth of the Montreal River; thence through the middle of the main channel of the said river Montreal to the headwaters thereof; thence in a direct line to the center of the channel between Middle and South islands in the Lake of the Desert; thence in a direct line to the southern shore of Lake Brule; thence along said southern shore and down the river Brule to the main channel of the Menominee River; thence down the center of the main channel of the same to the center of the most usual ship channel of the Green Bay of Lake Michigan' (see Thorpe, F. N., 'The Federal and State Constitutions', House Doc. 357, 59th Congress, 2nd Session, Vol. 4, 1909, p. 1945).

"The line through Lake Superior is thus described in the Wisconsin enabling act of 1846 (9 Stat. L. 56-57. Thorpe, F. N., op. cit., vol. 7, p. 4072) : 'thence down the main channel of the Montreal River to the middle of Lake Superior; thence through the center of Lake Superior to the mouth of the St. Louis River.'

"The straight parts of the boundary were surveyed and marked, in 1847, from a point where the Balsam River and the Pine River unite to form the Montreal, S. 74° 27' E. to the Lake of the Desert, a distance of 50 miles 67 chains 6 links. The southern part of the line begins at the lower end of Lake Brule and runs N. 59° 38' W. for 13 miles 37 chains 66 links to an intersection with the former line in the Lake of the Desert (See General Land Office files, Boundaries, No. 39, 1 and 2).

"Suit was commenced by Michigan in the United States Supreme Court in October, 1923, for a redetermination of the Michigan-Wisconsin boundary, the claim being made that the surveys of 1840-1847 were not in accord with the descriptions. The change from the previously accepted boundary to that proposed by Michigan would have resulted in a loss to Wisconsin of about 255,000 acres of land, but the court by decree dated March 1, 1926, (270 U. S. 295; 272 U. S. 398) confirmed Wisconsin's title to the disputed area, principally because

'The rule, long settled and never doubted by this court, is that long acquiescence by one State in the possession of territory by another and in the exercise of sovereignty and dominion over it is conclusive of the latter's title and rightful authority.'

"A resurvey of the Michigan-Wisconsin line was completed in 1929 by commissioners representing the two States. There are now 160 concrete monuments on this 65-mile line. This work was executed in accordance with the Supreme Court decree of November 22, 1926.

"The boundary from Lake Brule to the mouth of the Menominee is practically that described in the enabling act and

'follows the channels of the Brule and Menominee wherever they are free from islands; * * * wherever islands are encountered above Quinnesec Falls the line follows the channel nearest the Wisconsin mainland, so as to throw all such islands into Michigan; and * * * wherever islands are encountered below Quinnesec Falls the line follows the channel nearest the Michigan mainland, so as to throw all such islands into Wisconsin.'

"Through Green Bay the line was fixed as claimed by Wisconsin and includes in that State Washington, Detroit, Plum, Rock, and some smaller islands.*

"On March 3, 1847, a supplementary act for the admission of Wisconsin was passed by Congress, in which the western boundary of the proposed State was described as 'follows:

'That the assent of Congress is hereby given to the change of boundary proposed in the first article of said constitution, to wit: leaving the boundary line prescribed in the act of Congress entitled 'An Act to enable the People of Wisconsin Territory to form a Constitution and State Government, and for the Admission of such State into the Union,' at the first rapids in the river St. Louis; thence in a direct line southwardly to a point fifteen miles east of the most easterly point of Lake St. Croix; thence due south to the main channel of the Mississippi River or Lake Pepin; thence down the said main channel, as prescribed in said act.'

"This act had it been accepted would have given the State an area considerably less than it now has.

"The first constitution, completed December 6, 1846, (The full text of this constitution is given by Quaife, M. M., Wisconsin State Hist. Soc. Coll., Vol. 27, 1919, pp. 732-753), accepted the boundaries as described in the enabling act of August 6, 1846, but proposed that Congress consent to the change as described in the later act above referred to. This constitution was rejected by popular vote April 5, 1847 (14,119 ayes, 20,231 noes)—not, however, because of unsatisfactory boundaries.

"A second constitution dated February 1, 1848, with the boundaries also as described in the act of August 6, 1846, was accepted by the people but with the proviso (art. 2, sec. 1) that, if Congress approved, the boundary line should run southwesterly from the foot of the rapids of the St. Louis River to the mouth of the Rum River, thence down the Mississippi River as previously described. This boundary would have added materially to the area of the State had it been accepted by Congress.

"Congress accepted the constitution dated February 1, 1848, without action on the proviso and by act approved May 29, 1848 (9 Stat. L. 233), admitted Wisconsin as a state.

---

*This paragraph is quoted exactly as it was printed in 1930 (see p. 425).

"The admission of Wisconsin to statehood left an area of more than 30,000 square miles west of the St. Croix River, east and north of the Mississippi River, practically without a government. The settlers organized a temporary government and elected a Delegate to Congress who was admitted as the representative of the "Territory of Wisconsin". This area became a part of the Territory of Minnesota by congressional act of March 3, 1849.

"The State of Minnesota in 1916 instituted a suit against Wisconsin in the United States Supreme Court in order to have that part of the State boundary line from St. Louis Bay up the St. Louis River to the falls near Fond du Lac finally determined. The court handed down an opinion March 8, 1920 (252 U. S. 273), and on October 11, 1920, (254 U. S. 14) appointed commissioners to survey and mark the line. The survey was made on the ice during the winter of 1920-21, and the commissioners' report was confirmed by the court February 27, 1922, (258 U. S. 149). The line surveyed was 18.4 miles in length and was almost entirely over water. Rectangular coordinates were computed for each angle, and suitable reference marks were established on shore.

"The meridian boundary between Wisconsin and Minnesota from the St. Louis River to the St. Croix River was surveyed and marked in 1852 under the General Land Office.*

"   *   *   *

"The northern boundary (of Illinois, i. e., the southern boundary of Wisconsin) was surveyed and marked in 1831-32 by commissioners representing the United States and Illinois. The position on the east side of the Mississippi of a point in latitude 42° 30' having been found by observation, a stone about 7 feet long and of an estimated weight of 5 tons was set in the ground on the high-water line. The stone was marked 'Illinois' on its south side and 'Michigan latitude 42° 30' N.' on its north side. (The Mississippi River Commission later located this stone or one on the State line near it and determined its latitude as 42° 30' 29.3".) From this point the line was run east to the fourth principal meridian of the General Land Office, where a large mound of earth was erected,

---

*The publication quoted fails to say that the western boundary of Wisconsin was fixed in 1848 in accordance with the provisions of the Act of Aug. 6, 1846 (U. S. Statutes at Large, Vol. 9, pp. 56-57). From the mouth of the St. Louis River the boundary runs "up the main channel of said river to the first rapids in the same, above the Indian village, according to Nicollet's map; thence due south to the main branch of the River St. Croix; thence down the main channel of said river to the Mississippi; thence down the centre of the main channel of that river to the north-west corner of the State of Illinois; thence due east with the northern boundary of the State of Illinois to the place of beginning" [i. e., the northeast corner of Illinois].

and was continued east to the Rock River. Observations then taken showed that the line was 54″ too far north. An offset was taken the proper distance to the south, and a post was set on the east bank of the river, 81 miles 31 chains 9 links from the Mississippi, from which the line was extended (with frequent astronomic observations) to Lake Michigan, where an oak post 12 inches square and 9 feet long was set 5 feet in the ground at a point about 1 chain from the lake shore. Recent observations show that this end of the marked line is about half a mile south of the parallel of 42° 31′. The total length of the boundary as measured is 144 miles 48 chains 80 links. A post was also set on the east bank of the Fox River 125 miles 9 chains 10 links from the initial point. There is a signed copy of the report and notes in the files of the General Land Office (Boundaries, No. 22. See U. S. Geol. Survey Bulls. 310, 551, and 644 for latitude and longitude of points on this line). The line west of the Rock River was later rerun and placed in a corrected position."

Certain geographical relationships of these boundaries have been discussed in this book (see pp. 43, 171, 422, and 447). The territory and state of Wisconsin have at various times been attached to Indiana, Illinois, and Michigan. Wisconsin has included Minnesota, Iowa, and the Dakotas. It has possessed and lost the sites of Chicago, St. Paul, and Minneapolis, the iron mines of Minnesota, the iron and copper mines of Michigan, the coal and lead mines of northern Illinois, and the rich corn and wheat lands of Iowa, of northern Illinois, and of the Red River Valley in Minnesota and the Dakotas.

APPENDIX C

AREAS AND POPULATIONS OF COUNTIES IN WISCONSIN

WITH NAMES OF COUNTY SEATS, AND DISTANCES OF EACH FROM MADISON—THE STATE CAPITAL—AND FROM MILWAUKEE —THE LARGEST CITY

An excellent publication containing much geographical information about the boundaries of Wisconsin counties and the origin of their names is Louise P. Kellogg's "Organization, Boundaries, and Names of Wisconsin Counties" (Proc. Wis. Hist. Soc., Vol. 57, 1909, pp. 184-231). See also Wisconsin Statutes, 1911, pp. 3-16.

The following table is quoted from the 15th Census and the 1923 edition of the Official Railroad Map of Wisconsin, prepared under the direction of the Railroad Commission of Wisconsin.

County	Land Area in Square Miles	1930 Population	County Seat	Distance From	
				Madison	Milwaukee
Adams	684	8,003	Friendship	126	125
Ashland	1,082	21,054	Ashland	259	367
Barron	885	34,301	Barron	238	296
Bayfield	1,503	15,006	Washburn	372	380
Brown	529	70,249	Green Bay	156	112
Buffalo	687	15,330	Alma	186	250
Burnett	860	10,233	Grantsburg	339	397
Calumet	324	16,848	Chilton	160	78
Chippewa	1,039	37,342	Chippewa Falls	193	250
Clark	1,218	34,165	Neillsville	154	211
Columbia	778	30,503	Portage	37	93
Crawford	579	16,781	Prairie du Chien	98	180
Dane	1,202	112,737	Madison	....	82
Dodge	897	52,092	Juneau	59	57
Door	469	18,182	Sturgeon Bay	213	170
Douglas	1,337	46,583	Superior	336	371
Dunn	869	27,037	Menomonie	209	266
Eau Claire	638	41,087	Eau Claire	183	240
Florence	497	3,768	Florence	290	263
Fond du Lac	726	59,883	Fond du Lac	90	63
Forest	1,017	11,118	Crandon	256	229
Grant	1,169	38,469	Lancaster	86	168
Green	593	21,870	Monroe	37	105
Green Lake	360	13,913	Green Lake	117	90
Iowa	781	20,039	Dodgeville	47	129
Iron	792	9,933	Hurley	407	325
Jackson	990	16,468	Black River Falls	127	184
Jefferson	552	36,785	Jefferson	36	52
Juneau	802	17,264	Mauston	73	129
Kenosha	282	63,277	Kenosha	115	33
Kewaunee	337	16,037	Kewaunee	193	149
La Crosse	481	54,455	La Crosse	133	197
Lafayette	642	18,649	Darlington	107	138

County	Land Area in Square Miles	Population 1930	County Seat	Distance From Madison	Distance From Milwaukee
Langlade	875	21,544	Antigo	212	185
Lincoln	902	21,072	Merrill	197	209
Manitowoc	602	58,674	Manitowoc	159	77
Marathon	1,554	70,629	Wausau	177	187
Marinette	1,415	33,530	Marinette	205	178
Marquette	457	9,388	Montello	60	116
Milwaukee	235	725,263	Milwaukee	82	
Monroe	937	28,739	Sparta	108	171
Oconto	1,118	26,386	Oconto	185	158
Oneida	1,183	15,899	Rhinelander	259	231
Outagamie	646	62,790	Appleton	126	99
Ozaukee	233	17,394	Port Washington	107	25
Pepin	236	7,450	Durand	212	270
Pierce	563	21,043	Ellsworth	274	331
Polk	935	26,567	Balsam Lake	296	353
Portage	812	33,827	Stevens Point	108	159
Price	1,279	17,284	Phillips	323	268
Racine	324	90,217	Racine	105	23
Richland	590	19,525	Richland Center	59	141
Rock	716	74,206	Janesville	39	71
Rusk	925	16,081	Ladysmith	267	263
St. Croix	735	25,455	Hudson	249	306
Sauk	842	32,030	Baraboo	37	119
Sawyer	1,320	8,878	Hayward	291	349
Shawano	1,158	33,516	Shawano	185	152
Sheboygan	521	71,235	Sheboygan	134	52
Taylor	991	17,685	Medford	287	226
Trempealeau	748	23,910	Whitehall	169	226
Vernon	821	28,537	Viroqua	143	207
Vilas	934	7,294	Eagle River	270	243
Walworth	560	31,058	Elkhorn	69	57
Washburn	835	11,103	Shell Lake	272	329
Washington	431	26,551	West Bend	116	34
Waukesha	549	52,358	Waukesha	62	20
Waupaca	759	33,513	Waupaca	186	131
Waushara	646	14,427	Wautoma	145	118
Winnebago	459	76,622	Oshkosh	108	81
Wood	809	37,865	Wisconsin Rapids	133	160

The total population of the 71 Wisconsin counties in 1930 was 2,939,006. In 1840 it was only 30,945, and in 1836, 11,000 people. The average number of inhabitants per square mile in 1930 was 53.2 as compared with 47.6 in 1920. The state now ranks thirteenth in population in the United States. The density of population in each county in 1930 and the relation to the geographic provinces of the state is shown in Figure 68. The center of population in 1930 still remained in eastern Marquette County. In 1920 it was 2.6 miles south of Neshkoro.

APPENDIX D

AREAS OF STATE PARKS, FORESTS, INDIAN RESERVATIONS, MILITARY RESERVATIONS, PUBLIC LANDS AND EDUCATIONAL LANDS IN WISCONSIN

Wisconsin State Parks. The establishment and maintenance of a system of state parks is rightly accorded a definite and important place in the administration of the conservation program in Wisconsin. The basic principle underlying the administration of the conservation program is proper land utilization. Certain areas in Wisconsin, or in any state, are better adapted to state park use than to any other.

There are three principal reasons for the establishment of state parks. They are:

1. To perpetuate some spot of intense historic interest of significance to the entire state.
2. To preserve some area of outstanding beauty or wonder.
3. To provide recreation grounds of unusual importance.

When two or more of these reasons can be fulfilled by one area, there is all the more reason to establish and maintain it as a state park.

Wisconsin today has 14 state parks varying in size from two to 36,000 acres, and comprising in all approximately 50,000 acres. These parks are well distributed throughout the state with the exception of the southeastern district which is the area of heaviest population and in which are located but two comparatively small parks.

Practically every type of scenic beauty peculiar to Wisconsin or the Middle West is represented in at least one of the state parks and scenes in some of them remind one of the Rocky Mountains or the rugged coast of northern New England.

Places of outstanding scenic value are contained within Wisconsin's state parks. Among these are Devil's Lake, Rib Mountain, the highest known point in the state; Manitou Falls, the highest waterfall in the state; the dalles of the St. Croix; Trempealeau mountain; some high bluffs at the junction of the Wisconsin and Mississippi rivers; a section of beautiful pine-covered dune land on the Lake Michigan shore; Government Bluff and rolling wooded country in Door county; and sections of the beautiful lake country of northern Wisconsin.

Three parks in the southern part of the state are interesting primarily because of their historic interest. They are First Capitol State Park near Belmont; Tower Hill State Park, the site of the first shot tower in Wisconsin, near Spring Green; and Cushing Memorial State Park, the ancestral home of the three Wisconsin Cushings, in Waukesha County.

Wisconsin was the first state to establish any area and name it a state park. "The State Park", an area of some 50,000 acres of forest lands located in what was then part of Lincoln county, was created by legislative act in 1878. It existed for 19 years, but met a sad fate in 1897 when by legislative act the lands were sold to lumber companies. Part of this same area was later repurchased and is now contained in Northern State Forest.

Just thirty years ago the state made a new beginning in acquiring state parks and since that time 16 areas have been set aside as state parks.

The appointing of a committee in 1899 by Governor Edward Scofield to investigate park possibilities of the St. Croix river region in Polk County was the first official act in the new movement. Acquisition of lands in this area began in 1900 and a park was established here in the same year in co-operation with the State of Minnesota which established a park on the opposite side of the river. The two areas, one on either side of the river, are known as Interstate Park.

The legislature of 1907 created the first State Park Board which reported to Governor James O. Davidson in favor of establishing a state park system. During the next few years two additional parks were acquired, Devil's Lake State Park in Sauk County and Peninsula State Park in Door County.

In 1913, the State Board of Forestry co-operated with the State Park Board to develop and improve the parks. One forester and several rangers were furnished by the State Board of Forestry to locate and construct roads and trails within the parks and prepare maps of the areas.

On July 1, 1915, the State Park Board, the State Board of Forestry, the Fisheries Commission, and the State Game Warden Department were consolidated to form the Conservation Commission. Since 1915 administration of the constantly increasing number of state parks has come under the jurisdiction of the Conservation Commission.

State parks answer a very definite purpose in the complicated life of modern America. They provide playgrounds and vacation lands for all of the people. They furnish places for the casual afternoon picnic as well as for the tourist who wishes to stay overnight or for a week. Some of Wisconsin's parks offer unexcelled opportunities for the camper who goes out for several weeks to live the life of several generations ago.

In addition, each state park might be called a "conservation area" for here ruthless destruction of natural resources and natural beauties is not tolerated. Nature is permitted to regulate her own affairs in most state parks, and the educational value of such areas to the people of the state is inestimable.

WISCONSIN STATE PARKS

Name of Park	Location (County)	Size (Acres)	How acquired	Year Established	Highway
Interstate	Polk	580	Purchase	1900	35, 8, 87
Peninsula	Door	3,400	Purchase	1910	17
Devil's Lake	Sauk	1,400	Purchase	1911	12, 113, 159
Cushing Memorial	Waukesha	8	Gift	1915	18
Nelson Dewey	Grant	1,650	Purchase	1917	35, 60, 18
Perrot	Trempealeau	910	Gift	1918	167
Pattison	Douglas	660	Gift	1920	35
Tower Hill	Iowa	60	Gift	1922	11
First Capitol	Lafayette	2	Gift	1924	118, 80
Rib Mountain	Marathon	160	Gift	1927	51, 29
Potawatomi	Door	1,100	Purchase	1928	17, 78
Terry Andrae	Sheboygan	112	Gift	1928	141
American Legion	Oneida	36,000	Purchase	1929	47
Copper Falls	Ashland	520	Purchase	1929	13, 77

Publicly Owned Forests. Under present plans approximately three million acres of land in Wisconsin will be developed in governmentally owned and established forests. Over 1,000,000 acres are included in state forest purchase areas, of which the state now owns approximately 300,000 acres. The federal government has been authorized to purchase not to exceed 1,000,000 acres of Wisconsin land for national forests. Eight counties have established county forests on lands obtained through tax delinquency.

Indian Reservations. The Federal Indian Reservations are as follows, (see Report of the Commissioner of Indian Affairs to the Secretary of the Interior, Washington, 1930, and General Data concerning Indian Reservations, 1930.)

Reservation	County	Area in Acres	Number of Indians (1930)	Tribe
La Pointe	Ashland	129,898	1,171	Chippewa
Red Cliff	Bayfield	14,142	584 (1928)	Chippewa
La Court D'Oreilles	Sawyer	69,590	1,532(a)	Chippewa
Lac du Flambeau	Vilas-Iron	70,119	827	Chippewa
Menominee	Shawano-Oconto	231,680	1,928	Menominee
Stockbridge	Shawano	8,920	599 (1910)	Stockbridge & Munsee
Oneida	Outagamie-Brown	65,553	3,046	Oneida
Rice Lake Band			221	Chippewa
Tomah School (b)			1,378	Winnebago
Potawatomi(c)	Forest	14,556	419	

(a) Includes Hayward School.
(b)The Hayward, Tomah, and Carter schools are not within any of the Indian Reservations.
(c) Not a reservation.

The total area of Indian lands in 1929 was 604,458 acres, of which 330,874 were allotted, and 273,584 acres were unallotted.

The total Indian population of the state in 1930 was 11,548.

Many geographical relationships of the Indians in Wisconsin in early days are found in F. J. Turner's "Character and Influence of the Fur Trade in Wisconsin" (Proc. Wis. Hist. Soc., Vol. 36, 1889, pp. 52-98; reprinted as "Character and Influence of the Indian Trade in Wisconsin," Johns Hopkins University Studies, Vol. 9, 1891, 75 pp). See, also, J. G. Shea's "Indian Tribes in Wisconsin," (Collections Wis. Hist. Soc., Vol. 3, 1857, pp. 125-138.)

Military Reservations. The State military reservation at Camp Douglas in Juneau County covers 1200 acres.

The Federal Military reservation, at Camp McCoy in Monroe County between Sparta and Tomah, covers 14,111 acres.

Educational Lands. In the lands held for educational purposes are included the school lands in 1931 amounting to 12,318 acres, the normal school lands (formerly swamp lands) amounting to 156 acres and agricultural college lands amounting to 40 acres.

APPENDIX E
WISCONSIN MAPS

The State as a Whole. The following list describes the more important maps of the whole state and tells where they may be obtained.

Name of Map	Scale	Size in Inches	Cost	Where Obtainable
Base map of Wisconsin (1911)	8 miles to 1 inch	43 by 45	.25	U. S. Geol. Survey, Wash., D. C.
Base map of Wisconsin (1911)	16 miles to 1 inch	21 by 24	.05	U. S. Geol. Survey, Washington, D. C.
Geological and Road Map of Wisconsin (1911) a	6 miles to 1 inch	51 by 59	1.00	Bureau of Purchases, Madison
Geological Map of Wisconsin (1929)	16 miles to 1 inch	21 by 19	.30	Wis. Geol. Survey, Madison
State Trunk Highway Map of Wisconsin	6 miles to 1 inch	51 by 59	2.00	Bureau of Purchases, Madison
Relief Model of Wisconsin (1910)c	7 miles to 1 inch	42 by 46	d	Univ. of Wis., Madison
Geological Model of Wisconsin (1910)	7 miles to 1 inch	42 by 46	d	Univ. of Wis., Madison
Model of Wisconsin showing Glacial Deposits (1930)	7 miles to 1 inch	42 by 46	d	Univ. of Wis., Madison
Relief Map of Wisconsin (1915)	16 miles to 1 inch	20 by 21	------	Pl. I in this bulletin. Also issued separately
Railroad map of Wisconsin (1923)	10 miles to 1 inch	33 by 36	Out of Print	Railroad Commission of Wisconsin, Madison
Land Office Map of Wisconsin (1912)	12 miles to 1 inch	25 by 29	.25	U. S. Land Office, Wash., D. C.
Post Office Map of Wisconsin and Michigan	9 miles to 1 inch	46 by 60	1.60	U. S. Post Office, Wash., D. C.
Political Map of Wisconsin	16 miles to 1 inch	19 by 26	.25	Rand, McNally & Co., Chicago
Political Map of Wisconsin	12 miles to 1 inch	28 by 38	1.00	Rand, Mc Nally & Co., Chicago
Political Map of Wisconsin	10 miles to 1 inch	38 by 42	2.50	A. J. Nystrom & Co., Chicago
Blackboard Outline Map of Wisconsin	6 miles to 1 inch	48 by 60	2.60	A. J. Nystrom & Co., Chicago
Political Map of Wisconsin	10 miles to 1 inch	29 by 35	2.50	National Map Co., Indianapolis, Ind.

(a) Earlier colored geological maps of the state were published as follows: For places of publication of these maps see Appendix G.

Author	Date	Scale of Map
D. D. Owen	1851	20 miles to one inch
I. A. Lapham	1855	25 miles to one inch
James Hall	--------	22 miles to one inch
I. A. Lapham	1869	15 miles to one inch
I. A. Lapham	1874	36 miles to one inch
T. C. Chamberlin	1878	24 miles to one inch
R. D. Irving	1879	20 miles to one inch
T. C. Chamberlin	1881	15 miles to one inch

(b) Cloth backed and mounted. The same on paper 30c.
(c) Earlier models were published as follows; (1) in 1882 by F. H. King (editions issued to show separately (a) geology, (b) glacial deposits, (c) forests, and (d) soils.) (2) in 1906 by E. C. Case (for information apply to Central Scientific Co., Chicago.)
(d) For prices and descriptions, address Business Manager, Univ. of Wisconsin, Madison.

Relief models of two portions of the state have been made and issued by the University of Wisconsin. Photographs of these models appear as Plates XXX and XXXI in this book. These are:

The Lake Superior Region, 10 miles to 1 inch, model 70 by 44 inches, shows topography in parts of Wisconsin, Michigan, Minnesota and Canada.

The Baraboo District, 2 inches to 1 mile, model 44 by 60 inches, shows topography and geology.

For information address the Business Manager, University of Wisconsin, Madison, Wisconsin.

The following maps of part or all of Wisconsin are on more specialized subjects. Some of them were made many years ago and are not easily obtained at present, but may be consulted in the larger libraries:

Name of Map (a)	Date	Scale	Size in inches	Published as
Quaternary Formations	1881	15 miles to 1 inch	23 by 27	Atlas Pl. II: Geol. of Wis.
Native Vegetation	1882	15 miles to 1 inch	23 by 27	Atlas Pl. IIA: Geol. of Wis.
Soils	1882	15 miles to 1 inch	23 by 27	Atlas Pl. IIB: Geol. of Wis.
Rainfall and Temperature	1882	15 miles to 1 inch	23 by 27	Atlas Pl. IIC: Geol. of Wis.
Geological Map, showing Quarries	1898	20 miles to 1 inch	16 by 20	Pl. I, Bull. 4, Wis. Geol. and Nat. Hist. Survey.
Forests of Northern Wisconsin	1898	24 miles to 1 inch	9 by 15	Pl. I, Bull. 1, Wis. Geol. and Nat. Hist. Survey.
Clay and Shale Map	1901	21 miles to 1 inch	16 by 20	Pl. I, Bull. 7, Wis. Geol. and Nat. Hist. Survey.
Wisconsin Clays	1906	21 miles to 1 inch	16 by 20	Pl. II, Bull. 15, Wis. Geol. & Nat. Hist. Survey.
Soils	1908	55 miles to 1 inch	16 by 7	Pl. II, Bull. 20, Wis. Geol. and Nat. Hist. Survey
Creameries and Cheese Factories	1910	9 miles to 1 inch	43 by 31	Bull. 210, Wis. Agr. Exp. Sta.
Geology of the Lake Superior Region	1911	16 miles to 1 inch	35 by 23	Pl. 1, Monograph 52, U. S. Geol. Survey.
Swamp Lands	1915	25 miles to 1 inch	13 by 15	Pl. 1, Bull. 45, Wis. Geol. and Nat. Hist. Survey.
Geological Map, showing Artesian Water Conditions	1915	16 miles to 1 inch	19 by 21	Pl. 1, Bull. 35, Wis. Geol. and Nat. Hist. Survey.
Soils North Half of Wisconsin	1921	6 miles to 1 inch	29 by 49	Pl. in Bull. 55, Wis. Geol. and Nat. Hist. Survey.
Geological Map, showing Location of Limestone and Clay Analyses	1924	16 miles to 1 inch	19 by 21	Pl. 6, Bull. 66, Wis. Geol. and Nat. Hist. Survey.
General Soils Map of Wisconsin	1926	10 miles to 1 inch	31 by 36	Pl. in Bull. 68, Wis. Geol. and Nat. Hist. Survey
Index to Atlas Sheets, Wisconsin	1929	16 miles to 1 inch	19 by 21	U. S. Geol. Survey, Washington, D. C. (Free).

(a) The maps listed in this table are all good-sized wall maps, usually in colors. For smaller black and white maps showing geographical distributions see the following:

Farm animals, crops, etc., in Wisconsin in 1905—Bull. 26, Wis., Geol. and Nat. Hist. Survey, 1913:

Ibid., Wisconsin Blue Book, 1915 ; Ibid., in Merrill's Industrial Geography of Wisconsin, 1911.

Farm animals, crops, etc., in 1910—Unbound pamphlet, Wis. Agr. Exp. Sta., 1915.

Population, improved land, etc., Abstract, U. S. Census, 1930 with Supplement for Wisconsin.

Population, crops, farm animals, gaging stations, judicial circuits, Wisconsin senate districts and federal congressional districts, Wisconsin Blue Book, Madison, 1929, especially maps on pp. 52-74, 93, 484, 516, 616.

Name of Quadrangle	Number on Figure	Name of Quadrange	Number on Figure
Alma	64	Mondovi	70
Baraboo	37	Monroe	8
Bayview	21	Muskego	20
Black River Falls	68	Neenah	58
Blair	67	Neshkoro	55
Blanchardville	13	New Glarus	14
Blue Mounds	30	North Bend	59
Briggsville	43	Oconomowoc	24
Brodhead	7	Portage	42
Cross Plains	29	Port Washington	41
Delavan	4	Poynette	38
Denzer	36	Prairie du Chien	33
Durand	71	Racine	1
Eagle	19	Richland Center[a]	31
Elkader (Iowa-Wis.)[a]	12	Ripon	56
Evansville	15	Robbins[d]	75
Ferryville	34	St. Croix Dalles (Wis.-Minn.)	74
Fond du Lac	57	Shopiere	5
Fountain City	61	Silver Lake	2
Galesville	60	South Wayne	9
Gays Mills	35	Sparta	51
Geneva-Racine[b]		Stoddard	48
Gilmanton	65	Stoughton	16
Gogebic Iron Range[c]	78	Strum	69
Hartford	39	Sun Prairie	27
Hillsboro	45	Superior	79
Iron River (Mich.-Wis.)	77	The Dells	44
Janesville	6	Three Lakes[d]	76
Kendall	53	Tomah	52
Koshkonong	17	Viroqua	47
La Crosse	50	Wabasha	63
La Crescent	49	Waterloo	26
La Farge	46	Watertown	25
Lake Geneva	3	Waukesha	23
Lancaster[a]	11	Wausau Special[a]	73
Madison	28	Wauzeka	32
Marathon Special[a]	72	West Bend	40
Mauston	54	Whitehall	66
Milwaukee	22	Whitewater	18
Milwaukee Special	22[a]	Winnebago Special[e]	57 & 58
Minneiska	62		
Mineral Point[a]	10		

[a] Scale of 1:125,000.
[b] Bayview, Eagle, Lake Geneva, Muskego, Racine, and Silver Lake sheets, on the scale of 1:62,500, have been reduced and form the Geneva-Racine double sheet, on the scale of 1:125,000. It sells for 20 cents, or 12 cents when included in wholesale orders.
[c] Advance sheets on scale of 1:24,000. Contour interval 10 feet.
[d] Aerial base maps. Scale 1:48,000.
[e] Neenah and Fond du Lac sheets form the Winnebago Special, price 20 cents.

U. S. Geological Survey Maps. The United States Geological Survey, in cooperation with the Wisconsin Geological Survey, has published the quadrangles outlined in Figure 197, showing topography. Parts of these quadrangles are reproduced in Figure 15, etc. They are sold for 10 cents each, or $3 for 50 maps, on application to The Director, U. S. Geological Survey, Washington, D. C., or from The University Cooperative Co., 702 State St., Madison, or the Caspar Krueger Dory Co., 772 N. Water St., Milwaukee. They may also be purchased from the postmasters in certain towns. The names of the sheets or quadrangles thus far published are listed on p. 496. All of them are on the scale of 1:62,500, or about an inch to the mile, except as noted. Each sheet is 16½ by 20 inches. A large index map of these quadrangles, or atlas sheets, may be obtained, free, by writing The Director, U. S. Geological Survey, Washington, D. C.

On the index map (Fig. 197), these quadrangles bear the arbitrarily-chosen numbers given in the alphabetical list on p. 496. In ordering these sheets or quadrangles it is necessary to refer to them by name rather than by number and to pay in advance.

The topographic sheets of the United States Geological Survey are described as follows:

"The features shown on these atlas sheets or maps may be classed in three groups—(1) *water,* including seas, lakes, rivers, canals, swamps, and other bodies of water; (2) *relief,* including mountains, hills, valleys, and other elevations and depressions; (3) *culture* (works of man), such as towns, cities, roads, railroads, and boundaries. The conventional signs used for these features are explained below. Variations appear on some earlier maps.

"All water features are printed in *blue,* the smaller streams and canals in full blue lines and the larger streams, lakes, and the sea in blue water-lining. Intermittent streams—those whose beds are dry at least three months in the year—are shown by lines of dots and dashes.

"Relief is shown by contour lines in *brown.* A contour on the ground passes through points that have the same altitude. One who follows a contour will go neither uphill nor downhill but on a level. The contour lines on the map show not only the shapes of the hills, mountains and valleys but also their elevations. The line of the sea coast itself is a contour line, the datum or zero of elevation being mean sea level. The contour at, say, 20 feet above sea level would be the shore line if the sea were to rise or the land to sink 20 feet. On a gentle slope this contour is far from

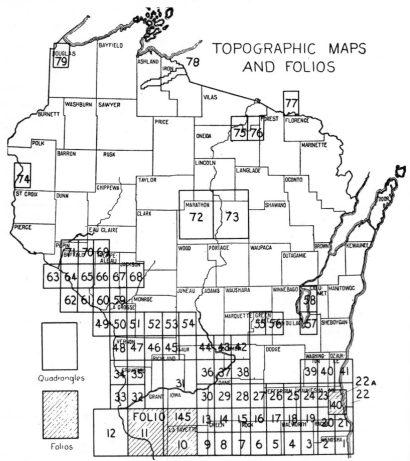

Fig. 197. Index map showing the location of areas in Wisconsin where there are topographic maps by the United States Geological Survey. The numbers on this map correspond to those in the list on page 496, where the names of these quadrangles are given. Ruled areas are covered by folios with geological as well as topographic maps.

the present coast; on a steep slope it is near the coast. Where successive contour lines are far apart on the map they indicate a gentle slope; where they are close together they indicate a steep slope; and where they run together in one line they indicate a cliff.

"The manner in which contour lines express altitude, form, and grade is shown in the Figure 198.

"The sketch represents a river valley between two hills. In the foreground is the sea, with a bay that is partly inclosed by a hooked sand bar. On each side of the valley is a terrace into which small streams have cut narrow gullies. The hill on the right has a rounded summit and gently sloping spurs separated by ravines.

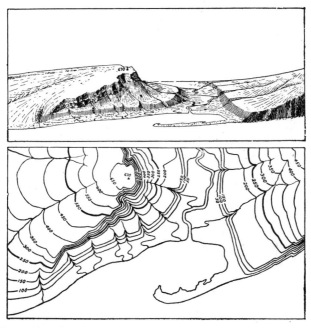

Fig. 198. Perspective sketch of hills, valley, cape, bay, and cliffs; map showing the same features by contour lines. Contour interval 50 feet. (U. S. Geol. Survey.)

The spurs are truncated at their lower ends by a sea cliff. The hill on the left terminates abruptly at the valley in a steep scarp. It slopes gradually back away from the scarp and forms an inclined table-land, which is traversed by a few shallow gullies. On the map each of these features is indicated, directly beneath its position in the sketch, by contour lines.

"The contour interval, or the vertical distance in feet between one contour and the next, is stated at the bottom of each map. This interval differs according to the character of the area mapped; in a flat country it may be as small as 5 feet; in a mountainous region it may be 250 feet. Certain contour lines, every fourth or fifth one, are made heavier than the others and are accompanied by figures stating elevation above sea level. The heights of many points, such as road corners, summits, surfaces of lakes, and bench marks, are also given on the map in figures, which ex-

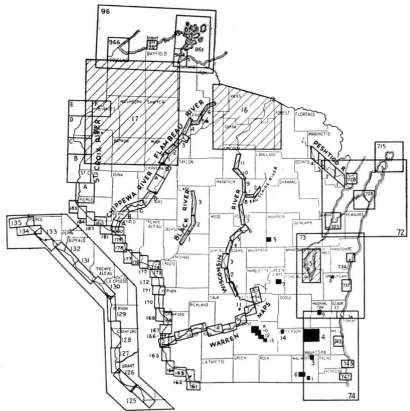

Fig. 199. Index map showing the location of areas in Wisconsin where there are hydrographic maps—charts and maps of the coast, lakes, and rivers. The numbers on this map correspond to those in the lists on pages 502 to 506.

press the elevations to the nearest foot only. More exact elevations of bench marks, as well as geodetic coordinates of triangulation stations, are published in bulletins issued by the Geological Survey. A bulletin pertaining to any state may be had on application.

"The works of man are shown in black, in which color all lettering also is printed. Boundaries, such as those of a State, county, city, land grant, township, or reservation, are shown by continuous or broken lines of different kinds and weights. Public and through roads are shown by fine double lines; private and poor roads by dashed double lines; trails by dashed single lines."

Within the state the U. S. Geological Survey in cooperation with the Wisconsin Geological Survey, issued profile river maps in 1907, supplementary to Water Supply Paper 207. The scale is 1:24,000 or about $2\frac{1}{2}$ inches to the mile. The contour interval is 5 or 10 feet on the land and 1 or 5 feet on river surfaces. The stock of these maps is now deposited with the Wisconsin Geological Survey and may be purchased for 5 cents per sheet. These profile river maps cover parts of the rivers listed below.

River	Number of sheets	Area covered
Wisconsin	11	Kilbourn to Tomahawk
Eau Claire	2	Mouth to Dells
Peshtigo	4	Peshtigo to Strong Falls
Black	3	Black River Falls to Soo Line
Flambeau	6	Mouth to Turtle River

The U. S. Geological Survey has republished these maps in Water Supply Paper 417. In addition, the following sheets are in-included in this report.

Profile of Chippewa River, mouth to Chippewa Falls—1 sheet.

Plan and profile of Chippewa River, Chippewa Falls
   to Flambeau, Wisconsin ....4 sheets.

New river plan and profile sheets were published in 1929. Scale
   1:24,000.

Chippewa River from mouth to Chippewa Falls. Contour interval on land 10 feet, on river surface 5 feet—4 plan sheets,
   2 profile sheets.

Chippewa River from Flambeau River to Chippewa Reservoir.
   Contour interval 5 feet—4 plan sheets, 1 profile sheet.

St. Croix River from mouth to a point eleven miles above Dan-
   bury. Contour interval 5 feet—6 plan sheets.

The United States Geological Survey has also issued detailed geological (Figs. 197, 202) and glacial (Figs. 203, 204) maps of parts of Wisconsin in recent years (p. 533), see Monographs 5, 19, 52, and 53, Folios 140 and 145, Bulletins 273 and 294. Water

Supply Papers 207, etc., and Professional Papers 34, 106, and 154A. For information address The Director, U. S. Geological Survey, Washington, D. C.

U. S. Lake Survey Charts. The United States Lake Survey (War Department, Corps of Engineers, Survey of the Northern and Northwestern Lakes) has published colored charts of the coasts of Lakes Superior and Michigan, as outlined in Figure 199. Part of one of these charts is reproduced as Figure 195. They may be purchased from The U. S. Lake Survey Office, Old Custom House, Detroit, Mich., at the prices indicated in the table. In ordering these charts it is necessary to refer to them by number.

Chart No.	Name of U. S. Lake Survey Chart	Scale	Price	Remarks
9	Lake Superior	1:500,000	.40	Shows the whole lake
96	Wisconsin portion of Lake Superior	1:120,000a	.40	Includes inset map of Port Wing.
961	Apostle Islands	1:60,000a	.40	
964	Ashland Harbor	1:15,000	.20	Includes Washburn Harbor.
966	Superior Harbor	1:18,000	.40	Includes Duluth Harbor.
7	Lake Michigan	1:500,000	.40	Includes Green Bay
70	North End of Lake Michigan	1:240,000	.40	North of Kewaunee
71	North end of Green Bay	1:120,000	.40	
72	South end of Green Bay	1:120,000	.40	Includes part of Lake Michigan north of Kewaunee
⌊73	Lake Winnebago and Lower Fox River	1:120,000	.40	Includes Lake Michigan from Kewaunee to Port Washington
74	Lake Michigan coast from Kewaunee to Waukegan, Ill.	1:120,000	.40	Includes inset map of Port Washington
715	Entrance to Green Bay	1:40,000	.40	Contour map of Washington Island
723	Menominee-Marinette Harbor	1:15,000	.20	
725	Head of Green Bay	1:25,000	.20	Includes Fox River below DePere
728	Sturgeon Bay Canal	1:30,000	.20	Includes inset map of Sturgeon Bay
734	Manitowoc Harbor	1:8,000	.20	
737	Sheboygan Harbor	1:10,000	.20	
743	Milwaukee Harbor	1:12,000	.40	
745	Racine Harbor	1:15,000	.20	Includes Racine Bay
747	Kenosha Harbor	1:10,000	.20	

aThe scale of 1:120,000 is about one and nine-tenths miles to the inch: the scale of 1:60,000 is nearly one mile to one inch, and the other scales are proportional.

Mississippi River Commission Maps. The land close to the Mississippi River in Wisconsin is mapped on excellent black and white sheets in two series, as shown in Figure 199. Part of one of the smaller scale maps is reproduced as Figure 45, on the scale of 1:63,360, or 1 inch to the mile. The Mississippi River Commission maps may be purchased from the Mississippi River Commission (Corps of Engineers, U. S. Army), Vicksburg, Miss. The eleven sheets of the small-scale map of the Mississippi valley in

Wisconsin sell for 10 cents each. Their dimensions are 15 by 24 inches. The twenty-five sheets of the larger-scale map sell for 27 cents each. Their dimensions are 25 by 39 inches.

The whole river in Wisconsin is also covered by a set of 4 sheets—Map of the Alluvial Valley of the Upper Mississippi River, from the Falls of the St. Anthony to the Mouth of the Ohio River. These maps cost 20 cents per sheet or 70 cents for the set. The two northern sheets of this map cover the Mississippi River in Wisconsin, as well as the Wisconsin and Fox rivers eastward to Lake Winnebago. The map shows (a) lands subject to overflow, and (b) bluff lines defining the limit of the alluvial valley. The scale is 5 miles to 1 inch.

The charts on the scale of 1:20,000 have 5-foot contours on the bottom lands and terraces, and 20-foot contours on the bluffs. This is the larger and better series. The maps on the scale of

City or village	Chart on scale of 3⅛ inches to the mile	Map on scale of one inch to the mile
Alma	179	132
Bagley	165	127
Bay City	182	133
Cassville	164	126
Cochrane	178	131
De Soto	169	128
Diamond Bluff	184	134
Ferryville	168	128
Fountain City	176	131
Genoa	170	129
Glen Haven	165	126
Hager	183	133
La Crosse	172	129, 130
Lynxville	167	128
Maiden Rock	182	133
Midway	173	130
Nelson	180	132
North La Crosse	172	130
Onalaska	173	130
Pepin	181	132
Prairie du Chien	166	127
Prescott	185	135
Stockholm	181	133
Stoddard	171	129
Trempealeau	174	130
Victory	170	129
Wyalusing	165	127

1: 63,360 show topography graphically by hachures — short lines running straight up and down the slopes (see Fig. 55). The sheet of the Mississippi River Commission maps covering the neighborhood of each of the principal cities and villages along the Mississippi River in this state is indicated in the list on page 503. In ordering these maps it is necessary to refer to them by number.

The lower Wisconsin River is not shown in detail on any set of maps now easily available, except in the portion covered by U. S. Geological Survey maps (Fig. 197). A set of maps covering the whole of the lower Wisconsin was made long ago by the Engineer Department of the United States Army and may be consulted in the larger libraries. These are the 8 sheets of the Warren Survey of 1867, on the scale of 1⅜ inches to the mile. They extend from the mouth of the Wisconsin River at Prairie du Chien to the Fox River portage in Columbia County. The positions of the several sheets are shown in Figure 199, (see Warren, G. K., bibliography, page 208).

Wisconsin Geological Survey Maps. The maps issued by the state geological survey are of two sorts, one showing topography, the other geology, the latter being sometimes combined with topography.

The present Geological and Natural History Survey has published ten sheets of hydrographic maps of the principal lakes of southern and eastern Wisconsin. These were prepared under the direction of L. S. Smith, and cover the areas outlined in Figure 199.

Number	Name	Size of sheet in inches	Scale in inches per mile	Contour interval, in feet
1	Lake Geneva (Out of print)	17.5 by 10.8	2	10
2	Elkhart Lake	15.5 " 13.1	5	10
3	Lake Beulah	22.5 " 20.0	6	10
4	Oconomowoc—Waukesha Lakes	29.8 " 19.1	2	10
5	The Chain of Lakes, Waupaca	21.7 " 20.6	6	10
6	Delavan and Lauderdale Lakes	22.5 " 16.8	4	10
7	Green Lake	26.0 " 17.8	3.2	20
8	Lake Mendota	23.7 " 19.5	6	5
9	Big Cedar Lake	18.0 " 13.5	2.9	10
10	Lake Monona	17.6 " 17.3	4	5

In all of these maps the depth of the lakes is indicated by contour lines, and by tints in all except No. 1. They are sent on receipt of 10 cents each and may be had either mounted in a manila cover, or unmounted. Address *The State Geologist*, Madison, Wis. These

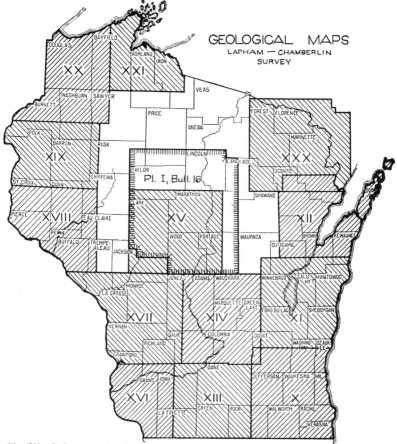

Fig. 200. Index map showing the location of the areas in Wisconsin which are covered by the atlas sheets of the Wisconsin Geological Survey of 1873-1879. These maps show geology without topography and are all on the scale of 3 miles to 1 inch. The numbers are those of the original sheets. The one exception (marked Pl. I, Bull, 16) is a geological map on the same scale published by the Wisconsin Geological and Natural History Survey in 1907.

maps, on a reduced scale, and five new sheets in addition, are included in Bulletin 27. Details concerning some of the shorelines of the inland lakes appear as black and white maps in Bulletin 8, which may be obtained for 50 cents.

In Bulletin 27 the maps on sheets 4 and 6 are split up into separate hydrographic maps of the following lakes:

Oconomowoc map:
    Nagawicka, Nashotah, and Nemahbin Lakes.
    Silver, Crooked, Otis, and Genesee Lakes.
    Pewaukee Lake.
    North, Pine, and Beaver Lakes.
    Mouse, Garvin, Okauchee, and Oconomowoc Lakes.
    Fowler Lake, and Lac la Belle.
Delavan-Lauderdale map:
    Delavan Lake.
    Lauderdale Lake.
The new sheets include the following:
    11. Hydrographic map of Devils Lake by F. T. Thwaites
    12. Lake Waubesa   15. Lake Winnebago
    13. Lake Kegonsa   16. Northeastern Lake District
    14. Rock Lake   17. Northwestern Lake District

The Geological and Natural History Survey has also issued 15 large scale topographic maps of the lead and zinc district, as outlined in Figure 201. Part of one of these maps is reproduced as Figure 63. These maps are in two uniform series, the first printed in 1906 as a supplement to Bulletin 14 by U. S. Grant. The bulletin is out of print but the maps may be obtained for 20 cents. The second series of maps, issued in 1909 by W. O. Hotchkiss and Edward Steidtmann—topography and geology combined—may be obtained for 6 cents additional. All these maps are on the scale of 4 inches to 1 mile with contour interval of 10 feet. The maps are listed below:

Plate	Name	Supplementary Plates
I	Highland Sheet	1. Cuba Sheet
III	Dodgeville Sheet	2. Big Patch—Elk Grove
V	Mineral Point Sheet	Sheet
VII	Mifflin Sheet	3. Ipswich Sheet
IX	Platteville Sheet	4. East M e e k e r s Grove
XI	Hazel Green—Benton Sheet	Sheet
XIII	Shullsburg Sheet	5. East M i n e r a l Point
XV	Meekers Grove Sheet	Sheet
XVII	Potosi Sheet	6. Montfort Sheet

County topographic highway maps have been published by the State Geological Survey. The scale is one inch to the mile, the contour interval 20 feet. The type of road surfacing is shown by red overprint. State and county trunk highways are indicated by symbols. The following maps are available: Green County—1926, Vernon County—1928, Trempealeau County—1929, La Crosse County—1931. Map of Buffalo County is in press. These maps are sold for 25 cents each, and may be obtained from the County Highway Commissioner or from the State Geological Survey.

Three areas in the state have state survey maps showing topography combined with geology. These maps are no longer available for distribution, but copies may be consulted in the larger libraries. The first of these areas is the eastern portion of the lead and zinc district in Lafayette and Green counties where parts of a more extensive series of maps, made in 1876 by Moses Strong, bridge over the gap between the Mineral Point and Brodhead

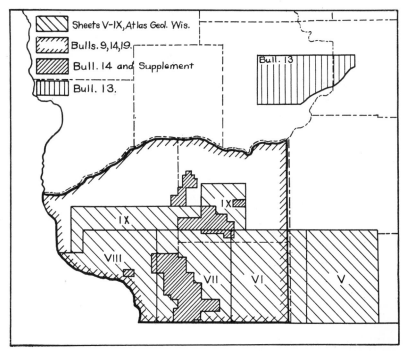

Fig. 201. Index map showing the location of the areas in southwestern Wisconsin where there are geological maps combined with topography. These are all maps by the state geological surveys. For list of maps in Bulletin 14 and supplement (fine oblique ruling) see p. 506.

Quadrangles (Fig. 197) of the U. S. Geological Survey. The maps are sheets V and VI of the Atlas accompanying the Geology of Wisconsin. They are on the scale of 1 inch to one mile with very satisfactory 50 foot contours. The areas covered are outlined in Figure 201.

Another area where topographic maps were made by the State Geological Survey is in the Penokee Range of Ashland and Iron counties (Fig. 202). Here large-scale maps, made in 1877 by R.

D. Irving, show the topography very well indeed from Penokee Gap to the Montreal River. The sheets are numbers XXIV, XXV, and XXVI of the Atlas accompanying the Geology of Wisconsin. They are on the scale of 3⅗ inches to one mile with contour interval of 20 feet. Sheet XXIII, a large scale topographic map of Penokee Gap, is reproduced in this book as Figure 152. New topographic maps of the Gogebic Range in Wisconsin are now available (Fig. 197).

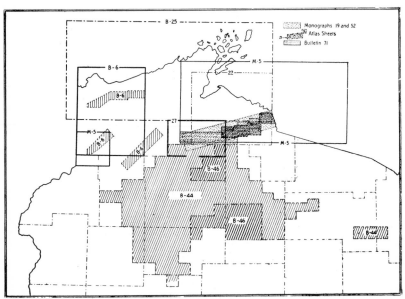

Fig. 202. Index map showing the location of the areas in northern Wisconsin where there are geological maps. The numbers correspond to those of the original publications. The areas marked Bull. 6, Bull. 25, and Bull. 44 are covered by maps in bulletins of the Wisconsin Geological and Natural History Survey; the areas marked Monograph 5 and Monographs 19 and 52 by United States Geological Survey monographs; the areas marked 22 to 27 by sheets in the Atlas of the Geology of Wisconsin.

A generalized topographic map of eastern Wisconsin includes the area not yet mapped by the United States Geological Survey. It was compiled in 1876 by T. C. Chamberlin and is Plate IV of the Atlas accompanying the Geology of Wisconsin. The scale is 12 miles to 1 inch and the contour interval is 100 feet.

Maps showing the glacial geology of parts of Wisconsin cover the areas outlined in the index sheets (Figs. 203, 204). Some of these were published by the state geological surveys, some by the U. S. Geological Survey, and some are in geological periodicals.

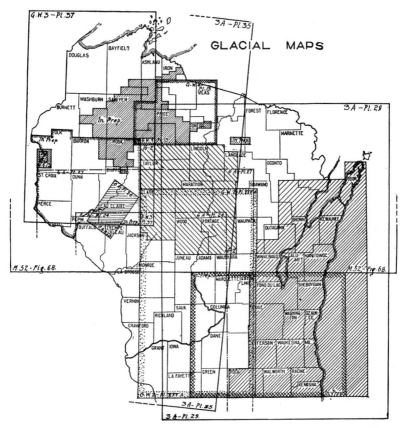

Fig. 203.   Index map with the location of the areas in Wisconsin where there are maps showing the glacial geology (see also Fig. 204).   GW3—Pl. 37 means Geology of Wisconsin, Vol. 3, 1880, Plate 37 facing p. 383; GW4—Pl. 13, Ibid., Vol. 4, 1882, Plate 13, facing p. 612; GW2—Pl. XXVA, Ibid., Vol. 2, 1877, Plate XXVA, facing p. 608; 3A—Pl. 35.   Third Annual Rept., U. S. Geol. Survey, 1883, Plate 35, facing p. 382; 3A—Pl. 29, Ibid., Plate 29, facing p. 316; 6A—Pl. 28, Ibid., Sixth Annual Rept., 1885, Plate 28, facing p. 306; 6A—Pl. 24, Ibid., Plate 24, facing p. 220; 6A—Pl. 27, Ibid., Plate 27, facing p. 259; M 52—Fig. 68, Ibid., Monograph 52, 1911, Fig. 68, p. 453; the same area is shown in Plates 1 and 2 of Professional Paper 154 A, U. S. Geol. Survey; J. G. 13, Journal of Geology, Vol. 13, 1905, Fig. 2, p. 243; A. G. 21, American Geologist, Vol. 21, 1898, Pl. 13, p. 146; 16—2, Bull. 16, Wis. Geol. and Nat. Hist. Survey, Plate 2; obliquely-ruled area in eastern Wisconsin, Geology of Wisconsin, Atlas Plate IV, 1876; two areas in northwestern Wisconsin marked In Prep., to be described in bulletins of the Wisconsin Geological and Natural History Survey; area in southeastern Wisconsin marked In Prep., Plate II in Professional Paper 106, U. S. Geol. Survey.   There is also a glacial map of the whole state, published as Atlas Plate IIA, Geology of Wisconsin, 1882.   A new glacial map was published on the relief map of the state in 1930.   The soils maps whose locations are shown on Figure 205 also give much information about glacial deposits, especially if used in connection with the key list on page 10 of the first edition of this book.   Maps showing glacial striae alone are not referred to on Figure 203.   Maps of the borders of the Glacial Great Lakes are given in Monographs 52 and 53, U. S. Geol. Survey, 1911 and 1915 (see Figures 115, 116, and 184 to 188 in this book).   This cut was not revised in 1931.

Much glacial and physiographic data may also be obtained from the county soil maps now covering nearly the whole state. In 1930 the detailed maps on the scale of 1 inch to the mile covered the areas outlined in Figure 205. There are reconnoissance maps of other parts of northern Wisconsin. These are on the scale of 3 miles to 1 inch. Bulletin 55 includes a map of the north half of the state, Bulletin 68 a general soil map of the state.

These maps and the accompanying bulletins may be obtained from Prof. A. R. Whitson, in charge of Soil Survey, Wisconsin Geological and Natural History Survey, Madison, or from the State Geologist.

Geological maps of 43 counties of Wisconsin were published in Hotchkiss and Steidtman's Bulletin 34, of the Geological and Natural History Survey, issued in 1914. These are colored maps on the scale of 6 miles to 1 inch, and show the locations of all quarries. The bulletin and maps may be obtained from the State Geologist for 50 cents.

A series of maps showing the roads, streams, and swamps in 87 townships in parts of Ashland, Bayfield, Washburn, Sawyer, Price, Oneida, Forest, Rusk, Barron, and Chippewa counties was published in 1915 in Bulletin 44 of the Wisconsin Geological and Natural History Survey. These are on the scale of $1\frac{5}{8}$ inches to the mile. Many of them contain leveled road profiles showing elevations. The bulletin and maps may be obtained from the State Geologist for $5.00.

Similar maps are available in Bulletin 46, 1929, price $1.00, and Bulletin 71, 1929, price $1.25.

The series of outline maps published in this Appendix (Figs. 197 to 205) is thought to refer to all maps of the state as a whole or of any parts of it that will be of use to the teacher or layman interested in the physical geography of Wisconsin. A very few maps showing details of the economic or structural geology of small areas are omitted. In such cases, however, the area is covered in some other map referred to in this Appendix. Those interested in the mineral resources of Wisconsin may wish to refer to some of these maps—such as the large-scale crevice maps of the lead and zinc region, and the blue prints of Florence County. Information concerning these latter maps may be obtained from the State Geologist. References to detailed economic maps of

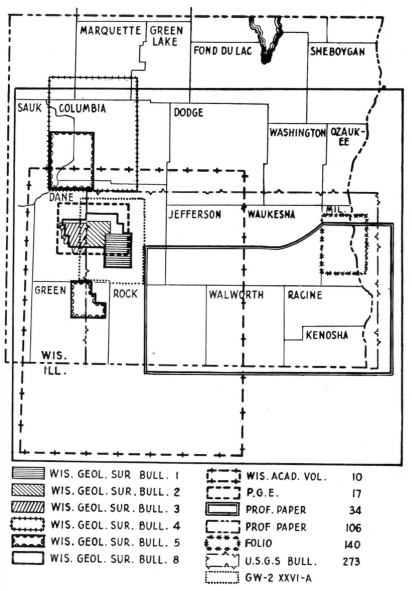

▤ WIS. GEOL. SUR BULL. I	⌐¬⌐⌐¬ WIS. ACAD. VOL.	10
▨ WIS. GEOL. SUR. BULL. 2	⌐ ⌐ ⌐ P.G.E.	17
▧ WIS. GEOL. SUR. BULL. 3	▭ PROF. PAPER	34
▨ WIS. GEOL. SUR. BULL. 4	⌐ ⌐ PROF PAPER	106
▨ WIS. GEOL. SUR. BULL. 5	✳✳✳ FOLIO	140
▭ WIS. GEOL. SUR. BULL. 8	⌐∧⌐ U.S.G.S BULL.	273
	⦙⦙⦙⦙⦙ GW-2 XXVI-A

Fig. 204. Index map of southeastern Wisconsin, with the areas covered by maps which show the glacial geology (see also Fig. 203). Prof. Paper 34 means Professional Paper 34, U. S. Geol. Survey, 1904, Plate XIV, facing p. 64; U. S. G. S. Bull. 273, Ibid., Bull. 273, 1905, Plate I facing p. 10; Folio 140, Ibid., Geologic Atlas of the United States, Milwaukee Folio; Wis. Bull. 5, Bulletin V, Wis. Geol. and Nat. Hist. Survey, 1900, Plate 37, facing p. 108; Wis. Bull. 8, Ibid., Bulletin VIII, 1910, folded map; Wis. Acad. Vol. 10, Wisconsin Academy of Sciences, Arts, and Letters, Vol. 10, 1895, Plates XIII to XVI; GW2—XXVI—A, Geology of Wisconsin, Vol. 2, 1877, Plate XXVI—A, facing p. 613; PGE 17, Manual of Physical Geography Excursions, Fig. 17, p. 116; the numerals refer to maps in theses (see bibliographies at ends of earlier chapters), 1 and 3 by F. T. Thwaites, 2 by O. W. Stromme, 4 by R. W. Ellie, 5 by E. S. Park.

parts of the Lake Superior iron district and the lead and zinc district of southwestern Wisconsin will be found in the books and articles listed at the end of several of the chapters in this book and in Appendix G.

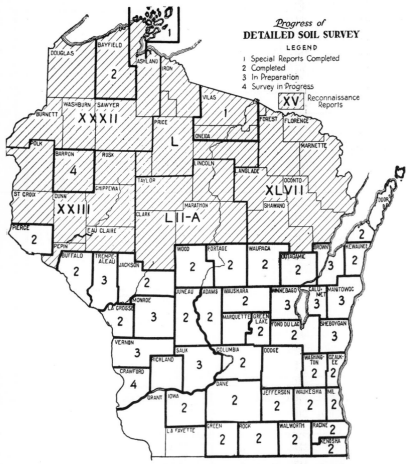

Fig. 205. Index map showing the location of the areas in Wisconsin where there are soils maps by the Wisconsin Geological and Natural History Survey. These cover counties or groups of counties. Soils maps of the Janesville Area, Viroqua Area, Racine County Area, and Portage County Area have been issued by the United States Department of Agriculture. The Roman numerals refer to Wis. Geol. Survey bulletin numbers.

The geological maps issued by the State Geological Survey cover a large part of the state. The sheets on the scale of 3 miles to 1 inch (Fig. 200) were made and issued by the Lapham-Chamberlin

Survey of 1873 to 1879. An area in North-Central Wisconsin has also been mapped more recently on this scale (Bull. 16, Wis. Geol. and Nat. Hist. Survey). The later maps are on various scales, the areas covered being indicated in Figures 201 and 202.

The territorial geological surveys between 1809 and 1848 (p. 531) issued only exploratory geological maps. The state geological survey of 1853 to 1862 issued reconnoissance maps of parts of the state, with a little detailed work in the lead and zinc district. The state geological survey of 1873 to 1879 issued good general maps of almost the whole state with detailed work in the iron country to the north as well as in the lead and zinc district. The present state survey since 1897 has issued detailed maps of all the areas it has investigated. For list of areas covered, see pp. 535-547. The maps and bulletins may be obtained from *The State Geologist,* Madison. For list of publications see pp. 536-547.

APPENDIX F

THE LAND SURVEY IN WISCONSIN

Three Systems of Describing Boundaries. There are three ways in which the location of lands may be made specific, so that one person may describe them to another. These are (a) by metes and bounds, (b) by the rectangular system of townships, ranges, and sections, (c) by latitude and longitude.

Metes and Bounds. The location of lands by metes and bounds is as follows. A point of beginning is taken arbitrarily, for example the bank of a river or lake, a prominent rock or hill, a tree, or some other natural feature. The property is then bounded in terms of lines run in stated compass directions for stated distances, often in relation to other natural features. The disadvantages of this system are, first, that the starting point may not be easy to identify after the lapse of some years. The bank of a river may be worn back; the tree may be cut down; one hill or rock may be confused with another; the boundary stakes or monuments may be destroyed. The second objection to this system is that it does not always provide boundaries which fit each other. Thus disputes and litigations over property lines often arise.

This system is found to work satisfactorily, however, in places where it has been long established, though it is less simple than the location by townships and sections. We sometimes use metes and bounds in delineating our city and village building lots in Wisconsin, though more often they are designated as Lot No. —— in Plat so-and-so. Metes and bounds are still used in the French claims of parts of this state (p. 285). This is the approved method throughout eastern United States from Ohio and Tennessee to the Atlantic coast.

Township and Range System. The land survey of the United States government, as adopted in 1785, locates public lands and rural property in relation to principal meridians, base lines, and correction lines, so that the land is divided into a rectangular system of townships and ranges.

In Wisconsin the ranges—north-south strips of land 6 miles wide —are determined by the 4th principal meridian (Fig. 206). The 1st principal meridian is in eastern Indiana. The 4th principal meridian is a north-south line extending from western Illinois, near St. Louis, Missouri, through southwestern Wisconsin near Platteville, west of Richland Center and northward to a point near the mouth of the Montreal River and through Outer Island in the Apostle archipelago. It is a short distance east of the geographical meridian of 90° 30′ west longitude. The 4th principal meridian

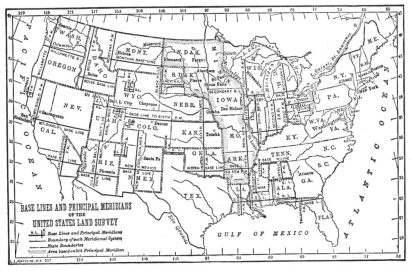

Fig. 206. Map to show the location of the 4th principal meridian, in relation to which government lands in Wisconsin are described. (From Johnson's Mathematical Geography.)

was surveyed about 1847. Lapham has quoted some of the difficulties of the work (Lapham, I. A., Latitude and Longitude of Places in Wisconsin, Collections Wis. Hist. Soc., Vol. 4, 1859, pp. 359-363). The lands were subdivided into townships and sections between 1834 and 1866.

The ranges are numbered eastward and westward from the 4th principal meridian (Fig. 207). There are 30 ranges east of the principal meridian in Wisconsin and 20 ranges west.

The townships are arranged in north-south tiers, beginning at the base line along the Illinois boundary. There are 53 townships in the longest Wisconsin ranges.

Thus we may locate any township in the state by reference to the township and range numbers. Madison, for example, is in township 7 north, range 9 east. This is usually written T. 7 N., R. 9 E.

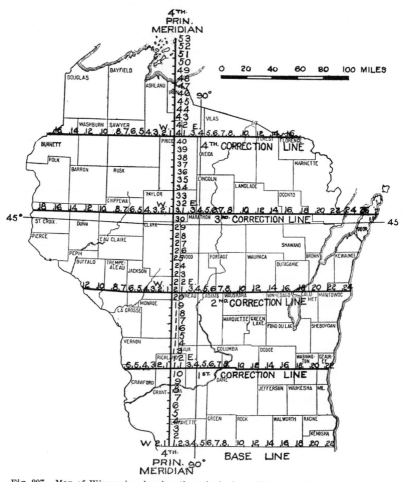

Fig. 207. Map of Wisconsin, showing the principal meridian, base line, correction lines, and numbering of ranges and townships.

Each township is supposed to be 6 miles square. The meridians converge, however, due to the earth's curvature. This necessitates allowances and corrections, as do the lack of coincidence of state boundaries with township boundaries, the presence of lakes and streams, and errors in surveying. Accordingly, no township is exactly six miles square. In order to avoid carrying errors clear across the state the government surveyors established correction lines. These are east-west lines parallel to the original base line in southern Wisconsin. There are four correction lines in Wisconsin (Fig. 207).

In a region of slight relief, as in northern or eastern Wisconsin, the highways are laid out along the township, range, and sec-

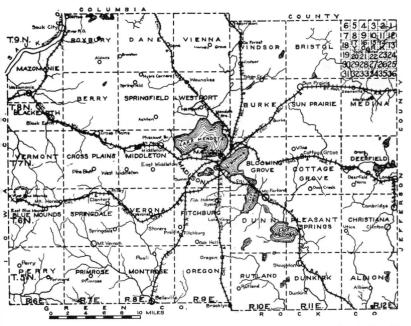

Fig. 208.   Map of Dane County, showing the civil towns, nearly all of which coincide with the government townships.

tion lines. Where the relief is greater, however, as among the hills of the Driftless Area, the roads are apt to follow ridge tops or valley bottoms, regardless of the land lines.

Townships Within Counties. When townships are combined into counties these, like the townships, are often rectangular. Dane County is typical (Fig. 208). It consists of 35 townships, 7 east and west by 5 north and south. The four western tiers lie slightly to the north of the three on the east, producing an offset between ranges 9 and 10. This is because town 1 in range 9 is more than

6 miles long, while town 1 in range 10 east is just 6 miles long. Accordingly the northern boundaries of all the townships in range 9 and the ranges to the west are a fraction of a mile farther north than the northern boundaries of the townships in range 10 east, although all the townships on each side are 6 miles long. This off-set appears only as far north as the first correction line, however, where the townships numbered 12, in ranges 10 and 11 east, are nearly 7 miles long. There are similar adjustments throughout

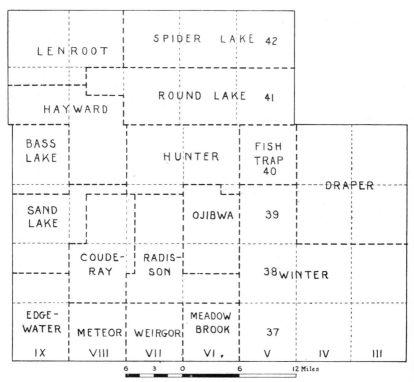

Fig. 209. Map of Sawyer County, showing the civil towns. The town of Winter contains seven government townships.

the state. The particular departure from the regular dimensions in townships 1, ranges 9 and 10, appears to be due to the error in determining the southern boundary of the state as already described (p. 43), for the Wisconsin-Illinois boundary is not a true east-west line.

Civil Towns and Congressional Townships. Another particular in which Dane County departs from regularity is that the towns of Black Earth and Mazomanie—Ts. 8, 9, N., R. 6 E.—are not square. This introduces the difference between the civil, or organized, or municipal town and the congressional township or government township. The civil town of Black Earth—a unit of present government—is only half the size of a congressional township. The civil town of Mazomanie is more than 6 miles long on the eastern side, but is cut off at the northwest by a natural boundary—the Wisconsin River.

Other counties furnish numerous variations of form and relationship of civil towns. Sawyer County (Fig. 209) is about the same size as Dane County, but it contains several very large civil towns. In southeastern Sawyer County is the civil town of Winter. This civil town contains about 252 square miles, made up of 7 congressional townships each containing 36 square miles. As Sawyer County becomes more densely populated the large town of Winter will doubtless be split into smaller and smaller units until it is only 6 miles square. Only a few years ago the civil town of Winter contained nine townships instead of seven.

The most irregular civil town in Wisconsin, so far as form is concerned, is Vaughn, Iron County. It contains the city of Hurley. This town is 16 miles long. It is 7 miles wide at the south. For 5 miles, however, it is only a half mile wide, and for three miles only a quarter mile wide. At one point, 3 miles south of Hurley, it is actually separated into two parts. It terminates at the north in a narrow point, where the Montreal River, along the Wisconsin-Michigan state line, furnishes its eastern boundary (Fig. 210).

Sections. Every township is divided into 36 sections, each 1 mile or 320 rods square. They are numbered as shown in the upper right hand corner of Figure 208. The capitol square at Madison is at the corner of sections 13, 14, 23, and 24 in township 7 north, range 9 east.

A section contains 640 acres. Each of the quarter-sections, therefore, contains 160 acres. As the townships are surveyed from the southeast corner and measured off to the north and west, the northern and western sections often contain a little more or a little less than 640 acres. The quarter-sections are, accordingly, irregular in some cases. There are aberrant 40 acre tracts, and smaller plats, usually known as fractions or lots.

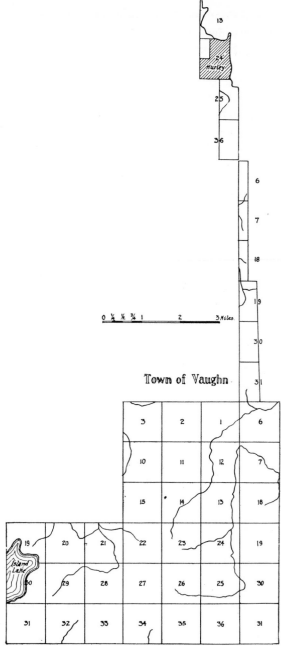

Fig. 210. The town of Vaughn, Iron County, which contains the city of Hurley.

The quarters of a section are designated by the points of the compass. Thus the tract of land originally purchased by the University of Wisconsin was the northwest quarter of section 23, T. 7 N., R. 9 E. It contains the part of the present campus including Chadbourne Hall.

A quarter of a quarter-section, of course, contains a fourth of 160 acres or 40 acres. These tracts, generally spoken of as "forties," are likewise designated in relation to the points of the compass. No one could misunderstand if the shaded area in Figure 211 were described as the northeast quarter of the southwest quarter of section 31, township 18 north, range 9 east. This may be abbreviated to the following: NE¼ of SW¼ of Sec. 31, T. 18 N., R. 9 E. It might be well, however, to add "of the 4th principal meridian." The word quarter is often omitted, a forty being spoken of as "the northeast of the southwest of section so-and-so."

The Marking of Township and Section Corners. The monuments or land bounds which mark the corners of townships and sections may be stakes, trees, stones, or mounds. Too often they are not permanent in character. The section post or stone is notched on the edges, the number of notches indicating the distance in miles from the boundary of the township. Each monument is also marked by several witness trees or stakes. Usually there are four witness trees for a section corner and two witness trees for the corner of a quarter section. The witness trees are blazed and inscribed with the number of the township, range, and section, as follows:

T 18 N
R   9 E
S 31

It is sometimes possible to read the marks on the witness trees when they are reblazed after a period of at least 60 years.

The French Land System. Allusion has already been made to the long, narrow farms (p. 285) which the French laid out along the Fox River near Green Bay and Depere and on the Mississippi at Prairie du Chien. Some of these farms were over 3 miles long and only 400 feet wide (Fig. 108). This narrow frontage was always on the river bank so that every *habitant* might have his share of the common highway of communication. As these French claims were already occupied when the government lands in Wis-

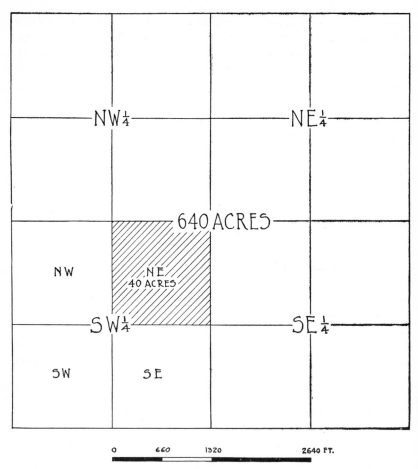

Fig. 211. Map showing a section of government land and the way its quarter-sections and forties are designated.

consin were surveyed, the tracts in question were not laid out in the township and range system.

Like the system of metes and bounds, this township and range system, or rectangular system, is open to certain objections. If we were establishing land lines now we should probably use some system similar to the one next described.

The Quadrangle System. A third method of locating lands is that used in describing the location of the topographic sheets of the United States Geological Survey. The Madison quadrangle, for example, is bounded by the parallels of 43° and 43° 15′ north latitude and by the meridians of 89° 15′ and 89° 30′ west longitude. This means that the area is nearly a quarter of the way around the world from Greenwich, England, and nearly half way from the equator to the north pole. These topographic sheets usually show both the parallels and meridians and the township and range lines.

Triangulation and Maps. The astronomical observations, triangulation, and spirit leveling, upon which the location of base maps and the contouring of topographic maps depends, began long ago. The earlier explorers made rough determinations of latitude and longitude at scattered points, while mapping the rivers and lakes. One of the best early maps of Wisconsin, based on more refined work, was made by Nicollet, who delineated the Mississippi region, (Nicollet, J. N., Report Intended to Illustrate a Map of the Hydrographic Basin of the Upper Mississippi River, Senate Document 237, 26th Congress, 2nd Session, Washington, 1843, 170 pp., and map,—for geographical positions and altitudes in Wisconsin determined in 1838 and 1839 see pp. 122-124, 128). A similar map of excellent quality is the work of Cram, who covered eastern Wisconsin (Cram, T. J., Internal Improvements in the Territory of Wisconsin, Senate Document 140, 26th Congress, 1st Session, Washington, 1840, 22 pp., and map). Astronomical positions along the coast of Lake Superior were first determined with accuracy by H. W. Bayfield. His chart was published by the British Admiralty in 1828. The Jesuit fathers published a general map of Lake Superior as early as 1672. The United States Coast and Geodetic Survey and United States Lake Survey have carried on extensive triangulations in connection with the more modern mapping of Lake Superior and Lake Michigan. Triangulation and other geodetic work in this state were performed by Davies nearly 40 years ago (Davies, J. E., Geodetic Survey, Geology of Wisconsin, Vol. 4, 1882, pp. 727-754, and Atlas Plates 41, 42). As the first step in this triangulation, a very accurate base line was measured in the Wisconsin valley near Spring Green. Triangulation along the Mississippi River was done by the Mississippi River Commission.

Triangulation and leveling by the U. S. Geological Survey, U. S. Coast and Geodetic Survey, U. S. Corps of Engineers (Lake Survey and Mississippi River Commission) are summarized in Marshall's "Results of Spirit Leveling in Wisconsin" (Marshall, R. B., Bull. 570, U. S. Geol. Survey, 1914, 82 pp; earlier edition, Ibid., Bull. 461, 1911, pp. 28-60). The positions of lines of levels, lines of primary traverse, and of triangulation stations are shown on the *Index to Atlas Sheets in Wisconsin,* published in 1929 by the U. S. Geological Survey.

APPENDIX G

CHRONOLOGICAL LIST OF FEDERAL AND STATE SUR-
VEY REPORTS, WITH A FEW OTHER EARLY PAPERS
ON THE GEOLOGY AND PHYSICAL GEOGRA-
PHY OF WISCONSIN

OBSERVATIONS BEFORE THE COMING OF GEOLOGISTS

There are incidental observations of geological and physiographic features in the narratives of the very earliest exploration
and settlement of what is now Wisconsin. We may fairly designate as the first contribution to the physical geography of Wisconsin the following papers written by Father André, a black-
robed Jesuit priest.

André, Louis. The Tide in the Bay des Puans, Jesuit Relations, 1671-72,
Thwaites' edition, Vol. 56, Cleveland, 1899, pp. 137-139; Remarkable
Facts Concerning the River (that Discharges into the Bay des Puans at
the Bottom of the Cove), Jesuit Relations, 1672-73, Ibid., Vol. 57, pp.
301-303.

Both of these papers deal with the fluctuations in level of Green Bay,
also discussed in 1673 by Father Marquette. Father André's papers
are quoted in full in Chapter XII of this book.

The first contributions to the geology of Wisconsin are perhaps
the following:

Marquette, Jacques. Jesuit Relations, 1673-77, Thwaites' edition, Vol. 59,
Cleveland, 1900, p. 107; Ibid., Hennepin's translation in "A New Discovery of a Vast Country in America," 1679-82, Part 1, London, 1698,
p. 326.

Marquette and Jolliet had crossed over from the Fox River to
the Wisconsin River at Portage in 1673, when Father Marquette
states that:

"After navigating about 30 leagues, we saw a spot presenting all
the appearances of an iron mine; and, in fact, one of our party
who had formerly seen such mines, assures us that The One which
we found is very good and very rich. It is Covered with three

feet of good soil, and is quite near a chain of rocks, the base of which is covered by very fine trees. After proceeding 40 leagues on This same route, we arrived at the mouth of our River; and, at 42 and a half degrees Of latitude, We safely entered Missisipi on The 17th of June."

Jolliet's map (see reference p. 182) shows "Mines de fer," to the northwest of what seems to be meant for Blue Mound. The locality mentioned by Marquette should be near Boscobel, where the Cambrian sandstone contains some iron. The iron in adjacent areas was mined in subsequent years, but no geologist is known to have mentioned the deposit described by Father Marquette.

Still earlier, as we learn from the Jesuit Relations, the copper deposits were noticed and commented upon, the first mention being over 270 years ago. Among others, Father Dablon took stock of the copper near Lake Superior, as quoted at the beginning of Chapter XVIII.

Dablon, Claude. Jesuit Relations, 1669-71, Thwaites' edition, Vol. 54, pp. 161-163; see also the following:—The Relation of 1659-60, Vol. 45, pp. 219-221; the Relation of 1670-72, Vol. 55, p. 99; Ibid., p. 103, the latter perhaps alluding to the diamonds in the glacial drift, as well as to copper.

The volumes of the Jesuit Relations contain much more geographical information regarding Wisconsin. The contemporary accounts published in the Collections and Proceedings of the State Historical Society of Wisconsin also constitute an undeveloped mine of geographical information.

Among the early French travelers whose writings contain incidental references to geographical features and mineral resources were Radisson and Groseilliers, Marquette and Jolliet, Hennepin, Perrot, and many others who traveled in Wisconsin in the seventeenth century.

Kellogg, Louise Phelps. French Regime in Wisconsin and the Northwest, Madison, 1925, Chapter XVI, "Early Mining in the Northwest."

In the eighteenth century came English travelers, including Carver, who noted and commented upon the simpler features of our physical geography.

Carver, Jonathan. Travels through the Interior Parts of North America in the years 1766, 1767, and 1768, London, 1778, 1781—reprinted as Travels in Wisconsin, from the third London edition, New York, 1838, 362 pp.

Lapham has collected some of the early maps of Wisconsin, a few of which were subsequently destroyed in the Chicago fire.

Lapham, I. A. Maps of Early Wisconsin Published between 1670 and 1823, 10 blueprint plates.

The bibliography in this Appendix includes the chief publications dealing with first-hand observations of the geology and physical geography of the area included within the present boundaries of the state of Wisconsin. The list includes all the publications seen by the author, but may not be absolutely complete. From this point on, these are all works by geologists and geographers, or by men somewhat skilled in geological observation. Periodical literature has not been included, except for the earliest years, the bibliography being mainly confined to portions of official reports and to books. Geological papers for later years are referred to in the bibliographies at the ends of the chapters in this book. Only a few reports on paleontology and economic geology are included. For physical geography and general geology the list is substantially complete.

Period of Exploratory Geological Surveys
1809-1848

Nuttall, Thomas. Observations of the Geological Structure of the Valley of the Mississippi, Journal of the Academy of Natural Sciences of Philadelphia, Vol. 2, Part 1, 1821, pp. 14-52.

This is a series of notes on a journey in 1809 which included Green Bay, the Fox River at Portage, the Wisconsin River to Prairie du Chien, and the Mississippi to St. Louis. The author, primarily a botanist, speaks of rounded gravels at Prairie du Chien, the contrast in the current of the Fox and the "Ouisconsin" Rivers, the "almost uninterrupted hills" on either side of the Wisconsin, the southward dip of the rocks, the pieces of native copper from Lake St. Croix which were shown to him by an Indian, and the copper at the western end of Lake Superior, the lead mines near Dubuque, the "adventitious granitic gravel and bolders throughout the western states," the marine origin of the sedimentary rocks with citation of fossils collected,—see pp. 16, 18, 22, etc.

Schoolcraft, H. R. Narrative Journal of Travels through the Northwestern Regions of the United States in the Year 1820, Albany, 1821, pp. 191-204, 321-383.

This report contains notes on topography and geology, in a journey along the coast of Lake Superior from the Montreal River to the St. Louis River, along the Mississippi from Lake St. Croix to Dubuque, up the Wisconsin River from Prairie du Chien to Green Bay, and on the coast of Lake Michigan from Green Bay to Chicago, on meteorology, (p.

204), on Lake Pepin, (pp. 324, 327-331), on Trempealeau Mountain (pp. 334-335), on the lead and zinc district of southwestern Wisconsin (pp. 342-354).

Keating, W. H.   Major Long's Second Expedition, Narrative of an Expedition to the Source of St. Peter's River, Lake Winnepeek, Lake of the Woods, etc., etc., Performed in the Year 1823, under the command of Stephen H. Long, 2 volumes, Philadelphia, 1824, 439, 459 pp.—especially Vol. 1, pp. 172-293.

The author was Professor of Mineralogy and Chemistry in the University of Pennsylvania. His book includes geological and geographical observations in Wisconsin during a journey overland from the Pecatonica River to Prairie du Chien and up the Mississippi to Minneapolis. On drift copper from near Milwaukee, (Vol. 1, p. 168); on the Driftless Area, (pp. 200, 263, 287-288); Platte Mounds, (p. 193); topography and geology of southwestern Wisconsin, including fossils, (pp. 194-209); Mississippi gorge and its rock formations, (pp. 236-237, 265-266, 293-294); Kickapoo Valley, (pp. 241-242); Trempealeau Bluffs, (pp. 271-272); Lake Pepin, (pp. 278-280). The account of the expedition also contains some very general statements about topography and drainage by Major Long, "Of the Country and Navigable Communications between Lake Michigan and the Mississippi River" (Vol. 2, pp. 212-220); Latitudes and longitudes in western Wisconsin determined in 1823 by J. E. Colhoun, (Vol. 2, pp. 401-404); Comparison of temperatures at Prairie du Chien and Green Bay in 1822 with those in other parts of United States, by Joseph Lovell and Major Long, (Vol. 2, pp. 417-448).

Schoolcraft, H. R.   Exploration of the St. Croix and Burntwood (or Brule) Rivers, Narrative of an Expedition through the Upper Mississippi to Itasca Lake, the Actual Source of the River; Embracing an Exploratory Trip through the St. Croix and Burntwood (or Broule) Rivers in 1832, New York, 1834, pp. 123-144, 157-159.

This report includes Remarks on the Lead Mine Country on the Upper Mississippi (pp. 294-307); on the Driftless Area, (p. 306).

Featherstonhaugh, G. W.   Report of a Geological Reconnoissance Made in 1835 from the Seat of Government by the way of Green Bay and the Wisconsin Territory to the Coteau de Prairie, an Elevated Ridge dividing the Missouri from the St. Peter's River, Senate Document 333, 24th Congress, 1st Session, Washington, 1836, 162 pp.,—see especially pp. 119-135; A Canoe Voyage up the Minnay Sotor, 2 volumes, London, 1847, on Wisconsin, and Mississippi Rivers, (Vol. 1, pp. 191-258, 270-273; Vol. 2, pp. 15-22, 28-33, 113).

Owen, David D.   Report of a Geological Exploration of Part of Iowa, Wisconsin, and Illinois in 1839, House Ex. Document 239, 26th Congress, 1st Session, Washington, 1840, 161 pp. Reprinted as Senate Ex. Document 407, 28th Congress, 1st Session, Washington, 1844, pp. 15-145.

This volume includes the report of John Locke described below, a report by E. Phillips on timber and soil, a series of woodcuts of Wisconsin scenery, a number of diagrams with geological sections on the border and perspective sketches above (much as in the modern block diagrams), township descriptions of southwestern Wisconsin, and a geological map of the region south of the Wisconsin River and west of Madison.

Locke, John, in Owen's report, Op. cit., 1840, pp. 116-159.

Report compares rocks of southwestern Wisconsin with those of Ohio, gives detailed sections with some physiographic interpretation, and summarizes instrumental observations on elevations, on terrestrial magnetism, and on meteorology.

Lapham, Increase A. A Geographical and Topographical Description of Wisconsin, Milwaukee, 1844, 256 pp; Wisconsin, Its Geography and Topography, Milwaukee, 1846, 202 pp.

Owen, David D. Preliminary Report containing Outlines of the Progress of the Geological Survey of Wisconsin and Iowa, up to October 11, 1847, Senate Ex. Document 2, 30th Congress, 1st Session, Washington, 1847, pp. 160-174.

Owen, David D. A Report of a Geological Reconnoissance of the Chippewa Land District of Wisconsin, and the Northern Part of Iowa, Senate Ex. Document 57, 30th Congress, 1st Session, Washington, 1848, 134 pp.

This volume includes a report by J. G. Norwood, many graphic illustrations of the block diagram type, and a large geological map of northern and western Wisconsin.

Norwood, J. G. In Owen's report, 1848, pp. 75-129.

This report includes a description of the Pillared Rocks in the Apostle Islands, General Observations of the Topography and Climate of Wisconsin, Reconnoissance of a Portion of St. Louis River, of a Portion of the District lying between Fond du Lac and the Falls of St. Anthony, and from the mouth of Montreal River, via Lac du Flambeau, and the Head Waters of the Wisconsin River to Prairie du Chien.

Jackson, C. T. Report on the Geological and Mineralogical Survey of the Mineral Lands of the United States in the State of Michigan, Senate Ex. Document 1, 31st Congress, 1st Session, Part 3, Washington, 1849, 935 pp., description of falls on Montreal River, (p. 417).

Foster, J. W., and **Whitney, J. D.** Report on the Geology and Topography of a Portion of the Lake Superior Land District in the State of Michigan, Part 1, Copper Lands, House Ex. Document 69, 31st Congress, 1st Session, Washington, 1850, 224 pp.

Includes Desor's report on the drift. For references to Wisconsin geology and geography see Montreal River, (pp. 24, 103-104); Menominee River, (pp. 30-31); Meteorological records at Fort Howard on Green Bay, (pp. 40, 42, 46, 47). Among the illustrations are drawings of the falls of the Montreal and Menominee Rivers.

Foster, J. W., and **Whitney, J. D.** Report on the Geology of the Lake Superior Land District, Part 2, The Iron Region, Senate Ex. Document 4, Special Session, 1851, Washington, 1851, 400 pp.

Includes Desor's Report on the Superficial Deposits, Lapham's Geology of Southeastern Wisconsin, Hall's discussion of the Paleozoic, and Whittlesey's Geology of Eastern Wisconsin and Observed Fluctuations of the Surfaces of the Lakes. Notes on Wisconsin geology and geography,—Menominee River, (p. 27); Geology in eastern Wisconsin, (pp. 153, 174, etc.); red clay of Fox River and well records at Fond du Lac, Wis., (pp. 175-176, 234, 393-395); origin of basin of Lake Michigan, (pp. 176-177); drift copper in southeastern Wisconsin, (p. 201); drift of Menominee valley including eskers, (pp. 234-237).

Lapham, Increase A.   On the geology of the Southeastern Portion of the State of Wisconsin,—in Foster and Whitney's Geology of the Lake Superior Land District, Senate Ex. Document 4, Special Session, 1851, Washington, 1851, pp. 167-173; Geological Formations of Wisconsin, Trans. Wis. State Agr. Soc., Vol. 1, 1851, pp. 122-128.

Owen, David D.   Abstract of an Introduction to the Final Report on the Geological Surveys made in Wisconsin, Iowa and Minnesota, in the years 1847-'48-'49 and '50, containing a Synopsis of the Geological Features of the Country, Proc. Amer. Assoc. Adv. Science, Vol. 5, 1851, pp. 119-131; On the Palaeontology of the Lowest Sandstones of the Northwest, Ibid., pp. 169-172; On the Number and Distribution of Fossil Species in the Paleozoic Rocks of Iowa, Wisconsin, and Minnesota, Ibid., pp. 235-239.

Owen, David D.   Report of a Geological Survey of Wisconsin, Iowa, and Minnesota, Philadelphia, 1852, 634 pp. and atlas.

This report contains a rather good geological map of Wisconsin, (see notice in Proc. Acad. Nat. Sci. Philadelphia, Vol. 6, 1854, pp. 189-191.)

Norwood, J. G., in Owen's 1852 report, pp. 213-418,—including Geology of the Northwest and West Portion of the Valley of Lake Superior (pp. 333-418).

Whittlesey, Charles, in Owen's 1852 report, pp. 419-473,—including Geological Report on that Portion of Wisconsin Bordering on the South Shore of Lake Superior.

Shumard, B. F., in Owen's 1852 report, pp. 475-634,—including Local Detailed Observations in the Valleys of the Minnesota, Mississippi, and Wisconsin Rivers.

Schoolcraft, H. R.   Thirty Years with the Indian Tribes, Philadelphia, 1851. For journey from southwestern Wisconsin to Portage in 1831, (pp. 352-396). This volume also contains notes and journals concerning Schoolcraft's travels in the Fox and Wisconsin valleys and on Lake Superior between 1825 and 1827.

Schoolcraft, H. R.   Summary Narrative of an Exploratory Expedition to the Sources of the Mississippi River in 1820; Resumed and Completed by the Discovery of its Origin in Itasca Lake, in 1832. By Authority of the United States. With Appendices, comprising the Original Report of the Copper Mines of Lake Superior, and Observations on the Geology of the Lake Basins, and the Summit of the Mississippi, Together with all Official Reports and Scientific Papers of Both Expeditions, Philadelphia, 1855, 588 pp.

Description of journey from Montreal River to the west end of Lake Superior in 1820, (pp. 102-109); journey from Prescott to Prairie du Chien, trip to lead mines of Grant County, and journey from Prairie du Chien to Chicago via the Wisconsin and Fox Rivers, Green Bay, and the west coast of Lake Michigan in 1820, (pp. 162-199).

For Observations on the Geology and Mineralogy of the Region Embracing the Sources of the Mississippi River, and the Great Lake Basins, during the Expedition of 1820 see the following: South coast of Lake Superior at Montreal River, Apostle Islands, St. Louis River, etc., (pp. 321-322, 325); Mississippi, Wisconsin, and Fox River Valleys and west coast of Lake Michigan, Lake St. Croix, Lake Pepin, Trempealeau, Prairie du Chien, Grant County, Portage, Green Bay, Milwaukee, etc., (pp. 332-336). Geological profiles of the Mississippi Valley and basins of the Great Lakes, with diagrams and views of scenery, were submitted with this report but not published.

For journey from Lake Superior to the Mississippi River via the Bad, Namakagon, and Chippewa Rivers in 1831, (pp. 540-544). For journey from southwestern Wisconsin to Portage, via Blue Mound and Madison in 1831, and second recorded observation of Driftless Area phenomena, (pp. 560-572). For journey from Prescott to Lake Superior via the St. Croix and Brule Rivers in 1832, (pp. 269-274). For latitudes and longitudes in western Wisconsin determined in 1836, (pp. 653-654).

PERIOD OF STATE SURVEYS BY DANIELS, PERCIVAL, LAPHAM, HALL AND WHITNEY, 1853-1862.

Daniels, Edward. First Annual Report on the Geological Survey of the State of Wisconsin, Madison, 1854, 83 pp.
——————————, Report of the Committee on Mining and Smelting, J. H. Earnest, (Chairman), March 8, 1854, 8 pp.

Percival, James G. Annual Report on the Geological Survey of the State of Wisconsin, Madison, 1855, 101 pp.

Lapham, Increase A. Geological Map of Wisconsin, New York, 1855.

Percival, James G. Report on the Iron of Dodge and Washington Counties, State of Wisconsin, Milwaukee, 1855, 13 pp.

Percival, James G. Annual Report on the Geological Survey of the State of Wisconsin, Madison, 1856, 111 pp.

Daniels, Edward. Annual Report of the Geological Survey of the State of Wisconsin for the Year Ending Dec. 31, 1857, Madison, 1858, 62 pp. This report includes Document "P," and Iron Ores of Wisconsin.

Hall, James. Report of the Commissioners of the Geological Survey, Madison, 1858,—in Governor's Message and Documents, 1859, 12 pp.

Hall, James. Report of the Superintendent of the Geological Survey, Exhibiting the Progress of the work January 1, 1861, Madison, 1861, 52 pp. This report includes Document O and Descriptions of New Species of Fossils from the Investigations of the Survey; To Accompany the Report of Progress made to His Excellency, Alexander W. Randall, on the 24th day of December, 1860. (A list of new species of fossils appeared as a 4 page pamphlet with the imprint—Albany, 1860).

Hall, James, and **Whitney, J. D.**   Report on the Geological Survey of the State of Wisconsin, Vol. 1, 1862, 448 pp.

Whitney, J. D.   The Upper Mississippi Lead Region (reprint of part of above, bound and with different title, 1862, pp. 273-420.

Hall, James.   Preliminary Notice of the Fauna of the Potsdam Sandstone, with Remarks upon the Previously Known Species of Fossils and Descriptions of Some New Ones, from the Sandstone of the Upper Mississippi Valley, Transactions Albany Institute, 1852, Vol. 5, 1867, pp. 93-195; 16th Annual Rept., N. Y. State Cabinet of Natural History, 1863, pp. 119-206.

Hall, James.   Geological Survey of the State of Wisconsin, 1859-1863, Paleontology, Part Third, Organic Remains of the Niagara Group and Associated Limestones, Albany, 1871, 94 pp.

   This appears to be a reprint of the report written in 1864 and entitled Account of Some New or Little Known Species of Fossils from Rocks of the Age of the Niagara Group, 20th Annual Rept., N. Y. State Cabinet of Natural History, 1867, p. 305.

Hall, James.   Geological Survey of Wisconsin. James Hall, Dirext., Geological Map of Wisconsin, Showing the Relations of its Geology with that of the Surrounding States, compiled from the work of the Geological Surveys of Wisconsin and Iowa and from the Surveys of Doctors D. D. Owen, Foster and Whitney, and Professor A. Winchell. (Date of publication not determined).

Lapham, Increase A.   Geological Map of Wisconsin, new edition, prepared mostly from original observations, Milwaukee, 1869.

PERIOD OF STATE SURVEYS BY LAPHAM, CHAMBERLIN, IRVING, STRONG, KING, AND OTHERS, 1873-1879

Lapham, Increase A.   Wisconsin Geological Survey, Report of Progress and Results for the Year 1873, Geology of Wisconsin, Vol. 2, 1877, pp. 5-44.

Lapham, Increase A.   Report of Progress and Results for the Year 1874, Ibid., pp. 45-66.   See also Synopsis of a Report on the Geology of East Central Wisconsin, Made to Dr. I. A. Lapham, Chief of the Geological Corps of Wisconsin, Jan. 1st, 1874, by T. C. Chamberlin; Synopsis of a Report on the Geology of the Lake Shore Region of Wisconsin, Made to Dr. I. A. Lapham, Chief of the Geological Corps, Jan. 1, 1875, by T. C. Chamberlin; report by F. H. King. The last three reports were published separately, together with Chamberlin's report to O. W. Wight, in a 16-page pamphlet, not dated.

Lapham, Increase A.   Geological Map of Wisconsin, 4 sheets, 6 miles to the inch, submitted with annual reports of 1873 and 1874, but probably not published.

Lapham, Increase A.   Geology, Walling's Atlas of the State of Wisconsin, 1876, pp. 16-19 (includes a geological section from Lake Superior through the Penokee Range and Iron Ridge to Lake Michigan, and a colored geological map, report dated July, 1874).

Wight, O. W. Report of Progress and Results for the Year 1875, Geology of Wisconsin, Vol. 2, 1877, pp. 67-89; Synopsis of a Brief Report of Progress and Results for the Year 1875, Made to Dr. O. W. Wight, Chief of the Geological Corps of Wisconsin, by T. C. Chamberlin, published separately.

Chamberlin, T. C. Annual Report of the Progress and Results of the Wisconsin Geological Survey for the Year 1876, Madison, 1877, 40 pp. (Includes reports of progress by Moses Strong, R. D. Irving, C. E. Wright, A. C. Clark and P. R. Hoy).

Chamberlin, T. C. Annual Report of the Wisconsin Geological Survey for the Year 1877, Madison, 1878, 93 pp. (Includes reports of progress by E. T. Sweet, R. D. Irving, C. E. Wright, L. C. Wooster, A. C. Clark, and R. P. Whitfield); Topography and Geology of Wisconsin, Snyder, Van Vechten & Co.'s Atlas, 1878, pp. 15, 148-151, including geological map of Wisconsin.

Chamberlin, T. C. Annual Report of the Wisconsin Geological Survey for the Year 1878, Madison, 1879, 52 pp. (Includes report of progress by T. B. Brooks.)

Chamberlin, T. C. Annual Report of the Wisconsin Geological Survey for the Year 1879, Madison, 1880, 72 pp. (Includes description of new fossils by R. P. Whitfield, and of new species of fungi by W. F. Bundy.)

Chamberlin, T. C., R. D. Irving, Moses Strong, E. T. Sweet, F. H. King, C. E. Wright, Raphael Pumpelly, A. A. Julian, Charles Whittlesey, T. B. Brooks, Arthur Wichmann, T. Sterry Hunt, L. C. Wooster, R. P. Whitfield, A. C. Clark, J. E. Davies, R. D. Salisbury, and others. Geology of Wisconsin, Survey of 1873-1879, 4 volumes and atlas (maps drawn by **W. J. L. Nicodemus** and **A. D. Conover**).

Vol. II, 1877, (second edition, 1878), 752 pp., Plates III to XVI of Atlas;

Vol. III, 1880, 741 pp., Plates XVII to XXX of Atlas;

Atlas Plate I, 1881, Geology; Plate II, 1881, Quaternary Formations; Plate IIA, 1882, Native Vegetation; Plate IIB, 1882, Soils; Plate IIC, 1882, Rainfall and Temperature;

Vol. IV, 1882, 754 pp., Plates XXXI to XLII of Atlas;

Vol. I, 1883, 701 pp.

Irving, R. D. The Mineral Resources of Wisconsin, Trans. Amer. Inst. Mining Engineers, Vol. 8, 1880, pp. 478-508, with geological map of Wisconsin on the scale of 20 miles to 1 inch.

King, F. H. Model of Wisconsin, showing topography and geology, River Falls, Wis., 1882.

PERIOD OF UNITED STATES GEOLOGICAL SURVEY WORK SINCE 1881, BY IRVING, VAN HISE, CHAMBERLIN, SALISBURY, LEITH, BUELL, ALDEN, GRANT, AND OTHERS

This work has resulted in the publication of the following major contributions, within which unusually complete references to other literature will be found:

Monographs:

    5. The Copper-Bearing Rocks of Lake Superior, by R. D. Irving, 1883, 464 pp.

  19. The Penokee Iron-Bearing Series of Wisconsin and Michigan, by R. D. Irving and C. R. Van Hise, 1892, 534 pp.

  47. A Treatise on Metamorphism, by C. R. Van Hise, 1904, 1286 pp.

  52. The Geology of the Lake Superior Region, by C. R. Van Hise and C. K. Leith, 1911, 641 pp.

Geologic Atlas of the United States—Folios:

 140. Milwaukee Special, by W. C. Alden, 1906.

 145. Lancaster-Mineral Point, by U. S. Grant and E. F. Burchard, 1907.

 — Sparta-Tomah, by F. T. Thwaites, W. H. Twenhofel, and Lawrence Martin, (in preparation).

Professional Papers:

  34. The Delavan Lobe of the Lake Michigan Glacier of the Wisconsin Stage of Glaciation and Associated Phenomena, by W. C. Alden, 1904, 106 pp.

 106. The Quaternary Geology of Southeastern Wisconsin, with a chapter on the Older Rock Formations, by W. C. Alden, 1918, 356 pp.

 154A. Moraines and Shorelines of the Lake Superior Basin, by Frank Leverett, 1928, 72 pp.

Water Supply Papers:

 156. Water Powers of Northern Wisconsin, by L. S. Smith, 1906, 145 pp. (See also other Water Supply Papers, and Bibliography in Water Supply Paper 347, 1918).

 417. Profile Surveys of Rivers in Wisconsin, prepared under the direction of W. H. Herron, 1916, 16 pp.

Topographic Maps:

  80 Quadrangles, up to 1931, see Appendix E and Fig. 197. See also Results of Spirit Leveling in Wisconsin, 1897 to 1914 inclusive, Bulletin 570, 1914, 86 pp.

Other U. S. Geological Survey Reports:

Copper-Bearing Rocks of Lake Superior, by R. D. Irving, 3rd Annual Rept., 1883, pp. 89-188.

Preliminary Paper on the Terminal Moraine of the Second Glacial Epoch, by T. C. Chamberlin, Ibid., pp. 291-402.

The Topographic Features of Lake Shores, by G. K. Gilbert, 5th Annual Rept., 1885, pp. 69-123.

Secondary Enlargements of Mineral Fragments in Certain Rocks, by R. D. Irving and C. R. Van Hise, Bulletin 8, 1884, 56 pp.

The Requisite and Qualifying Conditions of Artesian Wells, by T. C. Chamberlin, 5th Annual Rept., 1885, pp. 125-173.

A Preliminary Paper on an Investigation of Archean Formations of the Northwestern States, by R. D. Irving, Ibid., pp. 175-242.

Observations on the Junction between the Eastern Sandstone and the Keweenaw Series on Keweenaw Point, Lake Superior, by R. D. Irving and T. C. Chamberlin, Bulletin 23, 1885, 124 pp.

Preliminary Paper on the Driftless Area of the Upper Mississippi Valley, by T. C. Chamberlin and R. D. Salisbury, 6th Annual Rept., 1885, pp. 199-322.

The Rock Scorings of the Great Ice Invasions, by T. C. Chamberlin, 7th Annual Rept., 1888, pp. 147-248.

On the Classification of Early Cambrian and pre-Cambrian Formations, by R. D. Irving, Ibid., pp. 365-454.

The Penokee Iron-Bearing Series of Michigan and Wisconsin, by R. D. Irving and C. R. Van Hise, 10th Annual Rept., Part 1, 1890, pp. 341-507.

Principles of pre-Cambrian North American Geology, by C. R. Van Hise, 16th Annual Rept., Part 1, 1896, pp. 571-874.

Principles and Conditions of the Movements of Ground Water, by F. H. King, 19th Annual Report, Part 2, 1899, pp. 59-294.

The Iron-Ore Deposits of the Lake Superior Region, by C. R. Van Hise, 21st Annual Rept., Part 3, 1901, pp. 304-434.

The Drumlins of Southeastern Wisconsin, by W. C. Alden, Bulletin 273, 1905, 46 pp.

Rock Cleavage, by C. K. Leith, Bulletin 239, 1905, 216 pp.

Zinc and Lead Ores near Dodgeville, Wis., by E. E. Ellis, Bulletin 260, 1905, pp. 311-315.

Zinc and Lead Deposits of the Upper Mississippi Valley, by H. F. Bain, Bulletin 294, 1906, 155 pp.

Pre-Cambrian Geology of North America, by C. R. Van Hise and C. K. Leith, Bulletin 360, 1909, 939 pp., (especially on Wisconsin, pp. 178-196, 717-724).

The Physical Geography of the Lake Superior Region, and The Pleistocene, by Lawrence Martin, Monograph 52, 1911, pp. 85-117, 427-459.

Recent Discoveries of Clinton Iron Ore in Eastern Wisconsin, by F. T. Thwaites, Bulletin 540, 1915, 5 pp.

History of the Great Lakes, by F. B. Taylor, Monograph 53, 1915, pp. 316-469.

THE PRESENT WISCONSIN GEOLOGICAL AND NATURAL HISTORY SURVEY SINCE 1897.

Birge, E. A. Director, Wisconsin Geological and Natural History Survey, 1898-1919.

Hotchkiss, W. O. Director, Wisconsin Geological and Natural History Survey, 1919-1925.

Bean, E. F. Director, Wisconsin Geological and Natural History Survey, 1925-31.

Following is a list of the publications of the present Survey:

BULLETINS

The bulletins are issued in four series:

Scientific Series.—The bulletins so designated consist of original contributions to the geology and natural history of the state, which are of scientific interest rather than of economic importance.

Economic Series.—This series includes those bulletins whose interest is chiefly practical and economic.

Educational Series.—The bulletins of this series are primarily designed for use by teachers and in the schools.

Soil Series.—This includes the bulletins prepared by the Soils Division.

The following bulletins have been issued:

Bulletin No. I. Economic Series No. 1.

On the Forestry Conditions of Northern Wisconsin. Filibert Roth, Special Agent, United States Department of Agriculture. 1898. Pp. vi, 78; 1 map. *Out of Print.*

Bulletin No. II. Scientific Series No. 1.

On the Instincts and Habits of the Solitary Wasps. George W. Peckham and Elizabeth G. Peckham. 1898. Pp. iv, 241; 14 plates, of which 2 are colored; 2 figures in text. *Out of print.*

Bulletin No. III. Scientific Series No. 2

A Contribution to the Geology of the Pre-Cambrian Igneous Rocks of the Fox River Valley, Wisconsin. Samuel Weidman, Ph. D. 1898. Pp. iv, 63; 10 plates; 13 figures in text. *Out of print.*

Bulletin No. IV. Economic Series No. 2.

On the Building and Ornamental Stones of Wisconsin. Ernest Robertson Buckley, Ph. D. 1898. Pp. xxvi, 544; 69 plates, of which 7 are colored, and 1 map; 4 figures in text. Price $2.50. Ed. 5,000.

Bulletin No. V. Educational Series No. 1.

The Geography of the Region About Devil's Lake and the Dalles of the Wisconsin with some notes on its surface geology. Rollin D. Salisbury, A. M., Professor of Geographic Geology, University of Chicago, and Wallace W. Atwood, B. S., Assistant in Geology, University of Chicago. 1900. Pp. x, 151; 38 plates; 47 figures in text. *Out of Print.* Ed. 5,000.

Bulletin No. VI. Economic Series No. 3. Second Edition.

Preliminary Report on the Copper-Bearing Rocks of Douglas County, and parts of Washburn and Bayfield Counties, Wisconsin. Ulysses Sherman Grant, Ph. D., Professor of Geology, Northwestern University. 1901. Pp. vi, 83; 13 plates. Price 50 cents. Ed. 3,000, 1750.

Bulletin No. VII. Economic Series No. 4.

The Clays and Clay Industries of Wisconsin. Part 1. Ernest Robertson Buckley, Ph. D. 1901. Pp. xii, 304; 55 plates. Price 50 cents. Ed. 5,000.

Bulletin No. VIII. Educational Series No. 2.

The Lakes of Southeastern Wisconsin. N. M. Fenneman, Ph. D., Professor of General and Geographic Geology, University of Wisconsin. 1902. Pp. xv, 178; 36 plates; 38 figures in text. A second edition has been issued and is sold at the price of 50 cents. Ed. 2,000.

Bulletin No. IX. Economic Series No. 5.

Preliminary Report on the Lead and Zinc Deposits of Southwestern Wisconsin. Ulysses Sherman Grant, Ph. D., Professor of Geology, Northwestern University. 1903. Pp. viii, 103; 2 maps; 2 plates; 8 figures in text. *Out of print.* Ed. 2,000.

Bulletin No. X. Economic Series No. 6.

Highway Construction in Wisconsin. Ernest Robertson Buckley, Ph. D. 1903. Pp. xvi, 339; 106 plates, including 26 maps of cities. Price 50 cents. Ed. 5,000.

Bulletin No. XI. Economic Series No. 7.

Preliminary Report on the Soils and Agricultural Conditions of North Central Wisconsin. Samuel Weidman. 1903. Pp. viii, 67; 10 plates, including soil map. Second edition, 1908. *Out of print.* Ed. 6,000.

Bulletin No. XII. Scientific Series No. 3.

The Plankton of Lake Winnebago and Green Lake. Dwight Marsh, Ph. D., Professor of Biology, Ripon College. 1903. Pp. vi, 94; 22 plates. *Out of Print.* Ed. 1,500.

Bulletin No. XIII. Economic Series No. 8.

The Baraboo Iron-Bearing District of Wisconsin. Samuel Weidman. 1904. Pp. x, 190; 23 plates, including geological map. *Out of print.* Ed. 3,000.

Bulletin No. XIV. Economic Series No. 9.

Report on Lead and Zinc Deposits of Wisconsin. Ulysses Sherman Grant, Ph. D., Professor of Geology, Northwestern University. 1906. Pp. ix, 100; 8 plates; 10 figures in text; an atlas containing 18 maps. This report is out of print, but the 18 maps which formerly accompanied it are still available. Price 20 cents. Ed. 3,500.

A supplementary series of 6 maps including the Montfort, East Mineral Point, Ipswich, Big Patch, Elk Grove, Cuba City, and East Meeker's Grove sheets—by W. O. Hotchkiss and Edward Steidtmann—was issued in 1909. Price 6 cents.

Bulletin No. XV. Economic Series No. 10.

The Clays of Wisconsin and Their Uses. Heinrich Ries, Ph. D., Assistant Professor of Economic Geology, Cornell University. 1906. Pp. xii, 259; 30 plates, including 2 maps, 7 figures in text. Price 50 cents. Ed. 3,000.

Bulletin No. XVI. Scientific Series No. 4.

The Geology of North Central Wisconsin. Samuel Weidman. 1907. Pp. xxxi, 697; 86 plates, including 2 maps; 38 figures in text. Price $2.00. Ed. 3,000.

Bulletin No. XVII. Scientific Series No. 5.

The Abandoned Shore-Lines of Eastern Wisconsin. J. W. Goldthwait, Ph. D., Assistant Professor of Geology, Northwestern University. 1907. Pp. x, 134; 38 plates; 37 figures in text. Price 50 cents. Ed. 2,500.

Bulletin No. XVIII. Economic Series No. 11.

Rural Highways of Wisconsin. W. O. Hotchkiss. 1906. Pp. xiv, 135; 16 plates; 2 figures in text. Price 25 cents. Ed. 14,000, 8,000.

Bulletin No. XIX. Economic Series No. 12.

Zinc and Lead Deposits of the Upper Mississippi Valley. H. Foster Bain, Director of State Geological Survey of Illinois. Washington, D. C. 1907. Pp. xii, 155; 9 plates, including 5 maps; 45 figures in text. Price 25 cents.

This bulletin is a reprint of Bulletin No. 294 of the United States Geological Survey. *Only a small number of copies were reprinted for local use. It has not been sent out to libraries and exchanges.*

Bulletin No. XX. Economic Series No. 13.

The Water Powers of Wisconsin. L. S. Smith, C. E.; Engineer Wisconsin Geological and Natural History Survey; Engineer U. S. Geological Survey. Pp. xvi, 354; 54 plates; 17 figures in text. 1908. *Out of print.*

Bulletin No. XXI. Scientific Series No. 6.

The Fossils and Stratigraphy of the Middle Devonic of Wisconsin. Herdman F. Cleland, Professor of Geology, Williams College. 1911. Pp. 206; 55 plates. *Out of print.* Ed. 1,200.

Bulletin No. XXII. Scientific Series No. 7.

The Inland Lakes of Wisconsin; the Dissolved Gases of the Water and their Biological Significance. Edward A. Birge and Chancey Juday. 1911. Pp. xxi, 254; 10 plates; 142 figures in text; all diagrams of gases and plankton. *Out of print.* Ed. 1,500.

Bulletin No. XXIII. Economic Series No. 14.

Reconnoissance Soil Survey of Northwestern Wisconsin. Samuel Weidman, with the assistance of E. B. Hall and F. L. Musback. 1911. Pp. viii, 103; 15 plates including one map; 16 figures in text. *Out of print.* Ed. 8,000.

Bulletin No. XXIV. Soil Series No. 1 (Economic Series No. 15.)

Reconnoisance Soil Survey of Marinette County. Samuel Weidman and Percy O. Wood. 1911. Pp. 44; 4 plates; one map. *Out of print.* Ed. 3,000.

Bulletin No. XXV.   Scientific Series No. 8.

Sandstone of the Wisconsin Coast of Lake Superior. Fredrik Turville Thwaites. 1912. Pp. viii, 117; 23 plates; large map in pocket; 10 figures in text. Cloth bound. *Out of print.* Ed. 1,400.

Bulletin No. XXVI.   Educational Series No. 3.

The Geography and Industries of Wisconsin. R. H. Whitbeck. 1913. Pp. viii, 65; 23 plates; 46 figures in text. *Out of print.* Ed. 6,000.

Bulletin No. XXVII.   Scientific Series No. 9.

The Inland Lakes of Wisconsin. C. Juday. 1914. Pp. vi 137; 29 maps; 8 figures in text. Cloth bound. *Out of print.* Ed. 2,000.

Bulletin No. XXVIII.   Soil Series No. 2.

Soil Survey of Waushara County. A. R. Whitson, W. J. Geib, Guy Conrey and A. R. Kuhlman, of the Wisconsin Geological and Natural History Survey; and J. W. Nelson, of the United States Department of Agriculture. 1913. Pp. iv, 63; 3 plates, including one map. Paper bound. *Out of print.* Ed. 3,000.

Bulletin No. XXIX.   Soil Series No. 3.

Soil Survey of Waukesha County. A. R. Whitson, W. J. Geib and A. H. Meyer, of the Wisconsin Geological and Natural History Survey; and Percy O. Wood and Grove B. Jones, of the United States Department of Agriculture. 1914. Pp. iv, 82; 3 plates, including one map. Paper bound. *Out of print.* Ed. 4,000.

Bulletin No. XXX.   Soil Series No. 4.

Soil Survey of Iowa County. A. R. Whitson, W. J. Geib, T. J. Dunnewald and Emil Truog, of the Wisconsin Geological and Natural History Survey; and Clarence Lounsbury, of the United States Department of Agriculture. 1914. Pp. 61; 2 plates, including one map. Paper bound. *Out of print.* Ed. 3,000.

Bulletin No. XXXI.   Soil Series No. 5.

Soil Survey of the Bayfield Area. A. R. Whitson, W. J. Geib, L. R. Schoenemann and F. L. Musback, of the Wisconsin Geological and Natural History Survey; and Gustavus B. Maynadier, of the United States Department of Agriculture. 1914. Pp. 51; 4 plates, including one. map. Paper bound. *Out of print.* Ed. 3,000.

Bulletin No. XXXII.   Soil Series No. 6.

Reconnaissance Soil Survey of North Part of Northwest Wisconsin. F. L. Musback, T. J. Dunnewald, Carl Thompson and O. I. Bergh. 1914. Pp. vi, 92; 11 plates; 10 figures in text. Includes soil map. Paper bound. *Out of print.* Ed. 10,000.

Bulletin No. XXXIII.   Scientific Series No. 10.

The Polyporaceae of Wisconsin. J. J. Neuman. 1914. Pp. iii, 206; 25 plates. Cloth bound. Price 50 cents. Ed. 2,000.

Bulletin No. XXXIV.   Economic Series No. 16.

The Limestone Road Materials of Wisconsin. W. O. Hotchkiss and Edward Steidtmann. 1914. Pp. viii, 137; 2 figures in text; 41 plates and geological maps of counties. Cloth bound. Price 50 cents. Ed. 3,500.

Bulletin No. XXXV.   Economic Series No. 17.

The Underground and Surface Water Supplies of Wisconsin. Samuel Weidman, of the Wisconsin Geological and Natural History Survey; and A. R. Schultz, of the United States Geological Survey. 1915. Pp. xv, 664; 72 figures in text; 5 plates, and a colored geological map of the state. Scale: 1 inch= 16 miles. Cloth bound. Price $1.50. Ed. 3,000.

Bulletin No. XXXVI.   Educational Series No. 4.

The Physical Geography of Wisconsin. Lawrence Martin, formerly Associate Professor of Physiography and Geography, Universiy of Wisconsin. Second edition, 1932. Pp. xxiv, 610; 211 figures in text; 36 plates. Includes relief map of Wisconsin. Cloth bound. Price $1.25. Ed. 5,000.

Bulletin No. XXXVII.   Soil Series No. 7.

Soil Survey of Fond du Lac County. A. R. Whitson, W. J. Geib, L. R. Schoenemann and C. A. LeClair, of the Wisconsin Geological and Natural History Survey; and Guy Conrey and A. E. Taylor, of the United States Department of Agriculture. 1914. Pp. iv, 85; 5 plates; 2 figures in text. Includes soil map of the county. Paper bound. *Out of print.* Ed. 3,000.

Bulletin No. XXXVIII.   Soil Series No. 8.

Soil Survey of Juneau County. A. R. Whitson, W. J. Geib, L. R. Schoenemann, and C. A. LeClair, of the Wisconsin Geological and Natural History Survey; and E. B. Watson, of the United States Department of Agriculture. 1914. Pp. iv, 93; 5 plates; 2 figures in text. Includes soil map of the county. Paper bound. Price 5 cents. Ed. 4,000.

Bulletin No. XXXIX.   Soil Series No. 9.

Soil Survey of Kewaunee County. A. R. Whitson, W. J. Geib, and E. J. Graul, of the Wisconsin Geological and Natural History Survey, and A. H. Meyer, of the United States Department of Agriculture. 1914. Pp. v, 84; 3 plates; 3 figures in text. Includes soil map of the county. Paper bound. *Out of print.* Ed. 3,000.

Bulletin No. XL.   Soil Series No. 10.

Soil Survey of La Crosse County. A. R. Whitson, W. J. Geib, and T. J. Dunnewald, of the Wisconsin Geological and Natural History Survey; and Clarence Lounsbury, of the United States Department of Agriculture. 1914. Pp. 77; 5 plates; 2 figures in text. Includes soil map of the county. Paper bound. *Out of print.* Ed. 3,000.

Bulletin No. XLI. Economic Series No. 18.

A Study of Methods of Mine Assessments and Valuation. W. L. Uglow. 1914. Pp. v, 73; 12 plates. Cloth bound. Price 50 cents. Ed. 2,500.

Bulletin No. XLII. Educational Series No. 5.

Geography of the Fox-Winnebago Valley. R. H. Whitbeck, Associate Professor of Geography, University of Wisconsin. 1915. Pp. ix, 105; 28 plates. Cloth bound. Price 50 cents. Ed. 4,000.

Bulletin No. XLIII. Soil Series No. 11.

Soil Survey of Vilas and Portions of Adjoining Counties. A. R. Whitson, T. J. Dunnewald, assisted by W. C. Boardman, C. B. Post, and A. R. Albert. 1915. Pp. 77; 4 plates; 2 figures in text. Includes soil map of area. **Paper bound.** *Out of print.* Ed. 2,000.

Bulletin No. XLIV. Economic Series No. 19,

Mineral Land Classification in parts of Northwestern Wisconsin. W. O. Hotchkiss, assisted by E. F. Bean and O. W. Wheelwright. 1915. Pp. vi, 378; 39 figures in text, 9 plates. Cloth bound. Price $5.00. Ed. 3,500.

Bulletin No. XLV. Economic Series No. 20.

The Peat Resources of Wisconsin. Frederick William Huels. 1915. Pp. viii, 274; 20 figures in text, 22 plates. Cloth bound. Price 50 cents. Ed. 2,500.

Bulletin No. XLVI. Economic Series No. 21.

Mineral Lands of Part of Northern Wisconsin. W. O. Hotchkiss, E. F. Bean and H. R. Aldrich. 1929. Pp. xiii, 212; 51 figures; 21 plates. Paper bound. Price $1.00. Ed. 2,500.

Bulletin No. XLVII. Soil Series No. 12.

Reconnoissance Soil Survey of North Eastern Wisconsin. A. R. Whitson, W. J. Geib, Carl Thompson, Clinton B. Post, and A. L. Buser. 1916. Pp. iv, 87; 10 plates; 3 figures in text. Includes soil map of Forest, Florence, Marinette, Oconto, Shawano and Langlade counties. Paper bound. Price 5 cents. Ed. 8,000.

Bulletin No. XLVIII. Soil Series No. 13.

Soil Survey of Jefferson County. A. R. Whitson, W. J. Geib and O. J. Noer, of the Wisconsin Geological and Natural History Survey; and A. H. Meyer, of the United States Department of Agriculture. 1916. Pp. ix, 77; 4 plates; 2 figures in text. Includes map of the county. Paper bound. Price 5 cents. Ed. 4,000.

Bulletin No. XLIX. Soil Series No. 14.

Soil Survey of Columbia County. A. R. Whitson, W. J. Geib and Guy W. Conrey, of the Wisconsin Geological and Natural History Survey; and Arthur

E. Taylor, of the United States Department of Agriculture. 1916. Pp. iv, 84; 4 plates; 3 figures in text. Includes map of the county. Paper bound. Price 5 cents. Ed. 4,000.

Bulletin No. L. Soil Series No. 15.

Soil Survey of North Part of North Central Wisconsin. A. R. Whitson, W. J. Geib, T. J. Dunnewald, C. B. Post, W. C. Boardman, and A. R. Albert, of the Wisconsin Geological and Natural History Survey; and A. E. Taylor, L. R. Schoenemann, and Carl Thompson, of the United States Department of Agriculture. 1916. Pp. iv, 80; 10 plates; 3 figures in text. Includes soil map of Vilas, Oneida, Iron and Price counties. Paper bound. Price 5 cents. Ed. 8,000.

Bulletin No. 51. Scientific Series No. 11.

Inland Lakes of Wisconsin. Temperatures. E. A. Birge. *In preparation.*

Bulletin No. 52A. Soil Series No. 16.

Reconnoissance Soil Survey of South Part of North Central Wisconsin. A. R. Whitson, W. J. Geib, T. J. Dunnewald, C. B. Post, of the Wisconsin Geological and Natural History Survey; and A. E. Taylor, J. B. R. Dickey, and Carl Thompson, of the U. S. Department of Agriculture. 1918. Pp. 108; 4 plates; 2 figures in text. Includes soil map of Marathon, Lincoln, Taylor and Clark counties. Paper bound. Price 5 cents. Ed. 5,000.

Bulletin No. 52B. Soil Series No. 17.

Soil Survey of Wood County. A. R. Whitson, W. J. Geib, G. W. Conrey, C. B. Post, and W. C. Boardman, of the Wisconsin Geological and Natural History Survey. 1918. Pp. 86; 4 plates; 3 figures in text. Includes soil map of the county. Paper bound. Price 5 cents. Ed. 3,500.

Bulletin No. 52C. Soil Series No. 18.

Soil Survey of Portage County. A. R. Whitson, W. J. Geib, T. J. Dunnewald, L. P. Hanson, of the Wisconsin Geological and Natural History Survey; and Clarence Lounsbury and L. Cantrel, of the U. S. Department of Agriculture. 1918. Pp. iv, 79; 3 plates; 3 figures in text. Includes soil map of the county. Paper bound. Price 5 cents. Ed. 3,500.

Bulletin No. 52D. Soil Series No. 19.

Soil Survey of Door County. A. R. Whitson, W. J. Geib, and H. V. Gieb, of the Wisconsin Geological and Natural History Survey; and Carl Thompson, of the U. S. Department of Agriculture. 1919. Pp. 72; 3 plates; 3 figures in text. Includes soil map of the county. Paper bound. Price 5 cents. Ed. 3,000.

Bulletin No. 53A. Soil Series No. 20.

Soil Survey of Dane County. A. R. Whitson, W. J. Geib, and G. W. Conrey, of the Wisconsin Geological and Natural History Survey; and A. E. Taylor of the U. S. Department of Agriculture. 1917. Pp. iv, 86; 10 plates; 3 figures in text. Includes soil map of the county. Paper bound. Price 5 cents. Ed. 7,000.

Bulletin No. 53B. Soil Series No. 21.

Soil Survey of Rock County. A. R. Whitson, W. J. Geib, Guy Conrey, and W. M. Gibbs, of the Wisconsin Geological and Natural History Survey; and A. E. Taylor, of the U. S. Department of Agriculture. 1922. Pp. 80; 5 plates; 3 figures in text. Includes soil map of the county. Paper bound. Price 5 cents. Ed. 4,500.

Bulletin No. 53C. Soil Series No. 22.

Soil Survey of Green County. A. R. Whitson, W. J. Geib, T. J. Dunnewald, F. J. O'Connell, Walter Voskuil, Max J. Edwards, and Kenneth Whitson, of the Wisconsin Geological and Natural History Survey; and A. C. Anderson, of the U. S. Department of Agriculture. 1930. Pp. 82; 7 figures in text. Includes soil map of the county. Paper bound. Price 5 cents. Ed. 3,000.

Bulletin No. 54A. Soil Series No. 23.

Soil Survey of Buffalo County. A. R. Whitson, W. J. Geib, T. J. Dunnewald, and O. J. Noer, of the Wisconsin Geological and Natural History Survey; and Clarence Lounsbury and L. Cantrel, of the U. S. Department of Agriculture. 1917. Pp. iv, 76; 4 plates; 3 figures in text. Includes soil map of the county. Paper bound. Price 5 cents. Ed. 3,000.

Bulletin No. 54B. Soil Series No. 24.

Soil Survey of Jackson County. A. R. Whitson, W. J. Geib, T. J. Dunnewald, of the Wisconsin Geological and Natural History Survey; and A. L. Goodman, of the U. S. Department of Agriculture. 1923. Pp. 85; 5 plates; 2 figures in text. Includes soil map of the county. Paper bound. Price 5 cents. Ed. 3,500.

Bulletin No. 54C. Soil Series No. 25.

Soil Survey of Waupaca County. A. R. Whitson, W. J. Geib, Martin C. Ford, and Martin O. Tosterud, of the Wisconsin Geological and Natural History Survey; and Clarence Lounsbury, of the U. S. Department of Agriculture. 1921. Pp. 84; 3 plates; 2 figures in text. Includes soil map of the county. Paper bound. Price 5 cents. Ed. 3,000.

Bulletin No. 54D. Soil Series No. 26.

Soil Survey of Outagamie County. A. R. Whitson, W. J. Geib, Martin C. Ford, and Martin O. Tosterud, of the Wisconsin Geological and Natural History Survey; and H. V. Geib, of the U. S. Department of Agriculture. 1922. Pp. 77; 3 plates; 3 figures in text. Includes soil map of the county. Paper bound. Price 5 cents. Ed. 3,000.

Bulletin No. 55. Soil Series No. 27.

Report and map (General) of the North Half of Wisconsin. A. R. Whitson, T. J. Dunnewald, and Carl Thompson, of the Wisconsin Geological and Natural

History Survey. This report including map is sold for 25 cents and can only be obtained from the Director of Purchases, State Capitol, Madison. Ed. 15,000.

Bulletin No. 56A. Soil Series No. 28.

Soil Survey of Milwaukee County. A. R. Whitson, W. J. Geib, and T. J. Dunnewald, of the Wisconsin Geological and Natural History Survey. 1919. Pp. 63; 2 plates; 3 figures in text. Includes soil map of the county. Paper bound. *Out of Print.* Ed. 3,000.

Bulletin No. 56B. Soil Series No. 29.

Soil Survey of Racine and Kenosha Areas. W. J. Geib, H. W. Stewart, and Wm. Gibbs, of the Wisconsin Geological and Natural History Survey; and A. E. Taylor, of the U. S. Department of Agriculture. 1923. Pp. 94; 4 plates; 2 figures in text. Includes soil map. Paper bound. Price 5 cents. Ed. 4,000.

Bulletin No. 56C. Soil Series No. 30.

Soil Survey of Walworth County. A. R. Whitson, W. J. Geib, Vern C. Leaper, W. M. Pierre, of the Wisconsin Geological and Natural History Survey; and L. R. Schoenemann, and W. B. Cobb, of the U. S. Department of Agriculture. 1924. Pp. 98; 3 plates; 3 figures in text. Includes soil map of the county. Paper bound. Price 5 cents. Ed. 3,500.

Bulletin No. 57. Scientific Series No. 12.

The Phytoplankton of the Inland Lakes of Wisconsin. Part I. Gilbert M. Smith, Associate Professor of Botany, University of Wisconsin. 1920. Pp. 243; 51 plates. Cloth bound. Price $1.00. Ed. 2,500.
Part II. 1924. Pp. 227; 88 plates. Cloth bound. Price $1.00. Ed. 2,500.

Bulletin No. 58. Educational Series No. 6.

Geography of Southeastern Wisconsin. R. H. Whitbeck, Associate Professor of Geography, University of Wisconsin. 1920. Pp. 160; 3 plates; 100 figures in text. Cloth bound. Price 50 cents. Ed. 4,000.

Bulletin No. 59A. Soil Series No. 31.

Soil Survey of Sheboygan County. A. R. Whitson, W. J. Geib, A. H. Meyer, Geo. D. Scarseth, R. P. Bartholomew, and W. H. Pierre, of the Wisconsin Geological and Natural History Survey; and A. C. Anderson, of the U. S. Department of Agriculture. *In preparation.*

Bulletin No. 59B. Soil Series No. 32.

Soil Survey of Manitowoc County. A. R. Whitson, W. J. Geib, John B. Woods, Harold H. Hull, F. Reed Austin, of the Wisconsin Geological and Natural History Survey; and A. C. Anderson, of the U. S. Department of Agriculture. *In preparation.*

Bulletin No. 59C. Soil Series No. 33.

Soil Survey of Washington and Ozaukee Counties. A. R. Whitson, W. J. Geib, Julius Kubier, Wm. H. Pierre, Vern C. Leaper, and F. J. O'Connell, of the Wisconsin Geological and Natural History Survey; and A. C. Anderson, of the U. S. Department of Agriculture. 1926. Pp. 93; 6 plates; 2 figures in text. Includes soil map of the counties. Paper bound. Price 5 cents. Ed. 3,500.

Bulletin No. 60A. Soil Series No. 34.

Soil Survey of Pierce County. A. R. Whitson, W. J. Geib, Edward Templin, Howard Lathrope, E. H. Rohrbeck, W. H. Pierre, of the Wisconsin Geological and Natural History Survey; and Max J. Edwards and E. H. Bailey, of the U. S. Department of Agriculture. 1930. Pp. 71; 8 figures in text. Includes soil map of the county. Paper bound. Price 5 cents. Ed. 4,000.

Bulletin No. 60B. Soil Series No. 35.

Soil Survey of Monroe County. A. R. Whitson, W. J. Geib, Olaf Stockstad, R. P. Bartholomew, and H. D. Chapman, of the Wisconsin Geological and Natural History Survey; and Ernest H. Bailey, Max J. Edwards, and A. C. Anderson, of the U. S. Department of Agriculture. *In preparation.*

Bulletin No. 60C. Soil Series No. 36.

Soil Survey of Sauk County. A. R. Whitson, W. J. Geib, T. J. Dunnewald, Joseph Fudge, and Homer Chapman, of the Wisconsin Geological and Natural History Survey; and Ernest H. Bailey, of the U. S. Department of Agriculture. 1924-25. *In preparation.*

Bulletin No. 60D. Soil Series No. 37.

Soil Survey of Trempealeau County. A. R. Whitson, W. J. Geib, Joseph F. Fudge, and Burel S. Butman, of the Wisconsin Geological and Natural History Survey; and Max J. Edwards, of the U. S. Department of Agriculture. *In preparation.*

Bulletin No. 60E. Soil Series No. 38.

Soil Survey of Vernon County. A. R. Whitson, W. J. Geib, A. H. Meyer, and Harold G. Frost, of the Wisconsin Geological and Natural History Survey; and Joseph A. Chucka, of the U. S. Department of Agriculture. *Field work in progress.*

Bulletin No. 61A. Soil Series No. 39.

Soil Survey of Calumet County. A. R. Whitson, W. J. Geib, A. H. Meyer, Harold H. Hull, and Joseph A. Chucka, of the Wisconsin Geological and Natural History Survey. *In preparation.*

Bulletin No. 61B. Soil Series No. 40.

Soil Survey of Winnebago County. A. R. Whitson, W. J. Geib, Merritt Whitson, and Harold H. Hull, of the Wisconsin Geological and Natural History Survey; and A. C. Anderson, of the U. S. Department of Agriculture. *In preparation.*

Bulletin No. 61C. Soil Series No. 41.

Soil Survey of Green Lake County. A. R. Whitson, W. J. Geib, T. J. Dunnewald, H. D. Chapman, Kenneth Whitson, and F. J. O'Connell, of the Wisconsin Geological and Natural History Survey; and A. C. Anderson, Ernest H. Bailey, and Max J. Edwards, of the U. S. Department of Agriculture. 1929. Pp. 79; 2 plates; 2 figures in text. Includes soil map of the county. Paper bound. 5 cents. Ed. 2,500.

Bulletin No. 61D. Soil Series No. 42.

Soil Survey of Adams County. A. R. Whitson, W. J. Geib, T. J. Dunnewald, H. W. Stewart, and Oscar Magistad, of the Wisconsin Geological and Natural History Survey; and Julius Kubier, and F. J. O'Connell, of the U. S. Department of Agriculture. 1924. Pp. 83; 6 plates; 3 figures in text. Includes soil map of the county. *Out of print.* Ed. 3,500.

Bulletin No. 62A. Soil Series No. 43.

Soil Survey of Brown County. A. R. Whitson, Harold Bandoli and C. E. Born, of the Wisconsin Geological and Natural History Survey; and A. C. Anderson, of the U. S. Department of Agriculture. *In preparation.*

Bulletin No. 63. Educational Series No. 7.

Educational Collection of Wisconsin Rocks. Frederick Turville Thwaites. 1921. Pp. 33. Only a small number of copies of this bulletin were printed to be sent out with the rock collection to Wisconsin High Schools. *It has not been sent to libraries and exchanges.*

Bulletin No. 64. Scientific Series No. 13.

The Inland Lakes of Wisconsin. The Plankton. 1. Its Quantity and Chemical Composition. Edward A. Birge and Chancey Juday. 1922. Pp. ix, 222; 40 figures in text. Paper bound. Price 50 cents. Ed. 2,500.

Bulletin No. 65. Educational Series No. 8.

The Geography of Southwestern Wisconsin. W. O. Blanchard, Professor of Geography, University of Illinois. (Formerly with Department of Geography, University of Wisconsin.) 1924, Pp. 105; 1 plate; 81 figures in text. Cloth bound. 50 cents. Ed. 5,000.

Bulletin No. 66. Economic Series No. 22.

Limestones and Marls of Wisconsin, by Edward Steidtmann, Professor of Geology, Virginia Military Institute, Lexington, Va. (formerly with Department of Geology, University of Wisconsin) ; with a chapter on the economic possibilities of manufacturing cement in Wisconsin, by W. O. Hotchkiss and E. F. Bean. 1924. Pp. ix, 208; 6 plates; 19 figures in text. Cloth bound. Price 75 cents. Ed. 4,000.

Bulletin No. 67.   Educational Series No. 9.

A Brief Outline of the Geology, Physical Geography, Geography, and Industries of Wisconsin, by W. O. Hotchkiss and E. F. Bean. 1925. Pp. 60. Paper bound. Price 15 cents. Ed. 5,000.

Bulletin No. 68.   Soil Series No. 49.

General Soil Survey Report and Map of the State.   This report includes a description of the larger types of soil occurring in each important section of the state with a discussion of their origin, agricultural adaptation and requirements. A. R. Whitson. 1927. Pp. xi, 270; 32 plates; 20 figures in text. With cloth binding and map.   Price $1.00.   Ed. 5,000.

Bulletin No. 69.   Economic Series No. 23.

Molding Sands of Wisconsin, by David W. Trainer, Jr., of Cornell University. 1928. Pp. 103; 2 plates; 13 figures in text; 5 tables. Price 30 cents. Ed. 3,100.

Bulletin No. 70.   Scientific Series No. 14.

The Fresh Water Mollusca of Wisconsin, by Frank Collins Baker, B. S., Curator, Museum of Natural History, University of Illinois. 1928. Paper bound.
Part I.   Gastropoda. Pp. xx, 507; 28 plates; 202 figures. Price $3.00. Ed. 1,400.
Part II.   Pelecyopoda. Pp. vi, 495; 77 plates; 97 figures. Price $3.00. Ed. 1,400.

Bulletin No. 71.   Scientific Series No. 24.

Geology of the Gogebic Iron Range of Wisconsin, by H. R. Aldrich. 1929. Pp. x, 279; 32 figures; 16 plates. Cloth bound. Price $1.25. Ed. 3,500.

Bulletin No. 72A.   Soil Series No. 50.

Soil Survey of Bayfield County. A. R. Whitson, W. J. Geib, Charles E. Kellogg, M. Whitson, Burel Butman, Delmar S. Fink, M. H. Gallatin, H. H. Hull, of the Wisconsin Geological and Natural History Survey, in cooperation with the State Department of Agriculture. 1929. Pp. 44; 5 plates; 8 figures, including soil map of the county. Paper bound. Price 5 cents.   Ed. 2,000.

Bulletin No. 77A.   Soil Series No. 54.

Preliminary Study of the Profiles of the Principal Soil Types of Wisconsin. Charles E. Kellogg. 1930. Pp. 112; 6 plates; 11 figures. Paper bound. Price 5 cents. Ed. 1,500.

APPENDIX H

ALTITUDES OF CITIES AND VILLAGES ON AND NEAR THE RAILWAYS IN WISCONSIN, WITH A FEW ELEVATIONS OF RIVERS, LAKES AND HILLS[a]

By F. T. Thwaites

Revised by J. M. Hansell

INTRODUCTORY NOTES

The following table is based upon Gannett's "Dictionary of Altitudes" (Gannett, Henry, A Dictionary of Altitudes in the United States, Bull. 274, U. S. Geol. Survey, 1906, pp. 1042-1062. See also Macfarlane, James, An American Geological Railway Guide, New York, 1890, pp. 223-232). It has been revised and enlarged from information furnished by the chief engineers of nearly all the railways operating in the state. All but a few of the elevations given in the older list have been adjusted by checking one railway survey against another and by connections with the levels of the U. S. Geological Survey and of the Corps of Engineers of the

---

[a] The origins of geographical names in Wisconsin are discussed in the following publications.
L. M.

Brunson, A., Hathaway, J., and Calkins, H., Wisconsin Geographical Names, Collections Wis. Hist. Soc., Vol. 1, 1855, pp. 110-126.

Gannett, Henry, The Origin of Certain Place Names in the United States, Bull. 197, U. S. Geol. Survey, 1902, 280 pp.

Kellogg, Louise P., Organization, Boundaries, and Names of Wisconsin Counties, Proc. Wis. Hist. Soc., Vol. 57, 1909, pp. 184-231.

Schoolcraft, H. R., Thirty Years with the Indian Tribes, Philadelphia, 1851.

Stennett, W. H., History of the Origin of Place Names connected with the Chicago and North Western and Chicago, St. Paul, Minneapolis, and Omaha Railways, second edition, Chicago, 1908, 202 pp.

Thwaites, R. G., The Boundaries of Wisconsin, Collections Wis. Hist. Soc., Vol. 11, 1888, pp. 451-501.

Verwyst, Chrysostom, Geographical Names in Wisconsin, Minnesota, and Michigan, Having a Chippewa Origin, Collections Wis. Hist. Soc., Vol. 12, 1892, pp. 390-398.

U. S. Army (U. S. Lake Survey and Mississippi River Commission). A small number of elevations were obtained from Smith's "Water Powers of Wisconsin" (Smith, L. S., Wis. Geol. and Nat. Hist. Survey, 1908, Bull. 20) and the "Geology of Wisconsin" (Vol. 2, 1877, pp. 17-24, 429-433; Vol. 3, 1880, pp. 77-78). A number of dumpy-level and stadia elevations recently obtained by field parties of this Survey on the larger lakes and at inland villages in Burnett, Polk, and Washburn counties, with origins at railroad stations, have also been included. A few levels were run especially for this report. In addition, a considerable number of typographical errors in elevations have been eliminated, so that the list is virtually a new one.

All elevations are above mean sea level. Unless otherwise stated, the elevation printed is that of the track in front of the local railroad station. Since the publication of the first edition of this bulletin the stations and steel on several of the short lines of the State have been removed and some stations on the main lines abandoned. The elevations of these points have been retained in this revised edition because of their value in comparing local and regional altitudes, although specific reference points are lacking. The elevations are given to the nearest foot only and are not intended for exact engineering work. For the elevations and locations of permanent "Bench Marks" recourse should be had to R. B. Marshall's "Results of Spirit Leveling in Wisconsin, 1897 to 1914 inclusive," (Bull. 570, U. S. Geol. Survey, 1914, 82 pp.), and to the mimeographed briefs now in the files of this Survey containing this information for the quadrangles surveyed since the above mentioned bulletin was published.

Approximate elevations of other points in Wisconsin may be obtained from the topographic maps listed in Appendix E. It should be remembered that those given on maps of the U. S. Geological Survey made before 1904 are barometric, whereas those on later maps are based on rapid spirit leveling over the roads. Barometric elevations will be found in the volumes of the "Geology of Wisconsin" as follows: Vol. 2, 1877, pp. 107-127, 429-447, 650-651; Vol. 3, 1880, pp. 78-80, 312-314; Vol. 4, 1882, pp. 138-140, 747-748; and in Bull. 27, Wis. Geol. and Nat. Hist. Survey, 1914. A large number of unpublished barometric elevations are also on file in the Survey office.

A few of the elevations given on Plate I*of this bulletin are barometric. Some are very accurate, but in general these barometric elevations cannot be depended upon within 50 feet. Hand-leveled profiles will be found on the maps of Bulletin 44 of this Survey. A few of these were not referred to sea level and the railroad surveys were not then adjusted as they were for this table. For instance, the elevations in the vicinity of Crandon are 47 feet too low. In addition to the foregoing a large number of hand-level elevations are in the Survey files for which there are no published profiles. Accurately leveled elevations on the abandoned shore lines of Lake Michigan and Green Bay will be found in Bulletin 17. These are not tied to permanent bench marks.

KEY TO ABBREVIATIONS

A. & W.	Ahnapee & Western Railway (Kewaunee and Door Counties).
B. T.	Bayfield Transfer Railway, and connecting lines (Bayfield County).
B. M.	Bench marks—usually an iron post or metal tablet when on a stone building. Elevation given is to nearest foot only and given only where there is no railway station. See Bull. 570, U. S. Geol. Survey, for exact elevations.
C. B. & Q.	Chicago, Burlington & Quincy Railroad.
C. & N. W.	Chicago & Northwestern Railway.
C. M. St. P. & P.	Chicago, Milwaukee, St. Paul, and Pacific Railway.
C. St. P. M. & O.	Chicago, St. Paul, Minneapolis & Omaha Railway, or "Omaha Line."
C. V. & N.	Chippewa Valley and Northern Railway (Rusk and Sawyer counties).
D. S. S. & A.	Duluth, South Shore & Atlantic Railway.
F. & N. E.	Fairchild & Northeastern Railway (Clark and Eau Claire counties).
G. B. & W.	Green Bay & Western Railroad.
G. N.	Great Northern Railway (Douglas County).
I. C.	Illinois Central Railway.
K. G. B. & W.	Kewaunee, Green Bay & Western Railroad (Brown and Kewaunee counties).
M. P. & N.	Mineral Point & Northern Railway (Lafayette and Iowa counties).
M. R. C.	Mississippi River Commission.
N. P.	Northern Pacific Railway.
S. L.	"Soo Line"—Minneapolis, St. Paul & Sault Ste. Marie Railway.
S. M. & P.	Stanley, Merrill & Phillips Railway (Clark, Taylor and Rusk counties).
U. S. G. S.	United States Geological Survey.
U. S. L. S.	United States Lake Survey, U. S. Army. Corps of Engineers.
W. G. & N. H. S.	Wisconsin Geological and Natural History Survey.
W. & M.	Wisconsin and Michigan Railway (Marinette County).
W. & N.	Wisconsin and Northern Railway (Shawano, Langlade and Forest counties).
X	Grade crossing of railways. Elevations given on authority of only one of the lines.

The officers of each of the railways mentioned above furnished profiles and other information for use in determining these altitudes. Acknowledgment is gratefully made to the railways, as well as to the United States Geological Survey. The latter bureau gave the Wisconsin Geological and Natural History Survey permission to revise this dictionary of altitudes and supplied some topographic data from unpublished maps.

---

*Excluded from the 1965 reprinting.

ELEVATIONS

Locality	County	Authority	Elevation
			Feet
Abbotsford	Clark	S. L.	1,422
Ablemans	Sauk	C. & N. W.	882
Abrams	Oconto	C., M., St. P. & P.	678
Ackerville	Washington	C., M., St. P. & P.	1,056
Adams	Adams	C. & N. W.	956
Adell	Sheboygan	C., M., St. P. & P.	904
Afton	Rock	C. & N. W.	764
X—C. & N. W.		C., M., St. P. & P.	754
Agnew	Ashland	S. L.	796
Albany	Green	C., M., St. P. & P.	818
Albertville	Chippewa	S. L.	1,022
Alden, B. M.	Polk	U. S. G. S.	953
Alder	Ashland	C. & N. W.	700
Algoma	Kewaunee	A. & W.	590
Allen	Eau Claire	U. S. G. S.	963
Allen Grove	Walworth	C., M., St. P. & P.	918
Allenton	Washington	S. L.	949
Allenville	Winnebago	C. & N. W.	801
Allouez	Douglas	N. P.	637
		D., S. S. & A.	648
X—G. N.		D., S. S. & A.	651
Alma	Buffalo	M. R. C.	674
Alma Center	Jackson	G. B. & W.	971
Almena	Barron	S. L.	1,191
Almond	Portage	C. & N. W.	1,161
Altoona	Eau Claire	C., St. P., M. & O.	890
Alverno	Manitowoc	S. L.	730
Amberg	Marinette	C., M., St. P. & P.	899
Amery	Polk	S. L.	1,076
Amherst	Portage	S. L.	1,064
Amherst Junction	Portage	G. B. & W.	1,126
		S. L.	1,097
Amnicon	Douglas	N. P.	852
Anderson Mills, B. M.	Grant	U. S. G. S.	699
Aniwa	Shawano	C. & N. W.	1,411
Anson	Chippewa	C., St. P., M. & O.	954
Anston	Brown	C. & N. W.	743
Antigo	Langlade	C. & N. W.	1,499
Anton	Outagamie	S. L.	834
Appleton	Outagamie	C. & N. W.	723
		C., M., St. P. & P.	719
		S. L.	801
Appleton Junction	Outagamie	C. & N. W.	803
Arbor Vitae	Vilas	C., M., St. P. & P.	1,627
Arbutus	Marinette	C., M., St. P. & P.	1,061
Arcadia	Trempealeau	G. B. & W.	725
Arena	Iowa	C., M., St. P. & P.	738
Argyle, I. C. Sta.	Lafayette	U. S. G. S.	802
Arkansaw	Pepin	U. S. G. S.	738
Arkdale	Adams	C. & N. W.	919
Arlington	Columbia	C., M., St. P. & P.	1,047
Armstrong Creek	Forest	S. L.	1,427

ELEVATIONS

Locality	County	Authority	Elevation
			Feet
Arnott	Portage	G. B. & W.	1,147
Arpin	Wood	C., M., St. P. & P.	1,147
		C. & N. W.	1,152
Ashippun	Dodge	C. & N. W.	854
Ashland	Ashland	S. L.	671
Union Sta.		C. & N. W.	666
Ashland Jct.	Bayfield	C., St. P., M. & O.	644
Askeaton	Brown	C., M., St. P. & P.	749
Askenett	Shawano	W. & N.	1,118
Asylum	Outagamie	C. & N. W.	780
Athelstane	Marinette	C., M., St. P. & P.	937
Athens	Marathon	S. L.	1,359
Attica	Green	U. S. G. S.	843
Atwater	Dodge	C., M., St. P. & P.	931
Auburndale	Wood	S. L.	1,215
Augusta	Eau Claire	C., St. P., M. & O.	968
Avalon	Rock	C., M., St. P. & P.	953
Avoca	Iowa	C., M., St. P. & P.	698
Babcock	Wood	C., M., St. P. & P.	975
Badger Mills	Chippewa	C., M., St. P. & P.	816
		S. L.	895
Bagley	Grant	M. R. C.	626
Bagley Jct.	Marinette	C., M., St. P. & P.	646
Baldwin	St. Croix	C., St. P., M. & O.	1,136
Ballou	Ashland	S. L.	1,428
Balsam Lake, water surface	Polk	W. G. & N. H. S.	1,128
Bancroft	Portage	S. L.	1,089
Bangor	La Crosse	C. & N. W.	753
		C., M., St. P. & P.	743
Bannerman	Waushara	U. S. G. S.	818
Baraboo	Sauk	U. S. G. S.	864
Baraboo River		W. G. & N. H. S.	
Mouth	Columbia		783
Baraboo, Woolen Mills dam, crest	Sauk		843
Reedsburg, above dam			871
Elroy	Juneau		938
Bardwell	Walworth	C., M., St. P. & P.	870
R. R. X.		C., M., St. P. & P.	867
Barneveld, C. & N. W. Sta.	Iowa	U. S. G. S.	1,235
Barnum	Crawford	C., M., St. P. & P.	678
Barron	Barron	S. L.	1,120
Barronett	Barron	C., St. P., M. & O.	1,375
Bartel	Ozaukee	C. & N. W.	673
Barton	Washington	C. & N. W.	905
Basco	Dane	I. C.	905
Bateman	Chippewa	S. L.	939
Baxter	Marinette	W. & M.	707
Bay City	Washington	M. R. C.	689
Bayfield	Bayfield	C., St. P., M. & O.	617
Bay View	Milwaukee	C. & N. W.	594

ELEVATIONS

Locality	County	Authority	Elevation
			Feet
Bear Creek	Outagamie	C. & N. W.	816
Bear Skin	Oneida	C., M., St. P. & P.	1,526
Bear Trap	Ashland	C. & N. W.	643
Beaver	Marinette	C., M., St. P. & P.	674
Beaver Dam	Dodge	C., M., St. P. & P.	867
Beaver Dam Jct.	Dodge	C., M., St. P. & P.	917
Beaver Lake, water surface	Waukesha	U. S. G. S.	910
Beecher Lake	Marinette	C., M., St. P. & P.	958
Beetown, B. M.	Grant	U. S. G. S.	793
Beldenville	Pierce	C., St. P., M. & O.	976
Belgium	Ozaukee	C. & N. W.	737
Bell Center	Crawford	C., M., St. P. & P.	697
Belle Plaine	Shawano	C. & N. W.	847
Belleville	Dane	I. C.	864
Bellevue	Brown	C. & N. W.	759
Belmont	Lafayette	C., M., St. P. & P.	1,010
Beloit	Rock	C., M., St. P. & P.	743
Beloit Jct.		C., M., St. P. & P.	766
Belton	Milwaukee	C. & N. W.	770
Bena	Bayfield	N. P.	1,075
Benders Corners	Vernon	U. S. G. S.	1,289
Bennett	Douglas	C., St. P., M. & O.	1,185
Benoit	Bayfield	C., St. P., M. & O.	890
Benoit Lake, water surface	Burnett	W. G. & N. H. S.	974
Benson (Randall P. O.)	Burnett	N. P.	885
Benton	Lafayette	C. & N. W.	846
Berg Park	Douglas	D., S. S. & A.	1,115
Berlin	Green Lake	C., M., St. P. & P.	762
Berryville	Kenosha	C. & N. W.	626
Beverly	Sawyer	S. L.	1,305
Bibon	Bayfield	D., S. S. & A.	914
		C., St. P., M. & O.	935
Big Falls	Waupaca	C. & N. W.	913
Big McKenzie Lake, water surface	Burnett	W. G. & N. H. S.	986
Big Suamico	Brown	C. & N. W.	605
Binghampton	Outagamie	S. L.	822
Birch	Ashland	C. & N. W.	838
Birch Island Lake, water surface	Burnett	W. G. & N. H. S.	988
Birchwood, Union Station	Washburn	C., St. P., M. & O.	1,246
Birnamwood	Shawano	C. & N. W.	1,286
Black Creek	Outagamie	S. L.	793
X—S. L.		G. B. & W.	792
Black Earth	Dane	C., M., St. P. & P.	816
Blackhawk, B. M.	Sauk	U. S. G. S.	775
Black River		W. G. & N. H. S.	
La Crosse	La Crosse		628
Black River Falls, above dam	Jackson		763
Hatfield			838
F. & N. E. Bridge	Clark		1,094

ELEVATIONS

Locality	County	Authority	Elevation
			Feet
Black River (continued)		W. G. & N. H. S.	
S. L. Bridge	Clark		1,198
Black River Falls	Jackson	C., St. P., M. & O.	805
Blackwell Junction	Forest	C. & N. W.	1,522
Black Wolf	Winnebago	S. L.	788
Blair	Trempealeau	G. B. & W.	849
Blanchardville	Lafayette	I. C.	830
Blanchardville, I. C. Sta.	Lafayette	U. S. G. S.	831
Bloomer	Chippewa	C., St. P., M. & O.	1,006
Blueberry	Douglas	N. P.	1,136
Blue Mounds	Dane	U. S. G. S.	1,301
Blue Mounds, top west mound	Iowa	W. G. & N. H. S.	1,716
Blue River	Grant	C., M., St. P. & P.	677
Bluff Siding, X—C. & N. W. & C. B. & Q.	Buffalo	U. S. G. S.	660
Boardman	St. Croix	C., St. P., M. & O.	957
Boland	Fond du Lac	S. L.	834
Bolton	Vilas	C. & N. W.	1,622
Bonduel	Shawano	C. & N. W.	884
Bone Lake, water surface	Polk	W. G. & N. H. S.	1,148
Bonita	Oconto	C. & N. W.	872
Boscobel	Grant	C., M., St. P. & P.	673
Boulder Junction	Vilas	C., M., St. P. & P.	1,655
Bowler	Shawano	C. & N. W.	1,081
Boyceville	Dunn	S. L.	950
Boyd	Chippewa	S. L.	1,107
Boydtown, B. M.	Crawford	U. S. G. S.	653
Boylston	Douglas	G. N.	689
		S. L.	690
Brackett	Eau Claire	U. S. G. S.	949
Branch	Manitowoc	C. & N. W.	732
Brandon	Fond du Lac	C., M., St. P. & P.	997
Brantwood	Price	S. L.	1,696
Breed	Oconto	C. & N. W.	879
Bridgeport	Crawford	C., M., St. P. & P.	640
Bright	Clark	F. & N. E.	1,300
Brighton Beach	Winnebago	S. L.	749
Brill	Barron	C., St. P., M. & O.	1,192
Brillion	Calumet	C. & N. W.	825
Bristol	Kenosha	C. & N. W.	772
Brodhead	Green	C., M., St. P. & P.	790
Brookfield	Waukesha	C., M., St. P. & P.	835
Brooklyn	Green	C. & N. W.	979
Brooks	Adams	C. & N. W.	954
Brookside	Oconto	C. & N. W.	603
Brownlee	Washburn	S. L.	1,198
Browntown	Green	C., M., St. P. & P.	789
Bruce	Rusk	S. L.	1,093
Brule	Douglas	N. P.	994
Brunet	Chippewa	C., St. P., M. & O.	1,022
Bryant	Langlade	C. & N. W.	1,587
Buckbee	Waupaca	C. & N. W.	832

ELEVATIONS

Locality	County	Authority	Elevation
			Feet
Budd	Vernon	U. S. G. S.	1,207
Buffalo	Marquette	C. & N. W.	799
Buncombe	Lafayette	C. & N. W.	697
Burke	Dane	C., M., St. P. & P.	902
Burkhardt	St. Croix	C., St. P., M. & O.	928
Burlington	Racine	C., M., St. P. & P.	786
		S. L.	765
Burnette Jct.	Dodge	C. & N. W.	873
Burnside	Outagamie	C. & N. W.	857
Bushman	Marathon	S. L.	1,413
Butler	Milwaukee	C. & N. W.	736
Butternut	Ashland	S. L.	1,504
Byron	Fond du Lac	S. L.	1,058
Cable	Bayfield	C., St. P., M. & O.	1,372
Cadott	Chippewa	S. L.	979
Calamine	Lafayette	C., M., St. P. & P.	824
Caledonia	Racine	C., M., St. P. & P.	723
Calhoun	Waukesha	C. & N. W.	851
Callon	Marathon	C. & N. W.	1,237
Calumet Yard	Manitowoc	C. & N. W.	645
Calvary	Fond du Lac	C. & N. W.	945
Calvert	La Crosse	C. B. & Q.	643
Cambria	Columbia	C., M., St. P. & P.	861
Cameron, Union Station	Barron	C., St. P., M. & O.	1,099
Campbellsport	Fond du Lac	C. & N. W.	1,045
Camp Douglas	Juneau	C., St. P., M. & O.	935
Campia	Barron	S. L.	1,168
Camp Lake	Kenosha	S. L.	751
Camp No. 4	Oneida	C. & N. W.	1,687
Canton	Barron	S. L.	1,101
Carson	Iron	C. & N. W.	1,624
Carter	Forest	C. & N. W.	1,474
Cary	Wood	C., M., St. P. & P.	1,073
Caryville	Dunn	C., M., St. P. & P.	761
Casco	Kewaunee	A. & W.	712
Casco Jct.	Kewaunee	K. G. B. & W.	728
Cashton	Monroe	C., M., St. P. & P.	1,365
Cassian, junction switch	Oneida	C., M., St. P. & P.	1,518
Cassville	Grant	M. R. C.	620
Cataract, B. M.	Jackson	U. S. G. S.	847
Catawba	Price	S. L.	1,493
Cato	Manitowoc	C. & N. W.	839
Cayuga	Ashland	S. L.	1,457
Cecil	Shawano	C. & N. W.	810
Cedar	Iron	C. & N. W.	1,094
Cedarburg	Ozaukee	C., M., St. P. & P.	778
Cedar Falls	Dunn	C., M., St. P. & P.	869
Cedar Grove	Sheboygan	C. & N. W.	697
Cedarhurst	Clark	C., St. P., M. & O.	1,214
Cedar Lake	Washington	S. L.	1,030

ELEVATIONS

Locality	County	Authority	Elevation
			Feet
Cedarville	Marinette	C., M., St. P. & P.	832
Center Valley	Outagamie	S. L.	825
Centerville	Trempealeau	U. S. G. S.	738
Centuria	Polk	S. L.	1,227
Chaseburg, B. M.	Vernon	U. S. G. S.	704
Chelsea	Taylor	S. L.	1,525
Cheney	Columbia	C., M., St. P. & P.	813
Chester	Dodge	C. & N. W.	877
Chetek	Barron	C., St. P., M. & O.	1,048
Chicago Jct	Washburn	C., St. P., M. & O.	1,086
Chili	Clark	C., St. P., M. & O.	1,233
Chilton	Calumet	C., M., St. P. & P.	856
Chippewa Falls	Chippewa	C., St. P., M. & O.	859
		S. L.	833
X—C., St. P., M. & O.		S. L.	861
X—C., St. P., M. & O.		S. L.	906
Chippewa River		W. G. & N. H. S.	
Chippewa Falls, head of dam	Chippewa		839
Jim Falls, head			936
Brunett Falls, head			993
Flambeau River, mouth	Rusk		1,050
Bruce			1,064
East Branch			
Pelican Lake	Ashland		1,462
Glidden			1,509
Chittamo	Washburn	S. L.	1,098
City Point	Jackson	G. B. & W.	964
Clam Falls	Polk	W. G. & N. H. S.	1,045
Clam Lake, water surface	Burnett	W. G. & N. H. S.	946
Clarno, I. C. Sta	Green	U. S. G. S.	841
Clayton	Polk	C., St. P., M. & O.	1,201
Clayton	Winnebago	S. L.	885
Claywood	Oconto	C. & N. W.	831
Clear Lake	Polk	C., St. P., M. & O.	1,199
Clear Lake, water surface	Rock	U. S. G. S.	800
Cleghorn	Eau Claire	U. S. G. S.	963
Cleveland	Manitowoc	C. & N. W.	638
Clinton	Rock	C. & N. W.	945
Clintonville	Waupaca	C. & N. W.	824
Cloverdale	Juneau	C. & N. W.	925
Clyde	Kewaunee	K. G. B. & W.	631
Clyman	Dodge	C. & N. W.	901
Clyman Jct	Dodge	C. & N. W.	890
Cobb	Iowa	C. & N. W.	1,175
Cochrane	Buffalo	C. B. & Q.	683
Coda	Bayfield	N. P.	1,095
Colby	Marathon	S. L.	1,355
Coleman	Marinette	C., M., St. P. & P.	715
Colfax	Dunn	S. L.	949
Colgate	Washington	S. L.	970

ELEVATIONS

Locality	County	Authority	Elevation
			Feet
Collins	Manitowoc	S. L.	809
Coloma	Waushara	S. L.	1,042
Columbia	Clark	C., St. P., M. & O.	962
Columbus	Columbia	C., M., St. P. & P.	842
Combined Locks	Outagamie	C. & N. W.	665
Comfort	Dunn	C., St. P., M. & O.	848
Commonwealth	Florence	C. & N. W.	1,315
Comstock	Barron	C., St. P., M. & O.	1,282
Conover	Vilas	C. & N. W.	1,658
Conrath	Rusk	S. L.	1,128
Coolidge	Price	S. L.	1,499
Coon Valley, B. M.	Vernon	U. S. G. S.	753
Coral City, B. M.	Trempealeau	U. S. G. S.	831
Corinth	Marathon	S. L.	1,436
Cormier	Brown	C., M., St. P. & P.	602
Corning	Columbia	S. L.	778
Cottage Grove	Dane	U. S. G. S.	889
Couderay	Sawyer	C., St. P., M. & O.	1,260
County House	Douglas	D., S. S. & A.	694
County Line	Racine	C. & N. W.	714
Coxie	Clark	F. & N. E.	1,258
Cranberry Center	Juneau	C. & N. W.	927
Crandon	Forest	C. & N. W.	1,635
		W. & N.	1,622
Cranmoor	Wood	C., M., St. P. & P.	982
Crawford	Crawford	C. B. & Q.	641
Cream, B. M.	Buffalo	U. S. G. S.	775
Crivitz	Marinette	C., M., St. P. & P.	684
Cross Plains	Dane	U. S. G. S.	861
Cuba City	Grant	C. & N. W.	1,005
Cudahy	Milwaukee	C. & N. W.	714
Cullen	Oconto	C. & N. W.	604
Culver	Langlade	W. & N.	1,280
Cumberland	Barron	C., St. P., M. & O.	1,242
Curtiss	Clark	S. L.	1,372
Cushing	Polk	W. G. & N. H. S.	1,015
Cusson	Bayfield	D., S. S. & A.	1,105
Custer	Portage	S. L.	1,173
Cutter	Douglas	N. P.	737
Cylon	St. Croix	S. L.	1,060
Dale	Outagamie	S. L.	806
Dallas	Barron	S. L.	1,156
Dalton	Green Lake	C. & N. W.	824
Daly	Wood	C., M., St. P. & P.	965
Danbury	Burnett	S. L.	932
Dancy	Marathon	C., M., St. P. & P.	1,128
Dane	Dane	C. & N. W.	1,062
Darien	Walworth	C., M., St. P. & P.	946
Darlington	Lafayette	C., M., St. P. & P.	814
Darwin	Dane	C., M., St. P. & P.	884
Dauby	Bayfield	C., St. P., M. & O.	853

ELEVATIONS

Locality	County	Authority	Elevation
			Feet
Deanville	Dane	U. S. G. S.	884
Deckers	Ozaukee	C. & N. W.	736
Dedham	Douglas	G. N.	803
Deer Brook	Langlade	C. & N. W.	1,531
Deerfield	Dane	U. S. G. S.	861
Deer Lake, water surface	Polk	W. G. & N. H. S.	1,107
Deer Park	St. Croix	C., St. P., M. & O.	1,060
De Forest	Dane	C., M., St. P. & P.	943
Delavan	Walworth	C., M., St. P. & P.	938
Dells Mills	Eau Claire	C., M., St. P. & P.	807
Delta	Bayfield	D., S. S. & A.	1,056
Denmark	Brown	C. & N. W.	874
Denzer, B. M.	Sauk	U. S. G. S.	803
Depere	Brown	C., M., St. P. & P.	595
		C. & N. W.	605
Deronda	Polk	S. L.	1,072
De Soto	Vernon	M. R. C.	636
Devils Lake	Sauk	C. & N. W.	971
Devils Lake, water surface	Burnett	W. G. & N. H. S.	953
Dewey	Douglas	G. N.	819
Dexterville	Wood	C., M., St. P. & P.	992
Diamond Bluff	Pierce	C. B. & Q.	722
Dickeyville, B. M.	Grant	U. S. G. S.	955
Dillmans	Milwaukee	C. & N. W.	668
Disco, B. M.	Jackson	U. S. G. S.	969
Dodge	Trempealeau	U. S. G. S.	676
Dodgeville	Iowa	C. & N. W.	1,253
		I. C.	1,178
Donald	Taylor	S. L.	1,186
X—C., St. P., M. & O.		S. L.	1,188
Donges	Ozaukee	C. & N. W.	691
Dorchester	Clark	S. L.	1,422
Dotyville, B. M.	Fond du Lac	U. S. G. S.	1,041
Dousman	Waukesha	C. & N. W.	868
Dover	Racine	C., M., St. P. & P.	811
Downing	Dunn	S. L.	980
Downsville	Dunn	C., M., St. P. & P.	765
Doylestown	Columbia	C., M., St. P. & P.	948
Draper	Sawyer	C., St. P., M. & O.	1,472
Drephal	Outagamie	S. L.	792
Dresser Jct.	Polk	U. S. G. S.	954
Drexel (Kent P. O.)	Langlade	C. & N. W.	1,698
Drummond	Bayfield	C., St. P., M. & O.	1,299
Duck Creek	Brown	C. & N. W.	589
Dunbarton	Lafayette	C., M., St. P. & P.	986
Dundas	Calumet	C. & N. W.	819
Dunnville	Dunn	C., M., St. P. & P.	733
Duplainville	Waukesha	S. L.	851
X—C., M., St. P. & P.		S. L.	852
X—C., M., St. P. & P.		S. L.	821
Durand	Pepin	C., M., St. P. & P.	728

ELEVATIONS

Locality	County	Authority	Elevation
			Feet
Eagle	Waukesha	C., M., St. P. & P.	945
Eagle Jct.	Waukesha	C., M., St. P. & P.	942
Eagle Point	Chippewa	C., St. P., M. & O.	968
Eagle River	Vilas	C. & N. W.	1,638
Earl	Washburn	C., St. P., M. & O.	1,102
East Farmington, B. M.	Polk	U. S. G. S.	1,038
Eastman, B. M.	Crawford	U. S. G. S.	1,224
East Winona—X—C. B. & Q. with C. & N. W.	Buffalo	M. R. C.	659
Eau Claire—X—C., M., St. P. & P. with C., St. P., M. & O.	Eau Claire	U. S. G. S.	880
		S. L.	840
		C., M., St. P. & P.	791
		C., St. P., M. & O.	837
Eau Galle, B. M.	Dunn	U. S. G. S.	838
Eden	Fond du Lac	C. & N. W.	1,015
Edgar	Marathon	C. & N. W.	1,238
Edgerton	Rock	C., M., St. P. & P.	822
Edgewater	Sawyer	S. L.	1,268
Edmund	Iowa	C. & N. W.	1,202
Eidsvold	Clark	S. L.	1,137
Eland Junction	Shawano	C. & N. W.	1,237
Elba	Dodge	C., M., St. P. & P.	823
Elcho	Langlade	C. & N. W.	1,635
Elderon	Marathon	C. & N. W.	1,203
El Dorado	Fond du Lac	C. & N. W.	897
Eleva	Trempealeau	C., St. P., M. & O.	866
Eliot	Bayfield	D., S. S. & A.	1,161
Elk Creek	Trempealeau	U. S. G. S.	827
Elkhart	Sheboygan	C., M., St. P. & P.	946
Elkhorn	Walworth	C., M., St. P. & P.	996
Elk Mound	Dunn	C., St. P., M. & O.	929
Ellsworth	Pierce	C., St. P., M. & O.	1,069
Elm Grove	Waukesha	C., M., St. P. & P.	744
Elmhurst	Langlade	C. & N. W.	1,461
Elm Lake	Wood	G. B. & W.	1,018
Elmo	Grant	C. & N. W.	1,016
Elmwood	Pierce	C., St. P., M. & O.	870
Elroy	Juneau	C., St. P., M. & O.	961
Elton	Langlade	C. & N. W.	1,385
Embarrass	Waupaca	C. & N. W.	795
Emerald, Union Sta.	St. Croix	S. L.	1,147
Endeavor	Marquette	S. L.	781
Enderline	Bayfield	N. P.	1,126
Engle	Columbia	C. & N. W.	883
Engoe.	Bayfield	N. P.	734
Ettrick, B. M.	Trempealeau	U. S. G. S.	772
Evansville	Rock	C. & N. W.	897
Evergreen Lodge	Langlade	W. & N.	1,276
Exeland	Sawyer	S. L.	1,192
X—C., V. & N.		S. L.	1,189
Exeter, B. M.	Green	U. S. G. S.	874

ELEVATIONS

Locality	County	Authority	Elevation
			Feet
Fairchild	Eau Claire	C., St. P., M. & O.	1,066
		F. & N. E.	1,069
Fairplay, B. M.	Grant	U. S. G. S.	863
Fairview	Crawford	U. S. G. S.	1,212
Fairwater	Fond du Lac	C., M., St. P. & P.	952
Fall Creek	Eau Claire	C., St. P., M. & O.	937
Fall River	Columbia	C., M., St. P. & P.	866
Falun	Burnett	W. G. & N. H. S.	956
Fancher	Portage	G. B. & W.	1,132
Fennimore	Grant	C. & N. W.	1,196
Fenwood	Marathon	C. & N. W.	1,286
Ferryville	Crawford	M. R. C.	635
Fifield	Price	S. L.	1,454
Fisk	Winnebago	C., M., St. P. & P.	840
Fitchburg	Dane	I. C.	1,014
Flambeau River		W. G. & N. H. S.	
Mouth	Rusk		1,050
Ladysmith, above dam			1,115
Foot, Big Falls			1,178
Forks	Sawyer		1,284
Park Falls	Price		1,470
Florence	Florence	C. & N. W.	1,290
Fond du Lac	Fond du Lac	C. & N. W.	759
		C., M., St. P. & P.	754
		S. L.	749
X—C. & N. W.		S. L.	775
Footville	Rock	C. & N. W.	819
Forest City	Bayfield	C., St. P., M. & O.	914
Forest Jct.	Calumet	C., M., St. P. & P.	833
Forestville	Door	A. & W.	593
Fort Atkinson	Jefferson	C. & N. W.	794
Forward, B. M.	Dane	U. S. G. S.	963
Foster	Ashland	S. L.	1,259
Foster	Eau Claire	F. & N. E.	982
Fountain City	Buffalo	M. R. C.	662
Fowler Lake, water surface	Waukesha	U. S. G. S.	861
Foxboro	Douglas	G. N.	951
Fox Lake	Dodge	C., M., St. P. & P.	885
Fox Lake Jct.	Dodge	C., M., St. P. & P.	882
Fox Point	Milwaukee	C. & N. W.	689
Fox River	Kenosha	C. & N. W.	781
Fox River		W. G. & N. H. S.	
Menasha dam, crest	Winnebago		747
Appleton, upper lock, crest	Outagamie		738
Littlechute locks, crest			692
Grand Kaukauna locks, crest			655
Depere lock, crest	Brown		589
Francis Creek	Manitowoc	C. & N. W.	717
Franksville	Racine	C., M., St. P. & P.	731
Frederic	Polk	S. L.	1,210
Fredonia Sta.	Ozaukee	C., M., St. P. & P.	795

ELEVATIONS

Locality	County	Authority	Elevation
			Feet
Fremont	Waupaca	S. L.	778
Friendship, Adams Sta.	Adams	C. & N. W.	956
Fries Lake, water surface	Washington	U. S. G. S.	958
Friesland	Columbia	C. & N. W.	995
Gagen	Oneida	C. & N. W.	1,645
Galesville, B. M.	Trempealeau	U. S. G. S.	712
Galloway	Marathon	C. & N. W.	1,175
Gaslyn Lake, water surface	Burnett	W. G. & N. H. S.	973
Gays Mills	Crawford	C., M., St. P. & P.	703
Genesee Lake, water surface	Waukesha	U. S. G. S.	866
Genesee Sta.	Waukesha	C., M., St. P. & P.	906
Geneva Lake, water surface	Walworth	W. G. & N. H. S.	864
Genoa	Vernon	M. R. C.	639
Genoa City	Walworth	C. & N. W.	845
Germania	Iron	S. L.	1,508
Germantown	Washington	C., M., St. P. & P.	858
Gibraltar Bluff	Columbia	U. S. G. S.	1,240
Gibson	Milwaukee	C., M., St. P. & P.	675
Gifford	Waukesha	C., M., St. P. & P.	885
Gilbert	Lincoln	C., M., St. P. & P.	1,432
Gilberts	Eau Claire	F. & N. E.	1,056
Gile	Iron	C. & N. W.	1,496
		S. L.	1,473
Gillett	Oconto	C. & N. W.	798
Gillett Jct.	Oconto	C. & N. W.	846
Gills Landing	Waupaca	S. L.	759
Gilman	Taylor	S. L.	1,213
		S. M. & P.	1,217
Gilmanton, B. M.	Buffalo	U. S. G. S.	786
Girard Jct.	Marinette	C., M., St. P. & P.	1,041
Glenbeulah	Sheboygan	C. & N. W.	972
Glendale	Monroe	C. & N. W.	995
Glen Flora	Rusk	S. L.	1,271
Glen Haven	Grant	M. R. C.	625
Glenoak	Marquette	C. & N. W.	788
Glenwood	St. Croix	S. L.	1,028
Glidden	Ashland	S. L.	1,521
Goll	Marinette	W. & M.	714
Goodnow	Oneida	C., M., St. P. & P.	1,513
Goodrich	Taylor	S. L.	1,397
Goodrich Jct.	Marathon	S. L.	1,340
Gordon	Douglas	C., St. P., M. & O.	1,038
		S. L.	1,058
X—C. & N. W.		S. L.	1,028
Gorman	Clark	F. & N. E.	1,126
Gotham	Richland	C., M., St. P. & P.	701
Government Hill	Waukesha	U. S. G. S.	1,233
Grafton	Ozaukee	C., M., St. P. & P.	757
Grand Crossing	La Crosse	C. B. & Q.	644
Grand Marsh	Adams	C. & N. W.	1,010
Grand View	Bayfield	C., St. P., M. & O.	1,021

ELEVATIONS

Locality	County	Authority	Elevation
			Feet
Granite City	Waupaca	C. & N. W.	899
Granite Heights	Marathon	C., M., St. P. & P.	1,217
Granton	Clark	C., St. P., M. & O.	1,112
Grantsburg	Burnett	N. P.	905
Granville	Milwaukee	C. & N. W.	723
		C., M., St. P. & P.	742
Gratiot	Lafayette	C., M., St. P. & P.	797
Green Bay	Brown	C., M., St. P. & P.	590
		C. & N. W.	589
		G. B. & W.	588
X—C. & N. W.		G. B. & W.	588
X—C., M., St. P. & P.		K. G. B. & W.	586
X—C. & N. W.		C., M., St. P. & P.	591
X—G. B. & W.		C., M., St. P. & P.	610
X—C., M., St. P. & P.		G. B. & W.	606
Green Lake	Green Lake	C. & N. W.	805
Green Lake, water surface	Green Lake	U. S. G. S.	796
Greenleaf	Brown	C., M., St. P. & P.	725
Green Valley	Shawano	C. & N. W.	801
Greenville	Outagamie	C. & N. W.	817
Greenwood	Clark	F. & N. E.	1,136
		S. L.	1,136
Gresham	Shawano	W. & N.	935
Grimms	Manitowoc	C. & N. W.	846
Grimpo	Bayfield	D., S. S. & A.	1,126
Gull Lake, water surface	Washburn	W. G. & N. H. S.	1,095
Hager	Pierce	M. R. C.	719
Half Moon Lake, water surface	Polk	W. G. & N. H. S.	1,161
Hamilton	Fond du Lac	S. L.	927
Hammond	St. Croix	C., St. P., M. & O.	1,103
Hancock	Waushara	S. L.	1,088
Hannibal	Taylor	C., St. P., M. & O.	1,255
		S. M. & P.	1,255
Hanover	Rock	C., M., St. P. & P.	780
Hanover Jct., X—C. & N. W.	Rock	C., M., St. P. & P.	778
Hansen	Wood	C., M., St. P. & P.	1,062
Harker	Iowa	M. P. & N.	927
Harmon	Washburn	S. L.	1,183
Harrington	Waupaca	S. L.	816
Harrison	Lincoln	C. & N. W.	1,683
Harshaw	Oneida	C., M., St. P. & P.	1,509
Hartford	Washington	C., M., St. P. & P.	984
Hartland	Waukesha	C., M., St. P. & P.	924
Hartman	Columbia	C., M., St. P. & P.	803
Hatfield	Jackson	G. B. & W.	889
Hatley	Marathon	C. & N. W.	1,267
Haugen	Barron	C., St. P., M. & O.	1,224
Hawkins	Rusk	S. L.	1,365
Hawthorne	Douglas	C., St. P., M. & O.	1,158
Hay Creek	Eau Claire	F. & N. E.	**1,037**

ELEVATIONS

Locality	County	Authority	Elevation
			Feet
Hayton	Calumet	C., M., St. P. & P.	825
Hayward	Sawyer	C., St. P., M. & O.	1,196
Hazelhurst	Oneida	C., M., St. P. & P.	1,592
Hazen Corners	Crawford	U. S. G. S.	1,222
Headquarters	Bayfield	N. P.	1,253
Heafford Jct.	Lincoln	S. L.	1,492
Hegg, B. M.	Trempealeau	U. S. G. S.	837
Hersey	St. Croix	C., St. P., M. & O.	1,204
Hertel	Burnett	W. G. & N. H. S.	1,037
Hewitt	Wood	S. L.	1,255
Hewitts Crossing	Lafayette	M. P. & N.	862
High Bridge	Ashland	S. L.	981
High Cliff Jct.	Calumet	S. L.	794
Highland	Iowa	M. P. & N.	1,185
Highland Jct.	Lafayette	M. P. & N.	849
Hilbert Jct.	Calumet	S. L.	827
Hiles	Forest	C. & N. W.	1,650
Hiles Jct.	Oneida	C. & N. W.	1,638
Hillcrest	Douglas	S. L.	1,168
Hillsboro, B. M.	Vernon	U. S. G. S.	1,001
Hillsdale	Barron	S. L.	1,197
Hines	Douglas	C., St. P., M. & O.	1,111
Hixon	Oneida	C., M., St. P. & P.	1,626
Hixton	Jackson	G. B. & W.	929
Hoffman	Marathon	S. L.	1,323
Holcombe	Chippewa	C., St. P., M. & O.	1,038
Hollandale	Iowa	U. S. G. S.	851
Holmen, B. M.	La Crosse	U. S. G. S.	719
Holmes Jct.	Marinette	C., M., St. P. & P.	969
Holy Hill	Washington	U. S. G. S.	1,361
Homewood	Milwaukee	C., M., St. P. & P.	716
Honey Creek	Racine	S. L.	818
Horicon	Dodge	C., M., St. P. & P.	881
Hortonville	Outagamie	C. & N. W.	805
Hoyt	Iron	S. L.	1,574
Hoyt	Douglas	D., S. S. & A.	1,130
Hubbleton	Jefferson	U. S. G. S.	795
Hubertus	Washington	S. L.	999
Hudson	St. Croix	C., St. P., M. & O.	700
Hulls Crossing	Sheboygan	C. & N. W.	940
Humbird	Clark	C., St. P., M. & O.	1,022
Hunting	Shawano	C. & N. W.	934
Hurley	Iron	C. & N. W.	1,497
		S. L.	1,493
Hurley Jct.	Iron	C. & N. W.	1,493
Hustler	Juneau	C., St. P., M. & O.	931
Independence	Trempealeau	G. B. & W.	780
Ingram	Rusk	S. L.	1,301
Ino	Bayfield	N. P.	1,088
Iola	Waupaca	G. B. & W.	947
Irma	Lincoln	C., M., St. P. & P.	1,507

ELEVATIONS

Locality	County	Authority	Elevation
			Feet
Iron Belt	Iron	S. L.	1,548
Iron Ridge	Dodge	C., M., St. P. & P.	919
Iron River	Bayfield	N. P.	1,098
		D., S. S. & A.	1,114
Irvine	Chippewa	S. L.	833
Itasca	Douglas	C., St. P., M. & O.	650
Ives Sta.	Racine	C. & N. W.	645
Ixonia	Jefferson	C., M., St. P. & P.	868
Jackson	Washington	C. & N. W.	893
Janesville	Rock	C., M., St. P. & P.	801
X—C. & N. W.		C., M., St. P. & P.	802
		C. & N. W.	801
Jefferson:	Jefferson	C. & N. W.	792
Jefferson Jct.	Jefferson	U. S. G. S.	804
Jeffris	Lincoln	C. & N. W.	1,651
Jeffris Jct.	Lincoln	C. & N. W.	1,673
Jerome	Rusk	S. L.	1,163
Jewett	St. Croix	S. L.	1,028
Jim Falls	Chippewa	C., St. P., M. & O.	953
Joel	Polk	S. L.	1,146
Johnsons Creek	Jefferson	C. & N. W.	805
Jonesdale	Iowa	I. C.	901
Juda	Green	C., M., St. P. & P.	812
Jump River	Taylor	S. M. & P.	1,180
Junction City	Portage	S. L.	1,142
Juneau	Dodge	C. & N. W.	909
Kansasville	Racine	C., M., St. P. & P.	821
Kaukauna	Outagamie	C. & N. W.	709
Keelers	Oneida	C. & N. W.	1,599
Keesus	Waukesha	C. & N. W.	990
Keesus Lake, water surface	Waukesha	U. S. G. S.	958
Kegonsa Lake, water surface	Dane	U. S. G. S.	842
Kellner	Wood	C. & N. W.	1,033
Kellys	Marathon	C. & N. W.	1,238
Kempster	Langlade	C. & N. W.	1,626
Kendall	Monroe	C. & N. W.	1,015
Kennan	Price	S. L.	1,509
Kennedy	Price	C., St. P., M. & O.	1,472
Kenosha	Kenosha	C. & N. W.	612
Kent	Langlade	C. & N. W.	1,687
Kewaskum	Washington	C. & N. W.	951
Kewaunee	Kewaunee	K. G. B. & W.	586
Kickapoo Center, B. M.	Vernon	U. S. G. S.	754
Kickapoo River		W. G. & N. H. S.	
Mouth	Crawford		633
Readstown	Vernon		720
La Farge			777
Kiel	Manitowoc	C., M., St. P. & P.	918
Kimball	Iron	C. & N. W.	1,262

ELEVATIONS

Locality	County	Authority	Elevation
			Feet
Kimberly	Outagamie	C. & N. W	747
Kingston	Oconto	C. & N. W	903
Kinsman	Marinette	W. & M	712
Knapp	Dunn	C., St. P., M. & O	921
Knowles	Dodge	C., M., St. P. & P	1,044
Knowlton	Marathon	C., M., St. P. & P	1,121
Kodatz	Iowa	M. P. & N	872
Koepenick	Langlade	C. & N. W	1,692
Koshkonong	Rock	C. & N. W	821
Koshkonong Lake, water surface	Jefferson and Rock	U. S. G. S	777
Kurth	Clark	C., St. P., M. & O	1,081
Lac du Flambeau	Vilas	C. & N. W	1,632
La Crosse	La Crosse	C., M., St. P. & P	649
		C. B. & Q	654
X—C., M., St. P. & P		C. B. & Q	644
Ladysmith	Rusk	S. L	1,142
La Farge	Vernon	C., M., St. P. & P	795
Lake	Milwaukee	C., M., St. P. & P	735
Lake Beulah	Walworth	S. L	827
Lake Emily	Portage	G. B. & W	1,112
		S. L	1,124
Lake Geneva	Walworth	C. & N. W	892
Lake Mills	Jefferson	U. S. G. S	860
Lake Nebagamon	Douglas	D., S. S. & A	1,153
Lake Park	Calumet	S. L	782
Lake Shore Junction	Milwaukee	C. & N. W	644
Lakeside	Washburn	C., St. P., M. & O	1,115
Lakeside	Waukesha	C., M., St. P. & P	880
Lake Tomahawk	Oneida	C. & N. W	1,632
Lakewood	Oconto	C. & N. W	1,296
Lamont	Lafayette	U. S. G. S	1,070
Lampson	Washburn	W. G. & N. H. S	1,134
Lancaster, B. M	Grant	U. S. G. S	1,086
Lancaster Jct	Grant	C. & N. W	1,170
Land O' Lakes	Vilas	C. & N. W	1,708
Lannon	Waukesha	C., M., St. P. & P	889
Laona	Forest	C. & N. W	1,553
Laona Junction	Forest	C. & N. W	1,531
Larsen	Winnebago	C. & N. W	762
La Valle	Sauk	C. & N. W	901
Lebanon	Dodge	C. & N. W	905
Lehigh	Barron	S. L	1,223
Leland, B. M	Sauk	U. S. G. S	787
Lena	Oconto	C., M., St. P. & P	714
Lenawee	Bayfield	N. P	1,181
Leslie	Lafayette	C. & N. W	1,159
Levis	Jackson	C., St. P., M. & O	847
Lewis	Polk	S. L	1,059
Lewiston	Columbia	C., M., St. P. & P	811
Liberty, B. M	Vernon	U. S. G. S	768

ELEVATIONS

Locality	County	Authority	Elevation
			Feet
Liberty Bluff	Marquette	S. L	941
Liberty Pole	Vernon	U. S. G. S	1,201
Liberty Pole Hill	Green	U. S. G. S	1,102
Lima	Rock	C., M., St. P. & P	890
Lindels	Florence	C. & N. W	1,501
Linden	Iowa	M. P. & N	1,005
Linderman	Trempealeau	C., St. P., M. & O	949
Lindsey	Wood	C., M., St. P. & P	1,154
Lindwerm	Milwaukee	C. & N. W	633
Little Black	Taylor	S. L	1,416
Little Chute (South Side)	Outagamie	C. & N. W	714
(North Side)		C. & N. W	728
Little Lake Butte Des Morts, water surface	Winnebago	U. S. G. S	739
Little Rapids	Brown	C. & N. W	644
Little Suamico	Oconto	C. & N. W	594
Livingston	Grant	C. & N. W	1,095
Lodi	Columbia	C. & N. W	852
Lodi Mill, B. M	Sauk	U. S. G. S	734
Lohrville	Waushara	C. & N. W	805
Lomira	Dodge	S. L	1,019
London	Dane	U. S. G. S	868
Lone Rock	Richland	C., M., St. P. & P	710
Long Lake	Florence	C. & N. W	1,526
Loomis	Marinette	C., M., St. P. & P	687
Lower McKenzie Lake, water surface	Washburn	W. G. & N. H. S	981
Loyal	Clark	S. L	1,228
Lublin	Taylor	S. L	1,280
Luck	Polk	S. L	1,213
Lunds	Shawano	S. L	854
Lusk (Polley P. O.)	Taylor	S. L	
X—S. M. & P.			1,213
Luxemburg	Kewaunee	K. G. B. & W	800
Lyndhurst	Shawano	C. & N. W	950
Lynn	Clark	C., M., St. P. & P	1,137
Lynxville	Crawford	M. R. C	635
Lyons	Walworth	C., M., St. P. & P	801
Lytle	La Crosse	C. B. & Q	664
McAllister	Marinette	W. & M	615
McCartney	Grant	M. R. C	617
McCord	Oneida	S. L	1,493
McCoy	Monroe	C. & N. W	888
X—C. & N. W., C. M. St. P. & P		C. & N. W	865
C., M. St. P. & P. Sta		U. S. G. S	892
McDill	Portage	S. L	1,081
McFarland	Dane	C., M., St. P. & P	868
McMillan	Marathon	C. & N. W	1,270
McNaughton	Oneida	C. & N. W	1,570
Mackville	Outagamie	S. L	802

ELEVATIONS

Locality	County	Authority	Elevation
			Feet
Madison	Dane	C., M., St. P. & P.	859
		C. & N. W.	853
X—I. C.		C., M., St. P. & P.	863
East Madison Sta.		C., M., St. P. & P.	852
Madsen	Manitowoc	S. L.	812
Magenta	Chippewa	S. L.	895
X—S. L.		C., M., St. P. & P.	900
X—C., St. P., M. & O.		C., M., St. P. & P.	883
Magnolia	Rock	C. & N. W.	913
Maiden Rock	Pierce	M. R. C.	682
Malcolm	Langlade	C. & N. W.	1,628
Malta	Sawyer	S. L.	1,353
Manawa	Waupaca	G. B. & W.	827
Manitowish	Iron	C. & N. W.	1,591
Manitowoc	Manitowoc	C. & N. W.	595
		S. L.	587
Mann	Marathon	S. L.	1,288
Manson	Oneida	S. L.	1,523
Maple	Douglas	N. P.	1,160
Mapleton	Waukesha	C. & N. W.	890
Maplewood	Door	A. & W.	708
Marathon	Marathon	C. & N. W.	1,239
Marengo	Ashland	D., S. S. & A.	775
		S. L.	778
Maribel	Manitowoc	C. & N. W.	858
Marinette	Marinette	C. & N. W.	600
		C., M., St. P. & P.	610
Marion	Waupaca	C. & N. W.	857
Markesan	Green Lake	C., M., St. P. & P.	853
Marshall	Dane	C., M., St. P. & P.	866
Marshfield	Wood	C., St. P., M. & O.	1,271
		S. L.	1,285
X—C. & N. W.		S. L.	1,278
Marshland	Buffalo	U. S. G. S.	665
Martintown	Green	I. C.	777
Mason	Bayfield	C., St. P., M. & O.	957
Mather	Juneau	C., M., St. P. & P.	960
Mathews Lake, water surface	Washburn	W. G. & N. H. S.	994
Mattoon	Shawano	C. & N. W.	1,254
Mauston	Juneau	C., M., St. P. & P.	880
Mayhews	Walworth	C., M., St. P. & P.	865
Mayville	Dodge	C., M., St. P. & P.	920
Mazomanie	Dane	C., M., St. P. & P.	780
Meadow Valley	Juneau	C., M., St. P. & P.	960
Medary	La Crosse	C., M., St. P. & P.	657
Medford	Taylor	S. L.	1,410
Medina	Outagamie	C. & N. W.	817
Medina Junction	Winnebago	S. L.	759
Meehan	Portage	G. B. & W.	1,067
Melrose, B. M.	Jackson	U. S. G. S.	731
Melvina	Monroe	U. S. G. S.	863

ELEVATIONS

Locality	County	Authority	Elevation
			Feet
Mellen	Ashland	S. L.	1,242
Menasha	Winnebago	C., M., St. P. & P.	765
		S. L.	761
Menasha—Neenah	Winnebago	C. & N. W.	755
Mendota Sta.	Dane	W. G. & N. H. S.	897
Mendota Lake, water surface.	Dane	U. S. G. S.	849
Menominee River		W. G. & N. H. S.	
Grand Rapids, head	Marinette		669
Ross			672
Sturgeon Falls, head			817
Big Quinnesec Falls, foot			966
Twin Falls, head	Florence	C., M., St. P. & P.	1,100
Menomonee Falls	Waukesha	C., M., St. P. & P.	881
Menomonie	Dunn		803
		C., St. P., M. & O.	785
Menomonie Jct.	Clark		842
Mentor	Clark	F. & N. E.	1,085
Mequon	Ozaukee	C. & N. W.	675
Mercer	Iron		1,598
Meridean	Dunn	C., M., St. P. & P.	749
Merrill	Lincoln	U. S. G. S.	1,257
Merrillan	Jackson	C., St. P., M. & O.	938
Merrill Park	Milwaukee	C., M., St. P. & P.	586
Merrimac	Sauk	C. & N. W.	800
Merton	Waukesha	C., M., St. P. & P.	965
Michigan, Lake, ordinary water level		U. S. L. S.	581
low water		U. S. L. S.	579
Middle Inlet	Marinette	C., M., St. P. & P.	711
Middle McKenzie Lake, water surface	Burnett	W. G. & N. H. S.	985
Middle River	Douglas	D., S. S. & A.	1,054
Middleton	Dane	U. S. G. S.	931
Midway	Brown	C., M., St. P. & P.	647
Midway	La Crosse	C. & N. W.	648
Mikana	Barron	S. L.	1,175
Milan	Marathon	S. L.	1,445
Milladore	Wood	S. L.	1,194
Millston	Jackson	C., St. P., M. & O.	919
Milltown	Polk	S. L.	1,239
Millville, B. M.	Grant	U. S. G. S.	657
Milton	Rock	C., M., St. P. & P.	874
Milton Jct.	Rock	C. & N. W.	877
Milwaukee	Milwaukee	C. & N. W.	592
		C., M., St. P. & P.	586
Milwaukee County Hospital		C., M., St. P. & P.	741
Milwaukee River		W. G. & N. H. S.	
Thiensville dam, crest	Ozaukee		653
Grafton Flour Mill Dam, crest			732
Saukville			745
Fredonia dam, crest			781

ELEVATIONS

Locality	County	Authority	Elevation
Milwaukee River—(continued)			Feet
West Bend Dam, crest...........	Washington...........	....	902
Kewaskum Dam, crest...........	....	....	950
Mindoro..............................	La Crosse...............	U. S. G. S....................	787
Mineral Point......................	Iowa......................	C., M., St. P. & P.........	944
Minnesota Jct......................	Dodge....................	C. & N. W....................	924
Minocqua............................	Oneida...................	C., M., St. P. & P.........	1,600
Minong...............................	Washburn..............	C., St. P., M. & O........	1,069
Mississippi River.................		M. R. C....................	
Dubuque...........................	....	....	595
Mouth of Wisconsin.............	Crawford...............	....	615
La Crosse..........................	La Crosse...............	....	628
Mouth of Chippewa.............	Pepin....................	....	664
Prescott............................	Pierce....................	....	667
Mitterhofer.........................	Taylor....................	S. M. & P....................	1,193
Modena..............................	Buffalo..................	U. S. G. S....................	800
Mohle.................................	Wood....................	S. L.........................	1,250
Mondovi.............................	Buffalo..................	C., St. P., M. & O........	788
Monico Jct..........................	Oneida...................	C. & N. W....................	1,597
Monona Lake, water surface...	Dane.....................	U. S. G. S....................	845
Monroe, I. C. Sta.................	Green....................	U. S. G. S....................	1,045
C. M. St. P. & P. Sta..............		U. S. G. S....................	1,070
Montello..............................	Marquette..............	S. L.........................	772
Montfort..............................	Grant.....................	C. & N. W....................	1,119
Montfort Junction................	Iowa......................	C. & N. W....................	1,132
Monticello, I. C. Sta..............	Green....................	U. S. G. S....................	827
X—I. C. and C. M. St. P. & P..........		U. S. G. S....................	828
....	....	C., M., St. P. & P.........	823
Moquah..............................	Bayfield.................	N. P.........................	851
Morgan...............................	Shawano...............	W. & N....................	989
Morrisonville.......................	Dane.....................	C., M., St. P. & P.........	859
Morse.................................	Ashland..................	S. L.........................	1,497
Mosel.................................	Sheboygan.............	C. & N. W....................	638
Mosinee..............................	Marathon...............	U. S. G. S....................	1,154
Moundville..........................	Marquette..............	S. L.........................	800
Mountain............................	Oconto..................	C. & N. W....................	970
Mount Hope, B. M................	Grant.....................	U. S. G. S....................	1,088
Mount Horeb.......................	Dane.....................	C. & N. W....................	1,230
Mount Ida, B. M..................	Grant.....................	U. S. G. S....................	1,211
Mount Sterling.....................	Crawford...............	U. S. G. S....................	1,183
Mount Tabor.......................	Vernon..................	U. S. G. S....................	1,361
Mount Zion.........................	Crawford...............	U. S. G. S....................	1,195
Mukwonago.........................	Waukesha..............	S. L.........................	841
Murry.................................	Grant.....................	S. L.........................	1,158
Muscoda.............................	Bayfield.................	U. S. G. S....................	693
Muskeg..............................	Rusk.....................	N. P.........................	1,102
Muskego Lake, water surface.	Waukesha..............	U. S. G. S....................	770
Nagawicka...........................	Waukesha..............	C., M., St. P. & P.........	933
Nagawicka Lake, water surface.................	Waukesha..............	U. S. G. S....................	890
Nashotah............................	Waukesha..............	C., M., St. P. & P.........	937

ELEVATIONS

Locality	County	Authority	Elevation
			Feet
Nashotah Lake, water surface	Waukesha	U. S. G. S.	871
Nashville	Forest	C. & N. W.	1,707
Navarino	Shawano	S. L.	805
Necedah	Juneau	C. & N. W.	927
Lower X—C. & N. W.		C., M., St. P. & P.	905
Neda	Dodge	C., M., St. P. & P.	997
Neenah	Winnebago	S. L.	747
		C., M., St. P. & P.	753
Neenah-Menasha	Winnebago	C. & N. W.	755
Nehmabin Lakes, water surface	Waukesha	U. S. G. S.	870
Neillsville	Clark	C., St. P., M. & O.	997
Nekoosa	Wood	C., M., St. P. & P.	957
Nelson	Buffalo	M. R. C.	680
Nelsons	Portage	S. L.	1,052
Neopit	Shawano	W. & N.	1,077
Neshkoro	Marquette	C. & N. W.	803
Newald	Forest	C. & N. W.	1,562
New Amsterdam	La Crosse	U. S. G. S.	684
New Auburn	Chippewa	C., St. P., M. & O.	1,105
Newbold	Oneida	C. & N. W.	1,571
New Franken	Brown	K. G. B. & W.	810
New Glarus	Green	C., M., St. P. & P.	860
New Holstein	Calumet	C., M., St. P. & P.	936
New Lisbon	Juneau	C., M., St. P. & P.	892
New London	Waupaca	C. & N. W.	764
		G. B. & W.	761
New London Junction	Outagamie	C. & N. W.	762
New Richmond	St. Croix	S. L.	985
X—C., St. P., M. & O.		S. L.	988
		C., St. P., M. & O.	986
Newry	Vernon	C., M., St. P. & P.	1,337
Newton	Manitowoc	C. & N. W.	660
Newton	Marinette	W. & M.	716
Newton, B. M.	Vernon	U. S. G. S.	733
Nichols	Outagamie	S. L.	795
Norrie	Marathon	C. & N. W.	1,282
North Bend, B. M.	Jackson	U. S. G. S.	723
North Branch	Oconto	C. & N. W.	810
Northfield, B. M.	Jackson	U. S. G. S.	949
North Fond du Lac	Fond du Lac	S. L.	749
North Freedom	Sauk	C. & N. W.	885
North Greenfield	Milwaukee	C. & N. W.	734
North La Crosse	La Crosse	C., M., St. P. & P.	640
North Lake	Waukesha	C. & N. W.	925
		C., M., St. P. & P.	916
North Lake, water surface	Waukesha	U. S. G. S.	897
North Lowell	Dodge	C. & N. W.	815
Northline	St. Croix	C., St. P., M. & O.	876
Northport	Waupaca	G. B. & W.	781

ELEVATIONS

Locality	County	Authority	Elevation
			Feet
North Milwaukee..........................	Milwaukee.................	C., M., St. P. & P..........	650
North Prairie..............................	Waukesha................	C., M., St. P. & P..........	944
North Tomah..............................	Monroe.......................	C. & N. W....................	965
Norton..	Dunn..........................	S. L...........................	922
Norwalk, B. M............................	Monroe......................	U. S. G. S....................	1,030
Norway.......................................	Monroe.......................	C., M., St. P. & P..........	979
Norwegian..................................	Winnebago..............	S. L...........................	833
Nye..	Polk...........................	S. L...........................	968
Oak Center................................	Fond du Lac...........	C. & N. W....................	895
Oakdale......................................	Monroe.......................	C., M., St. P. & P..........	951
Oakwood....................................	Milwaukee...............	C., M., St. P. & P..........	686
Oconomowoc..............................	Waukesha................	C., M., St. P. & P..........	866
Oconomowoc Lake, water surface...................		U. S. G. S....................	862
Oconto.......................................	Oconto......................	C. & N. W....................	590
		C., M., St. P. & P..........	593
Oconto Falls..............................		C. & N. W....................	732
Oconto Junction.........................		C., M., St. P. & P..........	680
Oconto River.............................		W. G. & N. H. S..........	
Suring......................................	Oconto......................		791
One mile south of Mountain............................			916
Wabeno.......................................	Forest.......................		1,526
Odanah.......................................	Ashland....................	C. & N. W....................	613
Ogdensburg................................	Waupaca...................	G. B. & W....................	864
Ogema...	Price..........................	S. L...........................	1,551
Okauchee....................................	Waukesha................	C., M., St. P. & P..........	898
Okauchee Lake, water surface...................	Waukesha................	U. S. G. S....................	873
Omaha Jct..................................	Clark.........................	C., M., St. P. & P..........	1,175
Omro.........	Winnebago..............	C., M., St. P. & P..........	758
Onalaska....................................	La Crosse................	C. & N. W....................	679
		C., M., St. P. & P..........	670
		C. B. & Q...................	647
X—C., M., St. P. & P.		C. B. & Q...................	645
Oneida..	Outagamie................	G. B. & W....................	747
Ontario, B. M.............................	Vernon......................	U. S. G. S....................	872
Oostburg....................................	Sheboygan................	C. & N. W....................	701
Oregon..	Dane...........................	C. & N. W....................	948
Orfordville..................................	Rock...........................	C., M., St. P. & P..........	883
Ormsby.......................................	Langlade...................	C. & N. W....................	1,530
Osceola.......................................	Polk...........................	U. S. G. S....................	809
Oshkosh......................................	Winnebago..............	C. & N. W....................	761
		C., M., St. P. & P..........	750
		S. L...........................	762
X—C. & N. W........................		S. L...........................	754
X—C., M., St. P. & P.		S. L...........................	749
Osseo..	Trempealeau...........	C., St. P., M. & O........	955
Owego...	Clark.........................	F. & N. E....................	1,184
Owen..	Clark.........................	S. L...........................	1,244
		F. & N. E....................	1,246

ELEVATIONS

Locality	County	Authority	Elevation
			Feet
Oxford	Marquette	C. & N. W.	855
Packard	Marinette	W. & M.	714
Packwaukee	Marquette	S. L.	771
Packwaukee Junction	Marquette	S. L.	778
Padus	Forest	C. & N. W.	1,601
Palmyra	Jefferson	C., M., St. P. & P.	840
Paoli	Dane	U. S. G. S.	907
Pardeeville	Columbia	C., M., St. P. & P.	810
Park Falls	Price	S. L.	1,494
Parrish	Langlade	C. & N. W.	1,590
Parrish Jct	Langlade	C. & N. W.	1,696
Patzu	Douglas	S. L.	1,017
Pearson	Bayfield	N. P.	1,060
Pecks Sta	Walworth	C., M., St. P. & P.	864
Peebles	Fond du Lac	U. S. G. S.	815
Pelican	Oneida	C. & N. W.	1,605
Pelican Lake, water surface	Oneida	C. & N. W.	1,593
Pembine	Marinette	C., M., St. P. & P.	971
Pence	Iron	C. & N. W.	1,623
Pennington	Price	S. L.	1,504
Penokee	Ashland	S. L.	1,294
Pensaukee	Oconto	C. & N. W.	589
Pepin	Pepin	M. R. C.	683
Perote	Shawano	W. & N.	1,253
Peshtigo	Marinette	C. & N. W.	609
		W. & M.	610
Peshtigo Harbor	Marinette	W. & M.	589
Peshtigo River		W. G. & N. H. S.	
Peshtigo dam, crest	Marinette		602
Crivitz dam, head			680
High Falls, head of			850
North Crandon R. R. Crossing	Forest		1,620
Petersburg	Crawford	C., M., St. P. & P.	691
Pewaukee	Waukesha	C., M., St. P. & P.	852
Pewaukee Lake, water surface		U. S. G. S.	852
Phelps	Vilas	C. & N. W.	1,681
Phillips	Price	S. L.	1,456
Phipps (Lenroot P. O.)	Sawyer	C., M., St. P. & P.	1,233
Phlox	Langlade	W. & N.	1,255
Pigeon Falls, B. M.	Trempealeau	U. S. G. S.	872
Pike River	Bayfield	D., S. S. & A.	960
Pine Creek	Trempealeau	C. & N. W.	656
Pine Grove	Sheboygan	C. & N. W.	852
Pine Lake, water surface	Waukesha	U. S. G. S.	903
Pine River	Lincoln	C., M., St. P. & P.	1,232
Pittsville	Wood	C., M., St. P. & P.	1,030
Pittsville Jct		C., M., St. P. & P.	1,013
Plainfield	Waushara	S. L.	1,108
Platte Mounds, top west mound	Grant	U. S. G. S.	1,420

ELEVATIONS

Locality	County	Authority	Elevation
			Feet
Platteville	Grant	C., M., St. P. & P.	918
Pleasant Prairie	Kenosha	C. & N. W.	700
Pleasantville	Trempealeau	U. S. G. S.	902
Pleshek	Shawano	S. L.	814
Plover	Portage	G. B. & W.	1,076
Plum City, B. M.	Pierce	U. S. G. S.	816
Plummer	Iron	S. L.	1,461
Plymouth	Sheboygan	C. & N. W.	845
X—C. & N. W.		C., M., St. P. & P.	846
Point Sauk	Sauk	U. S. G. S.	1,620
Pokegama	Douglas	N. P.	690
Pokegama Lake, water surface	Washburn	W. G. & N. H. S.	1,016
Polar	Langlade	C. & N. W.	1,520
Poplar	Douglas	N. P.	984
Portage	Columbia	S. L.	785
		C., M., St. P. & P.	811
Port Edwards	Wood	C., M., St. P. & P.	967
Porterfield	Marinette	C., M., St. P. & P.	672
Porters Mills	Eau Claire	C., M., St. P. & P.	771
Portland, B. M.	Monroe	U. S. G. S.	1,312
Port Washington	Ozaukee	C. & N. W.	671
Poskin Lake	Barron	S. L.	1,176
Potosi, B. M.	Grant	U. S. G. S.	786
Potosi Sta.	Grant	M. R. C.	615
Potter	Calumet	S. L.	805
Potts Corners	Vernon	U. S. G. S.	836
Pound	Marinette	C., M., St. P. & P.	723
Powell	Iron	C. & N. W.	1,595
Poynette	Columbia	C., M., St. P. & P.	845
Prairie du Chien	Crawford	C. B. & Q.	643
X—C., M. St. P. & P. & C., B. & Q.		M. R. C.	641
Sta. C., M., St. P. & P.		M. R. C.	626
Prairie du Sac	Sauk	C., M., St. P. & P.	758
Pratt Junction	Oneida	C. & N. W.	1,607
Pray	Jackson	G. B. & W.	978
Prentice	Price	S. L.	1,542
Prentice Jct.		S. L.	1,543
Prescott	Pierce	C. B. & Q.	702
Presque Isle	Vilas	C. & N. W.	1,644
Preston	Grant	C. & N. W.	1,105
Price	Jackson	C., St. P., M. & O.	1,087
Princeton	Green Lake	U. S. G. S.	776
Puckaway Lake, water surface	Green Lake	U. S. G. S.	764
Pulaski	Shawano	C. & N. W.	797
Purdy	Buffalo	C. B. & Q.	673
Racine	Racine	C. & N. W.	629
X—C. & N. W.		C., M., St. P. & P.	626
Racine Jct.	Racine	C. & N. W.	626
Radisson	Sawyer	C., St. P., M. & O.	1,239

ELEVATIONS

Locality	County	Authority	Elevation
			Feet
Randolph	Dodge	C., M., St. P. & P.	959
Random Lake	Sheboygan	C., M., St. P. & P.	883
Ranney	Kenosha	C., M., St. P. & P.	682
Raspberry	Bayfield	B. T.	847
Readstown	Vernon	C., M., St. P. & P.	754
Red Cedar, C. M. St. P. & P. Sta.	Dunn	U. S. G. S.	758
Red Cedar Junction, C. M. St. P. & P. Sta.	Dunn	U. S. G. S.	730
Red Cedar River		W. G. & N. H. S.	
Mouth	Dunn		705
Menomonie dam, crest			804
Colfax			895
Cameron	Barron		1,068
Red Granite	Waushara	C. & N. W.	792
Red Granite Jct. (Bannerman)	Waushara	U. S. G. S.	818
Redmound	Vernon	U. S. G. S.	1,302
Reed, B. M.	Crawford	U. S. G. S.	724
Reedsburg	Sauk	C. & N. W.	880
Reeds Corners	Fond du Lac	U. S. G. S.	1,004
Reedsville	Manitowoc	C. & N. W.	824
Reeseville	Dodge	C., M., St. P. & P.	823
Requa	Jackson	C., St. P., M. & O.	1,017
Reserve	Sawyer	S. L.	1,308
Retreat	Vernon	U. S. G. S.	1,284
Rewey	Iowa	C. & N. W.	1,065
Rhinelander	Oneida	C. & N. W.	1,543
X—S. L.		C. & N. W.	1,553
		S. L.	1,557
Rib Falls	Marathon	C. & N. W.	1,235
Rib Falls Jct.	Marathon	C. & N. W.	1,193
Rib Mountain	Marathon	U. S. G. S.	1,940
Rib Lake	Taylor	S. L.	1,566
Rice Lake	Barron	S. L.	1,112
		C., St. P., M. & O.	1,143
Rice Lake, water surface	Burnett	W. G. & N. H. S.	974
Richardson	Polk	C., St. P., M. & O.	1,201
Richfield	Washington	C., M., St. P. & P.	968
Richland Center	Richland	C., M., St. P. & P.	736
Richwood	Dodge	C., M., St. P. & P.	838
Rickhard	Eau Claire	F. & N. E.	1,050
Ridgeland	Dunn	S. L.	1,080
Ridgeway	Iowa	C. & N. W.	1,170
Riley	Dane	C. & N. W.	945
Ringle	Marathon	C. & N. W.	1,327
Rio	Columbia	C., M., St. P. & P.	928
Rio Creek	Kewaunee	A. & W.	699
Ripley Lake, water surface	Jefferson	U. S. G. S.	836
Ripon, C. & N. W. Sta.	Fond du Lac	U. S. G. S.	914
		C., M., St. P. & P.	935
Ripon Jct.	Fond du Lac	C., M., St. P. & P.	941
Riton	Rock	C., M. St. P. & P.	777

ELEVATIONS

Locality	County	Authority	Elevation
			Feet
River Falls	Pierce	C., St. P., M. & O.	887
Roberts	St. Croix	C., St. P., M. & O.	1,056
Rock Falls	Dunn	U. S. G. S.	861
Rockfield	Washington	C. & N. W.	881
Rock Lake, water surface	Jefferson	U. S. G. S.	827
Rockland	La Crosse	C., M., St. P. & P.	762
		C. & N. W.	767
Rockmont	Douglas	C., St. P., M. & O.	983
		D., S. S. & A.	1,000
Rodell	Eau Claire	C., St. P., M. & O.	947
Rolling Ground	Crawford	U. S. G. S.	1,217
Rolling Prairie	Dodge	C., M., St. P. & P.	940
Romadka	Clark	C., M., St. P. & P.	1,210
Romeo	Marathon	S. L.	1,320
Rosendale	Fond du Lac	C. & N. W.	899
Rosholt	Portage	C. & N. W.	1,135
Ross	Vernon	U. S. G. S.	795
Round Lake, water surface	Polk	W. G. & N. H. S.	1,146
Rouse	Iron	S. L.	1,493
Royalton	Waupaca	G. B. & W.	821
Rubicon	Dodge	C., M., St. P. & P.	1,017
Rudolph	Wood	C., M., St. P. & P.	1,137
Rugby Jct.	Washington	S. L.	983
Rummeles	Vilas	C. & N. W.	1,668
Rush Lake Jct.	Winnebago	C., M., St. P. & P.	846
Rusk	Dunn	C., St. P., M. & O.	903
Russell	Bayfield	B. T.	857
Rutledge	Grant	C. B. & Q.	612
Saint Cloud	Fond du Lac	C. & N. W.	933
Saint Croix Falls	Polk	U. S. G. S.	922
Saint Croix River		W. G. & N. H. S.	
Prescott	Pierce		667
Saint Croix Falls, crest of dam	Polk		750
Namakagon River, mouth	Burnett		908
Saint Croix Lake	Douglas		1,010
Saint Francis	Milwaukee	C. & N. W.	654
Saint Joseph, B. M.	La Crosse	U. S. G. S.	1,294
Saint Marie	Green Lake	U. S. G. S.	786
Salem	Kenosha	C. & N. W.	779
Sanborn	Ashland	D., S. S. & A.	834
Sand Lake, water surface	Burnett	W. G. & N. H. S.	983
Sand Rock	Iron	C. & N. W.	1,624
Sarona	Washburn	C., St. P., M. & O.	1,294
Satuit	Oneida	C. & N. W.	1,567
Sauk City	Sauk	C., M., St. P. & P.	757
Saukville	Ozaukee	C., M., St. P. & P.	771
Saunders	Douglas	G. N.	678
Sauntry	Douglas	C., St. P., M. & O.	1,142
Savoy	Buffalo	C., M., St. P. & P.	699
Sawyer	Door	A. & W.	590

ELEVATIONS

Locality	County	Authority	Elevation
			Feet
Saxon	Iron	C. & N. W.	1,117
X—D. S. S. & A.		C. & N. W.	1,086
		D., S. S. & A.	1,114
Scandinavia	Waupaca	G. B. & W.	930
Schofield	Marathon	C., M., St. P. & P.	1,206
Scott	Shawano	W. & M.	1,025
Scott	Vilas	C. & N. W.	1,633
Scotts	Eau Claire	F. & N. E.	1,028
Scranton	Wood	G. B. & W.	996
Sedgwick	Ashland	D., S. S. & A.	855
Seeley	Sawyer	C., St. P., M. & O.	1,274
Seneca	Crawford	U. S. G. S.	1,263
Sevenmile Creek	Sheboygan	C. & N. W.	634
Seymour	Outagamie	G. B. & W.	791
Sharon	Walworth	C. & N. W.	987
Shawano	Shawano	C. & N. W.	824
Shawano Junction, X—S. L.	Shawano	C. & N. W.	823
Shawano Lake, water surface	Shawano	C. & N. W.	797
Shawtown	Eau Claire	C., M., St. P. & P.	777
Sheboygan	Sheboygan	C. & N. W.	589
Sheboygan Falls	Sheboygan	C. & N. W.	668
Sheldon	Rusk	S. L.	1,125
Shell Lake	Washburn	C., St. P., M. & O.	1,242
Shell Lake, water surface	Washburn	W. G. & N. H. S.	1,223
Shennington	Monroe	C. & N. W.	912
Shepley	Shawano	C. & N. W.	1,175
Sheppard	Jackson	C., St. P., M. & O.	879
Sheridan	Waupaca	S. L.	963
Sherry	Wood	S. L.	1,186
Sherry Jct	Langlade	C. & N. W.	1,651
Sherwood	Calumet	S. L.	834
Shilling	Clark	F. & N. E.	1,211
Shilo	Ashland	D., S. S. & A.	784
Shiocton	Outagamie	G. B. & W.	770
Shullsburg	Lafayette	C., M., St. P. & P.	930
Silver Lake	Kenosha	S. L.	757
X—C. & N. W.		S. L.	752
Silver Lake, water surface	Washburn	W. G. & N. H. S.	1,106
Silver Lake, water surface	Waukesha	U. S. G. S.	864
Silver Springs	Milwaukee	C. & N. W.	646
Siren	Burnett	S. L.	991
Slinger	Washington	S. L.	1,053
X—C., M., St. P. & P.		S. L.	1,054
Slow Bridge	Bayfield	N. P.	1,080
Sobieski	Oconto	C., M., St. P. & P.	668
Soldiers Grove	Crawford	C., M., St. P. & P.	726
Solon Springs	Douglas	S. L.	1,138
		C., St. P., M. & O.	1,085
Somers	Kenosha	C., M., St. P. & P.	697
Somerset	St. Croix	S. L.	922
Soperton	Forest	C. & N. W.	1,534
South Beaver Dam	Dodge	C. & N. W.	892

ELEVATIONS

Locality	County	Authority	Elevation
			Feet
South Chippewa	Chippewa	S. L.	865
X—C., St. P., M. & O.		S. L.	859
South Kaukauna	Outagamie	C. & N. W.	656
South Milwaukee	Milwaukee	C. & N. W.	674
South Oshkosh	Winnebago	S. L.	758
X—C., M., St. P. & P.		S. L.	750
		C. & N. W.	758
South Randolph	Dodge	C. & N. W.	915
South Range	Douglas	D., S. S. & A.	776
		C., St. P., M. & O.	766
South Wayne	Lafayette	C., M., St. P. & P.	789
Sparta	Monroe	C., M., St. P. & P.	789
Sparta Jct.	Monroe	C. & N. W.	798
Spencer	Marathon	S. L.	1,308
Spider	Bayfield	N. P.	1,147
Spirit Lake, water surface	Burnett	W. G. & N. H. S.	946
Split Rock	Shawano	C. & N. W.	963
Spokeville	Clark	S. L.	1,280
Spooner	Washburn	C., St. P., M. & O.	1,095
Spooner Lake, water surface	Washburn	W. G. & N. H. S.	1,090
Sprague	Juneau	C., M., St. P. & P.	939
Spring Brook	Washburn	C., St. P., M. & O.	1,090
Springfield	Walworth	C., M., St. P. & P.	852
Spring Green	Sauk	C., M., St. P. & P.	728
Spring Lake	Waushara	C. & N. W.	833
Spring Valley	Price	C., St. P., M. & O.	941
Springville, B. M.	Vernon	U. S. G. S.	1,092
Stanbury	Washburn	S. L.	1,153
X—C., St. P., M. & O.		S. L.	1,176
Stanley	Chippewa	S. L.	1,079
Stanton	St. Croix	C., St. P., M. & O.	1,060
Star Lake	Vilas	C., M., St. P. & P.	1,679
State Hospital	Winnebago	S. L.	772
Stella Jct.	Oneida	C. & N. W.	1,643
Stetsonville	Taylor	S. L.	1,447
Steuben	Crawford	C., M., St. P. & P.	670
Stevens Point	Portage	S. L.	1,085
X—G. B. & W.		S. L.	1,086
Stewart, B. M.	Green	U. S. G. S.	1,046
Stiles	Oconto	C., M., St. P. & P.	622
Stiles Junction	Oconto	C. & N. W.	656
Stinnett	Washburn	C., St. P., M. & O.	1,149
Stitzer, B. M.	Grant	U. S. G. S.	1,166
Stockbridge	Calumet	U. S. G. S.	828
Stockholm	Pepin	C. B. & Q.	688
Stockton	Portage	S. L.	1,144
Stoddard	Vernon	C. B. & Q.	643
Stone Lake	Sawyer	S. L.	1,306
Stoughton	Dane	C., M., St. P. & P.	859
Strader	Eau Claire	F. & N. E.	1,006
Stratford	Marathon	C. & N. W.	1,249
Strawbridge	Lafayette	C. & N. W.	**745**

ELEVATIONS

Locality	County	Authority	Elevation
			Feet
Strickland	Rusk	S. L.	1,298
Strum	Trempealeau	C., St. P., M. & O.	902
Sturgeon Bay	Door	A. & W.	590
Sturtevant	Racine	C., M., St. P. & P.	725
Sucker Lake, water surface	Polk	W. G. & N. H. S.	1,025
Sugar Bush	Outagamie	C. & N. W.	831
Sugar Grove	Vernon	U. S. G. S.	1,233
Sullivan	Jefferson	C. & N. W.	859
Summit	Brown	K. G. B. & W.	791
Summit	Dane	I. C.	1,005
Summit	Fond du Lac	C. & N. W.	980
Summit	Monroe	C. & N. W.	1,181
Summit	Polk	S. L.	1,028
Summit Lake	Langlade	C. & N. W.	1,723
Sunnyside	Bayfield	B. T.	817
Sunnyside	Douglas	S. L.	698
Sun Prairie	Dane	C., M., St. P. & P.	936
Sunset, B. M.	Marathon	U. S. G. S.	1,355
Superior	Douglas	C., St. P., M. & O.	629
Union Station		N. P.	635
Central Ave.		N. P.	676
Central Ave.		G. N.	672
X—D., S. S. & A.		S. L.	649
X—G. N.		S. L.	643
Superior, Lake, water surface		U. S. L. S.	602
Suring	Oconto	C. & N. W.	803
Sussex	Waukesha	C., M., St. P. & P.	917
		C. & N. W.	938
Sutherland	Bayfield	D., S. S. & A.	938
Sweden	Bayfield	C., St. P., M. & O.	1,161
Syene	Dane	C. & N. W.	897
Sylvania	Racine	C., M., St. P. & P.	759
Taycheedah	Fond du Lac	C. & N. W.	751
Taylor	Jackson	G. B. & W.	882
Teegarden	Dunn	C., St. P., M. & O.	884
Templeton	Waukesha	C., M., St. P. & P.	899
		S. L.	904
Tesch	Shawano	W. & N.	894
Theresa	Dodge	S. L.	940
Thetis	Dodge	C., M., St. P. & P.	999
Thiensville	Ozaukee	C., M., St. P. & P.	666
Thornton	Shawano	C. & N. W.	859
		W. & N.	858
Thorpe	Clark	S. L.	1,219
Three Lakes	Oneida	C. & N. W.	1,658
Tiffany	Rock	C. & N. W.	848
Tigerton	Shawano	C. & N. W.	1,024
Tioga	Clark	F. & N. E.	1,084
Tomah, C., M. St. P. & P. Sta.	Monroe	U. S. G. S.	958
Tomahawk	Lincoln	C., M., St. P. & P.	1,449
Tomahawk Jct.	Lincoln	S. L.	1,473
Tomahawk Lake	Oneida	C. & N. W.	1,632

ELEVATIONS

Locality	County	Authority	Elevation
			Feet
Tony	Rusk	S. L.	1,222
Topside	Bayfield	N. P.	1,152
Townsend	Oconto	C. & N. W.	1,355
Trade Lake	Burnett	W. G. & N. H. S.	931
Trade Lake, water surface	Burnett	W. G. & N. H. S.	909
Tramway	Dunn	C., St. P., M. & O.	880
Trego	Washburn	C., St. P., M. & O.	1,090
Tremble	Brown	C., M., St. P. & P.	626
Trempealeau	Trempealeau	C. & N. W.	712
C., B. & Q. Sta.		M. R. C.	654
Trevino—X—C., B. & Q. and			
C., M., St. P. & P.	Buffalo	U. S. G. S.	684
Trevor	Kenosha	S. L.	785
Trow	Clark	C., St. P., M. & O.	1,012
Troy Center	Walworth	C., M., St. P. & P.	891
Truax	Eau Claire	C., St. P., M. & O.	895
Truesdell	Kenosha	C., M., St. P. & P.	682
Tunnel	Juneau	C., St. P., M. & O.	1,077
Tunnel City	Monroe	C., M., St. P. & P.	1,057
		C. & N. W.	1,053
Turtle Lake	Barron	S. L.	1,267
Tuscobia	Barron	C., St. P., M. & O.	1,184
Twin Bluffs	Richland	U. S. G. S.	703
Twin Creek Jct.	Marinette	W. & M.	677
Two Rivers	Manitowoc	C. & N. W.	597
Two Rivers Jct.	Manitowoc	C. & N. W.	627
Ulao	Ozaukee	C. & N. W.	699
Underhill	Oconto	C. & N. W.	793
Union Center	Juneau	C. & N. W.	934
Union Grove	Racine	C., M., St. P. & P.	780
Unity	Marathon	S. L.	1,333
Upson	Iron	S. L.	1,500
Utley	Green Lake	C., M., St. P. & P.	898
Valders	Manitowoc	S. L.	812
Vallee	Rusk	S. M. & P.	1,232
Valley	Vernon	U. S. G. S.	888
Valley Junction	Monroe	C., St. P., M. & O.	929
Valley View	Eau Claire	F. & N. E.	1,074
Valton, B. M.	Sauk	U. S. G. S.	1,032
Van Buskirk	Iron	C. & N. W.	1,512
Van Dyne	Fond du Lac	S. L.	791
Van Ostrand	Langlade	C. & N. W.	1,315
X—W. & N.		C. & N. W.	1,286
		W. & N.	1,318
Veedum	Wood	C., M., St. P. & P.	1,019
Veefkind	Clark	S. L.	1,303
Vernon	Waukesha	S. L.	795
Verona, I. C. Sta.	Dane	U. S. G. S.	983
Vesper	Wood	C. & N. W.	1,086
X—S. L.		C., M., St. P. & P.	1,088

ELEVATIONS

Locality	County	Authority	Elevation
			Feet
Victory	Vernon	M. R. C.	638
Viola	Richland	C., M., St. P. & P.	768
Viroqua	Vernon	C., M., St. P. & P.	1,274
Viroqua Jct	Monroe	C., M., St. P. & P.	782
Wabeno	Forest	C. & N. W.	1,529
Wagner	Marinette	W. & M.	716
Walbridge	Douglas	N. P.	814
Waldo	Sheboygan	C., M., St. P. & P.	840
Wales	Waukesha	C. & N. W.	1,002
Walker	Wood	G. B. & W.	1,009
Walsh	Marinette	W. & M.	695
Walworth	Walworth	C., M., St. P. & P.	1,004
Warrens	Monroe	C., St. P., M. & O.	1,017
Wascott	Douglas	C., St. P., M. & O.	1,063
Washburn	Bayfield	C., St. P., M. & O.	656
		N. P.	645
Waterloo	Jefferson	C., M., St. P. & P.	823
Watertown	Jefferson	C. & N. W.	822
		C., M., St. P. & P.	822
Watertown Jct	Jefferson	C. & N. W.	820
Waubesa Lake, water surface	Dane	U. S. G. S.	844
Waukau	Winnebago	U. S. G. S.	846
Waukesha	Waukesha	S. L.	821
X—C., M. & St. P.		S. L.	820
		C., M., St. P. & P.	807
		C., M., St. P. & P.	802
Waunakee	Dane	U. S. G. S.	926
Waupaca	Waupaca	S. L.	868
Waupun	Fond du Lac	C., M., St. P. & P.	888
Wausau	Marathon	C. & N. W.	1,191
C., M. & St. P. Sta.		U. S. G. S.	1,216
Wausau Jct	Marathon	C. & N. W.	1,194
Wausaukee	Marinette	C., M., St. P. & P.	745
Wausaukee Jct	Marinette	C., M., St. P. & P.	741
Wautoma	Waushara	C. & N. W.	897
Wauwatosa	Milwaukee	C., M., St. P. & P.	652
Wauzeka	Crawford	C., M., St. P. & P.	644
Way	Douglas	S. L.	989
Webster	Burnett	S. L.	980
Weed Dam	Shawano	W. & M.	902
Weedens	Sheboygan	C. & N. W.	702
Weirgor	Sawyer	S. L.	1,203
Weiskisit	Langlade	W. & N.	1,201
Wentworth	Douglas	N. P.	941
Werley	Grant	C. & N. W.	738
West Bend	Washington	C. & N. W.	896
Westboro	Taylor	S. L.	1,504
Westby	Vernon	C., M., St. P. & P.	1,309
West End	Bayfield	B. T.	857
Westfield	Marquette	S. L.	856
West Greenville	Outagamie	C. & N. W.	898

ELEVATIONS

Locality	County	Authority	Elevation
			Feet
West Lima, B. M.	Richland	U. S. G. S.	1,291
West Prairie	Vernon	U. S. G. S.	1,201
West Rosendale	Fond du Lac	C. & N. W.	902
West Salem	La Crosse	C. & N. W.	749
		C., M., St. P. & P.	**741**
Weston	Dunn	C., St. P., M. & O.	877
Weyauwega	Waupaca	S. L.	778
Weyerhaeuser	Rusk	S. L.	1,200
Wheatland	Kenosha	S. L.	759
Wheeler	Dunn	S. L.	939
Whitcomb	Shawano	C. & N. W.	1,122
White Corners, B. M.	Crawford	U. S. G. S.	1,212
White Fish Bay	Milwaukee	C. & N. W.	656
Whitehall	Trempealeau	G. B. & W.	821
Whitelaw	Manitowoc	C. & N. W.	855
White River	Ashland	S. L.	731
Whitewater	Walworth	C., M., St. P. & P.	822
Whitson Jct.	Iowa	M. P. & N.	1,193
Highest point on line		M. P. & N.	1,233
Whittlesey	Taylor	S. L.	1,465
Wiehe	Douglas	N. P.	1,096
Wilcox	Marinette	C. & N. W.	617
Wild Rose	Waushara	C. & N. W.	997
Willard	Clark	F. & N. E.	1,178
Willo	Bayfield	D., S. S. & A.	1,165
Wilson	St. Croix	C., St. P., M. & O.	1,157
Wilton	Monroe	C. & N. W.	992
Winchester	Vilas	C. & N. W.	1,637
Windfall—X—C., St. P., M. & O.	Sawyer	S. L.	1,313
Windsor	Dane	C., M., St. P. & P.	899
Winegar	Vilas	C. & N. W.	1,644
Wingra Lake, water surface	Dane	U. S. G. S.	849
Winnebago Lake, water surface	Calumet	U. S. G. S.	747
Winneboujou	Douglas	D., S. S. & A.	1,024
Winneconne	Winnebago	C., M., St. P. & P.	754
Winona Jct.	La Crosse	C. & N. W.	656
Winter	Sawyer	C., St. P., M. & O.	1,360
Wiscona	Milwaukee	C. & N. W.	681
Wisconsin Dells	Columbia	C., M., St. P. & P.	893
Wisconsin & Northern Jct.	Forest	W. & N.	1,628
Wisconsin Rapids	Wood	C. & N. W.	1,005
X—G. B. & W.		G. B. & W.	1,017
		C., M., St. P. & P.	1,016
Wisconsin River	Crawford	W. G. & N. H. S.	
Mouth			604
Muscoda Bridge	Grant		669
Prairie du Sac	Sauk		740
Wisconsin Dells (Kilbourn), below dam	Columbia		815
Nekoosa, below dam	Wood		922

ELEVATIONS

Locality	County	Authority	Elevation
Wisconsin River—(continued)			Feet
Wisconsin Rapids, above dam			1,004
Knowlton	Marathon		1,098
Wausau, above dam	Marathon	W. G. & N. H. S	1,180
Upper dam, Merrill, above	Lincoln		1,251
Grandfather Falls, head			1,385
Tomahawk dam, above			1,431
Rhinelander dam, above	Oneida		1,558
Lac Vieux Desert	Vilas		1,650
Withee	Clark	S. L	1,270
Wittenberg	Shawano	C. & N. W	1,168
Wolf Creek	Polk	W. G. & N. H. S	820
Wolf River		W. G. & N. H. S	
Winneconne	Winnebago		747
New London	Waupaca		750
Lenox	Oneida		1,562
Wolf River Junction	Langlade	C. & N. W	1,511
Wonewoc	Juneau	C. & N. W	914
Woodboro	Oneida	S. L	1,600
Woodford, I. C. Sta	Lafayette	U. S. G. S	789
Woodhull	Fond du Lac	C. & N. W	871
Woodland	Dodge	C., M., St. P. & P	949
Woodman	Grant	C. & N. W	639
		C., M., St. P. & P	657
Woodruff	Vilas	C. & N. W	1,609
X—C., M., St. P. & P		C. & N. W	1,615
Woodville	Calumet	S. L	848
Woodville	St. Croix	C., St. P., M. & O	1,171
Woodworth	Kenosha	C. & N. W	751
Worcester	Price	S. L	1,605
Wrightstown	Brown	C. & N. W	657
Wyalusing	Grant	M. R. C	628
Wyeville	Monroe	C., St. P., M. & O	922
Wyocena	Columbia	C., M., St. P. & P	824
Yarnell	Sawyer	C., St. P., M. & O	1,431
Yellow Lake, water surface	Burnett	W. G. & N. H. S	923
Zachow	Shawano	C. & N. W	863
Zenda	Walworth	C., M., St. P. & P	987

INDEX

L

La Baye, 22, 284.
Labrador ice sheet, 85, 87, 89, 126, 235, 397, 434.
Lac Court Oreilles, 373.
Lac du Flambeau, 414.
La Crosse, 43, 93, 105, 131, 137, 153, 154, 156, 158, 166, 167, 168, 173, 196, 199, 330, 488, 503.
La Crosse and Southeastern Railway, 205.
La Crosse County, 46, 48, 136, 169.
La Crosse River, 133, 184, 196, 199.
   ridges and coulees near, 48.
   upland near, 50.
La Crosse terrace, 156.
Lac Vieux Desert, 20, 418, 422.
Ladysmith, 489.
Lafayette County, 59, 93.
Lagoons, 467, 471.
Lake Algonquin, see *Glacial Lake Algonquin.*
Lake Beaver, 20.
Lake Beulah, 281, 504.
Lake-bottom deposits, see *Lake clay.*
   sand, 363.
   sediments, see *Lake clay.*
Lake Buffalo, 281, 358.
Lake Chicago, see *Glacial Lake Chicago.*
Lake clay, 10, 123, 130, 135, 265-267, 280, 333, 337, 339, 363, 469.
Lake, deepest, 329.
Lake Delavan, 281.
Lake deposits, see *Lake clay.*
Lake Duluth, see *Glacial Lake Duluth.*
Lake Europe, 304.
Lake Geneva, 20, 281, 504.
Lake Hennepin, 118.
Lake Kegonsa, 20, 271, 275.
Lake Koshkonong, 20, 277.
Lake maps, 500.
Lake Mendota, 20, 271, 275, 504.
Lake Michigan, 19.
   depth of, 21.
   drainage entering, 287-291.
   fluctuations in present level, 306-311.
   glacier, 121, 396.
   harbors, 306.
   lobe, 88, 235, 260.
   lowland, 214.
   predecessors of, 294-302.
   present shore line, 302-306.
   reefs in, 305.
   river, 434.
   rock basin, 237-239.
   Wisconsin coast of, 294-314.
Lake Middleton, extinct, 195.
Lake Mills, 253.
Lake Monona, 20, 275, 504.

Lake Nemadji, see *Glacial Lake Nemadji.*
Lake Nipissing, see *Nipissing Great Lakes.*
Lake Pepin, 20, 104, 105, 131, 169-171, 172, 174, 180, 201, 202.
   state boundary at, 171-172.
   terraces near, 159.
Lake Plains, 23.
Lake Poygan, 20, 216, 281, 358.
Lake Puckaway, 20, 281, 358.
Lake Region, 23.
Lake Ripley, 278.
Lake St. Croix, 20, 81, 105, 171, 202-204.
   terraces near, 162.
Lake Shawano, 20, 281, 358.
Lakes, 20, 281, 355, 358, 400, 410, 412, 418.
Lakes, coastal, 304.
Lakes, filled, 279-291, 416.
Lakes in Northern Wisconsin, 20, 411, 412, 413, 418.
Lakes, Oconomowoc, 20, 278.
Lakes of the Mississippi bottomlands, 169-176.
Lakes, origin of, 416.
Lakes, temporary, 133.
Lake Superior, 20, 21, 22, 126, 446-477.
   canal to the Mississippi, 202.
   coast of, 446.
   depth of, 21.
   glacier, 88, 396.
   lobe, 87, 329, 332, 396, 434, 436, 452.
   present shore lines of, 464-477.
   region, geological map of, 495.
Lake Superior Highland, 23, 32, 367-405.
   See *Northern Highland.*
Lake Superior Lowland, 31, 36, 210, 429-444.
   age of, 432-433.
   boundaries of, 431.
   drainage of, 438-442.
   glacial lakes in, 452-463.
   glaciation of, 433-437.
   origin of, 436.
   rivers of, 438-442.
Lake Waubesa, 20, 275.
Lake Waumandee, 172.
Lake Wingra, 271, 276.
Lake Winnebago, 20, 281, 283.
Lake Winnebago-Rock River Lowland, 214, 217-224.
Lake Wisconsin, see *Glacial Lake Wisconsin.*
La Montagne qui trempe à l'eau, 150.
Lancaster, 60, 71, 488.
Lancaster peneplain, 73.